The Most Successful Way of JEE Preparation

DPP

Daily Practice Problems

JEE Main & Advanced

Oscillations & Waves

PHYSICS Vol-5

Er PRS Murthy (Raghava)
M.Tech, HOD Physics
Jain College, Bengaluru

arihant
ARIHANT PRAKASHAN (SERIES), MEERUT

DPP
Daily Practice Problems

arihant

ARIHANT PRAKASHAN (SERIES), MEERUT

Administrative & Production Offices

Corporate Office 'Ramchhaya' 4577/15, Agarwal Road, Darya Ganj, New Delhi -110002
Tele: 011- 47630600, 43518550; Fax: 011- 23280316

Head Office Kalindi, TP Nagar, Meerut (UP) - 250002
Tel: 0121-2401479, 2512970, 4004199; Fax: 0121-2401648

Sales & Support Offices

Agra, Ahmedabad, Bengaluru, Bareilly, Chennai, Delhi, Guwahati, Hyderabad, Jaipur, Jhansi, Kolkata, Lucknow, Meerut, Nagpur & Pune

ISBN : 978-93-13193-35-7

Published by Arihant Publications (I) Ltd.

Production Team

Publishing Managers	Mahendra Singh Rawat & Keshav Mohan	*Inner Designer*	Mazher Chaudhary
Project Head	Sorabh Chaudhary	*Page Layouting*	Sunil Sharma, Hemant
Cover Designer	Shanu Mahad	*Proof Reading*	Ashish

For further information about the products from Arihant, log on to www.arihantbooks.com or email to info@arihantbooks.com

/arihantpub /@arihantpub Arihant Publications /arihantpub

ABOUT DPP

Daily Practice Problems

Indian Institutes of Technology better known as IITs have always been the dream destination of each science group students who want to pursue their career as Engineers. These are the institutes which have got very high value not even in India but in the whole world. To get the admission in such institutes one need to face entrance Examinations; JEE Main & JEE Advanced, out of which JEE Main is considered as the preliminary exam & JEE Advanced is the final exam to get into the IITs. Approximately per year 12 lakh students enroll themselves for JEE Main and only the top 2.25 lakh in JEE Main merit can attempt for JEE Advanced, the fact clearly express the degree of toughness of JEE Main & JEE Advanced.

To have success in such an exam one needs to be very planned & systematic in its preparation. A very popular wise saying 'Rome was not built in a day' suggests that to achieve a big target like IITs we need to work continuously. For continuous work, one need to have proper planning that how much work he/she has to do on regular basis, i.e., on daily basis. Each student preparing for JEE has many books but unfortunately none of them gives the students a plan for daily studies, which is the most effective success key for JEE.

Keeping the daily practice needs of the students in mind we have come up with DPP Series (Daily Practice Problems) having approximately 26 books. The primary focus of this series is on achieving success through practice & proper planning. For that we have divided the whole syllabus of JEE Main & JEE Advanced; Physics, Chemistry & Mathematics, into different volumes (around 8 volumes of Mathematics & 9 volumes each of Physics & Chemistry), each volume has Daily Practice Problems Sheets having specific questions on various Topics of the individual chapter of the volume ensuring the complete Practice & Assessment of the topic.

DPP

Daily Practice Problems

In this way we have tried to give you an effective daily planning for JEE keeping in mind the daily hours that a student can devote to his practice, you are just required to practice one sheet of each subject on daily basis. This habit of daily practice will keep you away from exam stress and thoughts like; how much I have prepared/am I preparing properly or not/ do I need to study more on daily basis or what I am doing is ok, etc., which regularly tease your mind and can give you panic at any stage of your preparation.

Some special features that we have tried to incorporate in our Daily Practice Problems (DPP) series are discussed below

1. Micro Level coverage of all the topics of a Chapter/ Unit given in JEE Main & Advanced Syllabus.
2. Each DPP has various types of questions; Subjective Questions, Single/ Multiple Correct Options, Statements type, Comprehension based, Integer Value Answer type, Matching, etc., making you practice & ready for all the question formats of JEE.
3. New & innovative questions in each DPP challenging & improving your problem solving skills.
4. Along with Topical Coverage, Revisal DPPs for JEE Main & JEE Advanced with each Chapter, covering multidisciplinary questions proving comprehensive practice.
5. JEE Main & JEE Advanced Archive (collection of last 13 years' exams questions) with each chapter.
6. Complete Solutions for each DPP.

We hope that DPP series will prove to be very beneficial for the students preparing for JEE, by giving them daily practice in a systematic & planned manner. The suggestions/criticism from the students, teachers & parents about DPP series will be welcomed whole heartedly and we promise to update this series accordingly from time to time.

Dedicated To My Mother

Late Smt. **Pavani. Subbarathnamma,**
whose Lullaby I want to listen
atleast one more time before
my Departure

DPP
Daily Practice Problems

DAILY TASKS...

1 OSCILLATIONS

2 PROGRESSIVE WAVES

DPP
Daily Practice Problems

3 SOUND

4 SUPERPOSITION AND STANDING WAVES

1. Oscillations

DPP
DAILY PRACTICE PROBLEMS

DPP-1 Kinematics of SHM

Subjective Questions

Working Space

Direction *(Q. Nos. 1 and 2) These questions are subjective in nature, need to be solved completely on notebook.*

1 Can every oscillatory motion be treated as simple harmonic motion in the limit of small amplitude?

2 Describe the motion of a particle acted upon by a force

i. $F = -2(x-3)^3$ ii. $F = -2(x-3)^2$

iii. $F = -2\,(x-3)$

Only One Option Correct Type

Direction *(Q. Nos. 3-16) This section contains 14 multiple choice questions. Each question has four choices (a), (b), (c) and (d), out of which ONLY ONE is correct.*

3 A particle executes SHM with a time period of 4 s. The time taken by the particle to go directly from its mean position to half of its amplitude is

a. $\frac{1}{3}$ s b. $\frac{1}{2}$ s

c. $\frac{1}{4}$ s d. $\frac{2}{3}$ s

4 A particle is performing a linear simple harmonic motion. If the instantaneous acceleration and instantaneous velocity of the particle are *a* and *v*, respectively identify the correct graph among the following.

a. 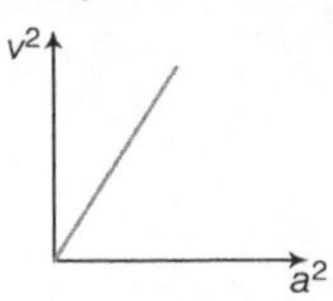

b. 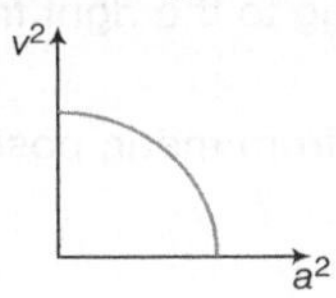

c. 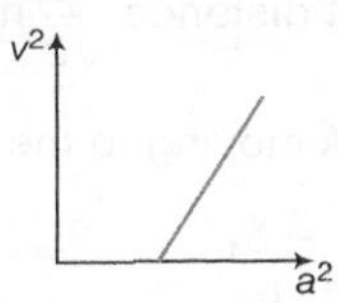

d. 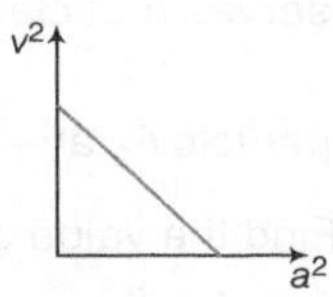

5 The maximum velocity of a body undergoing simple harmonic motion is 0.04 ms^{-1} and its acceleration at 0.02 m from the mean position is 0.06 ms^{-2}. Its amplitude and time period, respectively are

a. 3.21×10^{-2} m, 3.23 s b. 1.92×10^{-2} m, 2.82 s

c. 3.14×10^{-2} m, 3.82 s d. 2.31×10^{-2} m, 3.63 s

6 A man of mass 60 kg is standing on a platform executing SHM in vertical direction. The displacement from mean position of platform varies as $y = 0.5\sin(2\pi f t)$. The minimum value of *f*, for which the man will feel weightlessness at the highest point, is

a. $\frac{\sqrt{2g}}{\pi}$ b. $\sqrt{\frac{g}{2}} \cdot \frac{1}{\pi}$

c. $\sqrt{\frac{g}{2\pi}}$ d. $\sqrt{\frac{g}{3}} \cdot \frac{2}{\pi}$

Working Space

7 Figure shows the acceleration-displacement graph of a particle in SHM. The time period (in second) is

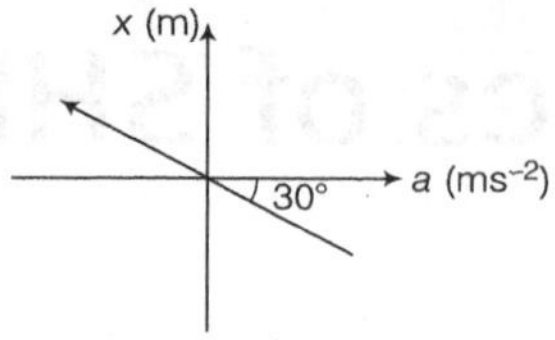

a. $\frac{2\pi}{3^{1/4}}$ b. 2π

c. $\sqrt{2}\,\pi$ d. $\frac{2\pi}{\sqrt{3}}$

8 A particle executes SHM in a straight line. The maximum speed of the particle during its motion is v_0. Its average speed in one complete oscillation is

a. $\frac{\sqrt{2}\,v_0}{\pi}$ b. $\frac{v_0}{2\pi}$

c. $\frac{2v_0}{\pi}$ d. $\frac{v_0}{\sqrt{2}\pi}$

9 The distance travelled by a particle executing SHM in 10 s, if the time period is 3 s, is

(it is given that the body starts from $\frac{A\sqrt{3}}{3}$ from equilibrium position, moves in positive direction at $t = 0$)

a. $\left(\frac{28-\sqrt{3}}{2}\right)A$ b. $\left(\frac{14-\sqrt{3}}{2}\right)A$

c. $\left(\frac{21-\sqrt{3}}{2}\right)A$ d. $\left(\frac{35-\sqrt{3}}{2}\right)A$

10 Two particles P_1 and P_2 are executing SHM along the same straight line, whose equations are given $x_1 = A\sin(\omega t + \delta_1)$ and $x_2 = A\sin(\omega t + \delta_2)$. An observer on the ground, at $t = 0$, observes particle P_1 at distance $\frac{A}{\sqrt{2}}$ moving to the right from mean position O while particle P_2 at $-\frac{\sqrt{3}}{2}A$ moving to the left from mean position O, as shown in figure. Find the value of $\delta_2 - \delta_1$.

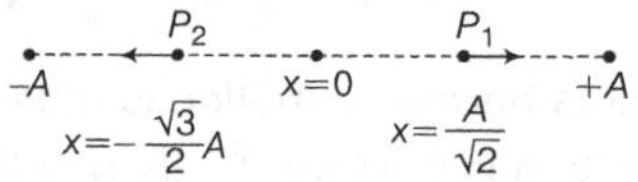

a. $\frac{7\pi}{12}$ b. $\frac{13\pi}{12}$

c. $\frac{5\pi}{6}$ d. π

11 Two particles are executing SHM with same amplitudes A and time period T. When one of the particles is located at right extreme, the other particle is located at mean position and moving towards left. From this instant, find the time after which both the particles will have same displacement from the equilibrium position.

a. $\frac{3T}{4}$ b. $\frac{5T}{8}$

c. $\frac{3T}{8}$ d. $\frac{2T}{7}$

Working Space

12 Two particles executing SHM with same angular frequency and amplitudes A and $2A$ on same straight line with same mean position cross each other in opposite direction at a distance $\frac{A}{3}$ from mean position. The phase difference between the two SHM's is

a. $\sin^{-1}\left(\frac{1}{3}\right) - \sin^{-1}\left(\frac{1}{6}\right)$

b. zero

c. $\frac{\pi}{2} - \sin^{-1}\left(\frac{1}{6}\right) - \sin^{-1}\left(\frac{1}{3}\right)$

d. $\pi - \sin^{-1}\left(\frac{1}{6}\right) - \sin^{-1}\left(\frac{1}{3}\right)$

13 Figure shows a spring-block system, hanging in equilibrium. The block of the system is pulled down by a distance d and imparted a velocity v_0 in the downward direction as shown in the figure. Find the time it will take to reach its mean position.

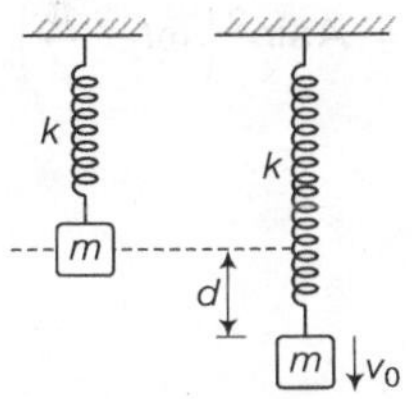

a. $\sqrt{\frac{m}{k}}\left[\pi - \cos^{-1}\left(\frac{d}{\sqrt{\frac{mv_0^2}{k} + d^2}}\right)\right]$

b. $\sqrt{\frac{m}{k}}\left[\frac{\pi}{4} - \cos^{-1}\left(\frac{d}{\sqrt{\frac{mv_0^2}{k} + d^2}}\right)\right]$

c. $\sqrt{\frac{m}{k}}\left[\frac{\pi}{2} - \sin^{-1}\left(\frac{d}{\sqrt{\frac{mv_0^2}{k} + d^2}}\right)\right]$

d. $\sqrt{\frac{m}{k}}\left[\pi - \sin^{-1}\left(\frac{d}{\sqrt{\frac{mv_0^2}{k} + d^2}}\right)\right]$

14 A spring-block pendulum is shown in the figure. The system is hanging in equilibrium. A bullet of mass $m/2$ moving with a speed u hits the block from down as shown in the figure. Find the amplitude of oscillation now.

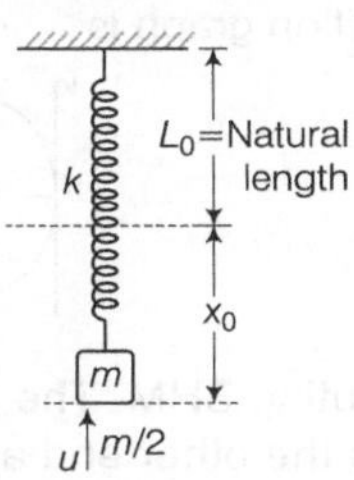

a. $\sqrt{\frac{mu^2}{8k} + \left(\frac{mg}{2k}\right)^2}$

b. $\sqrt{\frac{mu^2}{k} + \left(\frac{mg}{k}\right)^2}$

c. $\sqrt{\frac{mu^2}{6k} + \left(\frac{mg}{2k}\right)^2}$

d. $\sqrt{\frac{mu^2}{4k} - \left(\frac{mg}{2k}\right)^2}$

15 In the above problem, what is the time taken by the mass to reach its upper extreme position, after it is hit by the bullet?

a. $\sqrt{\frac{3m}{2k}}\sin^{-1}\left\{\frac{mg}{2k\sqrt{\frac{mu^2}{6k} + \left(\frac{mg}{2k}\right)^2}}\right\}$

b. $\sqrt{\frac{3m}{2k}}\cos^{-1}\left\{\frac{mg}{2k\sqrt{\frac{mu^2}{6k} + \left(\frac{mg}{2k}\right)^2}}\right\}$

c. $\sqrt{\frac{3m}{2k}}\sin^{-1}\left\{\frac{mg}{k\sqrt{\frac{mu^2}{6k} + \left(\frac{mg}{2k}\right)^2}}\right\}$

d. $\sqrt{\frac{3m}{2k}}\cos^{-1}\left\{\frac{mg}{k\sqrt{\frac{mu^2}{6k} + \left(\frac{mg}{2k}\right)^2}}\right\}$

Working Space

16 Two particles execute SHM of the same amplitude and frequency along the same straight line. They pass each other when going in opposite directions, each time their displacement is half of their amplitude. The phase difference between them is

a. $\frac{2\pi}{3}$ b. $\frac{\pi}{2}$

c. $\frac{3\pi}{2}$ d. $\frac{3\pi}{4}$

One or More than One Options Correct Type

Direction (Q. Nos. 17-22) *This section contains 6 multiple choice questions. Each question has four choices (a), (b), (c) and (d), out of which ONE or MORE THAN ONE are correct.*

17 A particle moves along X-axis according to the law $x = A\sin^2\left(\omega t - \frac{\pi}{4}\right)$. Its velocity is v. Choose the correct option(s).

a. Its amplitude is $\frac{A}{2}$

b. Its position-time graph is

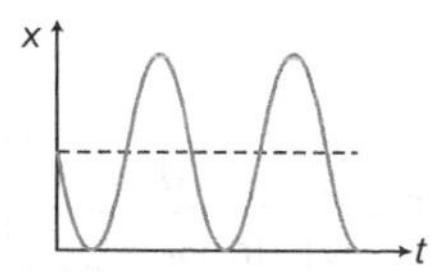

c. Its position-time graph is

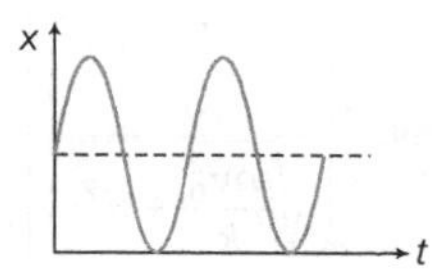

d. Its velocity-position graph is

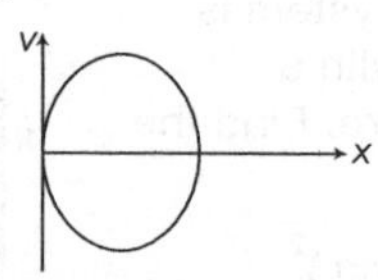

18 A particle is executing SHM. The ratio of mean velocity during motion from one end of the path to the other end and the maximum velocity is r_1. The ratio of average acceleration during motion from one extremity to the centre and maximum acceleration is r_2. Then,

a. $r_1 = \frac{2}{\pi}$ b. $r_2 = \frac{1}{\pi}$

c. $r_1 = \frac{1}{2\pi}$ d. $r_2 = \frac{2}{\pi}$

19 A particle executes simple harmonic motion about the point $x = 0$. At time $t = 0$, it has displacement $x = 2$ cm and zero velocity. Frequency of its motion is $0.25\ s^{-1}$. Then,

a. its amplitude is 2 cm

b. its maximum speed is $3.14\ cm\,s^{-1}$

c. its displacement at $t = 3$s is zero

d. its velocity at $t = 3$s is $3.14\ cm\ s^{-1}$

Working Space

20 An object of mass 0.8 kg is attached to one end of a spring and the system is set into simple harmonic motion. The displacement x of the object as a function of is shown in the figure. Then, choose the correct statement (s).

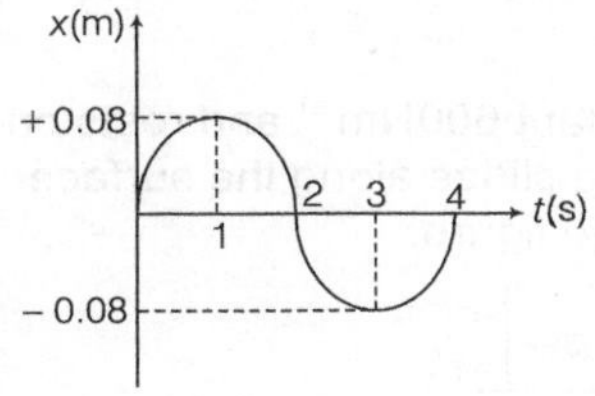

a. Its angular frequency is 1.57 rad s^{-1}
b. The spring constant is 1.97 Nm^{-1}
c. The speed of the object at $t = 1$ s is zero
d. The magnitude of the acceleration of the object at $t = 1$ s is 0.197 ms^{-2}

21 The equation of motion of a particle started at $t = 0$ is given by $x = 5\sin\left(20t + \frac{\pi}{3}\right)$, where x is in metre and t in second. Then,

a. the particle first comes to rest at $t = \frac{\pi}{30}$ s
b. the particle first has zero acceleration at $t = \frac{\pi}{15}$ s
c. the particle first has maximum speed at $t = \frac{\pi}{30}$ s
d. the particle has frequency $\frac{10}{\pi} s^{-1}$

22 Which of the following expressions represent simple harmonic motion?
a. $y = A\sin(\omega t + \delta)$
b. $y = B\cos(\omega t + \delta)$
c. $y = A\tan(\omega t + \delta)$
d. $y = A\sin\omega t\cos\omega t$

Statement Type

Direction (Q. Nos. 23 and 24) *This section is based on Statement I and Statement II. Select the correct answer from the codes given below.*

a. Both Statement I and Statement II are correct and Statement II is the correct explanation of Statement I
b. Both Statement I and Statement II are correct but Statement II is not the correct explanation of Statement I
c. Statement I is correct but Statement II is incorrect
d. Statement II is correct but Statement I is incorrect

23 **Statement I** An oscillatory motion is necessarily periodic.
Statement II A simple harmonic motion is necessarily periodic.

24 **Statement I** When a boy sitting on a swing stands up, the time period of swing will increase.
Statement II In standing position of the boy, the effective length of the swing decreases.

Comprehension Type

Working Space

Direction *(Q. Nos. 25-30) This section contains 2 paragraphs, each describing theory, experiments, data, etc. Six questions related to the paragraphs have been given. Each question has only one correct answer among the four given options (a), (b), (c) and (d).*

Passage I

A 2 kg mass is attached to a spring of force constant $600 Nm^{-1}$ and rests on a smooth horizontal surface. A second mass of 1 kg slides along the surface towards the first at a speed $6 ms^{-1}$ as shown in the figure.

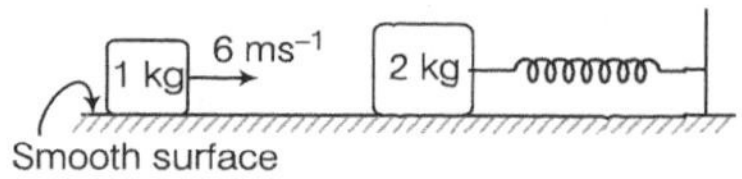

25 If the masses make perfectly inelastic collision, the amplitude and period of oscillations, respectively, are

a. 14.1 cm, 0.44 s
b. 12.414 cm, 0.22 s
c. 12.732 cm, 0.35 s
d. 15.2 cm, 0.28 s

26 If the masses make perfectly elastic collision, then the amplitude and period of oscillations, respectively, are

a. 28 cm, 0.48 s
b. 23 cm, 0.36 s
c. 20 cm, 0.42 s
d. 25 cm, 0.28 s

27 The equations of motion for both cases, respectively, are (assuming collision happens at $t = 0$)

a. $x = (14.1 \text{ cm})\sin(14.28t)$ and $x = (23 \text{ cm})\sin(17.45t)$
b. $x = (12.414\text{cm})\sin(12.21t)$ and $x = (20\text{cm})\sin(14.45t)$
c. $x = (15.2\text{cm})\sin(13.23t)$ and $x = (25\text{cm})\sin(15.25t)$
d. $x = (12.732\text{cm})\sin(12.82t)$ and $x = (28\text{cm})\sin(16.35t)$

Passage II

A block of mass 2 kg hangs without oscillating at the end of a spring of spring constant $500 Nm^{-1}$, which is attached to the ceiling of an elevator. The elevator is moving upwards with an acceleration $\frac{g}{3}$. At time $t = 0$, the acceleration suddenly ceases. (Given, $g = 10 ms^{-2}$)

28 What is the angular frequency of oscillation of the block after the acceleration ceases?

a. 12.72 rad s^{-1}
b. 13.62 rad s^{-1}
c. 14.53 rad s^{-1}
d. 15.81 rad s^{-1}

29 By what amount is the spring stretched during the time when the elevator is accelerating?

a. 4.2 cm
b. 6.7 cm
c. 5.3 cm
d. 3.7 cm

30 When the acceleration ceases, the amplitude of oscillation and epoch, as observed by an observer in the elevator, respectively, are

a. 2.1 cm, $\frac{\pi}{2}$
b. 1.3 cm, $\frac{3\pi}{2}$
c. 2.8 cm, π
d. 1.7 cm, $\frac{2\pi}{3}$

DPP-2 Energy in SHM

Only One Option Correct Type

Working Space

Direction *(Q. Nos. 1-15) This section contains 15 multiple choice questions. Each question has four choices (a), (b), (c) and (d), out of which ONLY ONE is correct.*

1 A particle of mass m is executing oscillations about the origin on the X-axis with amplitude A. Its potential energy is given as $U(x) = \beta x^4$, where β is a positive constant. The x-coordinate of the particle, where the potential energy is one-third of the kinetic energy, is

a. $\pm \frac{A}{2}$ b. $\pm \frac{A}{\sqrt{2}}$ c. $\pm \frac{A}{3}$ d. $\pm \frac{A}{\sqrt{3}}$

2 For a particle executing simple harmonic motion, the displacement x is given by $x = A\cos \omega t$. Identify the graphs that represent variation of potential energy (U) as a function of time (t) and displacement (x).

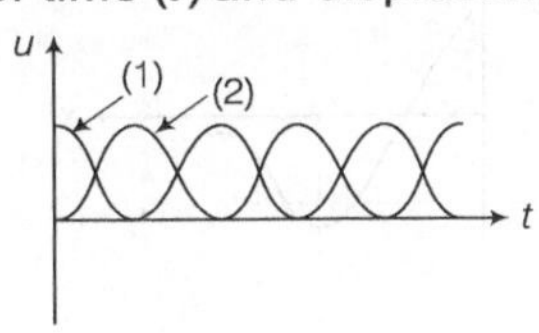

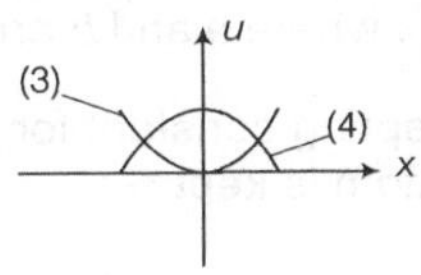

a. (1),(3) b. (2), (4)
c. (2), (3) d. (1), (4)

3 A body is executing a simple harmonic motion such that its potential energy is U_1 at x and U_2 at y. When the displacement is $x + y$, the potential energy will be

a. $U_1 + U_2$ b. $\sqrt{U_1^2 + U_2^2}$
c. $U_1 + U_2 + 2\sqrt{U_1 U_2}$ d. $\sqrt{U_1 U_2}$

4 Three springs of constants k_1, k_2 and k_3 are connected end to end and are used to hang a mass M from the roof. The elastic potential energy stored in the array is

a. $\frac{M^2g^2}{2}\left(\frac{1}{k_1}+\frac{1}{k_2}+\frac{1}{k_3}\right)$ b. $M^2g^2\left(\frac{1}{k_1}+\frac{1}{k_2}+\frac{1}{k_3}\right)$

c. $2M^2g^2\left(\frac{1}{k_1}+\frac{1}{k_2}+\frac{1}{k_3}\right)$ d. $\frac{M^2g^2}{3}\left(\frac{1}{k_1}+\frac{1}{k_2}+\frac{1}{k_3}\right)$

5 A particle is executing linear SHM. The average kinetic energy and average potential energy, over a period of oscillation, respectively are K_{av} and U_{av}. Then,

a. $K_{av} = \frac{U_{av}}{2}$ b. $U_{av} = \frac{K_{av}}{2}$ c. $K_{av} = U_{av}$ d. $U_{av} = \frac{K_{av}}{3}$

6 A linear harmonic oscillator of force constant 2×10^6 Nm^{-1} and amplitude 0.01 m has a total mechanical energy 160 J. Among the following statements, which are correct?

i. Maximum PE is 100 J ii. Maximum KE is 100 J
iii. Maximum PE is 160 J iv. Minimum PE is zero

a. Both (i) and (iv) b. Both (ii) and (iii)
c. Both (i) and (ii) d. Both (ii) and (iv)

Working Space

7 A particle free to move along the X-axis has potential energy given by $U(x) = k[1 - e^{-x^2}]$ for $-\infty \le x \le +\infty$, where k is a positive constant of appropriate dimensions. Then,

a. at points away from the origin, the particle is in unstable equilibrium
b. for any finite non-zero value of x, there is a force directed away from the origin
c. if its total mechanical energy is $\frac{k}{2}$, it has its minimum kinetic energy at the origin
d. for small displacements from $x = 0$, the motion is simple harmonic

8 From the variation of potential energy in the direction of small oscillation of a simple pendulum, find the effective spring constant for the simple pendulum, where m is mass of the bob, l is length of the simple pendulum.

a. $\frac{mg}{l}$ b. $\frac{mg}{2l}$ c. $\frac{2mg}{l}$ d. $\frac{mg}{\sqrt{2}l}$

9 The effective spring constant of a system whose potential energy varies with x as, $U = U_0 \sin ax$, is

a. $\frac{U_0a^2}{2}$ b. $U_0a^{3/2}$
c. U_0a^2 d. U_0a

10 The potential energy between two atoms in a diatomic molecule varies with x as $U = \frac{a}{x^{12}} - \frac{b}{x^6}$, where a and b are positive constants. Find the equivalent spring constant for the oscillation of one atom if the other atom is kept stationary.

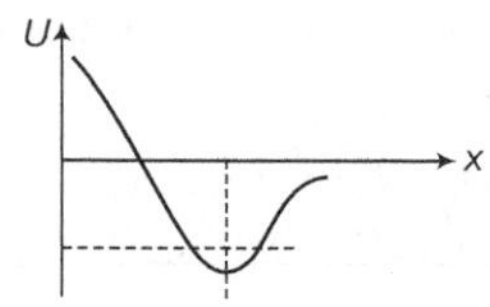

a. $\frac{9b^2}{2a}\left(\frac{4b}{a}\right)^{1/3}$ b. $\frac{9b^2}{a}\left(\frac{b}{2a}\right)^{1/3}$
c. $\frac{9b^2}{a}\left(\frac{2b}{a}\right)^{1/3}$ d. $\frac{9b^2}{a}\left(\frac{b}{a}\right)^{1/3}$

11 Find the effective spring constant for the oscillation of a particle in a spring like system whose potential energy varies as $U = \frac{a}{x^2} - \frac{b}{x}$, where a and b are positive constants.

a. $\frac{b^4}{4a^3}$ b. $\frac{4b^4}{a^3}$
c. $\frac{b^4}{2a^3}$ d. $\frac{b^4}{8a^3}$

12 A particle of mass m moves in the potential energy U shown in the figure. The period of the motion, if the total energy of the particle is E_0, is

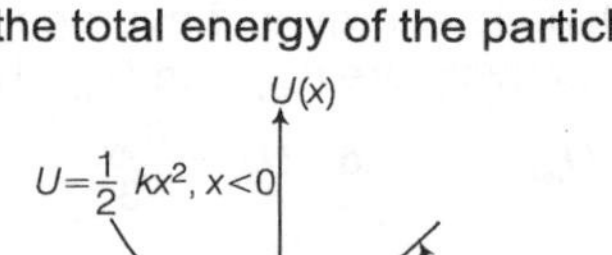

a. $\pi = \sqrt{\frac{m}{k}} + 4\sqrt{\frac{2E_0}{mg^2}}$ b. $2\pi\sqrt{\frac{m}{k}}$
c. $\pi\sqrt{\frac{m}{k}} + 2\sqrt{\frac{2E_0}{mg^2}}$ d. $2\sqrt{\frac{2E_0}{mg^2}}$

Working Space

13 The masses in the figure can slide on the frictionless horizontal surface. Now m_1 and m_2 are pushed to the left, so that the spring is compressed a distance L. What will be the amplitude of the oscillation of m_1 after the spring system is released ? Note that only m_1 is fixed to the spring.

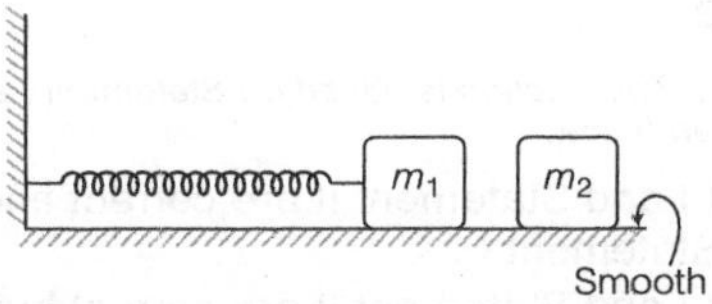

a. $\sqrt{\frac{m_1}{m_2}}L$ b. $\sqrt{\frac{m_1}{m_1+m_2}}L$

c. $\sqrt{\frac{m_2}{m_1}}L$ d. $\sqrt{\frac{m_1+m_2}{m_1}}L$

14 A point particle of mass 0.1 kg is executing SHM of amplitude 0.1 m. When the particle passes through the mean position, its kinetic energy is 0.008 J. If epoch is 45°, then the equation of motion of this particle is

a. $0.1\cos\left[(4s^{-1})t+\frac{\pi}{4}\right]$ b. $0.1\sin\left[(2s^{-1})t+\frac{\pi}{4}\right]$

c. $0.1\cos\left[(2s^{-1})t+\frac{\pi}{4}\right]$ d. $0.1\sin\left[(4s^{-1})t+\frac{\pi}{4}\right]$

15 A 2 kg particle undergoes SHM according to $x = 1.5\sin\left(\frac{\pi}{4}t+\frac{\pi}{6}\right)$, where x is in metre and t in second. The total mechanical energy of the particle is

a. 1.39 J b. 1.72 J

c. 2.39 J d. 2.72 J

One or More than One Options Correct Type

Direction (Q. Nos. 16 and 17) *This section contains 2 multiple choice questions. Each question has four choices (a), (b), (c) and (d), out of which ONE or MORE THAN ONE are correct.*

16 The energy of a particle executing simple harmonic motion is given by $E = Ax + Bv^2$, where x is the displacement from mean position $x = 0$ and v is the velocity of the particle at x and A and B are constants. Then, choose the correct statement(s).

a. Amplitude of SHM is $\sqrt{\frac{2E}{A}}$

b. Maximum velocity of the particle during SHM is $\sqrt{\frac{E}{B}}$

c. Time period of motion is $2\pi\sqrt{\frac{B}{A}}$

d. Displacement of the particle is proportional to the velocity of the particle

17 A man applies a force F on a spring-block system shown, towards right when the block is at rest and spring is relaxed. If F is constant, then

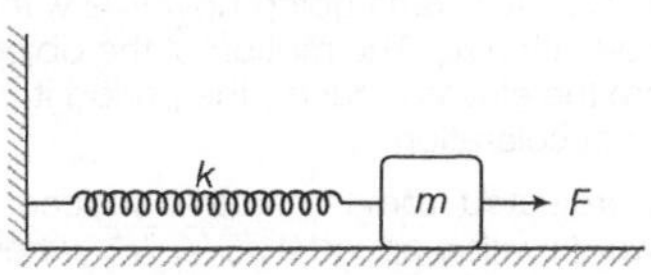

a. the time period of the block is $2\pi\sqrt{\frac{M}{k}}$

b. the amplitude is $\frac{F}{k}$

c. When the block passes through equilibrium position, the energy stored in the spring and the kinetic energy of the block respectively, are $\frac{F^2}{2k}$ and $\frac{F^2}{2k}$

d. The block does not oscillate

Working Space

Statement Type

Direction (Q. Nos. 18-20) *This section is based on Statement I and Statement II. Select the correct answer from the codes given below.*

a. Both Statement I and Statement II are correct and Statement II is the correct explanation of Statement I
b. Both Statement I and Statement II are correct but Statement II is not the correct explanation of Statement I
c. Statement I is correct but Statement II is incorrect
d. Statement II is correct but Statement I is incorrect

18 **Statement I** In SHM, the phase differences between displacement, velocity and velocity, acceleration and displacement, acceleration are $\frac{\pi}{2}, \frac{\pi}{2}$ and π, respectively.

Statement II In SHM, total mechanical energy is conserved.

19 **Statement I** In SHM of a particle with angular frequency ω, kinetic energy and potential energy vary simple harmonically with angular frequency 2ω.

Statement II Kinetic energy and potential energy vary periodically but not simple harmonically with angular frequency 2ω.

20 **Statement I** In SHM, potential energy is zero at mean position and maximum at extreme position.

Statement II In SHM, kinetic energy is maximum at mean position and zero at extreme position.

Matching List Type

Direction (Q. Nos. 21 and 22) *Choices for the correct combination of elements from Column I and Column II are given as options (a), (b), (c) and (d), out of which one is correct.*

21 Column I describes some situations in which a small object moves. Column II describes some characteristics of these motions. Match the Column I with Column II and mark the correct option from the codes given below.

	Column I		Column II
i.	The object moves on the X-axis under a conservative force in such a way that its speed and position satisfy $v = k_1\sqrt{k_2 - x^2}$, where k_1 and k_2 are positive constants	p.	The object executes simple harmonic motion
ii.	The object moves on the X-axis in such a way that its velocity and its displacement from the origin satisfy $v = -kx$, where k is a positive constant	q.	The object does not change its direction
iii.	The object is attached to one end of a massless spring of given spring constant. The other end of the spring is attached to the ceiling of an elevator. Initially, everything is at rest. The elevator starts going upwards with a constant acceleration a_0. The motion of the object is observed from the elevator during the period it maintains this acceleration	r.	The kinetic energy of the object keeps on decreasing
iv.	The object is projected from the earth's surface vertically upwards with a speed $2\sqrt{GM_e/R_e}$, where M_e and R_e are mass and radius of the earth. Neglect forces from objects other than earth	s.	The object can change its direction only once

Codes

	i	ii	iii	iv
a.	p	q,r	p	q,r
b.	s	p	r	q,s
c.	r	s	p	q
d.	q	s	p,q	r

Working Space

22 In a simple harmonic motion of a particle, potential energy at mean position is 4 J and at extreme position is 20 J. Given that amplitude of oscillation is A, where U is potential energy and K is kinetic energy and $x = 0$ is mean position. Match the Column I with Column II and choose the correct option from the codes given below.

	Column I		Column II
i.	$(U)_{\text{at } x=\frac{A}{2}}$	p.	18 J
ii.	$(K)_{\text{at } x=\frac{A}{4}}$	q.	16 J
iii.	$(K)_{\text{at } x=0}$	r.	8 J
iv.	$(K)_{\text{at } x=\frac{A}{2}}$	s.	None of the above

Codes

	i	ii	iii	iv
a.	p	q	r	s
b.	q	q	p	s
c.	s	s	q	s
d.	q	p	r	p

Comprehension Type

***Direction** (Q. Nos. 23-26) This section contains a paragraph, describing theory, experiments, data, etc. Four questions related to the paragraph have been given. Each question has only one correct answer among the four given options (a), (b), (c) and (d).*

Passage

A single conservative force $F(x)$ acts on a 2 kg particle that moves along the X-axis. The potential energy $U(x)$ is given by $U(x) = 30 + (x - 3)^2$, where x is in metre. At $x = 5$m, the particle has a kinetic energy of 25 J.

23 What is the mechanical energy of the system?

a. 69 J
b. 49 J
c. 39 J
d. 59 J

24 The least value of x and the greatest value of x between which the particle can move, respectively, are

a. -1.39m, 7.39m
b. -3.19m, 9.19m
c. -2.39m, 8.39 m
d. -3.39m, 9.39m

25 The maximum kinetic energy of the particle and the value of x at which it occurs respectively, are

a. 39 J, 2m
b. 29 J, 3 m
c. 49 J, 2 m
d. 42 J, 1.5 m

26 For what finite value of x does $F(x) = 0$?

a. 3 m
b. 4 m
c. 2 m
d. 2.5 m

One Integer Value Correct Type

Working Space

***Direction** (Q. Nos. 27-29) This section contains 3 questions. When worked out will result in an integer from 0 to 9 (both inclusive).*

27 A particle is moving on X-axis and has potential energy $U = 2 - 20x + 5x^2$ joule, where x is position. The particle is released at $x = -3$. If the mass of the particle is 0.1 kg, then the maximum velocity (in ms^{-1}) of the particle is 25β. If amplitude is 5 m, then find the value of β.

28 A particle of mass 4 kg moves in simple harmonic motion and its potential energy U varies with position x as shown in the figure. The time period of oscillation of the particle is $\frac{2\pi}{\alpha}$ second. Find the value of α.

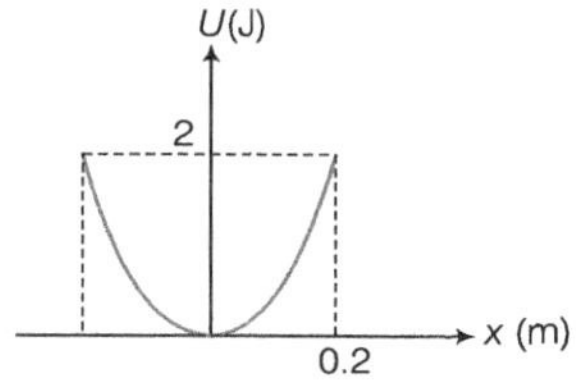

29 A particle moves under the force $F(x) = x^2 - 8x$, where x is in metre. For small displacements from the origin, the force constant in the simple harmonic approximation is $k\,Nm^{-1}$. Find the values of k.

DPP-3 Simple Pendulum and Spring-Block System

Only One Option Correct Type

Working Space

Direction *(Q. Nos. 1-15) This section contains 15 multiple choice questions. Each question has four choices (a), (b), (c) and (d), out of which ONLY ONE is correct.*

1 In the figure is shown a small block B of mass m resting on a smooth horizontal floor and the block is attached to an ideal spring (of force constant k). The spring is attached to vertical wall W_1. At a distance d from the block, right side of it, is present the vertical wall W_2. Now, the block is compressed by a distance $\frac{5d}{3}$ and released. It starts oscillating. If the collisions of the block with wall W_2 are perfectly elastic, the time period of oscillation of the block is

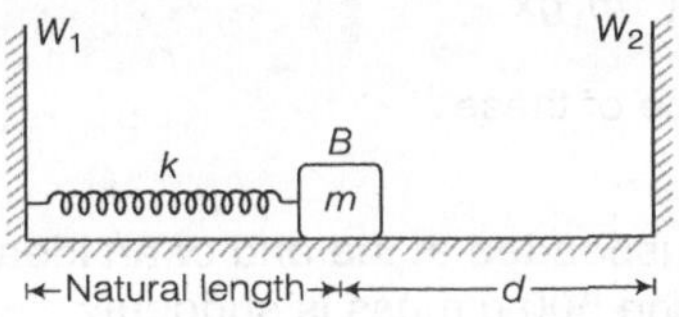

a. $\sqrt{\frac{m}{k}}\left[\pi + 2\sin^{-1}\left(\frac{3}{5}\right)\right]$
b. $\sqrt{\frac{m}{k}}\left(\pi + \sin^{-1}\left(\frac{3}{5}\right)\right)$
c. $\sqrt{\frac{m}{k}}\left[\frac{\pi}{2} + 2\sin^{-1}\left(\frac{5}{3}\right)\right]$
d. $\sqrt{\frac{m}{k}}\left[\frac{\pi}{2} + \sin^{-1}\left(\frac{3}{5}\right)\right]$

2 Two simple pendulums of lengths 1 m and 25 m, respectively, are both given small displacements in the same direction at the same instant. If they are in phase after the shorter pendulum has completed n oscillations, n is equal to

a. $\frac{4}{3}$ b. $\frac{5}{4}$ c. $\frac{7}{5}$ d. $\frac{8}{7}$

3 A block of mass M is kept on a smooth surface and touches the two springs as shown in the figure, but not attached to the springs. Initially, the springs are in their natural lengths (i.e. l_0 and $2l_0$). Now, the block is shifted $\frac{l_0}{2}$ from the given position leftwards in such a way that it compresses the left springs by $\frac{l_0}{2}$ and then released. The time period of oscillation of the block will be

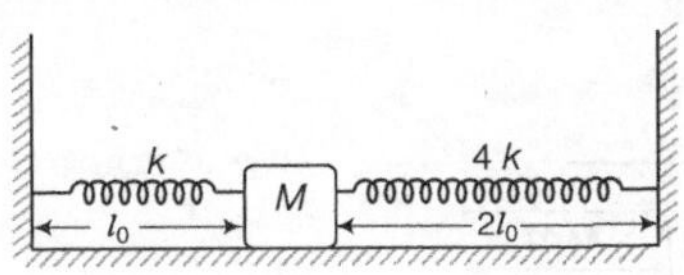

a. $\pi\sqrt{\frac{M}{3k}}$
b. $\frac{\pi}{2}\sqrt{\frac{M}{k}}$
c. $\frac{3\pi}{2}\sqrt{\frac{M}{k}}$
d. $2\pi\sqrt{\frac{M}{5k}}$

4 A ball is suspended by a thread of length L at the point O on an inclined wall as shown in the figure. The inclination of the wall with the vertical is α. The thread is displaced through a small angle β away from the vertical and the ball is released. If $\alpha < \beta$, find the period of oscillation of pendulum. Assume that any impact between the wall and the ball is elastic.

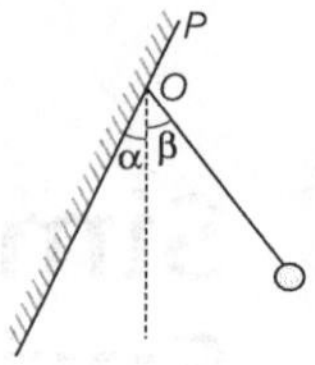

Working Space

a. $2\pi\sqrt{\frac{L}{g}}$

b. $2\sqrt{\frac{L}{g}}\left[1-\cos^{-1}\left(\frac{\alpha}{\beta}\right)\right]$

c. $2\pi\sqrt{\frac{L}{g}}\left[1-\cos^{-1}\left(\frac{\alpha}{\beta}\right)\right]$

d. $2\pi\sqrt{\frac{L}{g}}-2\sqrt{\frac{L}{g}}\cos^{-1}\left(\frac{\alpha}{\beta}\right)$

5 In the above problem if $\alpha > \beta$, find out the time period.

a. $2\pi\sqrt{\frac{L}{g}}$

b. $\sqrt{\frac{L}{g}}\left[1-\cos^{-1}\left(\frac{\alpha}{\beta}\right)\right]$

c. $2\pi\sqrt{\frac{L}{g}}\left[1-\cos^{-1}\left(\frac{\alpha}{\beta}\right)\right]$

d. $2\pi\sqrt{\frac{L}{g}}-2\sqrt{\frac{L}{g}}\cos^{-1}\left(\frac{\alpha}{\beta}\right)$

6 A mass m_1 is suspended from an ideal spring. An additional mass m_2 stretches the spring further by a distance x. The combined mass will oscillate with a period

a. $2\pi\sqrt{\frac{m_2 g}{(m_1+m_2)}}$

b. $2\pi\sqrt{\frac{m_1+m_2}{m_1 g x}}$

c. $2\pi\sqrt{\frac{(m_1+m_2)x}{m_2 g}}$

d. None of these

7 Three masses of 500 g, 300 g and 100 g are suspended at the end of an ideal spring as shown and are in equilibrium. When the 500 g mass is suddenly removed, the system oscillates with a period of 2 s. When 300 g mass is also removed, it will oscillate with the period.

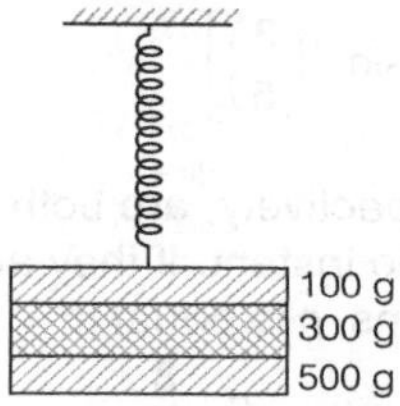

a. 1.75 s

b. 1.25 s

c. 1.5 s

d. 1 s

8 In the arrangement shown, all surfaces are smooth and spring is ideal. If block of mass m is given slight displacement, calculate the time period of oscillation of the system.

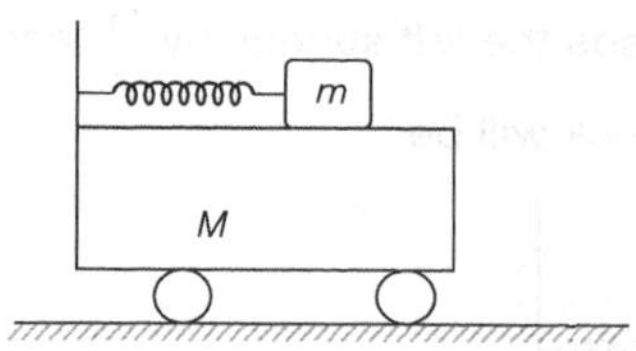

a. $2\pi\sqrt{\frac{M+m}{kMm}}$

b. $2\pi\sqrt{\frac{Mm}{(M+m)k}}$

c. $2\pi\sqrt{\frac{kMm}{(M+m)}}$

d. $2\pi\sqrt{\frac{M-m}{k(M+m)}}$

Working Space

9 A block of mass m is suspended from a spring as shown in the figure. It executes SHM of time period T. The amplitude of the SHM is A. The spring is never in compressed state during the oscillation. The minimum force exerted by spring is never in compressed state during the oscillations. The minimum force exerted by spring on the block is

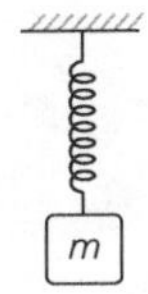

a. $mg - \frac{4\pi^2}{T^2}mA$

b. $mg + \frac{4\pi^2}{T^2}mA$

c. $mg - \frac{7\pi^2}{4T^2}mA$

d. $mg + \frac{7\pi^2}{4T^2}mA$

10 A block of mass m, attached by an ideal spring, to a fixed position P on a smooth inclined wedge of mass M, oscillates with amplitude A and angular frequency ω. The wedge is located on a rough horizontal surface. The angle of the wedge is 60°. If coefficient of friction is μ, then the force of friction acting on the wedge is

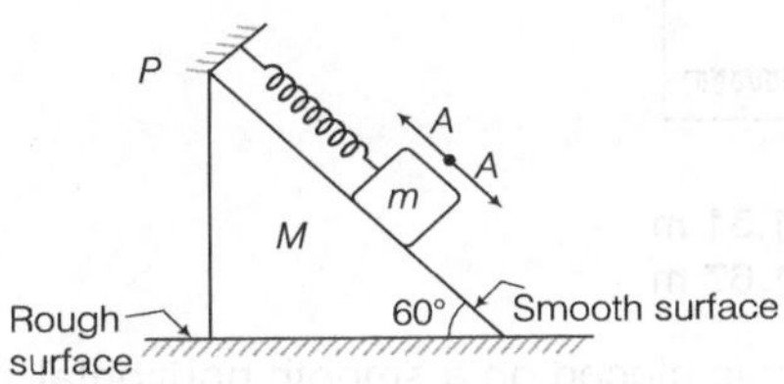

a. $\mu(M+m)g$

b. $\frac{1}{2}m\omega^2 A\sin\omega t$

c. $\mu(M+m)\omega^2 A\sin\omega t$

d. $\mu\left[(M+m)g + \frac{\sqrt{3}}{2}m\omega^2 A\sin\omega t\right]$

11 A block of mass m is suspended by means of an ideal spring of force constant k from the ceiling of a car, which is moving along a circular path of radius r with acceleration a_0. The time period of oscillation of the block, when it is displaced along the spring, will be

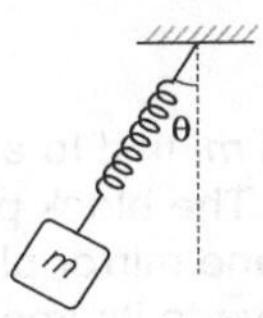

a. $2\pi\sqrt{\frac{mg+ma_0}{k}}$

b. $2\pi\sqrt{\frac{m}{k\sqrt{g^2+a_0^2}}}$

c. $2\pi\sqrt{\frac{m}{k}}$

d. $2\pi\sqrt{\frac{m}{\sqrt{k^2+g^2+a_0^2}}}$

Working Space

12 Two blocks of masses m_1 and m_2 are kept on a smooth horizontal surface. A spring of mass m and natural length L_0 connects the two blocks as shown in the figure. At $t = 0$, m_1 and m_2 are given velocities v_1 and v_2 as shown. Just after $t = 0$, what is the value of x, so that the point P remains at rest?

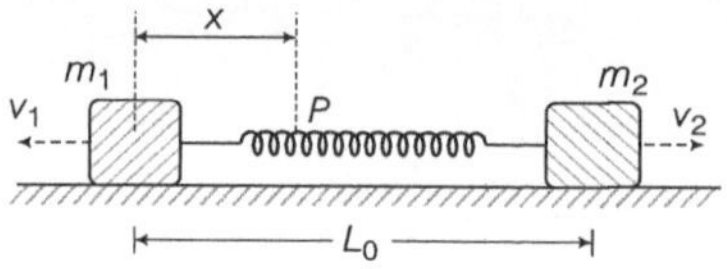

a. $\dfrac{L_0 v_1}{v_1 + v_2}$ b. $\dfrac{L_0 v_2}{v_1 + v_2}$

c. $\dfrac{L_0 m_1}{m_1 + m_2}$ d. $\dfrac{L_0 m_2}{m_1 + m_2}$

13 A block of mass m kg moving with kinetic energy 3 J on a smooth horizontal surface hits the free end of a spring, whose other end is attached to a rigid wall as shown in the figure. If restoring force applied by the spring is given by $F(x) = -2x - x^3$ (in N), then maximum compression in the spring is approximately.

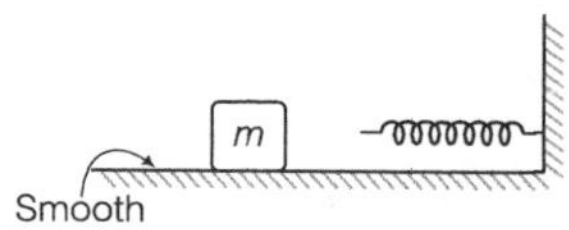

a. 2.2 m b. 1.51 m

c. 1.41 m d. 1.67 m

14 An assembly of identical spring-mass system is placed on a smooth horizontal surface as shown in the figure. Initially, the spring is released. The left mass is displaced to left while the right mass is displaced to the right and released. The resulting collision is elastic. The time period of the oscillations of the system is

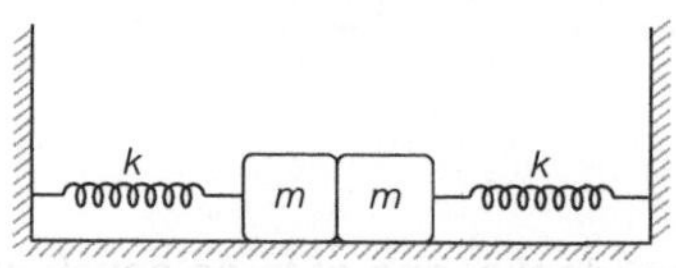

a. $2\pi\sqrt{\dfrac{2m}{k}}$ b. $\pi\sqrt{\dfrac{m}{k}}$

c. $2\pi\sqrt{\dfrac{m}{k}}$ d. $\pi\sqrt{\dfrac{2m}{k}}$

15 An insect of negligible mass is sitting on a block of mass m, tied to an ideal spring as shown in the figure. The force constant of spring is k. The block performs simple harmonic motion with amplitude A in front of a plane mirror placed as shown in the figure. The maximum speed of insect relative to its image will be

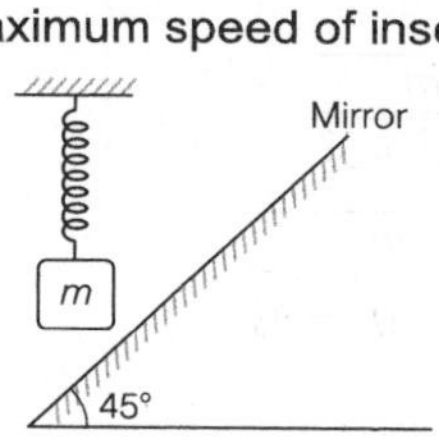

a. $A\sqrt{\dfrac{k}{m}}$ b. $\dfrac{A}{\sqrt{2}}\sqrt{\dfrac{k}{m}}$

c. $A\sqrt{2}\sqrt{\dfrac{k}{m}}$ d. $A\sqrt{\dfrac{m}{k}}$

Working Space

One or More than One Options Correct Type

Direction *(Q. Nos. 16-18) This section contains 3 multiple choice questions. Each question has four choices (a), (b), (c) and (d), out of which ONE or MORE THAN ONE are correct.*

16 Consider a simple pendulum of a very large length L. For small amplitudes, its time period is T. Given, radius of earth is $R_e = 6400$ km, then choose the correct option(s).

a. $T = 2\pi\sqrt{\dfrac{1}{g\left(\dfrac{1}{L}+\dfrac{1}{R_e}\right)}}$

b. When $L = R_e$, $T \simeq 59.8$ min

c. When $L \to \infty$, $T \simeq 84.6$ min

d. When $L = R_e$, $T \simeq 69.8$ min

17 In the shown figure, a block of mass m connected to an ideal spring (of force constant k) in different ways. Time period is denoted by T. Choose the correct option is.

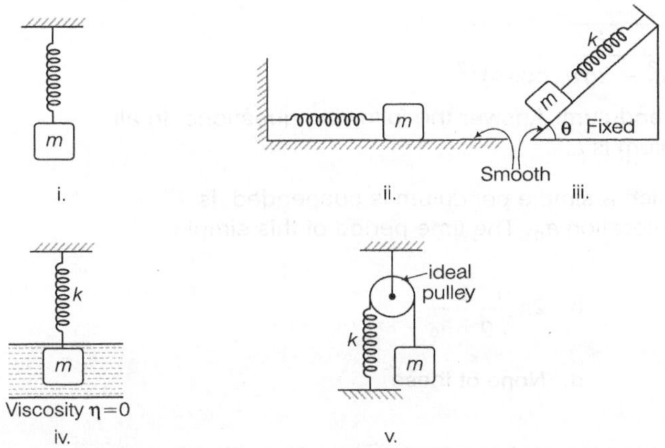

a. $T = 2\pi\sqrt{\dfrac{m}{k}}$ only in (i), (ii), (iii)

b. $T = 2\pi\sqrt{\dfrac{m}{k}}$ only in (i), (ii), (v)

c. $T = 2\pi\sqrt{\dfrac{m}{k}\cos\theta}$ in (iii)

d. $T = 2\pi\sqrt{\dfrac{m}{k}}$ in all cases

18 A spring-block system undergoes SHM on a smooth horizontal surface. The block is given some positive charge and a uniform electric field to the right is switched ON. As a result,

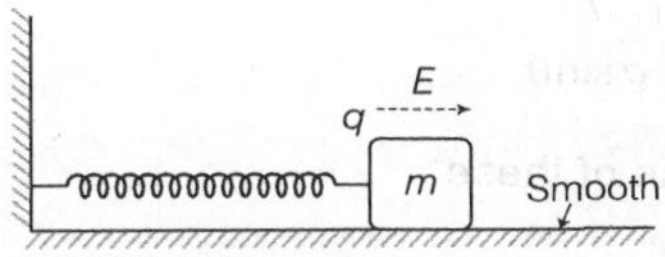

a. the time period of oscillation will decrease

b. the time period of oscillation will increase

c. the time period of oscillation will remain unaffected

d. the mean position of SHM will shift to the right

Comprehension Type

Working Space

Direction *(Q. Nos. 19-25) This section contains 2 paragraphs, each describing theory, experiments, data, etc. Seven questions related to the paragraphs have been given. Each question has only one correct answer among the four given options (a), (b), (c) and (d).*

Passage I

Consider a simple pendulum, whose point of suspension is accelerating at an angle ϕ with the vertical. The net force acting on the bob (in accelerating frame) is

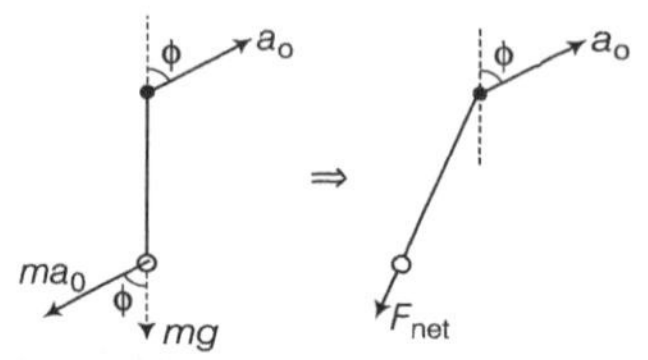

$$F_{net} = \sqrt{(mg)^2 + (ma_0)^2 + 2(mg)(ma_0)\cos\phi}$$

$$\therefore \quad a_{net} = \sqrt{g^2 + a_0^2 + 2ga_0\cos\phi}$$

$$T = 2\pi\sqrt{\frac{L}{a_{net}}} = 2\pi\sqrt{\frac{L}{(g^2 + a_0^2 + 2ga_0\cos\phi)^{1/2}}}$$

where, L is length of the simple pendulum. Answer the following questions. In all cases, length of the simple pendulum is L.

19 The container, from the roof of which a simple pendulum is suspended, is accelerating horizontally with acceleration a_0. The time period of this simple pendulum is

a. $2\pi\sqrt{\frac{L}{\sqrt{g^2 + a_0^2}}}$ b. $2\pi\sqrt{\frac{L}{g + a_0}}$

c. $\frac{\pi}{2}\sqrt{\frac{L}{g^2 - a_0^2}}$ d. None of these

20 A simple pendulum is suspended from the roof of a room. The bob (with mass m) of pendulum has net charge q. A vertically downward field E is switched ON. The time period of this simple pendulum is

a. $2\pi\sqrt{\frac{L}{\sqrt{g^2 + \left(\frac{Eq}{m}\right)^2}}}$ b. $2\pi\sqrt{\frac{L}{\sqrt{g^2 - \left(\frac{Eq}{m}\right)^2}}}$

c. $2\pi\sqrt{\frac{L}{g + \frac{Eq}{m}}}$ d. None of these

21 A container, from the roof of which a simple pendulum is suspended, is moving down on a smooth inclined plane. The inclined plane makes an angle θ with the horizontal. The time period of simple pendulum is

a. $2\pi\sqrt{\frac{L}{g}}$ b. $2\pi\sqrt{\frac{L}{g\sin\theta}}$

c. $2\pi\sqrt{\frac{L}{g\cos\theta}}$ d. None of these

Working Space

22 A simple pendulum is suspended in a horizontal electric field E. The mass of the bob is m and charge on it is q. The time period of the pendulum is

a. $2\pi\sqrt{\dfrac{L}{\sqrt{g^2+\left(\dfrac{Eq}{m}\right)^2}}}$ b. $2\pi\sqrt{\dfrac{L}{g^2+\left(\dfrac{Eq}{m}\right)^2}}$

c. $2\pi\sqrt{\dfrac{L}{g+\dfrac{Eq}{m}}}$ d. None of these

Passage II

A block is attached to a spring and is placed on a horizontal smooth surface as shown in the figure. Initially, the spring is unstretched. Now, the spring is given an initial compression $2d$ and the block is released from rest. Collisions of the block with wall W_2 are elastic.

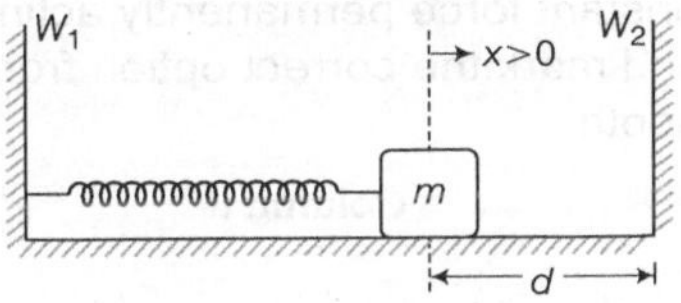

23 The time period of motion of the block is

a. $\dfrac{2\pi}{3}\sqrt{\dfrac{m}{k}}$ b. $\dfrac{3\pi}{2}\sqrt{\dfrac{m}{k}}$

c. $\dfrac{4\pi}{3}\sqrt{\dfrac{m}{k}}$ d. $\dfrac{\pi}{2}\sqrt{\dfrac{m}{k}}$

24 Its equation of motion is

a. $x=-2d\cos(\omega t+\pi)$ for $0<t<\dfrac{T}{2}$ b. $x=-d\cos\omega t$ for $0<t<\dfrac{T}{2}$

c. $x=-2d\cos\left(\omega t+\dfrac{4\pi}{3}\right)$ for $\dfrac{T}{2}<t<T$ d. $x=-2d\cos\left(\omega t+\dfrac{\pi}{3}\right)$ for $\dfrac{T}{2}<t<T$

25 Position-time (x-t) graph for one time period is (Treat the position of block in unstretched position as origin)

a. 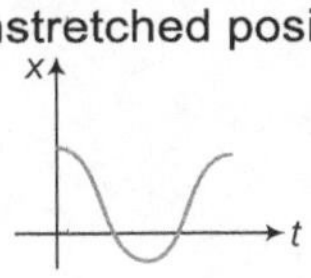

b. 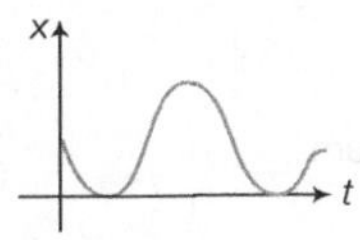

c. 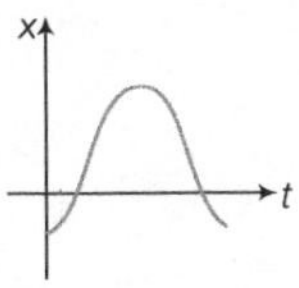

d. 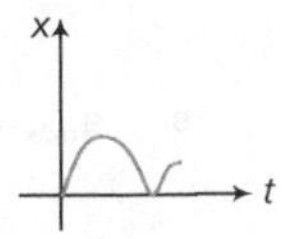

Matching List Type

Direction *(Q. Nos. 26 and 27) Choices for the correct combination of elements from Column I and Column II are given as options (a), (b), (c) and (d), out of which one is correct.*

26 A spring-block pendulum (horizontal configuration) executes SHM in such a way that the block has velocity v when it crosses the mean position. Now, the changes are made in such a way that the velocity while crossing the mean position gets doubled, without changing the mass of block. In Column I, some statements (in complete) are given and corresponding completions are given in Column II.

Working Space

Match the Column I with Column II and mark the correct option from the codes given below.

	Column I		Column II
i.	The frequency of oscillation will change by a factor of	p.	2
ii.	The amplitude of oscillation will change by a factor of	q.	$\sqrt{2}$
iii.	The magnitude of maximum acceleration will change by a factor of	r.	1
iv.	Maximum PE increases by a factor of	s.	4

Codes

	i	ii	iii	iv		i	ii	iii	iv
a.	r	p	p	s	b.	p	p	r	q
c.	p	q	r	s	d.	r	s	p	q

27 Column I shows spring-block system, with a constant force permanently acting on the block. Match the Column I with Column II and mark the correct option from the codes given below. Horizontal surfaces are smooth.

	Column I		Column II
i.	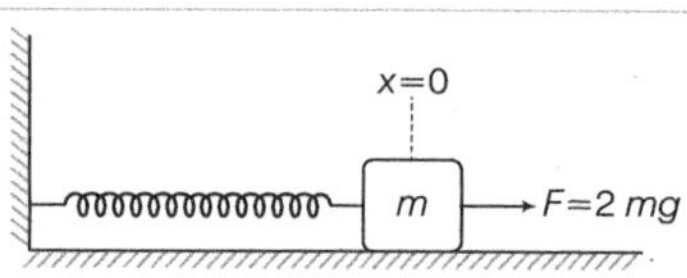 Spring is initially relaxed when force is applied.	p.	Time period is $T = 2\pi\sqrt{\frac{m}{k}}$
ii.	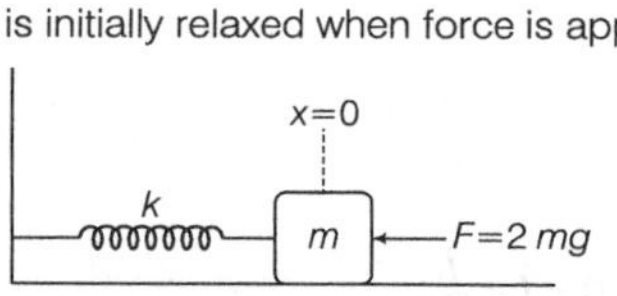 Spring is initially relaxed when force is applied.	q.	Amplitude is $A = \frac{2mg}{k}$
iii.	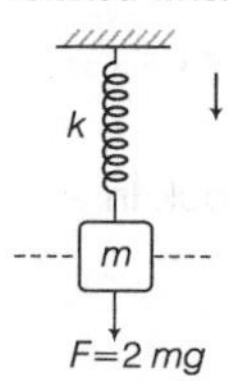 Before force is applied block is in equilibrium position.	r.	$v_{max} = 2g\sqrt{\frac{m}{k}}$
iv.	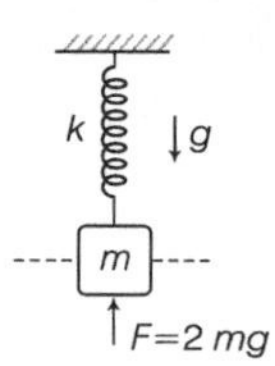 Before force is applied block is in equilibrium position.	s.	$a_{max} = 2g$
		t.	If spring acquires natural length, velocity of block at this instant is zero.

Codes

	i	ii	iii	iv
a.	p,q,r,s,t	p,q,r,s,t	p,q,r,s	p,q,r,s
b.	p,q,r,s	p,q,r,s,t	p,q,r,s,t	p,q,r,s
c.	p,q,r,s,t	p,q,r,s	p,q,r,s	p,q,r,s,t
d.	p,q,r,s,t	p,q,r,s,t	p,q,r,s,t	p,q,r,s

One Integer Value Correct Type

Direction (*Q. Nos.* 28-30) *This section contains 3 questions. When worked out will result in an integer from 0 to 9 (both inclusive).*

28 A sphere of mass M and radius R is on a smooth fixed inclined plane in equilibrium as shown in the figure. If now the sphere is displaced through a small distance along the plane, what will be the angular frequency (in rads^{-1}) of the resulting SHM? (Given, $k = \frac{4M}{3}$)

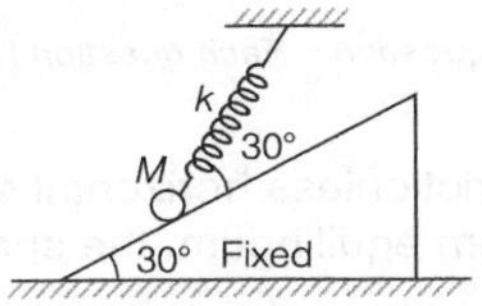

29 In the figure is shown, a plate of mass 60 g at rest and in equilibrium. A particle of mass 30 g is released from a height $\frac{4.5mg}{k}$ from the plate. The particle sticks to the plate. Spring has force constant 100 Nm^{-1}. Neglecting the duration of collision, the time from the collision of the particle and the plate to the moment when the spring has maximum compression is $(10\beta\pi)$ millisecond. Find the value of β.

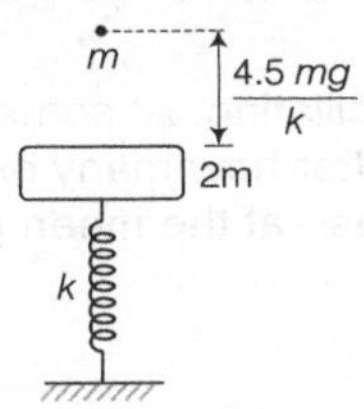

30 A spring is stretched by 0.2 m when a mass of 0.5 kg is suspended. The time period of the spring when a mass of 0.25 kg is suspended and put to oscillation is $\frac{\sqrt{2}\pi}{\psi}$ s. Find the value of ψ. (Given, $g = 9.8$ ms^{-2})

DPP-4 Time Period in SHM

Only One Option Correct Type

Working Space

Direction *(Q. Nos. 1-12) This section contains 12 multiple choice questions. Each question has four choices (a), (b), (c) and (d), out of which ONLY ONE is correct.*

1 A block is executing simple harmonic motion on a frictionless horizontal surface with an amplitude 0.1 m. At a point 0.06 m away from equilibrium, the speed of the block is 0.36 ms^{-1}. Its time period is

a. 1.12 s
b. 1.23 s
c. 1.40 s
d. 1.51 s

2 A particle of mass 5 g lies in a potential field $V = (8x^2 + 200)$ erg g^{-1}. Its time period is

a. 1.57 s
b. 3.14 s
c. 4.71 s
d. 2.23 s

3 Two pendulums of length 100 cm and 121 cm start oscillating. At some instant, the two are at the mean position in the same phase. After how many oscillations of the longer pendulum will the two be in the same phase at the mean position again.

a. 11
b. 10
c. 21
d. 20

4 The maximum displacement of the particle executing SHM is 1 cm and maximum acceleration is $(1.57)^2$ cms^{-2}. Its time period is

a. 0.25 s
b. 4 s
c. 1.57 s
d. 3.14 s

5 On a smooth inclined plane a body of mass m is attached between two springs. The other ends of the springs are fixed to firm supports. If each spring (ideal) has a force constant k, the period of oscillation of the body is (The springs are parallel to inclined plane)

a. $2\pi\sqrt{\frac{m}{2k}}$
b. $2\pi\sqrt{\frac{2m\sin\theta}{k}}$
c. $2\pi\sqrt{\frac{m\sin\theta}{2k}}$
d. $2\pi\sqrt{\frac{2m}{k}}$

6 The mass and diameter of a planet are twice those of the earth. The time period of a simple pendulum on this planet, if it is a second's pendulum on earth, is

a. $\sqrt{2}$ s
b. $\frac{1}{\sqrt{2}}$ s
c. $\frac{1}{2}$ s
d. $2\sqrt{2}$ s

Working Space

7 A small block is connected to one end of a massless spring of unstretched length 4.9 m. The other end of the spring is fixed as shown in the figure. The system lies on a horizontal frictionless surface. The block is stretched by 0.2 m and released from rest at $t = 0$. It then executes SHM with angular frequency $\omega = \frac{\pi}{3} \text{ rad s}^{-1}$. Simultaneously at $t = 0$, a small pebble is projected with a speed v from point P at an angle 45° as shown in the figure. Point P is at a horizontal distance of 10 m from O. If the pebble hits the block at $t = 1$ s, the value of v is (Given, $g = 10 \text{ ms}^{-2}$)

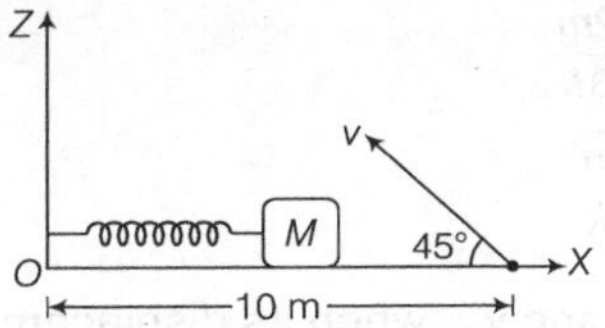

a. $\sqrt{50} \text{ ms}^{-1}$
b. $\sqrt{51} \text{ ms}^{-1}$
c. $\sqrt{52} \text{ ms}^{-1}$
d. $\sqrt{53} \text{ ms}^{-1}$

8 One end of a long metallic wire of length L_0 is tied to the ceiling. The other end is tied to a massless spring of spring constant k. A mass m hangs freely from the free end of the spring. The area of cross-section and the Young's modulus of the wire are A and Y respectively. If the mass is slightly pulled down and released, it will oscillate with a time period T equal to

a. $2\pi\sqrt{\frac{m}{k}}$
b. $2\pi\sqrt{\frac{mYA}{kL_0}}$
c. $\pi\sqrt{\frac{m(YA + kL_0)}{kYA}}$
d. $2\pi\sqrt{\frac{m(YA + kL_0)}{kYA}}$

9 A pendulum has time period T for small oscillations. Now, an obstacle is situated below the point of suspension O at a distance $\frac{3L_0}{4}$. The pendulum is released from rest. Throughout the motion, the moving string makes small angle with vertical. Time after which the pendulum returns back to its initial position is

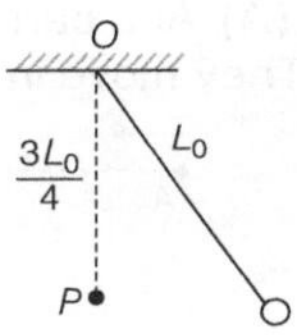

a. T
b. $\frac{4}{3}T$
c. $\frac{2}{3}T$
d. $\frac{3}{4}T$

10 A block of mass m is suspended by different springs of force constants as shown in the figure. Let time periods in these four situations are T_1, T_2, T_3 and T_4 respectively. Then,

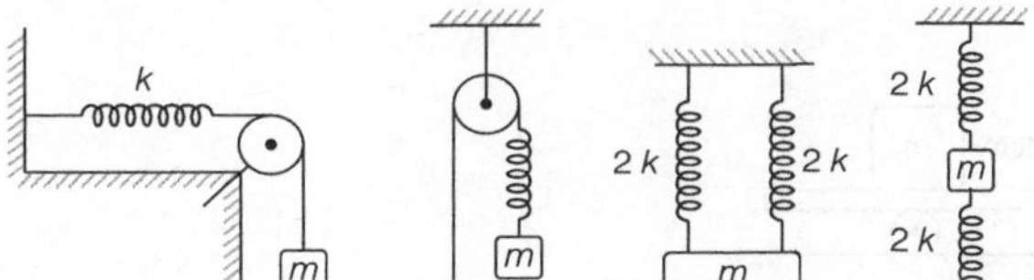

a. $T_1 = T_2 = T_3$
b. $T_1 = T_2 = T_4$
c. $T_1 = T_2$ and $T_3 = T_4$
d. $T_1 = T_3$ and $T_2 = T_4$

Working Space

11 A mass m is attached to four springs of spring constants $2k$, $2k$, k and k as shown in the figure. The mass is present on a smooth horizontal surface. It is displaced slightly leftwards and released. The time period (T) of oscillation is

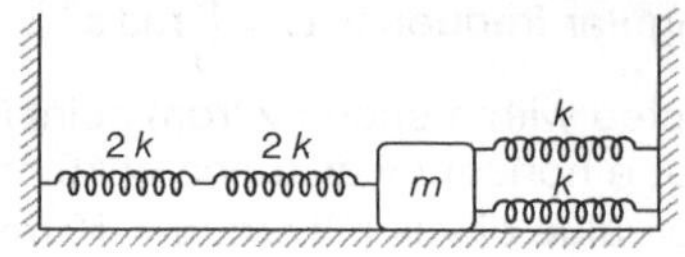

a. $2\pi\sqrt{\frac{m}{3k}}$ b. $2\pi\sqrt{\frac{2m}{3k}}$

c. $2\pi\sqrt{\frac{3m}{2k}}$ d. $\pi\sqrt{\frac{3m}{2k}}$

12 A particle is executing SHM. Its velocities are v_1 and v_2, when its displacements from the the mean position are x_1 and x_2, respectively. Its time period is

a. $2\pi\sqrt{\frac{x_1^2 + x_2^2}{v_1^2 + v_2^2}}$ b. $2\pi\sqrt{\frac{x_2^2 - x_1^2}{v_1^2 - v_2^2}}$

c. $2\pi\sqrt{\frac{x_1^2 - x_2^2}{v_1^2 - v_2^2}}$ d. None of these

One or More than One Options Correct Type

Direction (Q. Nos. 13-15) *This section contains 3 multiple choice questions. Each question has four choices (a), (b), (c) and (d), out of which ONE or MORE THAN ONE are correct.*

13 Two particles undergo SHM along the same line with the same time period (T) and equal amplitudes (A). At a particular instant, one particle is at $x = -A$ and the other is at $x = 0$. They move in the same direction. They will cross each other at

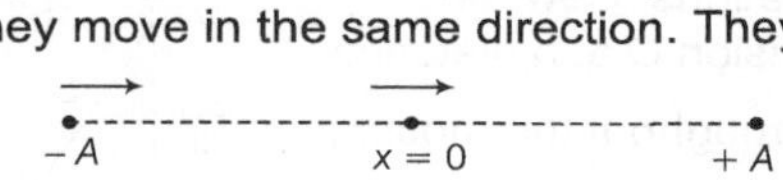

a. $t = \frac{4T}{3}$ b. $t = \frac{3T}{8}$

c. $x = \frac{A}{2}$ d. $x = \frac{A}{\sqrt{2}}$

14 A block of mass m is attached to a massless spring of force constant k, the other end of which is fixed from the wall of a truck as shown in the figure. The block is placed on the smooth horizontal surface of the truck and initially the spring is unstretched. Suddenly the truck starts moving towards left with a constant acceleration a_0. As seen from the truck,

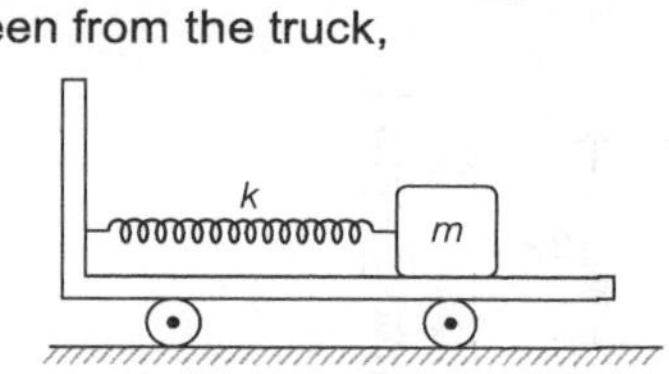

a. the particle will execute SHM

b. the time period of oscillation will be $2\pi\sqrt{\frac{m}{k}}$

c. the amplitude of oscillation will be $\frac{ma_0}{k}$

d. the energy of oscillation will be $\frac{m^2a_0^2}{k}$

15 Figure shows systems consisting of a massless pulley, a spring of force constant k and a block of mass m. If the block is slightly displaced vertically down from its equilibrium position and released, the periods of its vertical oscillations in cases (i), (ii) and (iii) respectively are T_1, T_2 and T_3. Then,

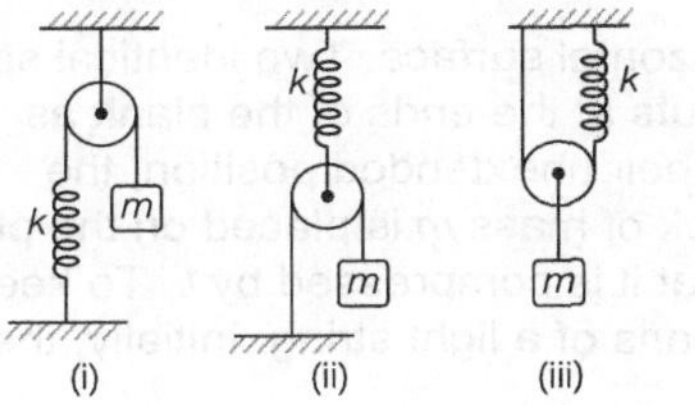

a. $T_1 = 2\pi\sqrt{\frac{m}{k}}$

b. $T_2 = 2\pi\sqrt{\frac{4m}{k}}$

c. $T_3 = 2\pi\sqrt{\frac{m}{2k}}$

d. $T_3 = 2\pi\sqrt{\frac{m}{4k}}$

Statement Type

Direction *(Q. Nos. 16-19) This section is based on Statement I and Statement II. Select the correct answer from the codes given below.*

a. Both Statement I and Statement II are correct and Statement II is the correct explanation of Statement I
b. Both Statement I and Statement II are correct but Statement II is not the correct explanation of Statement I
c. Statement I is correct but Statement II is incorrect
d. Statement II is correct but Statement I is incorrect

16 **Statement I** If a hole were drilled through the centre of earth, i.e. along the diameter and a ball is dropped into the hole at one end, it will not get out of the other end of the hole.

Statement II It will execute SHM and will be seen at the other end after 43.2 min from the time it is dropped.

17 **Statement I** If the length of a simple pendulum is increased by 4%, then the corresponding decrease in time period will be 2%.

Statement II $T \propto \sqrt{L}$

18 **Statement I** The bob of a simple pendulum is a hollow ball full of water. If a fine hole is made at the bottom of the ball, then the time period will vary till the ball gets empty and then becomes constant.

Statement II The time period of simple pendulum does not depend on the mass of the bob.

19 **Statement I** The time period of a hard spring is less as compared to that of a soft spring.

Statement II The time period is inversely proportional to the square root of the spring constant.

Working Space

Comprehension Type

Direction (Q. Nos. 20-25) *This section contains 2 paragraphs, each describing theory, experiments, data, etc. Six questions related to the paragraphs have been given. Each question has only one correct answer among the four given options (a), (b), (c) and (d).*

Passage I

A plank of mass M is placed on a smooth horizontal surface. Two identical springs each of stiffness k are rigidly connected to struts at the ends of the plank as shown in the figure. When the springs are in their unextended position, the distance between their free ends is $3L$. A block of mass m is placed on the plank and pressed against one of the springs, so that it is compressed by L. To keep the block at rest it is connected to the strut by means of a light string. Initially, the system is at rest. Now, the string is burnt.

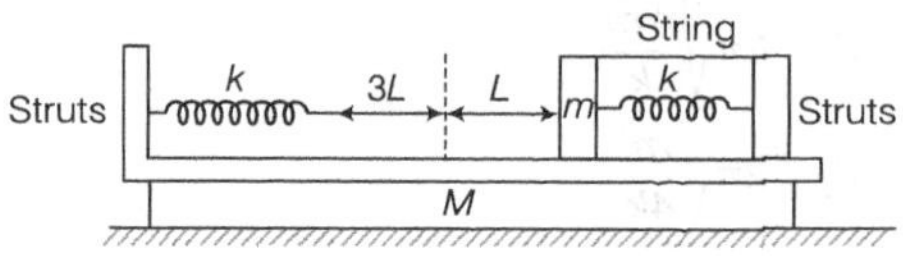

20 Maximum displacement of the plank is

a. $\dfrac{mL}{m+M}$ b. $\dfrac{5mL}{m+M}$

c. $\dfrac{3mL}{m+M}$ d. $\dfrac{2mL}{m+M}$

21 What is the relative speed of block with respect to the plank, immediately after right side spring regains its natural length?

a. $\left[\sqrt{\left(\dfrac{M+m}{Mm}\right)k}\right]L$ b. $\left[\sqrt{\left(\dfrac{M+m}{2Mm}\right)k}\right]L$

c. $\left[\sqrt{2\left(\dfrac{M+m}{Mm}\right)}\right]L$ d. None of these

22 Time period of oscillation of block is

a. $(2\pi+3)\sqrt{\dfrac{2Mm}{k(M+m)}}$ b. $(\pi+6)\sqrt{\dfrac{2Mm}{k(M+m)}}$

c. $(2\pi+6)\sqrt{\dfrac{Mm}{k(M+m)}}$ d. None of these

Passage II

In the shown arrangement, both the springs are in their natural lengths. The coefficient of friction between m_2 and m_1 is μ. There is no friction between m_1 and horizontal surface. If the blocks are displaced slightly, then they together perform simple harmonic motion.

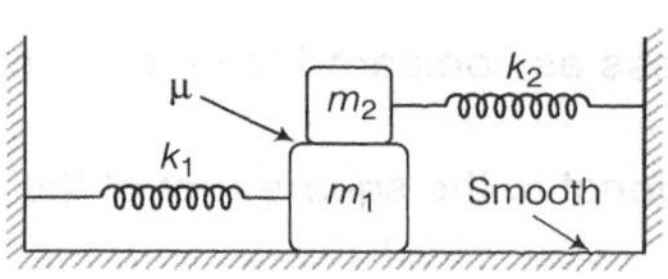

23 The frequency of such oscillations is

a. $\dfrac{1}{2\pi}\sqrt{\dfrac{k_1+k_2}{m_1+m_2}}$ b. $\dfrac{1}{\pi}\sqrt{\dfrac{k_1+k_2}{m_1+m_2}}$

c. $\dfrac{1}{2\pi}\sqrt{\dfrac{m_1+m_2}{k_1+k_2}}$ d. $\dfrac{1}{\pi}\sqrt{\dfrac{m_1+m_2}{k_1+k_2}}$

Working Space

24 If the frictional force on block m_2 is to act in the direction of its displacement from mean position, the condition is

a. $\frac{k_1}{k_2} > \frac{m_1}{m_2}$

b. $\frac{k_1}{k_2} < \frac{m_1}{m_2}$

c. $k_1 k_2 > m_1 m_2$

d. $k_1 k_2 < m_1 m_2$

25 If the condition obtained in the above problem is met, the maximum amplitude of their oscillations is

a. $\frac{\mu(m_1 - m_2)m_2 g}{m_2 k_1 - m_1 k_2}$

b. $\frac{\mu(m_1 + m_2)m_2 g}{m_1 k_1 - m_2 k_2}$

c. $\frac{\mu(m_1 + m_2)m_2 g}{m_1 k_2 - m_2 k_1}$

d. $\frac{\mu(m_1 + m_2)m_1 g}{m_2 k_1 + m_2 k_1}$

One Integer Value Correct Type

Direction *(Q. Nos. 26-29) This section contains 4 questions. When worked out will result in an integer from 0 to 9 (both inclusive).*

26 A mass m attached to a spring oscillates with a period of 3s. If the mass is increased by 1 kg, the period increases by 1s. The initial mass m is $\frac{\beta}{7}$ kg. Find the value of β.

27 A highly rigid cubical block B_1 of small mass m and side L is fixed rigidly onto another cubical block B_2 of same dimensions and of low modulus of rigidity η, such that the lower face of B_1 completely covers the upper face of B_2. The lower face of B_2 is rigidly held on a horizontal surface. A small force F is applied perpendicular to one of the side faces of B_1. After force is withdrawn, block B_1 executes small oscillations, the time period of which is $\pi\sqrt{\frac{\alpha m}{\eta L}}$. Find the value of α.

28 An object suspended from a spring exhibits oscillations of time period T. Now the spring is cut in two equal halves and the same object is suspended with two halves as shown in the figure. The new time period of oscillation is $\frac{T}{\beta}$. Find the value of β.

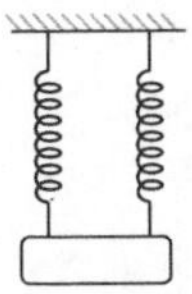

29 The period of small oscillations of a pendulum, that is bob suspended by a thread $L = 20$ cm in length, if it is located in a liquid whose density is 3 times less than that of bob is $\frac{5+x}{10}$ second (approximately). Find the value of x.

DPP-5 Time Period : Physical Systems-I

Only One Option Correct Type

Direction *(Q. Nos. 1-12) This section contains 12 multiple choice questions. Each question has four choices (a), (b), (c) and (d), out of which ONLY ONE is correct.*

Working Space

1 An air chamber of volume V, has a long neck of cross-sectional area A. A ball of mass m is fitted smoothly in the neck. The bulk modulus of air is B. If the ball is pressed down slightly and released, the time period of its oscillation is

a. $2\pi\sqrt{\frac{mV}{2BA^2}}$ b. $\pi\sqrt{\frac{2mV}{BA^2}}$

c. $2\pi\sqrt{\frac{mV}{BA^2}}$ d. $\frac{\pi}{2}\sqrt{\frac{m}{BA^2}}$

2 A tube of uniform cross-sectional area A is fixed to a horizontal surface in such a way that its two arms make angles θ_1 and θ_2 with the horizontal surface as shown in the figure. The tube contains water of mass m and density ρ. Now, air is blown into one of the arms and the water column starts oscillating with small amplitude. The period of oscillation of water column is (consider water to be ideal)

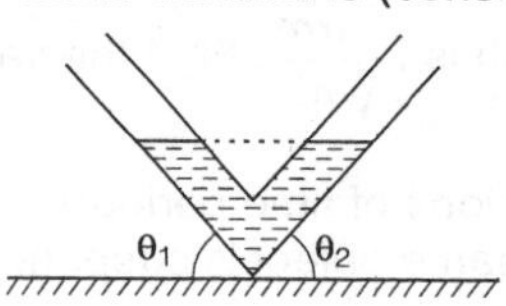

a. $2\pi\sqrt{\frac{m}{2\rho gA(\sin\theta_1+\sin\theta_2)}}$

b. $2\pi\sqrt{\frac{m}{\rho gA(\sin\theta_1+\sin\theta_2)}}$

c. $\pi\sqrt{\frac{2m}{\rho gA(\sin\theta_1+\sin\theta_2)}}$

d. None of these

3 A fixed U-tube of cross-sectional area A contains water in it as shown in the figure. The length of water column in one tube is L. The period of oscillations of water column is (consider water to be ideal)

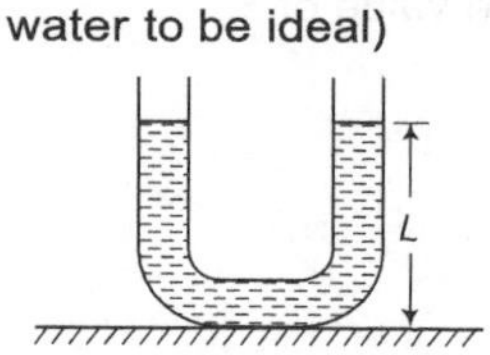

a. $2\pi\sqrt{\frac{L}{g}}$ b. $2\pi\sqrt{\frac{\sqrt{A}}{g}}$

c. $2\pi\sqrt{\frac{A}{Lg}}$ d. $2\pi\sqrt{\frac{gA}{L}}$

Working Space

4 A solid pole of cross-sectional area A and mass m is floating in a liquid of density ρ. The length of the pole immersed in the liquid is L in the equilibrium position. Now, the pole is pressed down slightly and released, its time period of oscillation is

a. $2\pi\sqrt{\dfrac{L}{g}}$ b. $2\pi\sqrt{\dfrac{\sqrt{A}}{g}}$

c. $2p\sqrt{\dfrac{A}{Lg}}$ d. $2\pi\sqrt{\dfrac{gA}{L}}$

5 Two identical springs, each of stiffness k are welded to each other at point P. The other two ends are fixed to the edge of a smooth horizontal tube as shown. A particle of mass m is welded at P. The entire system is horizontal. The period of oscillation of the particle in the direction of x is

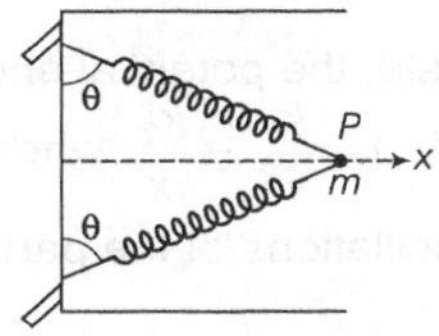

a. $\dfrac{2\pi}{\sin\theta}\sqrt{\dfrac{2m}{k}}$ b. $\dfrac{2\pi}{\sin\theta}\sqrt{\dfrac{m}{k}}$

c. $\dfrac{\pi}{2\sin\theta}\sqrt{\dfrac{m}{k}}$ d. $\dfrac{\pi}{\sin\theta}\sqrt{\dfrac{2m}{k}}$

6 A particle of mass m is attached to a network of four springs as shown in the figure. The entire system is on a horizontal smooth surface. The frequency of oscillation of the particle along Y-direction is

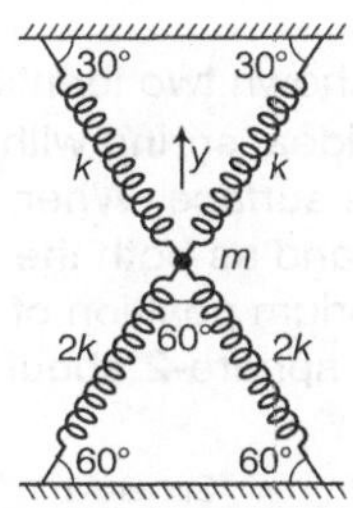

a. $\dfrac{1}{2\pi}\sqrt{\dfrac{5k}{3m}}$ b. $\dfrac{1}{2\pi}\sqrt{\dfrac{7k}{2m}}$

c. $\dfrac{1}{2\pi}\sqrt{\dfrac{9k}{4m}}$ d. $\dfrac{1}{\pi}\sqrt{\dfrac{4k}{3m}}$

7 A smooth piston of mass m, area of cross-section A is in equilibrium in a gas jar when the pressure of the gas is p_0. When the piston is disturbed slightly, the angular frequency of oscillation of the piston is
(Assume adiabatic change of state of the gas and γ is ratio of specific heats of the gas)

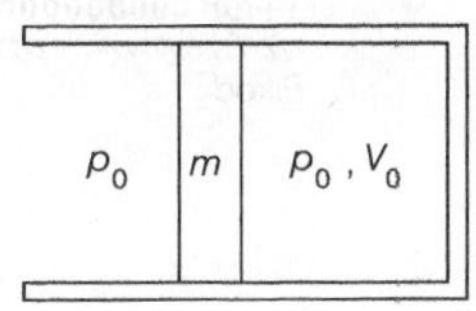

a. $\sqrt{\dfrac{\gamma p_0 A^2}{3m V_0}}$ b. $\sqrt{\dfrac{\gamma p_0 A^2}{\sqrt{2} m V_0}}$

c. $\sqrt{\dfrac{\gamma p_0 A^2}{m V_0}}$ d. $\sqrt{\dfrac{2\gamma p_0 A^2}{m V_0}}$

8 Two point charges each with charge $+q$ are fixed at points $(0, d)$ and $(0, -d)$ on Y-axis of a coordinate system as shown in the figure. Another small particle of mass m and charge $-q_0$ is placed at the origin of the system, where it stays in equilibrium. If the mass m is slightly displaced along $+x$-direction by a small distance, its time period of oscillations is $\left[k = \dfrac{1}{4\pi\varepsilon_0}\right]$

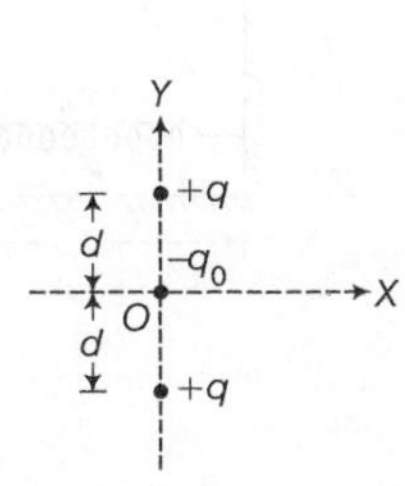

a. $\dfrac{\pi}{2}\sqrt{\dfrac{md^3}{2kqq_0}}$ b. $\pi\sqrt{\dfrac{md^3}{kqq_0}}$

c. $2\pi\sqrt{\dfrac{md^3}{2kqq_0}}$ d. None of these

Working Space

9 A particle of mass m is located in a unidimensional potential field, where the potential energy of the particle depends on the coordinate x as $U(x) = U_0(1 - \cos\beta x)$, where U_0 and β are constants. Find the time period of small oscillations that the particle performs about the equilibrium position. (Solve the problem by Taylor's method)

a. $4\pi\sqrt{\dfrac{m}{U_0\beta^2}}$ b. $2\pi\sqrt{\dfrac{m}{U_0\beta^2}}$

c. $2\pi\sqrt{\dfrac{m}{2U_0\beta^2}}$ d. $\pi\sqrt{\dfrac{m}{U_0\beta^2}}$

10 In a given force field, the potential energy of a particle is given as a function of its x-coordinate as $U(x) = \dfrac{k_1}{x^2} - \dfrac{k_2}{x}$, where k_1 and k_2 are positive constants. Find the period of small oscillations of the particle about its equilibrium position in the field.

a. $\pi\sqrt{\dfrac{4mk_1^3}{k_2^4}}$ b. $\dfrac{\pi}{2}\sqrt{\dfrac{8mk_1^3}{k_2^4}}$

c. $4\pi\sqrt{\dfrac{2mk_1^3}{k_2^4}}$ d. $2\pi\sqrt{\dfrac{2mk_1^3}{k_2^4}}$

11 In the figure are shown two identical spheres, each of mass m and charge $+Q$ connected by an ideal spring with force constant k. Sphere-1 is fixed to the horizontal smooth surface. When the spheres were uncharged, the spring was in its natural length and as both the spheres are charged, the spring length gets doubled in equilibrium position of the spheres. Find the time period of small oscillations of the sphere-2 about the new equilibrium position, after charging of the spheres.

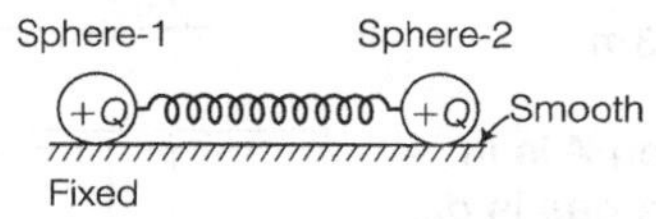

a. $2\pi\sqrt{\dfrac{m}{k}}$ b. $\dfrac{\pi}{3}\sqrt{\dfrac{2m}{k}}$

c. $\pi\sqrt{\dfrac{m}{2k}}$ d. $2\pi\sqrt{\dfrac{m}{2k}}$

12 In the spring-block system, the spring has mass m_s. Assume that the extension of the spring is uniform over its length and its linear mass density λ is uniform. Find the time period of the system for small oscillations.

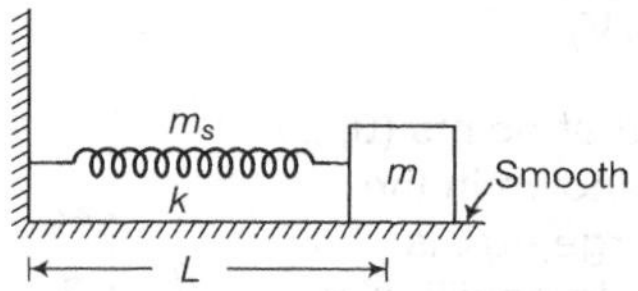

a. $2\pi\sqrt{\dfrac{m + \frac{m_s}{3}}{k}}$ b. $2\pi\sqrt{\dfrac{m + \frac{2m_s}{3}}{k}}$

c. $2\pi\sqrt{\dfrac{m + \frac{m_s}{2}}{k}}$ d. $2\pi\sqrt{\dfrac{m}{k}}$

Working Space

One or More than One Options Correct Type

Direction (Q. Nos. 13-15) *This section contains 3 multiple choice questions. Each question has four choices (a), (b), (c) and (d), out of which ONE or MORE THAN ONE are correct.*

13 A light rod is supported by two springs fixed to ceiling. A mass m hangs from centre of the rod with a third spring as shown in the figure. The natural frequency of the system when spring 1 and 2 share the weight mg equally is f_1 and the natural frequency of the system when the bar is assumed to stay horizontal all the time is f_2. Then, (Given : springs 1 and 2 have equal relaxed long lengths)

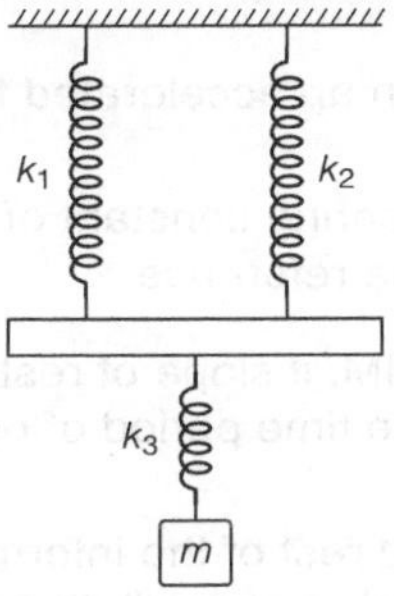

a. $f_1 = \frac{1}{2\pi}\sqrt{\frac{4k_1k_2k_3}{m(4k_1k_2 + k_1k_3 + k_2k_3)}}$

b. $f_2 = \frac{1}{2\pi}\sqrt{\frac{k_1k_3 + k_2k_3}{m(k_1 + k_2 + k_3)}}$

c. $f_1 = \frac{1}{2\pi}\sqrt{\frac{3k_1k_2k_3}{m(4k_1k_2 + k_1k_3 + k_2k_3)}}$

d. $f_2 = \frac{1}{2\pi}\sqrt{\frac{k_1k_2 + k_2k_3}{2m(k_1 + k_2 + k_3)}}$

14 In the figure is shown a vessel filled with a liquid whose density varies with height h from the bottom as $\rho = \rho_0\left(4 - \frac{3h}{h_0}\right)$, where ρ_0 is a constant and h_0 is the height of the liquid in the container. A solid block of small dimensions of density $\frac{5}{2}\rho_0$ and mass m is released from the bottom of the tank. Then,

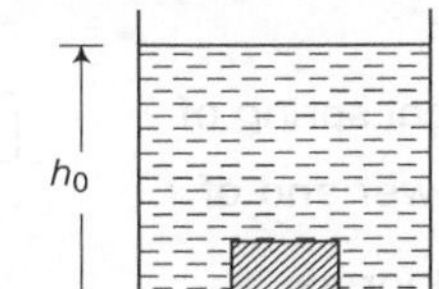

a. the block will perform SHM

b. the frequency of oscillation is $\frac{1}{2\pi}\sqrt{\frac{3g}{4h_0}}$

c. the frequency of oscillation is $\frac{1}{2\pi}\sqrt{\frac{6g}{5h_0}}$

d. the block will not perform SHM, but only periodic motion

15 A cone made of a material of relative density $\left[s = \frac{27}{64}\right]$ and height 4 m floats with its apex downward in water. The submerged height of cone in water is h. If the cone is slightly displaced vertically from its equilibrium position, its time period of oscillation is T. Then, (Given, $g = 10\,\text{ms}^{-2}$)

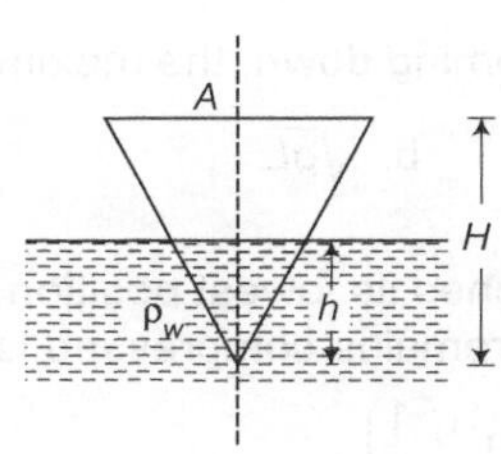

a. $h = 3$m

b. $h = 2.5$m

c. $T = 2.98$s

d. $T = 1.98$s

Working Space

Statement Type

Direction (Q. Nos. 16-18) *This section is based on Statement I and Statement II. Select the correct answer from the codes given below.*

a. Both Statement I and Statement II are correct and Statement II is the correct explanation of Statement I
b. Both Statement I and Statement II are correct but Statement II is not the correct explanation of Statement I
c. Statement I is correct but Statement II is incorrect
d. Statement II is correct but Statement I is incorrect

16 **Statement I** Time period of spring pendulum is same in an accelerated frame and in an inertial frame of reference.

Statement II Mass of the bob of spring pendulum and spring constant of the spring are independent of the acceleration of the frame of the reference.

17 **Statement I** For a particle of mass 1 kg, executing SHM, if slope of restoring force (Y-axis) *vs* displacement (X-axis) graph is -1, then the time period of oscillation is 6.28 s.

Statement II If 1 kg mass is replaced by 2 kg mass and rest of the information remains same as in the above statement, then the period of oscillation will still be 6.28 s.

18 **Statement I** In simple harmonic motion, the graph between velocity and displacement is ellipse.

Statement II In simple harmonic motion, phase difference between velocity and displacement is π.

Comprehension Type

Direction (Q. Nos. 19-21) *This section contains a paragraph, describing theory, experiments, data, etc. Three questions related to the paragraph have been given. Each question has only one correct answer among the four given options (a), (b), (c) and (d).*

Passage

A small block of mass m is fixed at upper end of a vertical spring of spring constant $k = \dfrac{4mg}{L}$ and natural length $10L$. The lower end of spring is free and is at a height L from fixed horizontal floor as shown in the figure. The spring is initially unstretched and the spring-block system is released from rest in the shown position.

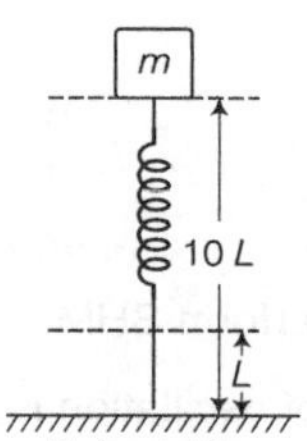

19 At the instant, the speed of block is maximum, the magnitude of force exerted by the spring on the block is

a. mg
b. $\dfrac{mg}{2}$
c. zero
d. None of these

20 As the block is coming down, the maximum speed attained by the block is

a. $\sqrt{\dfrac{3}{2}gL}$
b. $\sqrt{gL}$
c. $\sqrt{3gL}$
d. $\dfrac{3}{2}\sqrt{gL}$

21 Till the block reaches its lowest position for the first time, the time duration for which the spring remains compressed is

a. $\dfrac{\pi}{4}\sqrt{\dfrac{L}{g}} + \sqrt{\dfrac{L}{4g}}\sin^{-1}\left(\dfrac{1}{3}\right)$
b. $\pi\sqrt{\dfrac{L}{2g}} + \sqrt{\dfrac{L}{4g}}\sin^{-1}\left(\dfrac{2}{3}\right)$
c. $\pi\sqrt{\dfrac{L}{2g}} + \sqrt{\dfrac{L}{4g}}\sin^{-1}\left(\dfrac{1}{3}\right)$
d. $\dfrac{\pi}{2}\sqrt{\dfrac{L}{2g}} + \sqrt{\dfrac{L}{4g}}\sin^{-1}\left(\dfrac{2}{3}\right)$

Working Space

One Integer Value Correct Type

Direction *(Q. Nos. 22-25) This section contains 4 questions. When worked out will result in an integer from 0 to 9 (both inclusive).*

22 Three identical ideal springs, each of force constant k are joined to three identical balls (each of mass m) as shown in the figure. O is centroid of the triangle. Initially, each of the springs is in its natural length. Now, all the three balls are simultaneously given small displacements of equal magnitude along the directions shown in the figure. The oscillation frequency of balls is $\frac{1}{2\pi}\sqrt{\frac{\beta k}{2m}}$. Find the value of β.

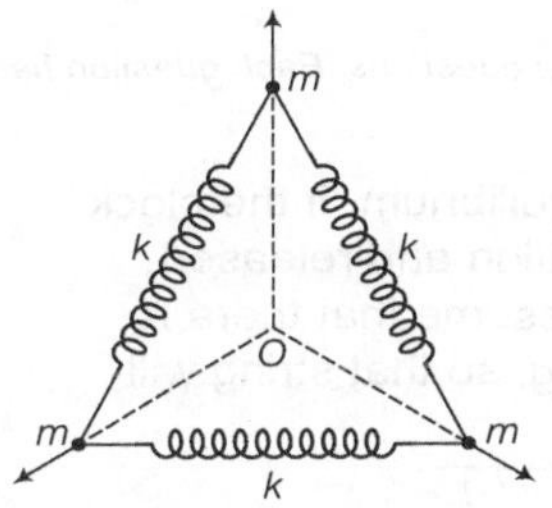

23 Two identical blocks (each of mass m) are connected by a spring in its natural length with force constant k. The system is kept on a smooth horizontal surface. At $t = 0$, the left side block is given a sharp impulse J towards the right and the blocks begin to slide along the table. If ω is the angular frequency of the oscillations, the velocity of left block as a function of time is $\frac{2J}{\alpha m}(1+\cos\omega t)$. Find the value of α.

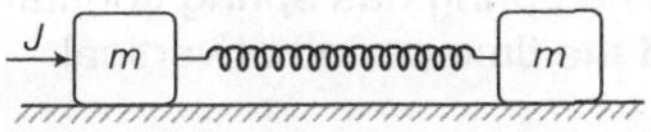

24 A pendulum clock is mounted on an elevator, which starts from rest and goes up with constant acceleration a_0, where $\frac{a_0}{g} = \frac{2}{5}$. At a height h, the acceleration of the elevator reverses in direction, its magnitude remaining constant. After a time $\frac{\psi}{5}\sqrt{\frac{2h}{a_0}}$, from the start the clock will show right time again. Find the value of ψ.
(Given, $\sqrt{7} = 2.6$, $\sqrt{3} = 1.7$ and $\sqrt{5} = 2.2$)

25 A vertical pole of length L, density ρ, area of cross-section A floats in two immiscible liquids of densities $2.25\,\rho$ and 4ρ. In equilibrium position, the bottom end is at the interface of the liquids. When the pole is slightly depressed and released, it performs oscillations. The time period of oscillations is $\frac{c}{6}\pi\sqrt{\frac{L}{g}}$. Find the value of c.

DPP-6 Time Period : Physical Systems-II

Only One Option Correct Type

Direction (*Q. Nos. 1-15*) *This section contains 15 multiple choice questions. Each question has four choices (a), (b), (c) and (d), out of which ONLY ONE is correct.*

Working Space

1 In the shown figure, a pulley-block system is in equilibrium. If the block is displaced down slightly from its equilibrium position and released, find the time period of oscillation of the system. Assume that there is sufficient friction present between pulley and string, so that string will not slip over pulley surface.

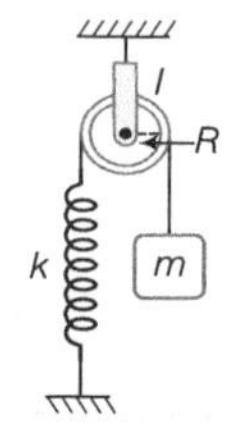

a. $\frac{\pi}{2}\sqrt{\frac{m+\frac{I}{R^2}}{k}}$ b. $2\pi\sqrt{\frac{m+\frac{I}{R^2}}{k}}$

c. $\pi\sqrt{\frac{m+\frac{I}{R^2}}{k}}$ d. $2\pi\sqrt{\frac{m+\frac{I}{R^2}}{\sqrt{2}k}}$

2 The pulley shown in the figure has moment of inertia I about its axis (perpendicular to plane of paper) and mass M. The spring has spring constant k and the string does not slip over the pulley. Find the time period of vertical oscillations of its centre of mass.

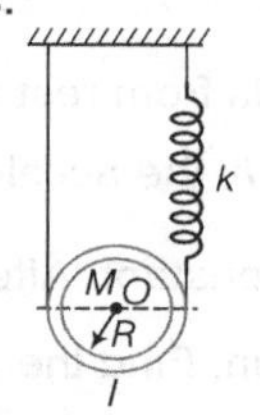

a. $2\pi\sqrt{\frac{M+\frac{I}{R^2}}{k}}$ b. $2\pi\sqrt{\frac{M+\frac{I}{R^2}}{2k}}$ c. $\pi\sqrt{\frac{M+\frac{I}{R^2}}{2k}}$ d. $2\pi\sqrt{\frac{M+\frac{I}{R^2}}{4k}}$

3 In the figure is shown, a pulley-block system in which block of mass M is hanging on one side of pulley and on the other side a particle P (of mass m) is welded on pulley. The moment of inertia of the pulley is I and the system is in equilibrium when the particle is at an angle ϕ from the vertical. If the system is slightly disturbed from its equilibrium position, find the time period of its oscillations.

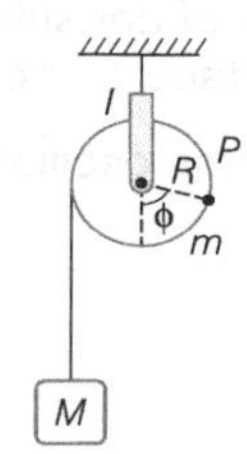

a. $2\pi\sqrt{\frac{R\left(M+m+\frac{I}{R^2}\right)}{mg\cos\phi}}$ b. $\pi\sqrt{\frac{R\left(M+m+\frac{I}{R^2}\right)}{2mg\cos\phi}}$

c. $2\pi\sqrt{\frac{R\left(M+m+\frac{I}{R^2}\right)}{2mg\cos\phi}}$ d. $\pi\sqrt{\frac{R\left(M+m+\frac{I}{R^2}\right)}{mg\cos\phi}}$

Working Space

4 A particle of mass m is allowed to oscillate on a smooth parabola $x^2 = 4ay$, $a > 0$, about the origin O (see figure). For small oscillations, find the angular frequency (ω).

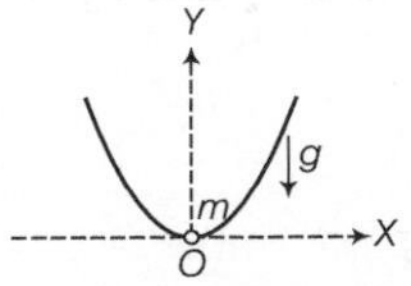

a. $\sqrt{\frac{g}{2a}}$ b. $\sqrt{\frac{g}{a}}$

c. $\sqrt{\frac{2g}{a}}$ d. $2\sqrt{\frac{g}{a}}$

5 A uniform horizontal plank is resting symmetrically in a horizontal position on two cylindrical drums, which are spinning in opposite directions about their horizontal axes with equal angular speed. The distance between the axes is $2L$ and the coefficient of friction between the plank and cylinder is μ. If the plank is displaced slightly from the equilibrium position along its length and released, find out the time period of oscillations of the plank.

a. $2\pi\sqrt{\frac{L}{\mu g}}$ b. $\pi\sqrt{\frac{L}{\mu g}}$

c. $\frac{2}{3}\pi\sqrt{\frac{L}{\mu g}}$ d. $\frac{\pi}{2}\sqrt{\frac{L}{\mu g}}$

6 A block of mass m hangs by means of a string which goes over a pulley (of mass m and moment of inertia I) as shown in the figure. The string does not slip on the pulley. The frequency of small vertical oscillations of the block is

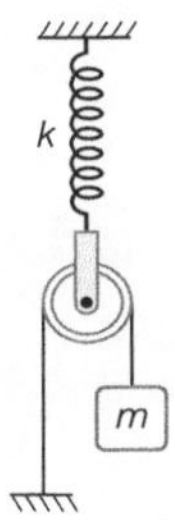

a. $\frac{1}{\pi}\sqrt{\frac{2k}{5m + \frac{I}{R^2}}}$ b. $\frac{1}{2\pi}\sqrt{\frac{k}{\frac{5m}{2} + \frac{I}{2R^2}}}$

c. $\frac{1}{2\pi}\sqrt{\frac{k}{5m + \frac{I}{R^2}}}$ d. None of these

7 Find the period of small oscillations in a horizontal plane performed by a ball of mass $m = 40$g fixed at the middle of a horizontally stretched string $l = 1$m in length. The tension of the string is assumed to be constant and equal to $T = 10$N.

a. 0.1 s b. 0.2 s

c. 0.3 s d. 0.4 s

8 A solid uniform cylinder of mass m performs small oscillations in horizontal plane if slightly displaced from its mean position shown in the figure. Initially springs are in natural lengths and cylinder does not slip on ground during oscillations due to friction between ground and cylinder. Force constant of each spring is k. The time period of oscillations is

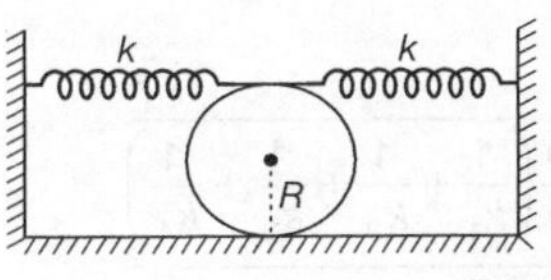

a. $\pi\sqrt{\frac{3m}{k}}$ b. $\frac{\pi}{4}\sqrt{\frac{3m}{k}}$

c. $\frac{\pi}{2}\sqrt{\frac{3m}{k}}$ d. $2\pi\sqrt{\frac{m}{2k}}$

Working Space

9 A pendulum inside an elevator has a time period T_1, when the elevator is at rest, T_2, when the elevator accelerates up, T_3, when the elevator accelerates down (both accelerations are equal in magnitude). T_1 in terms of T_2 and T_3 is

a. $\dfrac{\sqrt{2}\,T_2 T_3}{\sqrt{T_2^2 + T_3^2}}$ b. $\dfrac{2T_2T_3}{\sqrt{T_2^2 + T_3^2}}$

c. $\dfrac{\sqrt{2}\,T_2T_3}{\sqrt{2T_2^2 + T_3^2}}$ d. $\dfrac{2T_2 T_3}{\sqrt{T_2^2 + 2T_3^2}}$

10 An endless chord consists of two portions $2L$ and $2L'$ respectively, knotted together. The mass per unit length of each string is λ and λ'. It is placed in stable equilibrium (in vertical plane) over a small smooth peg and then slightly displaced vertically. The time period of oscillation is

a. $\pi\sqrt{\dfrac{\lambda L + \lambda' L'}{(\lambda' - \lambda)g}}$ b. $\dfrac{\pi}{2}\sqrt{\dfrac{\lambda L + \lambda' L'}{(\lambda' + \lambda)g}}$

c. $\pi\sqrt{\dfrac{\lambda L + \lambda' L'}{2(\lambda' - \lambda)g}}$ d. $2\pi\sqrt{\dfrac{\lambda L + \lambda' L'}{(\lambda' - \lambda)g}}$

11 In the figure shown pulleys, string and springs are ideal. The natural frequency of oscillation is

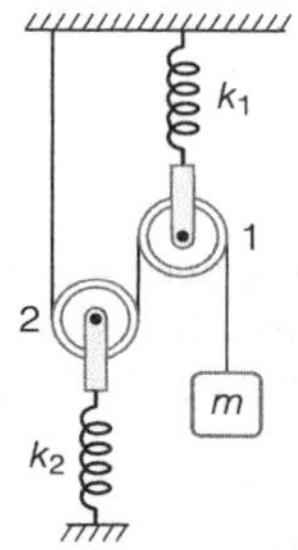

a. $\sqrt{\dfrac{k_1 k_2}{4m(k_1 + k_2)}}$ b. $\sqrt{\dfrac{k_1 k_2}{2m(k_1 + k_2)}}$

c. $\sqrt{\dfrac{2k_1 k_2}{m(k_1 + k_2)}}$ d. $\sqrt{\dfrac{4k_1 k_2}{m(k_1 + k_2)}}$

12 In the arrangement shown, pulleys, springs and string are ideal. k_1, k_2, k_3 and k_4 are respective force constants of the springs. The time period of small vertical oscillations of the block of mass m is

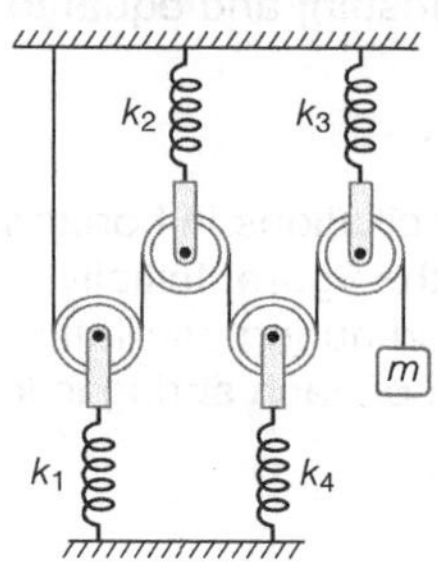

a. $2\pi\sqrt{m\left(\dfrac{1}{k_1} + \dfrac{1}{k_2} + \dfrac{1}{k_3} + \dfrac{1}{k_4}\right)}$ b. $2\pi\sqrt{\dfrac{m}{4}\left(\dfrac{1}{k_1} + \dfrac{1}{k_2} + \dfrac{1}{k_3} + \dfrac{1}{k_4}\right)}$

c. $2\pi\sqrt{4m\left(\dfrac{1}{k_1} + \dfrac{1}{k_2} + \dfrac{1}{k_3} + \dfrac{1}{k_4}\right)}$ d. $\pi\sqrt{4m\left(\dfrac{1}{k_1} + \dfrac{1}{k_2} + \dfrac{1}{k_3} + \dfrac{1}{k_4}\right)}$

13 In the arrangement shown spring, pulleys, strings are ideal. The natural frequency of the system for vertical oscillations is

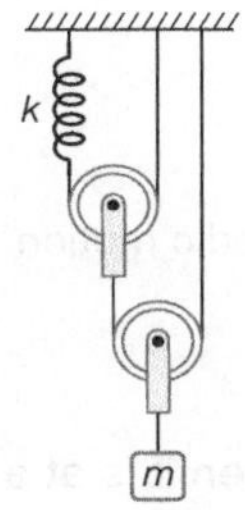

a. $\frac{1}{2\pi}\sqrt{\frac{8k}{m}}$ b. $\frac{2}{\pi}\sqrt{\frac{k}{m}}$

c. $\frac{1}{2\pi}\sqrt{\frac{2k}{m}}$ d. $\frac{2}{\pi}\sqrt{\frac{2k}{m}}$

14 A plank of mass m is kept at rest on two identical spheres, each of mass m_0 and radius R. The plank is connected to two ideal springs, which is turn are attached to walls as shown in the figure. Find the frequency of small (horizontal) oscillations (in the plane of the figure) of the plank. Assume pure rolling at all contact surfaces.

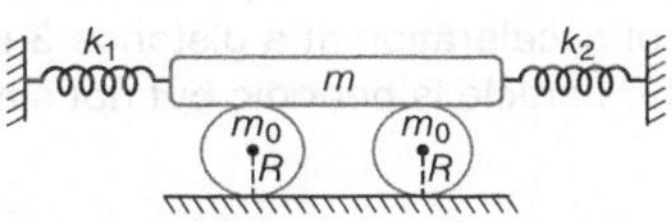

a. $\frac{1}{2\pi}\sqrt{\frac{k_1+k_2}{m+\frac{7}{5}m_0}}$ b. $\frac{1}{\pi}\sqrt{\frac{k_1+k_2}{m+\frac{7}{10}m_0}}$

c. $\frac{1}{2\pi}\sqrt{\frac{k_1+k_2}{m+\frac{7}{10}m_0}}$ d. $\frac{1}{\pi}\sqrt{\frac{k_1+k_2}{m+\frac{7}{5}m_0}}$

15 A ring of mass m can freely slide on a smooth vertical rod. The ring is symmetrically attached to two springs as shown in the figure. Equilibrium position of the ring is shown in the figure. If the ring is slightly displaced vertically, its time period is

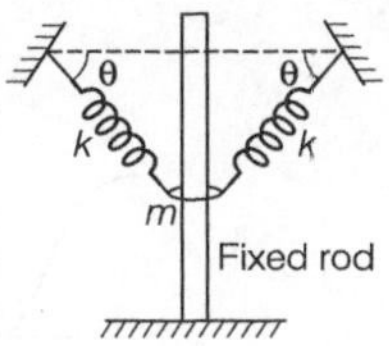

a. $2\pi\sqrt{\frac{2m}{k\sin^2\theta}}$ b. $\frac{\pi}{2}\sqrt{\frac{2m}{k\sin^2\theta}}$

c. $2\pi\sqrt{\frac{m}{k\sin^2\theta}}$ d. $\pi\sqrt{\frac{2m}{k\sin^2\theta}}$

One or More than One Options Correct Type

Direction (Q. Nos. 16-18) *This section contains 3 multiple choice questions. Each question has four choices (a), (b), (c) and (d), out of which ONE or MORE THAN ONE are correct.*

16 A particle is suspended by two ideal strings as shown in the figure. Now mass (m) is given a small displacement perpendicular to the plane of the triangle formed. Choose the correct statement(s). (Period of oscillation is T)

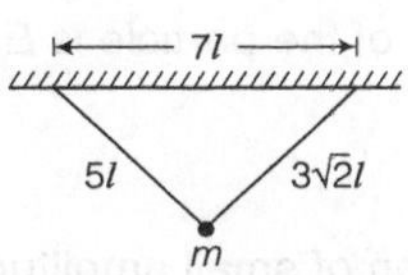

a. $T = 2\pi\sqrt{\frac{3\sqrt{3}\,l}{g}}$

b. $T = 2\pi\sqrt{\frac{3l}{g}}$

c. T is independent of m

d. If the distance between the suspension points is kept constant and the lengths of the strings are doubled, then time period gets doubled

Working Space

17 A particle slides back and forth between two inclined smooth planes joined smoothly at the bottom. The particle

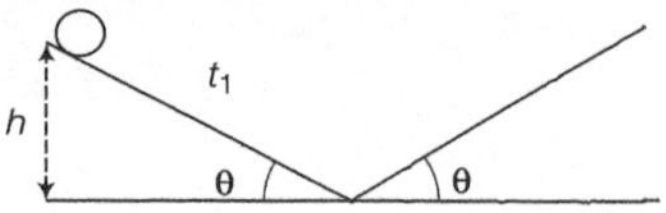

a. executes SHM
b. executes periodic motion
c. $T = \sqrt{\frac{32h}{g}\sin\theta}$
d. $T = \frac{4}{\sin\theta}\sqrt{\frac{2h}{g}}$

18 The speed v of a particle moving along a straight line, when it is at a distance x from a fixed point on the line is given by $v^2 = 144 - 9x^2$. Select the correct statement (s).

a. The motion of the particle is SHM with $T = \frac{2\pi}{3}$ unit
b. The maximum displacement of the particle from the fixed point is 4 unit
c. The magnitude of acceleration at a distance 3 unit from the fixed point is 27unit
d. The motion of the particle is periodic but not simple harmonic

Comprehension Type

Direction (Q. Nos. 19-23) This section contains 2 paragraphs, each describing theory, experiments, data, etc. Five questions related to the paragraphs have been given. Each question has only one correct answer among the four given options (a), (b), (c) and (d).

Passage I

When a particle of mass m moves on the X-axis in a potential of the form $V(x) = kx^2$, it performs simple harmonic motion. The corresponding time period is proportional to $\sqrt{\frac{m}{k}}$, as can be seen easily using dimensional analysis.

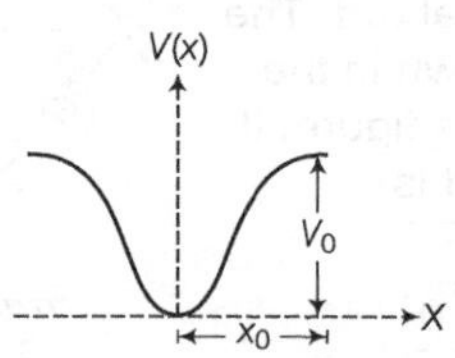

However, the motion of a particle can be periodic even when its potential energy increases on both sides of $x = 0$ in a way different from kx^2 and its total energy is such that the particle does not escape to infinity. Consider a particle of mass m moving on the X-axis. Its potential energy is given by $V(x) = Cx^4$ $(C > 0)$ for $|x|$ near the origin and becomes a constant equal to V_0 for $|x| > x_0$ (see figure). Based on this information, answer the following questions.

19 If the total energy of the particle is E_T, it will perform periodic motion only if

a. $E_T < 0$
b. $E_T > 0$
c. $V_0 > E_T > 0$
d. $E_T > V_0$

20 For periodic motion of small amplitude A, the periodic time T of this particle is proportional to

a. $A\sqrt{\frac{m}{C}}$
b. $\frac{1}{A}\sqrt{\frac{m}{C}}$
c. $A\sqrt{\frac{C}{m}}$
d. $\frac{1}{A}\sqrt{\frac{C}{m}}$

21 The acceleration of this particle for $|x| > x_0$ is

a. proportional to V_0

b. proportional to $\frac{V_0}{mx_0}$

c. proportional to $\sqrt{\frac{V_0}{mx_0}}$

d. zero

Passage II

A disc of mass m and radius R is attached to a spring of force constant k at its centre as shown in the figure. At $x = 0$, the spring is unstretched. The disc is moved to $x = +A$ and then released (at $t = 0$). There is no slipping between the disc and the ground. Let f be the force of friction on the disc from the ground. Now, answer the following questions.

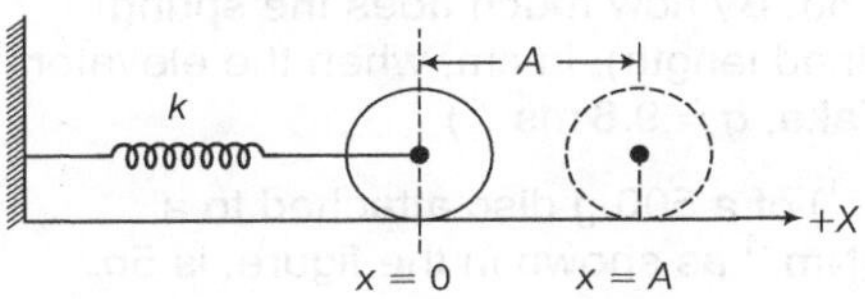

22 The best graph (among the following) that represents the variation of f with time t is

a. 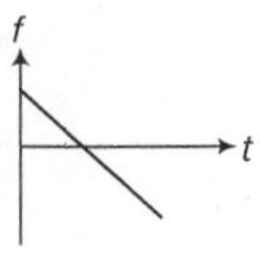

b. 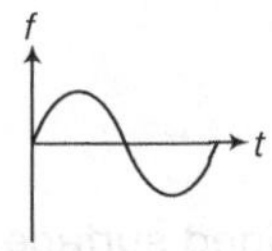

c. 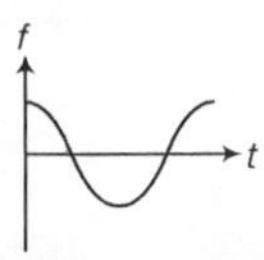

d. 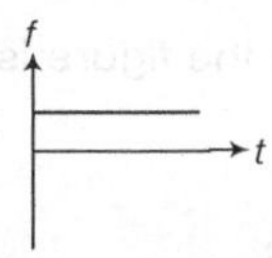

23 If $k = 10\ \text{Nm}^{-1}$, $m = 2$ kg, $R = 1$ m and $A = 2$ m, then the linear speed of the disc at the mean position is

a. $\sqrt{\frac{40}{3}}\ \text{ms}^{-1}$

b. $\sqrt{20}\ \text{ms}^{-1}$

c. $\sqrt{\frac{10}{3}}\ \text{ms}^{-1}$

d. $\sqrt{\frac{50}{3}}\ \text{ms}^{-1}$

One Integer Value Correct Type

Direction *(Q. Nos. 24-29) This section contains 6 questions. When worked out will result in an integer from 0 to 9 (both inclusive).*

24 A solid cylinder of mass $m = 4$ kg is attached to a horizontal spring with force constant $k = 3\text{Nm}^{-1}$. The cylinder can roll without slipping along the horizontal floor (see figure). The centre of mass of the cylinder executes simple harmonic motion with a period $T = 2\pi\sqrt{\beta}$. Find the value of β.

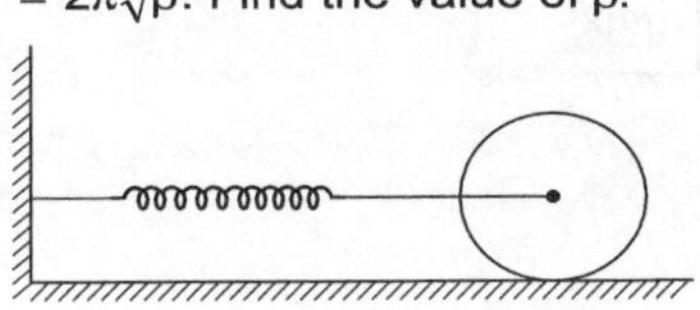

Working Space

25 The period of oscillations of mercury of mass m = 200 g poured into a bent tube (see figure), whose right arm makes an angle 30° with the vertical and whose left arm is vertical, is T seconds. The cross-sectional area of the tube is A = 0.5 cm^2. Find the value of $10T$. Neglect viscosity.

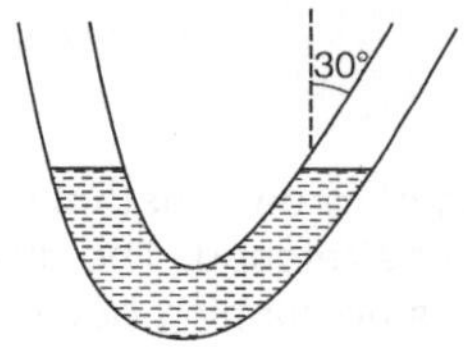

26 A spring of spring constant 1200 Nm^{-1} is hanging from the ceiling of an elevator and a 5 kg object is attached to the lower end. By how much does the spring stretch (in equilibrium, relative to its unstrained length), in cm, when the elevator is accelerating upward at a_0= 2.2 ms^{-2}? (Take, g = 9.8 ms^{-2})

27 The angular frequency of vibration (in rads^{-1}) of a 500 g disc attached to a support with a spring of force constant 300Nm^{-1} as shown in the figure, is $5p$. Find the value of p.

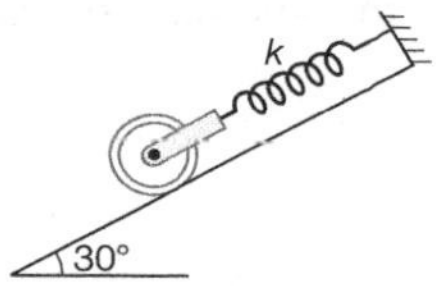

Assume that the disc does not slip on the inclined surface.

28 The natural frequency of the system,shown in the figure is $\frac{1}{\pi}\sqrt{\frac{qk}{m}}$. The pulleys and springs are ideal. Find the value of q.

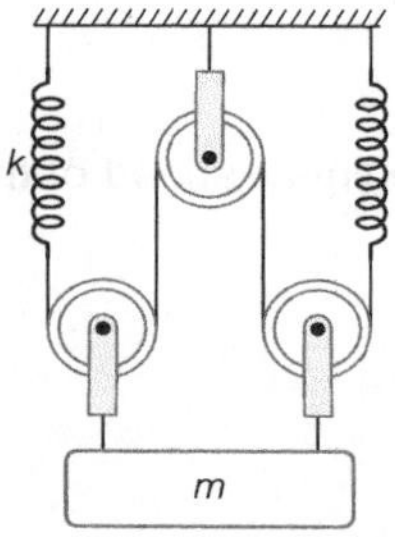

29 In the figure shown, threads, spring and pulleys are ideal. Given, m_1 = m and m_2 = m. The angular frequency of the system shown is $\sqrt{\frac{4k}{\beta m}}$. Find the value of β.

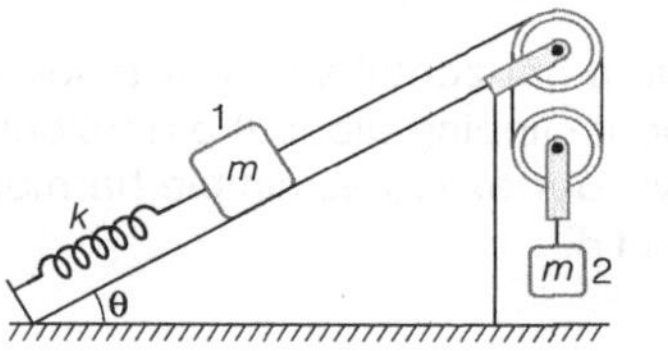

DPP-7 Angular SHM

Only One Option Correct Type

Working Space

***Direction** (Q. Nos. 1-11) This section contains 11 multiple choice questions. Each question has four choices (a), (b), (c) and (d), out of which ONLY ONE is correct.*

1 A simple pendulum of length L and mass m has a spring of force constant k, connected to it at a distance h below its point of suspension, as shown in the figure. For small amplitudes, the frequency of oscillations of the system is

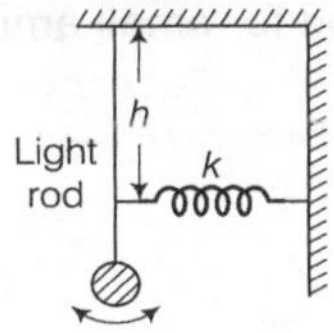

a. $\frac{1}{2\pi}\sqrt{\frac{mgL + 2kh^2}{mL^2}}$ b. $\frac{1}{2\pi}\sqrt{\frac{mgL + kh^2}{mL^2}}$

c. $\frac{1}{2\pi}\sqrt{\frac{mgL + kh^2}{2mL^2}}$ d. $\frac{1}{2\pi}\sqrt{\frac{mgL + 2kh^2}{2mL^2}}$

2 A cylinder of radius r and mass m rests on a curved path of radius R as shown in the figure. It is slightly displaced to its left. Thus, the cylinder makes oscillations about the mean position. The period of oscillations is (the cylinder rolls without slipping)

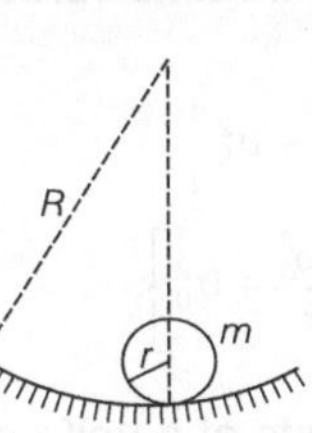

a. $\pi\sqrt{\frac{3(R-r)}{g}}$ b. $\pi\sqrt{\frac{4(R-r)}{g}}$

c. $\pi\sqrt{\frac{5(R-r)}{g}}$ d. $\pi\sqrt{\frac{6(R-r)}{g}}$

3 The natural period of vibration of a 50 N disc (semicircular) is (The centre of mass of the semicircular disc lies at a distance of $\frac{4r}{3\pi}$ for the centre. Assume pure rolling.)

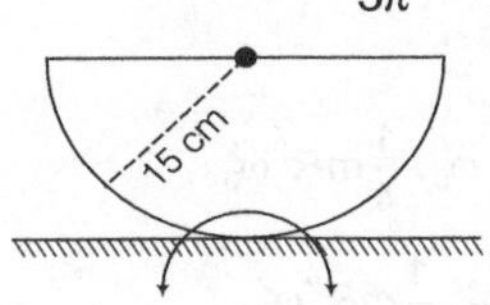

a. 0.71 s b. 0.96 s

c. 0.85 s d. 0.52 s

Working Space

4 Find the frequency of small oscillations of a thin uniform vertical rod of mass m and length L hinged at the point O as shown in the figure. The stiffness of each spring is k. Mass of the springs is negligible. The figure shows equilibrium position.

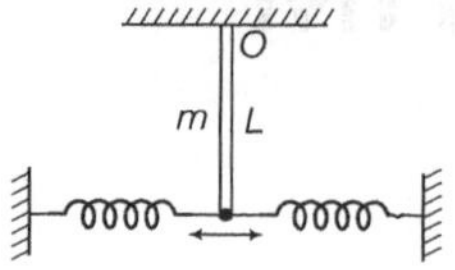

a. $\frac{1}{2\pi}\sqrt{\frac{3mg + 12kL}{2mL}}$

b. $\frac{1}{\pi}\sqrt{\frac{3mg + 12kL}{mL}}$

c. $\frac{1}{2\pi}\sqrt{\frac{3mg + 12kL}{mL}}$

d. $\frac{1}{2\pi}\sqrt{\frac{2mg + 7kL}{2mL}}$

5 A T-bar of uniform cross-section and mass m is supported in a vertical plane by a hinge O and a spring of force constant k at P. The period for small amplitude rotational oscillations in the plane of the figure is

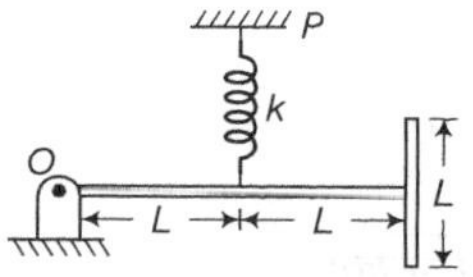

a. $2\pi\sqrt{\frac{3m}{k}}$

b. $2\pi\sqrt{\frac{2m}{3k}}$

c. $3\pi\sqrt{\frac{m}{k}}$

d. $\pi\sqrt{\frac{3m}{k}}$

6 A uniform rod of mass m and length L performs small oscillations about a horizontal axis passing through its upper end. Find the mean kinetic energy of the rod during its oscillation period, if at $t = 0$ it is deflected from its vertical by an angle θ_0 and imparted an angular velocity ω_0.

a. $\frac{1}{4}mgL\left[\frac{L\omega_0^2}{3g} + \theta_0^2\right]$

b. $mgL\left[\frac{2L\omega_0^2}{3g} + \theta_0^2\right]$

c. $\frac{1}{8}mgL\left[\frac{2L\omega_0^2}{3g} + \theta_0^2\right]$

d. $\frac{1}{4}mgL\left[\frac{2L\omega_0^2}{3g} + \theta_0^2\right]$

7 In the figure is shown, a torsional pendulum that consists of a uniform disc of mass m and radius R attached to a thin rod of torsional constant C. Initially, the disc is imparted an angular speed ω_0. The amplitude and energy of small torsional oscillations of the disc, respectively are

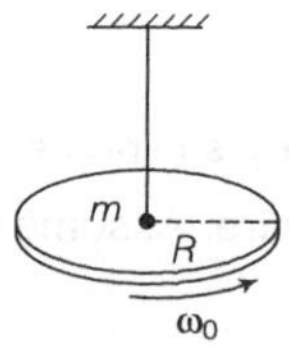

a. $2\sqrt{\frac{mR^2}{2C}}\,\omega_0, \frac{1}{2}mR^2\omega_0^2$

b. $\sqrt{\frac{3mR^2}{2C}}\,\omega_0, \frac{1}{4}mR^2\omega_0^2$

c. $\frac{1}{2}\sqrt{\frac{mR^2}{2C}}\,\omega_0, \frac{1}{2}mR^2\omega_0^2$

d. $\sqrt{\frac{mR^2}{2C}}\,\omega_0, \frac{1}{4}mR^2\omega_0^2$

Working Space

8 A uniform semicircular cylinder of radius R and mass m is displaced through a small angle θ from its equilibrium position. It rolls without slipping during oscillations. The time period of small oscillations is

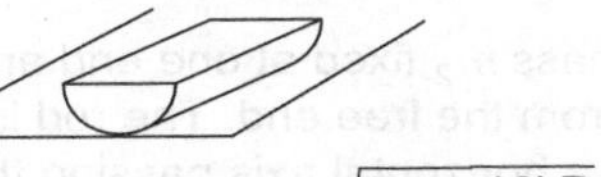

a. $2\pi\sqrt{\dfrac{(9\pi - 16)R}{2g}}$ b. $\pi\sqrt{\dfrac{(9\pi - 16)R}{2g}}$

c. $\dfrac{\pi}{2}\sqrt{\dfrac{(9\pi - 16)R}{g}}$ d. $\pi\sqrt{\dfrac{(9\pi - 16)R}{g}}$

9 A hole of radius $R/4$ is drilled through a cylinder of radius R to form a body of mass m as shown in the figure. The top point of the body is turned slightly and released. The body performs small oscillations about point O', without slipping. The time period is

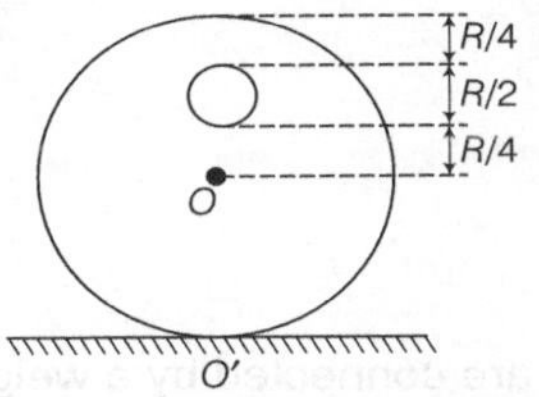

a. $41.4\sqrt{\dfrac{R}{g}}$ b. $73.2\sqrt{\dfrac{R}{g}}$

c. $24.1\sqrt{\dfrac{R}{g}}$ d. $37.5\sqrt{\dfrac{R}{g}}$

10 Carbon dioxide is a linear molecule. The carbon-oxygen bonds in this molecule are very much like springs. The figures show two possible ways of oscillations. In case (i) carbon atom is at rest. If mass of carbon atom is m and that of oxygen atom is M, the ratio of frequencies of oscillations in case (i) to case (ii) $\left(\text{i.e. } \dfrac{\omega_1}{\omega_2}\right)$ is

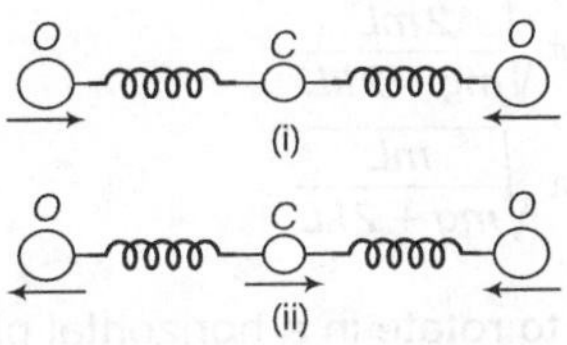

a. $\sqrt{\dfrac{m}{2M}}$ b. $\sqrt{\dfrac{m-2M}{m+2M}}$ c. $\sqrt{\dfrac{m}{m+2M}}$ d. $\sqrt{\dfrac{m+2M}{m-2M}}$

11 A uniform disc of mass m and radius R is pivoted smoothly at P. If a uniform thin ring of mass m and radius R is welded at the lowest point of the disc, find the period of SHM of the system (disc + ring).

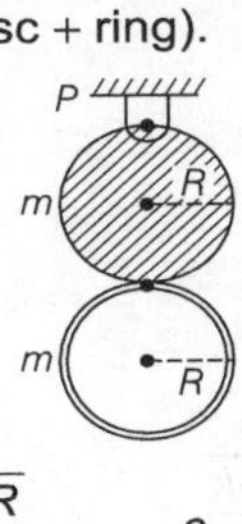

a. $\pi\sqrt{\dfrac{23R}{2g}}$ b. $2\pi\sqrt{\dfrac{23R}{g}}$ c. $\dfrac{\pi}{2}\sqrt{\dfrac{23R}{2g}}$ d. $2\pi\sqrt{\dfrac{23R}{2g}}$

Working Space

One or More than One Options Correct Type

Direction *(Q. Nos. 12-14) This section contains 3 multiple choice questions. Each question has four choices (a), (b), (c) and (d), out of which ONE or MORE THAN ONE are correct.*

12 A light rod of length l_2 has a small ball of mass m_2 fixed at one end and another ball of mass m_1 fixed on it at a distance l_1 from the free end. The rod is supported at its free end O and is free to rotate about a horizontal axis passing through O. The rod is slightly displaced from its equilibrium vertical position and released. Its frequency of oscillation about O is f and length of equivalent simple pendulum is L. Then,

a. $f = \frac{1}{2\pi}\left[\frac{m_1 l_1 + m_2 l_2}{m_1 l_1^2 + m_2 l_2^2}\cdot g\right]^{1/2}$

b. $f = \frac{1}{4\pi}\left[\frac{m_1 l_1 + m_2 l_2}{m_1 l_1^2 + m_2 l_2^2}\cdot g\right]^{1/2}$

c. $L = \frac{m_1 l_1^2 + m_2 l_2^2}{m_1 l_1 + m_2 l_2}$

d. $L = \frac{4(m_1 l_1^2 + m_2 l_2^2)}{m_1 l_1 + m_2 l_2}$

13 Two identical simple pendulums, each of length L are connected by a weightless spring as shown in the figure. The force constant of the spring is k. In equilibrium, the pendulums are vertical and the spring is horizontal and undeformed. They are deflected from their equilibrium positions through small and equal displacements in the same vertical plane (i) in the same directions and (ii) in opposite direction and released, In case (i), time period is T_1 and in case (ii) time period is T_2. Then,

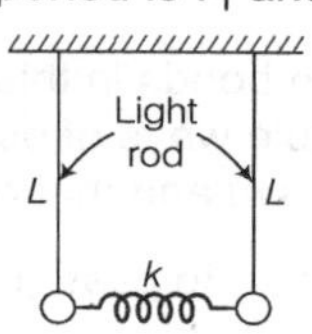

a. $T_1 = 2\pi\sqrt{\frac{2L}{g}}$

b. $T_2 = 2\pi\sqrt{\frac{2mL}{mg + 2kL}}$

c. $T_1 = 2\pi\sqrt{\frac{L}{g}}$

d. $T_2 = 2\pi\sqrt{\frac{mL}{mg + 2kL}}$

14 A long uniform rod of length L and mass m is free to rotate in a horizontal plane about a vertical axis through its one end O. A spring of force constant k is connected horizontally between one end (B) of the rod and a fixed wall (see figure). When the rod is in equilibrium, it is parallel to the wall. When the rod is rotated slightly in the direction shown (at B) and released, the time period of resultant oscillations is T. If amplitude of these oscillations is θ_0, then the maximum speed of displaced end of rod is v_{max}. Now,

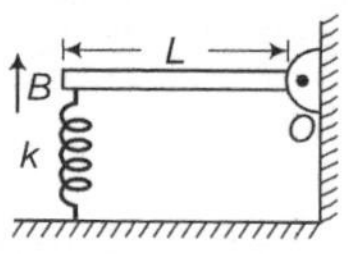

a. $T = 2\pi\sqrt{\frac{m}{5k}}$

b. $T = 2\pi\sqrt{\frac{m}{3k}}$

c. $v_{max} = L\theta_0\sqrt{\frac{3k}{m}}$

d. $v_{max} = \frac{L\theta_0}{2}\sqrt{\frac{5k}{m}}$

Comprehension Type

Direction *(Q. Nos. 15-19) This section contains 2 paragraphs, each describing theory, experiments, data, etc. Five questions related to the paragraphs have been given. Each question has only one correct answer among the four given options (a), (b), (c) and (d).*

Passage I

A uniform rod *AB* of mass *m* is suspended by the two identical strings of length *L* as shown in the figure. This rod is turned by a small angle in horizontal plane about the vertical axis passing through its centre of mass *C*. In this process, the strings are deviated by a small angle θ_0 from vertical. Then, the rod is released to start performing angular oscillations.

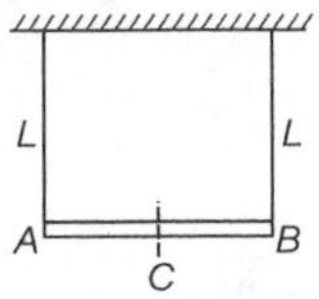

15 The period of oscillations is

a. $2\pi\sqrt{\frac{2L}{3g}}$ b. $2\pi\sqrt{\frac{L}{3g}}$

c. $\pi\sqrt{\frac{2L}{3g}}$ d. $\pi\sqrt{\frac{L}{3g}}$

16 The oscillation energy of the rod is

a. $\frac{1}{2}mgL\theta_0^2$ b. $mgL\theta_0^2$

c. $\frac{1}{4}mgL\theta_0^2$ d. $2mgL\theta_0^2$

Passage II

The pendulum in the figure consists of a uniform disc with radius 10 cm and mass 500 g, attached to a uniform rod with length 50 cm and mass 270 g. Answer the following questions.

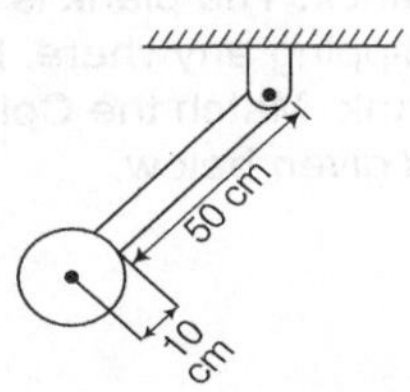

17 The rotational inertia of the pendulum about the pivot point is

a. 0.102 kg- m^2 b. 0.205 kg- m^2

c. 0.305 kg- m^2 d. 0.500 kg- m^2

18 What is the distance between the pivot point and centre of mass of the pendulum?

a. 30 cm b. 35.6 cm

c. 47.7 cm d. 49.3 cm

19 The period of oscillation is

a. 1.5 s b. 2 s

c. 3 s d. 4 s

Working Space

Matching List Type

Direction *(Q. Nos. 20 and 21) Choices for the correct combination of elements from Column I and Column II are given as options (a), (b), (c) and (d), out of which one is correct.*

20 A small block of mass m is connected to one end of a massless rod. A spring of force constant k is attached to the other end of the rod. The rod is pivoted at P. The block is displaced slightly downward and left free to oscillate. For small angular displacement θ of the block. Match the Column I with Column II and mark the correct option from the codes given below.

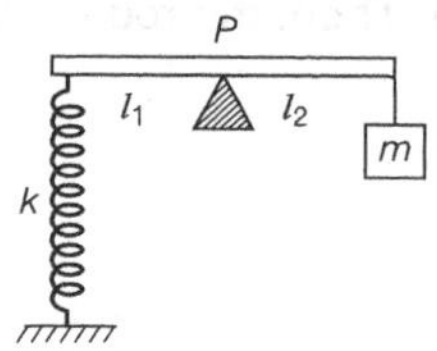

	Column I		Column II
i.	Restoring torque	p.	$\frac{1}{2}kl_1^2\theta^2$
ii.	Angular acceleration	q.	$kl_1^2\theta$
iii.	Energy of oscillations	r.	$kl_1^2\theta/ml_2^2$
iv.	Time period of oscillations	s.	$2\pi\sqrt{\frac{l_2^2 m}{l_1^2 k}}$

Codes

	i	ii	iii	iv
a.	q	r	p	s
b.	p	q	r	s
c.	s	p	r	q
d.	r	s	q	p

21 A uniform plank of mass m, free to move in the horizontal direction, is placed on the top of a solid cylinder of mass $2m$ and radius R. The plank is attached to a fixed wall with the help of a light spring of force constant k. The plank is slightly displaced towards spring and released. There is no slipping anywhere. For any displacement x, amplitude x_0 and velocity v of the plank. Match the Column I with Column II and mark the correct option from the codes given below.

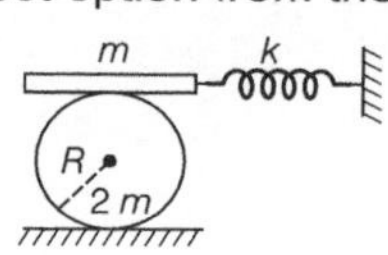

	Column I		Column II
i.	Restoring force on the plank	p.	$\frac{1}{2}kx_0^2$
ii.	Energy of oscillations of the system	q.	kx
iii.	Acceleration of the plank	r.	$\frac{7}{8}mv^2 + \frac{1}{2}kx^2$
iv.	Angular frequency	s.	$\frac{4kx}{7m}$
		t.	$\sqrt{\frac{4k}{7m}}$

Codes

	i	ii	iii	iv
a.	p,q	r	s	t
b.	q	p,r	s	t
c.	s	t	q	p
d.	p	q	r	s

Working Space

One Integer Value Correct Type

Direction (Q. Nos. 22-26) This section contains 5 questions. When worked out will result in an integer from 0 to 9 (both inclusive).

22 A rod of mass m and length L is pivoted at point O in a car whose acceleration towards left is a_0. The rod is free to rotate in vertical plane. In equilibrium state, the rod remains horizontal when other end is suspended by a spring of force constant k. The time period of small oscillations of rod is $T = \frac{2\pi}{D\sqrt{3}}$. Find the value of D. (Given, $k = 20 \text{Nm}^{-1}$, $a_0 = 10 \text{ms}^{-2}$, $m = 1 \text{kg}$, $L = 1 \text{m}$)

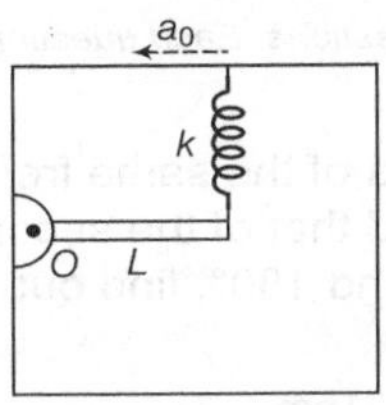

23 A uniform circular disc of radius R oscillates in a vertical plane about a horizontal axis. Its minimum time period is $\pi\sqrt{\frac{B\sqrt{2}R}{g}}$. Find the value of B.

24 The time period of small oscillations in vertical plane of a uniform rod with length L, pivoted at one end, is $\pi\sqrt{\frac{\psi L}{3g}}$. Find the value of ψ.

25 A body of mass 200 g oscillates about a horizontal axis at a distance of 20 cm from its centre of gravity. The length of equivalent simple pendulum is 35 cm. If its moment of inertia about point of suspension is $\frac{2\beta}{1000}$ kg-m^2. Find the value of β.

26 In the figure is shown, a solid uniform cylinder of radius R and mass m, which is free to rotate about a fixed horizontal axis O that passes through the centre of the cylinder. One end of an ideal spring of force constant k is fixed and the other end is hinged to the cylinder at point P. Distance OP is equal to $\frac{R}{2}$. An inextensible thread is wrapped round the cylinder and passes over a small, ideal pulley. A block of equal mass m and having cross-sectional area A is suspended from free end of the thread. The block is partially immersed in a non-viscous liquid of density ρ_L. In equilibrium, spring is horizontal and line OP is vertical. The frequency of small oscillations of the system is $\frac{1}{2\pi}\sqrt{\frac{k + 4\rho_L Ag}{Qm}}$. Find the value of Q.

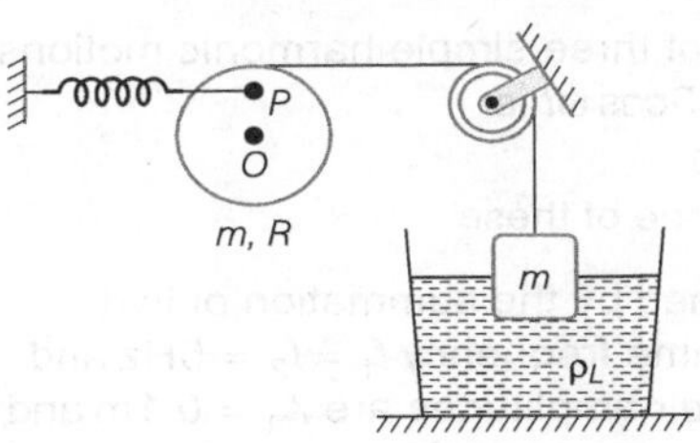

DPP-8 Superposition of SHM

Only One Option Correct Type

Working Space

Direction (Q. Nos. 1-10) *This section contains 10 multiple choice questions. Each question has four choices (a), (b), (c) and (d), out of which ONLY ONE is correct.*

1 A particle is subjected to two simple harmonic motions of the same frequency and direction. The amplitude of the first motion is 4 cm and that of the second is 3 cm. If the phase difference between them is 0°, 60°, 90° and 180°, find out corresponding resultant amplitude.

a. $(4, \sqrt{35}, 3, 1)$ cm
b. $(5, \sqrt{39}, 4, 2)$ cm
c. $(7, \sqrt{37}, 5, 1)$ cm
d. $(5, \sqrt{35}, 4, 2)$ cm

2 The displacement equation of the simple harmonic motion obtained by combining the motions $x_1 = 2\sin\omega t$, $x_2 = 4\sin\left(\omega t + \frac{\pi}{6}\right)$ and $x_3 = 6\sin\left(\omega t + \frac{\pi}{3}\right)$ is

a. $x = 9.25\sin[\omega t + \tan^{-1}(0.75)]$
b. $x = 11.25\sin[\omega t + \tan^{-1}(0.85)]$
c. $x = 10.25\sin[\omega t + \tan^{-1}(0.80)]$
d. $x = 9.25\sin[\omega t + \tan^{-1}(0.95)]$

3 Two linear SHMs of equal amplitudes and frequencies ω and 2ω are impressed on a particle along X and Y-axes respectively. The initial phase difference between them is $\frac{\pi}{2}$. The resultant path followed by the particle is

a. parabola
b. ellipse
c. circle
d. straight line

4 Three simple harmonic motions of equal amplitudes A and equal time periods along the same line combine. The phase of the second motion is 60° ahead of the first and phase of the third motion is 60° ahead of the second. The amplitude of resultant motion is

a. $3A$
b. $2\sqrt{2}A$
c. $\sqrt{3}A$
d. $2A$

5 The resultant amplitude due to superposition of three simple harmonic motions $x_1 = 3\sin\omega t$, $x_2 = 5\sin(\omega t + 37°)$ and $x_3 = -15\cos\omega t$ is

a. 18
b. 10
c. 12
d. None of these

6 The equation of the resulting oscillation obtained by the summation of two mutually perpendicular oscillations with the same frequency $f_1 = f_2 = 5$Hz and same initial phase $\delta_1 = \delta_2 = 60°$ is, (Given, their amplitudes are $A_1 = 0.1$m and $A_2 = 0.05$m)

a. $0.15\sin\left(10\pi t + \frac{\pi}{6}\right)$
b. $0.05\sin\left(10\pi t + \frac{2\pi}{3}\right)$
c. $0.112\sin\left(10\pi t + \frac{\pi}{3}\right)$
d. $0.313\sin\left(10\pi + \frac{\pi}{2}\right)$

Working Space

7 A particle simultaneously participates in two mutually perpendicular oscillations, $x = 5\sin\omega t$ and $y = 5\cos\omega t$. The trajectory of the particle is

a. ellipse
b. straight line
c. circle
d. parabola

8 The displacement x of a particle executing a certain periodic motion is given by $x = 4\cos^2\left(\frac{1}{2}t\right)\sin(1000t)$. This expression may be considered to be the superposition of n independent harmonic motions. Then, n is equal to

a. 6
b. 3
c. 4
d. 5

9 The displacement of a particle at time t is given by

$x = A_1\sin(-2\omega t) + A_2\sin^2\omega t$. Then,

a. the motion of the particle is SHM with an amplitude $\sqrt{A_1^2 + \frac{A_2^2}{4}}$

b. the motion of the particle is not SHM, but oscillatory with a time period of $T = \pi/\omega$

c. the motion of the particle is oscillatory with a time period $T = \frac{2\pi}{\omega}$

d. the motion of the particle is periodic but not SHM

10 A charged particle is deflected by two mutually perpendicular oscillating electric fields such that the displacement of the particle due to each one of them is given by $x = A\sin(\omega t)$ and $y = A\sin\left(\omega t + \frac{\pi}{6}\right)$ respectively. The trajectory followed by the charged particle is

a. a circle with equation $x^2 + y^2 = A^2$
b. a straight line with equation $y = \sqrt{3}x$
c. an ellipse with equation $x^2 + y^2 - xy = \frac{3}{4}A^2$
d. an ellipse with equation $x^2 + y^2 - \sqrt{3}xy = \frac{1}{4}A^2$

One or More than One Options Correct Type

***Direction** (Q. Nos. 11-14) This section contains 4 multiple choice questions. Each question has four choices (a), (b), (c) and (d), out of which ONE or MORE THAN ONE are correct.*

11 The function $x = A_1\sin^2\omega t + A_2\cos^2\omega t + A_3\sin\omega t\cos\omega t$ represents SHM for which of the options?

a. For all values of A_1, A_2 and A_3 $(A_3 \neq 0)$
b. $A_1 = A_2$, $A_3 = 2A_2$
c. $A_1 = -A_2$, $A_3 = 2A_2$
d. $A_1 = A_2$, $A_3 = 0$

12 A particle moves in the XY-plane, according to the equation $\vec{r} = A\cos\omega t\,\hat{i} + 2A\cos\omega t\,\hat{j}$. The motion of the particle is

a. on a straight line
b. on an ellipse
c. periodic
d. simple harmonic

Working Space

13 A 20 g particle is subjected to two simple harmonic motions simultaneously which are given by $x_1 = 2\sin 10t$ and $x_2 = 4\sin\left(10t + \frac{\pi}{3}\right)$, where x_1, x_2 are in metre and t in second. Choose the correct statement(s).

a. The displacement of the particle at time $t = 0$ will be $2\sqrt{3}$ m

b. Maximum speed of the particle will be $20\sqrt{7}$ ms^{-1}

c. Magnitude of maximum acceleration of the particle will be $200\sqrt{7}$ ms^{-2}

d. Energy of the resultant motion will be 28 J

14 Two SHMs $r_1 = A\sin\omega t$ and $r_2 = B\sin\omega t$ are superimposed on a particle. r_1 and r_2 are along the directions which make 37° with each other. Choose the correct options.

a. The particle will perform SHM

b. The path of the particle is straight line

c. The particle will perform only periodic motion but not SHM

d. The path is a closed curve but not a straight line

Comprehension Type

Direction *(Q. Nos. 15-18) This section contains 2 paragraphs, each describing theory, experiments, data, etc. Four questions related to the paragraphs have been given. Each question has only one correct answer among the four given options (a), (b), (c) and (d).*

Passage I

A particle is subjected to two simple harmonic motions given by $x_1 = 2\sin(100\pi t)$ and $x_2 = 2\sin\left(120\pi t + \frac{\pi}{3}\right)$ simultaneously, where x is in cm and t is in second.

15 The displacement of the particle at $t = 0.0125$ s is

a. −2.41cm

b. −3.732cm

c. −1.41cm

d. −2.732cm

16 The displacement of the particle at $t = 0.025$ s is

a. 1.73 cm

b. 0.73 cm

c. 3.27 cm

d. 0.27 cm

Passage II

Four SHMs happening on X-axis are represented by $x_1 = A\cos\omega t$, $x_2 = \frac{A}{2}\cos\left(\omega t + \frac{\pi}{2}\right)$, $x_3 = \frac{A}{4}\cos(\omega t + \pi)$ and $x_4 = \frac{A}{8}\cos\left(\omega t + \frac{3\pi}{2}\right)$. Now, these four SHMs are combined.

17 Amplitude of the resultant SHM is

a. $\frac{2\sqrt{5}}{7}A$

b. $\frac{4\sqrt{5}}{7}A$

c. $\frac{3\sqrt{5}}{8}A$

d. $\frac{4\sqrt{5}}{9}A$

18 The phase difference between the resultant and first SHM (i.e. x_1) is

a. $\tan^{-1}\left(\frac{2}{3}\right)$

b. $\tan^{-1}\left(\frac{1}{3}\right)$

c. $\tan^{-1}\left(\frac{3}{5}\right)$

d. $\tan^{-1}\left(\frac{1}{2}\right)$

Working Space

One Integer Value Correct Type

Direction (Q. No. 19) *This section contains 1 question. When worked out will result in an integer from 0 to 9 (both inclusive).*

19 If two SHM's are represented by

$$x_1 = 10\sin\left(4\pi t + \frac{\pi}{2}\right) \text{ and } x_2 = 5\sin(2\pi t) + 5\sqrt{8}\cos(2\pi t),$$

the ratio of their amplitudes is $\frac{2}{\beta}$. Find the value of β.

Subjective Questions

Direction (Q. Nos. 20-23) *These questions are subjective in nature, need to be solved completely on notebook.*

20 A particle oscillates according to the law $x = (1 + \cos^2 t + \sin^4 t)\sin 500t$. Expand this motion into its harmonic components and plot its spectrum.

21

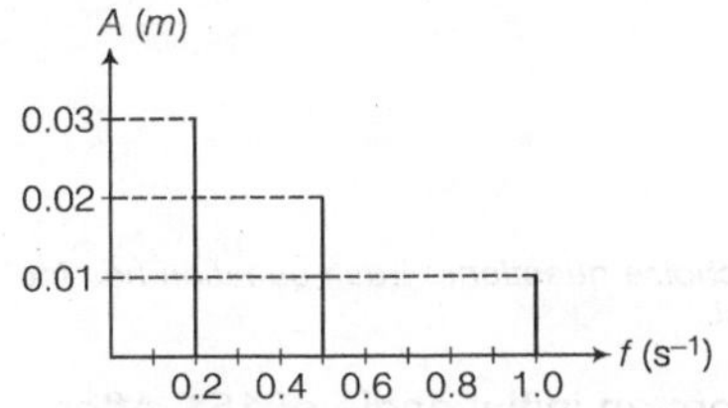

In the figure is shown a spectrum of a complex oscillation.

i. Write the equations of the constituent oscillations using the data of the graph.
ii. Draw a diagram of these oscillations assuming that the difference of phases between these oscillation is zero at the moment $t = 0$.

22 An oscillation is described by the equation $x = A\sin 2\pi f_1 t$, where A changes with time according to the law $A = A_0(1 + \cos 2\pi f_2 t)$, where A_0 is a constant.

i. Find the harmonic oscillations forming the oscillation.
ii. Draw a spectrum of the complex oscillations when $A_0 = 4\,\text{cm}$, $f_1 = 2\,\text{s}^{-1}$, $f_2 = 1\,\text{s}^{-1}$.

23 A particle oscillates according to the law $x = (1 + \cos^2 t + \cos^4 t)\sin 500t$. Expand this motion into its harmonic components and plot its spectrum.

DPP-9 Damped Oscillations and Resonance

Working Space

Subjective Questions

Direction (Q. Nos. 1-3) *These questions are subjective in nature, need to be solved completely on notebook.*

1 Explain damped harmonic oscillations and the equation of such oscillations.

2 Discuss three possible cases of damping oscillations.

3 Discuss driven oscillations.

Only One Option Correct Type

Direction (Q. Nos. 4-18) *This section contains 15 multiple choice questions. Each question has four choices (a), (b), (c) and (d), out of which ONLY ONE is correct.*

4 A pendulum with a length of 1m is released from an initial angle of 15°. After 1000 s, its amplitude has been reduced by friction to 5.5°. What is the value of $\frac{b}{2m}$? (Given, $\ln 2.73 = 1$)

a. $0.5 \times 10^{-3} s^{-1}$
b. $0.001 s^{-1}$
c. $0.002 s^{-1}$
d. $0.005 s^{-1}$

5 A 10.6 kg object oscillates at the end of a vertical spring that has a spring constant 20500 Nm^{-1}. The effect of air resistance is represented by the damping coefficient, $b = 3 Nsm^{-1}$. The frequency of damped oscillation is

a. 2.21 Hz
b. 9.23 Hz
c. 8.00 Hz
d. 7.00 Hz

6 In the above problem, by what percentage does the amplitude of the oscillation decrease in each cycle ? (Given, $e^{0.0202} = 1.0204$)

a. 1%
b. 2%
c. 3%
d. 4%

7 A 21.2 kg object oscillates at the end of a vertical spring that has a spring constant 20500 Nm^{-1}. The effect of air resistance is represented by the damping coefficient $b = 2 Nsm^{-1}$. Find the time interval that elapses while the energy of the system drops to 10% of its initial value (Given, $\ln 10 = 2.302$)

a. 34.4 s
b. 12.2 s
c. 24.4 s
d. 8.9 s

8 An unhappy 0.3 kg mouse, moving on the end of a spring with force constant $k = 2.5 Nm^{-1}$, is acted upon by a damping force $F_x = -bv_x$, where $b = 0.9 kgs^{-1}$. The frequency of oscillation of mouse is

a. 0.593 Hz
b. 0.393 Hz
c. 0.493 Hz
d. 0.293 Hz

Working Space

9 A simple pendulum has a period of 2 s and an amplitude of 2°. After 10 complete oscillations, its amplitude reduces to 1.5°. Find the value of $\frac{b}{2m}$.

[Given, ln(1.333) = 0.2877]

a. $0.0122\,s^{-1}$ b. $0.0144\,s^{-1}$
c. $0.0166\,s^{-1}$ d. $0.0100\,s^{-1}$

10 When a tuning fork (frequency 262 Hz) is struck, it loses half of its energy after 4s. The decay time (τ) and Q-factor for this tuning fork respectively are (Given, ln 2 = 0.693)

a. 4.77s, 10.5×10^3 b. 5.77s, 9.5×10^3
c. 6.77s, 11.5×10^3 d. 2.77s, 8.5×10^3

11 In the above problem, the fractional energy loss in a period or in a cycle is

a. 6.67×10^{-4} b. 5.51×10^{-4}
c. 4.81×10^{-4} d. 3.71×10^{-4}

12 In a spring-mass system with mass m = 0.5kg and k = 200Nm^{-1}, what is the value of damping coefficient for critical damping?

a. 20 kgs^{-1} b. 30kg s^{-1}
c. 10kgs^{-1} d. 25kgs^{-1}

13 A baby rejoices by crowing and bouncing up and down in her crib. Her mass is 12.5 kg and the crib mattress can be modeled as a light spring with force constant 4300Nm^{-1}. The baby soon learns to bounce with maximum amplitude and minimum effort by bending her knees at what frequency?

a. 1.95 Hz b. 3.95 Hz
c. 4.95 Hz d. 2.95 Hz

14 A 2 kg object attached to a spring moves without friction and is driven by an external force given by $F = 3\sin(2\pi t)$ newton. The force constant of the spring is 20Nm^{-1}. The amplitude of the motion is

a. 7.09 cm b. 5.09 cm
c. 6.09 cm d. 4.09 cm

15 In the above problem, the period is

a. 2 s b. 1 s
c. 1.5 s d. 0.5 s

16 A block weighing 40N is suspended from a spring that has a force constant of 200Nm^{-1}. The system is undamped and is subjected to a harmonic driving force of frequency 10Hz, resulting a forced-motion amplitude of 2 cm. The maximum value of driving force is

a. 118 N b. 218 N
c. 318 N d. 418 N

17 A 1000 kg car carrying four 82 kg people travels over a rough 'washboard' dirt road with corrugations 4 m apart, which cause the car to bounce on its spring suspension. The car bounces with maximum amplitude when its speed is 16 kmh^{-1}. The car now stops and the four people get out. By how much does the car body rise on its suspension doing to this decrease in mass?

a. 7 cm b. 8 cm
c. 5 cm d. 6 cm

Working Space

18 Find the limiting values of amplitude of a forced damped oscillator, when (i) $\omega_d << \omega_0$ and (ii) $\omega_d >> \omega_0$.

a. $\dfrac{F}{m\omega_0^2}, \dfrac{F}{m\omega_d\sqrt{\omega_d^2+\left(\dfrac{b}{m}\right)^2}}$

b. $\dfrac{2F}{m\omega_0^2}, \dfrac{F}{2m\omega_d\sqrt{\omega_d^2+\left(\dfrac{b}{m}\right)^2}}$

c. $\dfrac{F}{\sqrt{2}m\omega_0^2}, \dfrac{F}{m\omega_d\sqrt{2\left[\omega_d^2+\left(\dfrac{b}{m}\right)^2\right]}}$

d. $\dfrac{F}{2m\omega_0^2}, \dfrac{F}{2m\omega_d\sqrt{\omega_d^2+\left(\dfrac{b}{m}\right)^2}}$

Comprehension Type

Direction (Q. Nos. 19-21) *This section contains a paragraph, describing theory, experiments, data, etc. Three questions related to the paragraphs have been given. Each question has only one correct answer among the four given options (a), (b), (c) and (d).*

Passage

For the damped oscillation shown in the figure, $m = 250$g, $k = 85$ Nm^{-1} and $b = 70$g s^{-1}. (Given, ln 2 = 0.693)

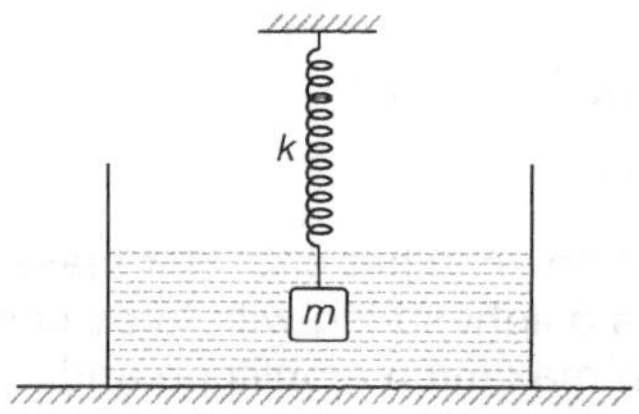

19 The period of motion is

a. 0.14 s
b. 0.34 s
c. 0.54 s
d. 0.74 s

20 How long does it take for the amplitude of the damped oscillation to drop to half its initial value?

a. 1s
b. 2.5 s
c. 5 s
d. 12 s

21 How long does it take for the mechanical energy to drop to half its initial value ?

a. 1s
b. 5 s
c. 3s
d. 2.5 s

SET - 1

Revisal Problems for JEE Main

Only One Option Correct Type

Direction *(Q. Nos. 1-31) This section contains 31 multiple choice questions. Each question has four choices (a), (b), (c) and (d), out of which ONLY ONE option is correct.*

Working Space

1 The figure depicts a circular motion. The radius of the circle, the period of revolution, the initial position and the sense of revolution are indicated in the figure.

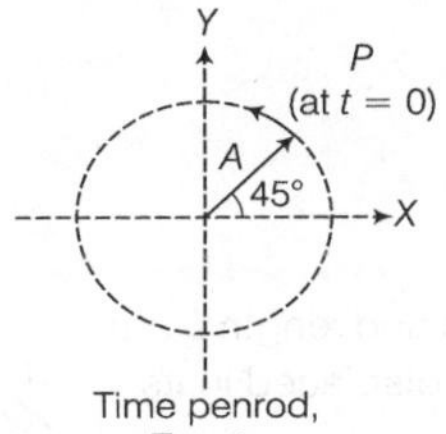

The simple harmonic motion of the x-projection of the radius vector of the rotating particle P can be shown as

a. $x(t) = A\cos\left(\frac{\pi}{2}t + \frac{\pi}{4}\right)$ b. $x(t) = A\cos\left(\frac{\pi}{4}t + \frac{\pi}{4}\right)$

c. $x(t) = A\sin\left(\frac{\pi}{2}t + \frac{\pi}{4}\right)$ d. $x(t) = A\sin\left(\frac{\pi}{4}t + \frac{\pi}{4}\right)$

2 A bent tube (see figure) of uniform cross-section area A has an ideal liquid of density ρ. The mass of the liquid oscillating in the tube is m. The time period of oscillation of the liquid is

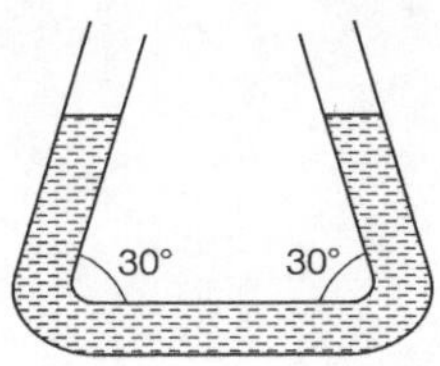

a. $2\pi\sqrt{\frac{m}{2\rho gA}}$ b. $2\pi\sqrt{\frac{2m}{\rho gA}}$

c. $2\pi\sqrt{\frac{m}{\rho gA}}$ d. $2\pi\sqrt{\frac{3m}{2gA}}$

3 Two blocks of masses m and $2m$ are connected to an ideal spring of force constant k. The spring is stretched by x_0 and the blocks are made to oscillate on a smooth horizontal surface. The amplitude of motion of block of mass m is

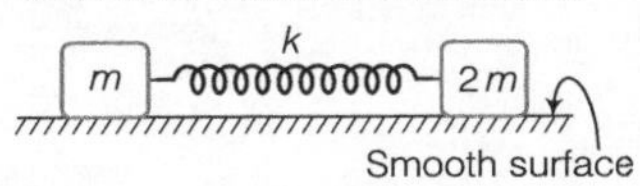

a. $\frac{x_0}{3}$ b. $\frac{2x_0}{3}$ c. x_0 d. $\frac{x_0}{6}$

4 A particle at the end of a spring executes simple harmonic motion with a period T_1, while the corresponding period for another spring is T_2. If the period of oscillation with the two springs in series is T, then

a. $T^{-1} = T_1^{-1} + T_2^{-1}$ b. $T = T_1 + T_2$

c. $T^{-2} = T_1^{-2} + T_2^{-2}$ d. $T^2 = T_1^2 + T_2^2$

Working Space

5 Two identical balls A and B, each of mass 0.1 kg are attached to two identical ideal springs. The spring-mass system is constrained to move inside a rigid smooth pipe bent in the form of a circle. The pipe is fixed in a horizontal plane. The centres of the balls can move in a circle of radius 0.06 m. Each spring has a natural length of 0.06π m and force constant $0.1\ \text{Nm}^{-1}$. Initially both the balls are displaced by an angle $\theta = \frac{\pi}{6}$ rad with respect to the diameter PQ of the circle and released from rest. The frequency of oscillation of the ball B (in Hz) is

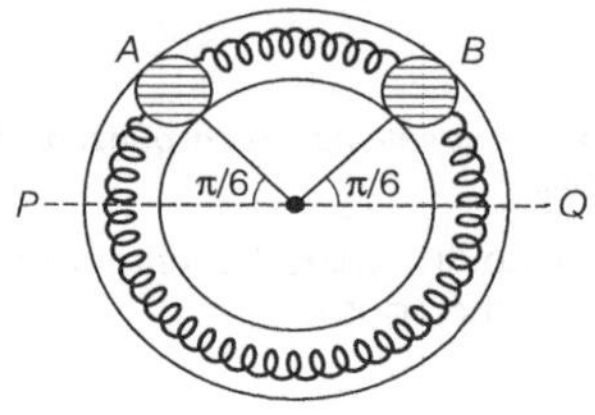

a. π b. $\frac{1}{\pi}$

c. 2π d. $\frac{1}{2\pi}$

6 A system of two identical rods (L-shaped) each of mass m and length L are resting on a peg P as shown in the figure. If the system is displaced in its plane by a small angle θ, find the period of oscillations.

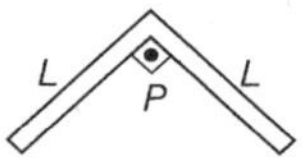

a. $2\pi\sqrt{\frac{\sqrt{2}}{3g}L}$ b. $2\pi\sqrt{\frac{2}{3g}L}$

c. $3\pi\sqrt{\frac{1}{3g}L}$ d. $2\pi\sqrt{\frac{2\sqrt{2}}{3g}L}$

7 In the figure shown, the springs are connected to the rod at one end and at the mid-point. The rod is hinged at its lower end. Rotational SHM of the rod (mass m and length L) will occur only if

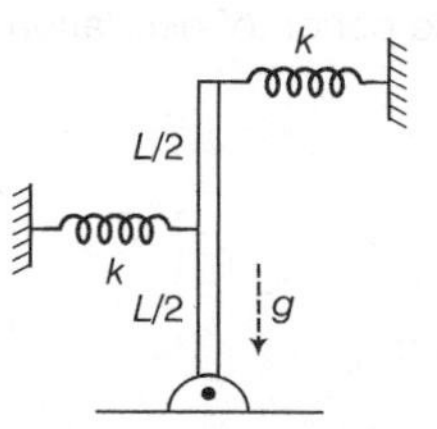

a. $k > 0$ b. $k > mg/3L$

c. $k > mg/4L$ d. $k > 2mg/5L$

8 Find the natural frequency for small vertical oscillations of the block of mass m shown in the figure. Spring is ideal. Pulley is a disc of mass M and radius R. There is no slipping.

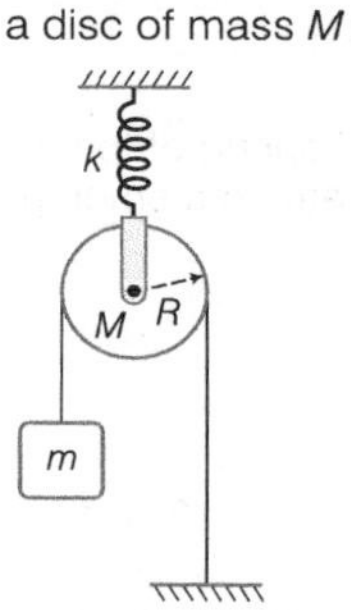

a. $\sqrt{\frac{k}{4m + 3M}}$ b. $\sqrt{\frac{k}{4m + \frac{3}{2}M}}$

c. $\sqrt{\frac{k}{2m + \frac{3}{2}M}}$ d. $\sqrt{\frac{k}{2m + 3M}}$

Working Space

9 The *v-t* graph of a particle in SHM is shown in the figure. Choose the correct statement.

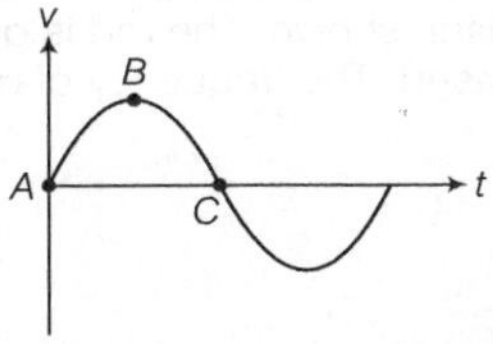

a. At *A* particle is at mean position and moving towards positive extreme
b. At *B* acceleration of particle is zero
c. At *C* acceleration of the particle is maximum and in positive direction
d. None of the above

10 The bob of a simple pendulum of period T is given a negative charge. It is allowed to oscillate above a positively charged plate, the new time period will be

a. equal to T b. more than T c. less than T d. infinite

11 In the figure shown, what can be maximum amplitude of the system so that there is no slipping between any of the blocks?

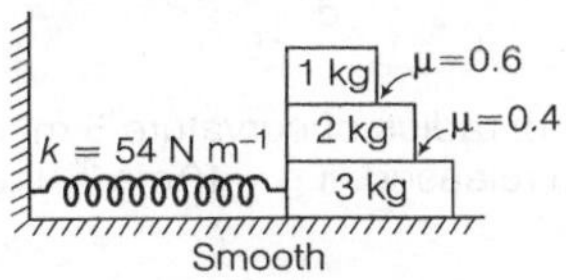

a. $\frac{2}{7}$ m b. $\frac{3}{4}$ m c. $\frac{4}{9}$ m d. $\frac{10}{3}$ m

12 A ball of mass m when dropped from certain height, strikes a wedge kept on smooth horizontal surface and moves horizontally just after impact as shown. If the ball strikes the ground at a distance R from its initial line of fall, then the amplitude of oscillation of wedge after being hit by the ball will be

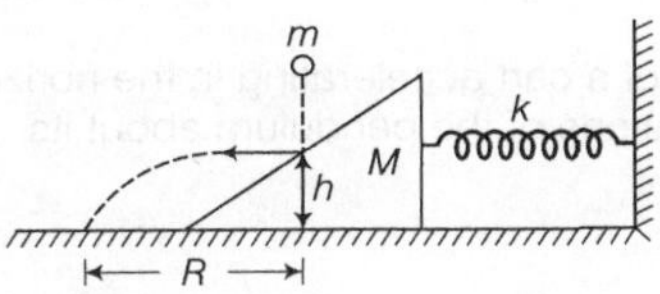

a. $\frac{mR}{M}\sqrt{\frac{Mg}{2kh}}$ b. $\frac{R}{m}\sqrt{\frac{kMg}{2h}}$ c. $R\sqrt{\frac{kg}{2Mh}}$ d. $R\sqrt{\frac{kg}{2mh}}$

13 The *x-t* graph of a particle undergoing simple harmonic motion is shown below. The acceleration of the particle at $t = \frac{4}{3}$ s is

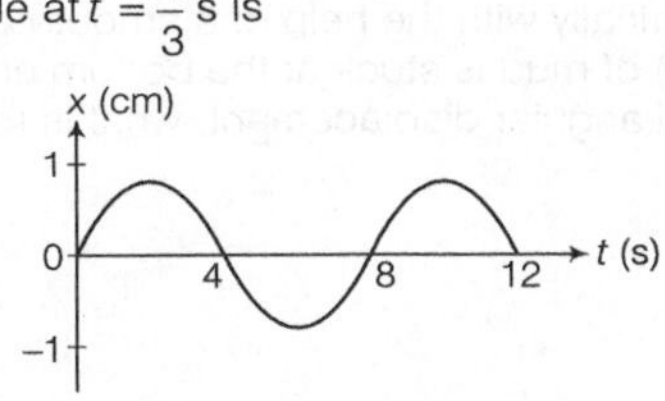

a. $\frac{\sqrt{3}}{32}\pi^2 \text{cm s}^{-2}$ b. $\frac{-\pi^2}{32}\text{cm s}^{-2}$ c. $\frac{\pi^2}{32}\text{cm s}^{-2}$ d. $-\frac{\sqrt{3}}{32}\pi^2 \text{cm s}^{-2}$

14 The mass m shown in the figure, oscillates in simple harmonic motion with amplitude A. The amplitude of the point P is

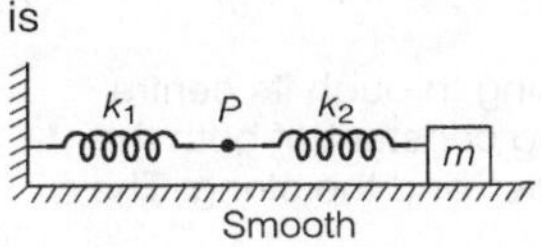

a. $\frac{k_1}{k_2}A$ b. $\frac{k_2}{k_1}A$

c. $\frac{k_1}{k_1 + k_2}A$ d. $\frac{k_2}{k_1 + k_2}A$

Working Space

15 A uniform rod of length L and mass m is pivoted at the centre. Its two ends are attached to two springs of equal spring constant k. The springs are fixed to rigid supports as shown in the figure and the rod is free to oscillate in a horizontal plane shown. The rod is gently pushed through a small angle θ in one direction and released. The frequency of oscillation is

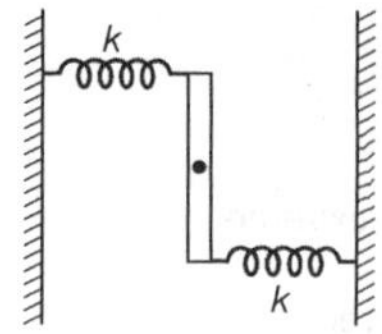

a. $\frac{1}{2\pi}\sqrt{\frac{2k}{m}}$ b. $\frac{1}{2\pi}\sqrt{\frac{k}{m}}$

c. $\frac{1}{2\pi}\sqrt{\frac{6k}{m}}$ d. $\frac{1}{2\pi}\sqrt{\frac{24k}{m}}$

16 The maximum tension in the string of a pendulum (during oscillation) is two times the minimum tension. Let θ_0 be the angular amplitude. Then the value of $\cos\theta_0$ is

a. $\frac{1}{2}$ b. $\frac{3}{4}$ c. $\frac{2}{3}$ d. $\frac{3}{5}$

17 A small sphere is placed on a concave smooth surface of radius of curvature 5 m. It is slightly pushed rightwards in the plane of the figure and released. if $g = 10\,\text{ms}^{-2}$, the time period of its oscillation is

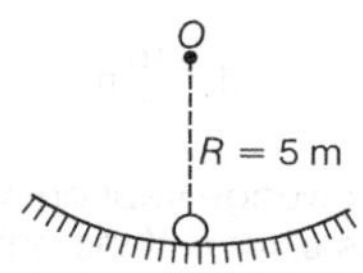

a. $\frac{\pi}{\sqrt{2}}$ s b. $\frac{\pi}{2}$ s c. π s d. $\pi\sqrt{2}$ s

18 A pendulum 50 cm long is suspended from the roof of a cart accelerating in the horizontal direction with $a_0 = 7\,\text{ms}^{-2}$. The period of small oscillations of the pendulum about its equilibrium angle is

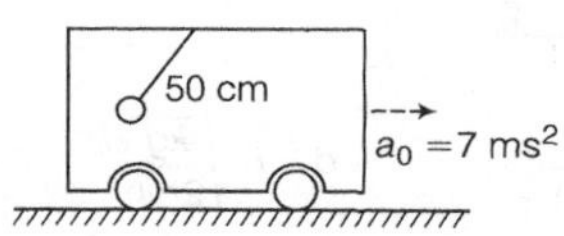

a. 12.8 s b. 14.2 s c. 13.7 s d. 8.4 s

19 A uniform disc of mass M and radius r is hanging vertically with the help of a smooth axle passing through its centre. A small quantity (mass m) of mud is stuck at the bottom end B near the rim of the disc. If the disc is now given small angular displacement, what is the equivalent length of simple pendulum?

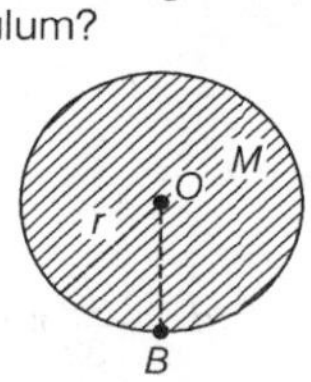

a. $\left(1+\frac{2M}{m}\right)r$ b. $\left(1+\frac{3M}{2m}\right)r$ c. $\left(1+\frac{M}{2m}\right)r$ d. $\frac{M}{m}r$

20 A 7 kg disc is free to rotate about a horizontal axis passing through its centre O (see figure). The radius of the disc is 10 cm and spring constant of both the springs is $600\,\text{Nm}^{-1}$. There is no slipping between the disc and the string. The time period of small oscillations of the disc is

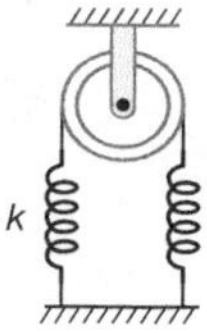

a. 0.12 s b. 0.34 s

c. 0.25 s d. 0.48 s

Working Space

21 A particle of mass m moves in a potential energy field given by $U = bx^2 - ax$, where a and b are positive constants. The frequency of oscillation of the particle is

a. $\frac{1}{2\pi}\sqrt{\frac{2b}{m}}$ b. $\frac{2}{\pi}\sqrt{\frac{b}{m}}$

c. $\frac{1}{\pi}\sqrt{\frac{b}{2m}}$ d. $\frac{1}{\pi}\sqrt{\frac{b}{m}}$

22 Position of a particle (executing SHM) as measured from mean position, at the end of three successive seconds is 1m, 5m, 5m. The time period of oscillations is

a. $\frac{\pi}{2\cos^{-1}\left(\frac{5}{7}\right)}$ b. $\frac{2\pi}{\sin^{-1}\left(\frac{3}{5}\right)}$

c. $\frac{\pi}{2\sin^{-1}\left(\frac{3}{5}\right)}$ d. $\frac{2\pi}{\cos^{-1}\left(\frac{3}{5}\right)}$

23 A particle of mass m free to move in the XY-plane is subject to a force whose components are $F_x = -kx$ and $F_y = -ky$, where k is a constant. The particle is released at $t = 0$, at the point (2, 3) at rest. The subsequent motion of the particle is

a. SHM along $2y - 3x = 0$ b. SHM along $2y + 3x = 0$

c. elliptical motion   d. circular motion

24 Acceleration due to gravity can be measured by using a pendulum. Suppose while doing the experiment you have made a tiny error dT in measuring T. The corresponding error in g would be (L is length of pendulum)

a. $\frac{g\sqrt{g}\,dT}{2\pi\sqrt{L}}$ b. $-\frac{g\sqrt{g}\,dT}{\pi\sqrt{L}}$

c. $-\frac{g\sqrt{g}\,dT}{2\pi\sqrt{L}}$ d. $\frac{2g\sqrt{g}\,dT}{\pi\sqrt{L}}$

25 Determine the time period of a simple pendulum, if the mass of the rod is not negligible. Mass of the rod is m_0 and mass of the bob is m. The length of the pendulum is L.

a. $\pi\sqrt{\frac{2\left(m+\frac{m_0}{2}\right)L}{\left(m+\frac{m_0}{3}\right)g}}$ b. $\pi\sqrt{\frac{\left(m+\frac{m_0}{3}\right)L}{\left(m+\frac{m_0}{2}\right)g}}$

c. $2\pi\sqrt{\frac{\left(m+\frac{m_0}{3}\right)L}{\left(m+\frac{m_0}{2}\right)g}}$ d. $2\pi\sqrt{\frac{\left(m+\frac{m_0}{2}\right)L}{\left(m+\frac{m_0}{3}\right)g}}$

26 A small particle is attached to a vertical wire which is under tension T. What will be the natural frequency of vibration of the mass, if it is displaced horizontally by a small distance and then released?

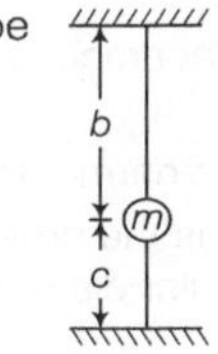

a. $\frac{1}{2\pi}\sqrt{\frac{2T}{m}\left(\frac{b+c}{bc}\right)}$ b. $\frac{1}{2\pi}\sqrt{\frac{T}{2m}\left(\frac{b+c}{bc}\right)}$

c. $\frac{1}{\pi}\sqrt{\frac{2T}{m}\left(\frac{b+c}{bc}\right)}$ d. $\frac{1}{2\pi}\sqrt{\frac{T}{m}\left(\frac{b+c}{bc}\right)}$

27 When a submerged body moves through a fluid, the particles of the fluid flow around the body and thus acquire kinetic energy. In the case of a sphere moving in an ideal fluid, the total kinetic energy acquired by the fluid is $\frac{1}{4}\rho V v^2$, where ρ is the density of the fluid, V the volume of the sphere and v is the velocity of the sphere. Consider a 0.5 kg hollow spherical shell of radius 8 cm which is held submerged in a tank of water (treat it is ideal fluid) by an ideal spring of force constant 500 Nm^{-1}. Neglecting fluid friction, determine the period of vibration of the shell when it is slightly displaced vertically and released.

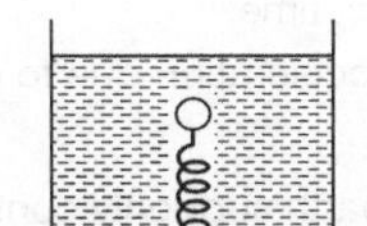

a. 0.278 s   b. 0.352 s

c. 0.431 s   d. 0.572 s

Working Space

28 In the above problem, if the tank is accelerated upward at a constant rate $8 ms^{-2}$, what is the time period?

a. 0.278 s
b. 0.352 s
c. 0.431 s
d. 0.572 s

29 Vertical displacement of a plank with a body of mass m on it is varying according to law $y = \sin \omega t + \sqrt{3} \cos \omega t$. The minimum value of ω for which the mass just breaks off the plank and the moment it occurs first time after $t = 0$, are respectively (assume y to be positive vertically upwards)

a. $\sqrt{\frac{g}{2}}, \frac{\sqrt{2}}{6} \frac{\pi}{\sqrt{g}}$
b. $\frac{g}{\sqrt{2}}, \frac{2}{3} \sqrt{\frac{\pi}{g}}$
c. $\sqrt{\frac{g}{2}}, \frac{\pi}{3} \sqrt{\frac{2}{g}}$
d. $\sqrt{2g}, \sqrt{\frac{2\pi}{3g}}$

30 The displacement of the motion of a particle is represented (in metre) by the equation $x = 0.4\left[\cos^2\left(\frac{\pi t}{2}\right) - \sin^2\left(\frac{\pi t}{2}\right)\right]$. The motion of the particle is

a. oscillatory but not SHM
b. SHM with amplitude 0.4 m
c. SHM with amplitude $0.4\sqrt{2}$ m
d. SHM with amplitude 0.8 m

31 Two SHMs are given by $y_1 = A \sin \omega t$ and $y_2 = \frac{A}{2} \sin \omega t + \frac{A}{2} \cos \omega t$. The ratio of their amplitudes is

a. 1
b. 2
c. $\frac{1}{\sqrt{2}}$
d. $\sqrt{2}$

Statement Type

***Direction** (Q. Nos. 32-35) This section is based on Statement I and Statement II. Select the correct answer from the codes given below.*

a. Both Statement I and Statement II are correct and Statement II is the correct explanation of Statement I
b. Both Statement I and Statement II are correct but Statement II is not the correct explanation of Statement I
c. Statement I is correct but Statement II is incorrect
d. Statement II is correct but Statement I is incorrect

32 **Statement I** In simple harmonic motion, the velocity is maximum when the acceleration is maximum.

Statement II Displacement and velocity of SHM differ in phase by $\frac{\pi}{2}$.

33 **Statement I** Simple harmonic motion is not a uniform motion.

Statement II SHM is the projection of uniform circular motion along any of the two mutually perpendicular diameters.

34 **Statement I** A watch (spring wound) falls from the top of a tower. The watch during motion shows correct time.

Statement II The acceleration due to gravity will have no effect on time period of the watch.

35 **Statement I** If the earth suddenly contracts, then duration of the day will decrease.

Statement II The angular velocity of the rotation of earth will decrease.

Revisal Problems for JEE Advanced

Only One Option Correct Type

Direction *(Q. Nos. 1-6) This section contains 6 multiple choice questions. Each question has four choices (a), (b), (c) and (d), out of which ONLY ONE option is correct.*

Working Space

1 In the block-springs system shown in the figure, all pulleys, springs and strings are ideal. The time period of small vertical displacement of the block is

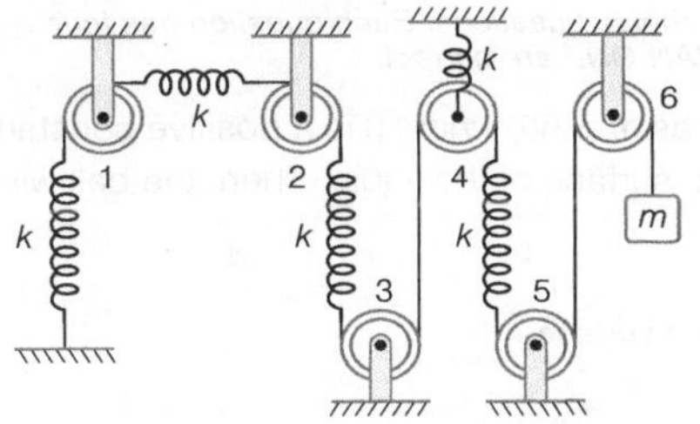

a. $2\pi\sqrt{\frac{m}{k}}$ b. $2\pi\sqrt{\frac{8m}{k}}$

c. $2\pi\sqrt{\frac{4m}{k}}$ d. None of these

2 In the arrangement shows pulleys and springs are ideal. Now the block is slightly displaced vertically, the time period of oscillations is

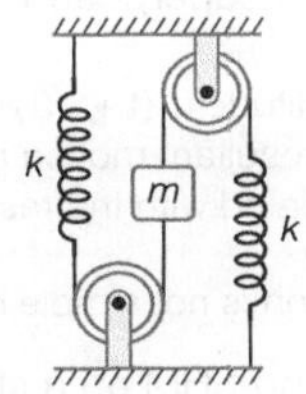

a. $\sqrt{2}\pi\sqrt{\frac{m}{k}}$ b. $2\pi\sqrt{\frac{m}{k}}$

c. $\pi\sqrt{\frac{m}{2k}}$ d. None of these

3 A system is shown in the figure. The time period for small oscillations of the two blocks will be

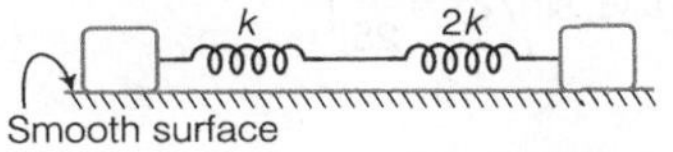

a. $2\pi\sqrt{\frac{3m}{k}}$ b. $2\pi\sqrt{\frac{m}{6k}}$

c. $2\pi\sqrt{\frac{3m}{2k}}$ d. $2\pi\sqrt{\frac{3m}{4k}}$

4 In the figure shown pulley is massless. Initially the blocks are held at a height such that spring is in its natural length. The amplitude and velocity amplitude of block B_1 respectively are (there is no slipping anywhere)

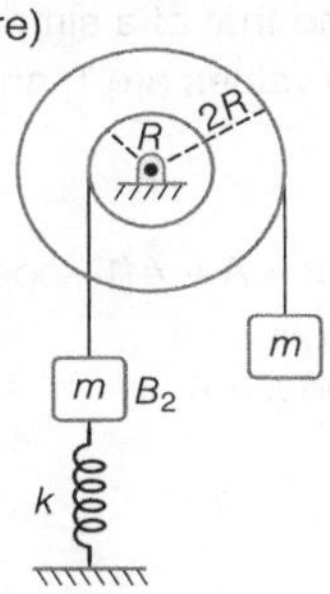

a. $\frac{2mg}{k}, 2g\sqrt{\frac{m}{5k}}$ b. $\frac{2mg}{k}, 2g\sqrt{\frac{m}{3k}}$ c. $\frac{mg}{2k}, 2g\sqrt{\frac{m}{3k}}$ d. $\frac{mg}{2k}\sqrt{\frac{m}{5k}}$

Working Space

5 In the above problem, frequency of oscillation of B_2 is

a. $\frac{1}{2\pi}\sqrt{\frac{k}{3m}}$ b. $\frac{1}{\pi}\sqrt{\frac{2k}{3m}}$

c. $\frac{1}{2\pi}\sqrt{\frac{k}{5m}}$ d. $\frac{1}{\pi}\sqrt{\frac{2k}{5m}}$

6 The period of the function $y = \sin\omega t + \sin 2\omega t + \sin 3\omega t$ is

a. $\frac{2\pi}{3\omega}$ b. $\frac{\pi}{2\omega}$

c. $\frac{2\pi}{\omega}$ d. $\frac{2\pi}{5\omega}$

One or More than One Options Correct Type

Direction *(Q. Nos. 7-11) This section contains 5 multiple choice questions. Each question has four choices (a), (b), (c) and (d), out of which ONE or MORE THAN ONE are correct.*

7 The density of a liquid ρ_L, varies with the depth h as $\rho_L = \beta h$, where β is a positive constant. A small ball of density ρ_0 is released from the free surface of the liquid. Then, the ball will

a. execute SHM of amplitude $\frac{\rho_0}{\beta}$

b. have its mean position at a depth $\frac{\rho_0}{2\beta}$ below the free surface

c. sink to a maximum depth of $\frac{2\rho_0}{\beta}$

d. All of the above

8 Three simple harmonic motions happening on the same line, having the same amplitude and same period are superposed. If each has a phase greater than the previous by 45°, then

a. the resultant amplitude is $(1+\sqrt{2})$ times individual amplitude
b. the phase of the resultant motion relative to the first is 90°
c. the energy associated with the resulting motion is $(3+2\sqrt{2})$ times the energy associated with any single motion
d. the resulting motion is not simple harmonic

9 A particle is executing SHM on a straight line. P_1 and P_2 are two points at which its velocity is zero. It starts from P_1 and passes through a certain point P $(P_1P < P_2P)$ at successive intervals of $\frac{1}{2}$ s and $\frac{3}{2}$ s with a speed 3ms^{-1}. Then, $\left(\text{Take, } \sin 72^\circ = \frac{24}{25}\right)$

a. maximum speed of particle is $\frac{50}{24}\text{ ms}^{-1}$

b. maximum speed of particle is $\frac{75}{24}\text{ ms}^{-1}$

c. ratio $\frac{P_1P}{P_2P}$ is $\frac{3}{5}$

d. ratio $\frac{P_1P}{P_2P}$ is $\frac{3}{4}$

10 Time period of spring-block system on surface of earth is T_1 and that of a simple pendulum is T_2. At a height $h = R$, (R is radius of earth) the corresponding values are T_1' and T_2', then

a. $T_1' > T_1$ b. $T_1' = T_1$
c. $T_2' > T_2$ d. $T_2' < T_2$

11 The x-t equation of a particle moving along X-axis is given by $x = A + A(1-\cos\omega t)$, then the

a. particle oscillates simple harmonically between points $x = 2A$ and $x = A$
b. velocity of particle is maximum at $x = 2A$
c. time taken by the particle in travelling from $x = A$ to $x = 3A$ is $\frac{\pi}{\omega}$
d. time taken by the particle in travelling from $x = A$ to $x = 2A$ is $\frac{\pi}{2\omega}$

Working Space

Comprehension Type

Direction (Q. Nos. 12-20) *This section contains 3 paragraphs, each describing theory, experiments, data, etc. Nine questions related to the paragraphs have been given. Each question has only one correct answer among the four given options (a), (b), (c) and (d).*

Passage I

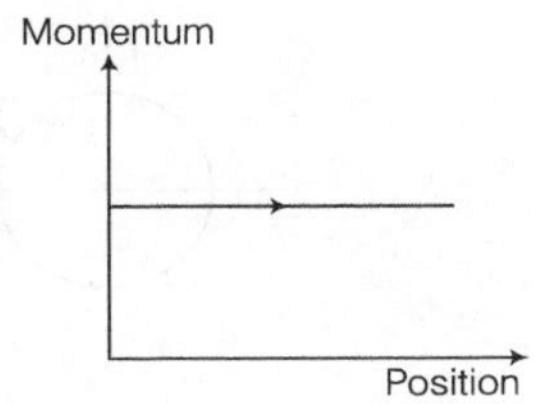

Phase space diagrams are useful tools in analysing all kinds of dynamical problems. They are especially useful in studying the changes in motion as initial position and momentum are changed. Here we consider some simple dynamical systems in one-dimension. For such systems, phase-space is a plane in which position is plotted along horizontal axis and momentum is plotted along vertical axis. The arrow on the curve indicates the time flow.

For example, the phase-space diagram for a particle moving with a constant velocity is a straight line as shown in the figure. We use the sign convention in which position or momentum upwards (or to right) is positive and downwards (or to left) is negative. Based on the above facts, answer the following question, where p is momentum and x is position.

12 The phase-space diagram for a ball thrown vertically up from the ground is

a. 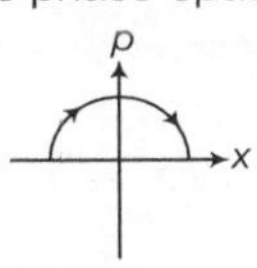

b. 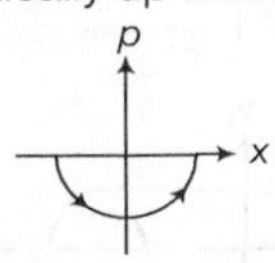

c. 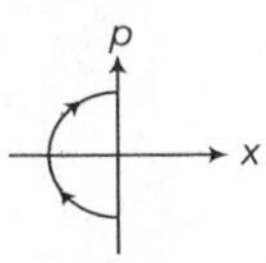

d. 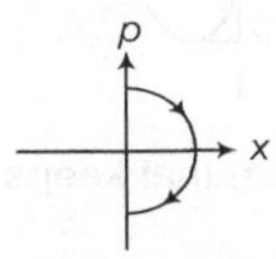

13 The phase-space diagram for simple harmonic motion is a circle centred at a origin. In the figure, the two circles represent the same oscillator but for different initial conditions and E_1 and E_2 are the total mechanical energies respectively. Then,

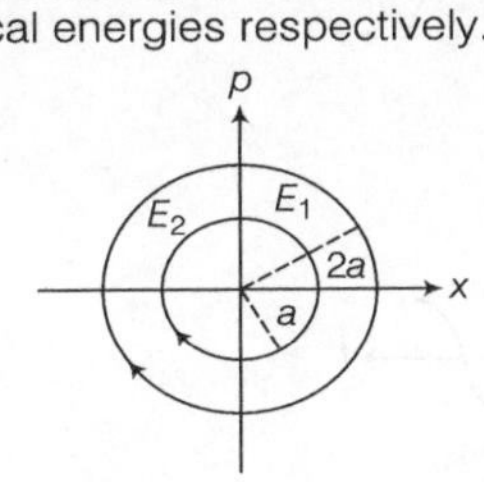

a. $E_1 = \sqrt{2}\, E_2$ b. $E_1 = 2E_2$ c. $E_1 = 4\, E_2$ d. $E_1 = 16\, E_2$

14 Consider the spring-mass system, with the mass submerged in water, as shown in the figure. The phase space diagram for one cycle of this system is

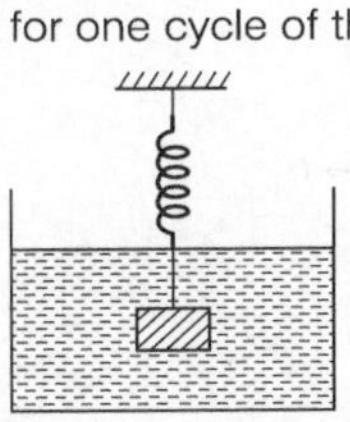

a. 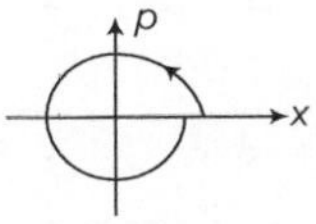

b. 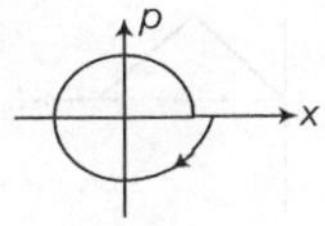

c. 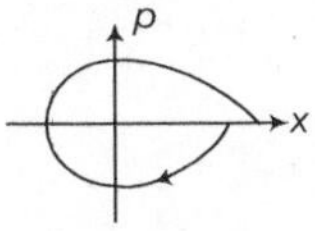

d. 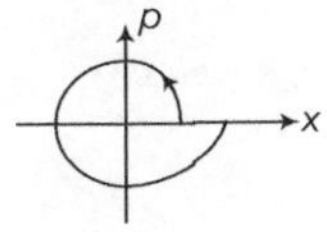

Passage II

An object is fixed to the edge of a disc that is rotating with uniform circular motion. At time $t = 0$, the position and the velocity of the object are shown in the figure. The object travels around the disc for a full rotation.

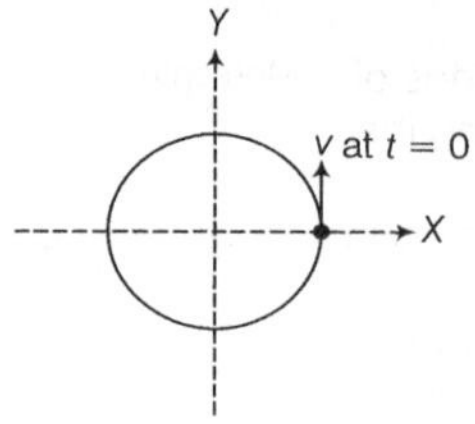

15 The *x*-component of the velocity, when plotted against time is represented by

a. 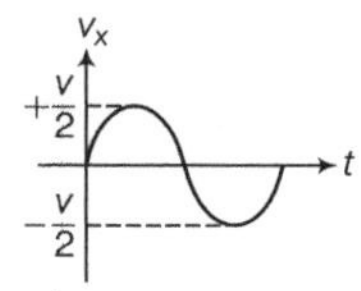

b. 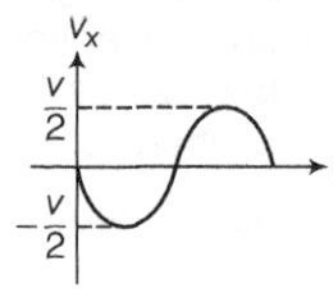

c. 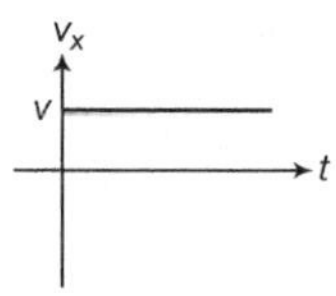

d. 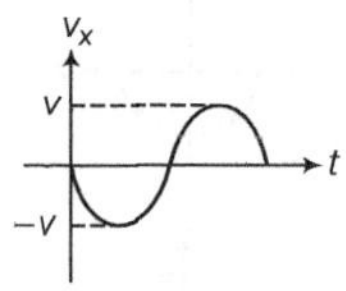

16 The *y*-component of the force (plotted against time) that keeps the object moving in the circle is represented by

a. 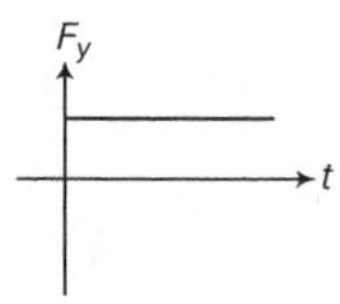

b. 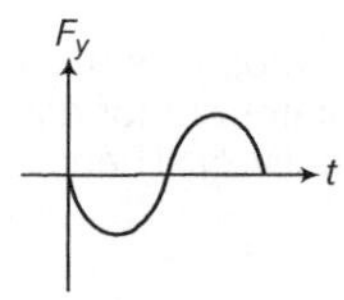

c. 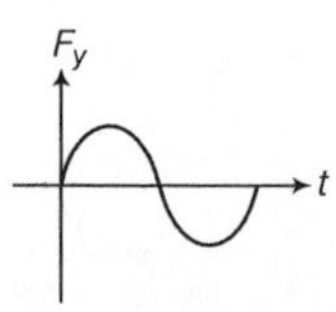

d. 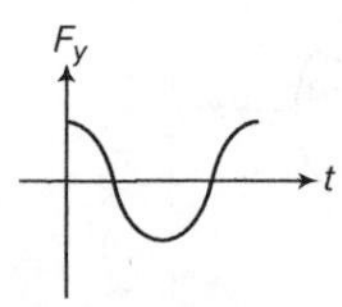

17 The graph of the angle of the position of the object (that it makes with *X*-axis) plotted against the time *t* is best represented by

a. 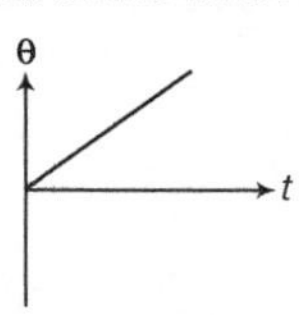

b. 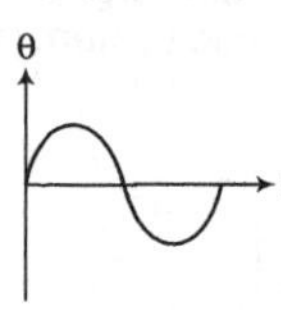

c. 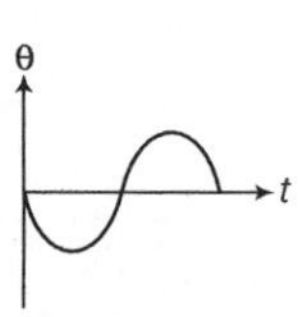

d. 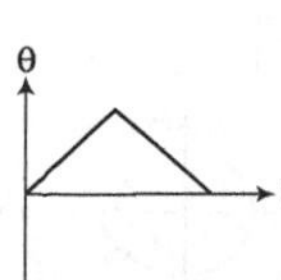

Passage III

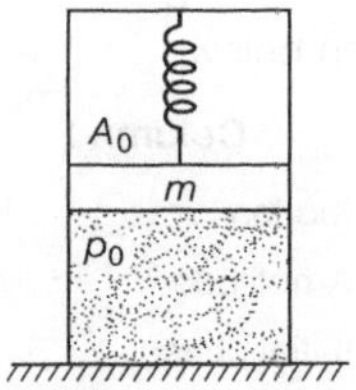

A non-conducting piston of mass m and area A_0 divides a non-conducting, closed cylinder as shown in the figure. Piston having mass m is connected with top wall of cylinder by a spring of force constant k. The upper part of the cylinder is evacuated and the lower part contains an ideal gas at pressure p_0 in equilibrium position. Assume the adiabatic constant for the gas to be γ, the equilibrium length of each part to be L_0 and neglect friction.

18 Assuming $p_0A_0 > mg$, the compression in the spring at equilibrium position is

a. zero
b. $\dfrac{2p_0A_0 - mg}{k}$
c. $\dfrac{p_0A_0 - mg}{k}$
d. $\dfrac{p_0A_0 - mg}{2k}$

19 The angular frequency for small oscillations is

a. $\sqrt{\dfrac{kL_0 + \gamma p_0A_0}{2mL_0}}$
b. $\sqrt{\dfrac{2kL_0 + \gamma p_0A_0}{mL_0}}$
c. $\sqrt{\dfrac{k}{m}}$
d. $\sqrt{\dfrac{kL_0 + \gamma p_0A_0}{mL_0}}$

20 If the spring is now removed and the upper part of the cylinder is opened, then the angular frequency for small oscillations (Assume pressure of the gas at equilibrium is p_1 and length of gas column is L_1) is

a. $\sqrt{\dfrac{\gamma p_1A_0}{mL_1}}$
b. $\sqrt{\dfrac{2\gamma p_1A_0}{mL_1}}$
c. $\sqrt{\dfrac{\gamma p_1A_0}{4mL_1}}$
d. $\sqrt{\dfrac{\gamma p_1A_0}{2mL_1}}$

Matching List Type

Direction *(Q. Nos. 21-23) Choices for the correct combination of elements from Column I and Column II are given as options (a), (b), (c) and (d), out of which one is correct.*

21 A block of mass $m = 2$ kg is connected to a spring of spring constant $k = 3200\,\text{Nm}^{-1}$ and placed on a frictionless horizontal surface as shown in the figure. Initially it is compressed by a distance 10 cm. It is then released at $t = 0$.

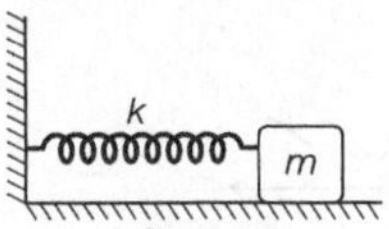

Match the Column I with Column II and mark the correct option from the codes given below.

	Column I		Column II (all in seconds)
i.	Time taken by the mass to move first 5 cm is	p.	$\dfrac{\pi}{160}$
ii.	Time taken by the mass to move next 5 cm is	q.	$\dfrac{\pi}{240}$
iii.	Time after which kinetic energy and potential energy become equal for first time is	r.	$\dfrac{\pi}{120}$
iv.	Time at which kinetic energy becomes one-fourth of its maximum value is	s.	$\dfrac{\pi}{40}$

Codes

	i	ii	iii	iv
a.	r	q	p	q
b.	p	q	r	s
c.	s	r	q	p
d.	q	p	r	s

22 If $y = A\sin\omega t + A\sin\left(\omega t + \frac{2\pi}{3}\right)$, match the following columns and mark the correct option from the codes given below.

	Column I		Column II
i.	Motion	p.	is periodic but not SHM
ii.	Amplitude	q.	is SHM
iii.	Initial phase	r.	A
iv.	Maximum velocity	s.	$\frac{\pi}{3}$
		t.	ωA
		u.	$\omega A\sqrt{2}$

Codes

	i	ii	iii	iv
a.	q	r	s	t
b.	p	u	s,t	q
c.	q,t	p	r,s	u
d.	p	q	r	u

23 The potential energy function for the force between two atoms in a diatomic molecule can approximately be expressed as $U(x) = \frac{a}{x^{12}} - \frac{b}{x^6}$, where a and b are positive constants and x is the distance between the atoms. Match the Column I with Column II and mark the correct option from the codes given below.

	Column I		Column II
i.	The graph between potential energy (on Y-axis) *vs* x (on X-axis) will be	p.	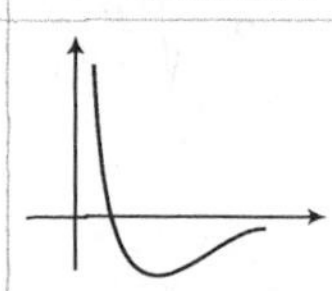
ii.	The graph between force of the atoms (on Y-axis) *vs* X (on X-axis) will be	q.	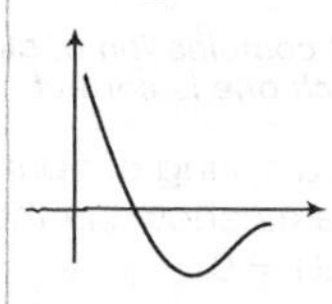
iii.	The dissociation of energy of the molecule is (assuming initially molecule is at rest at equilibrium)	r.	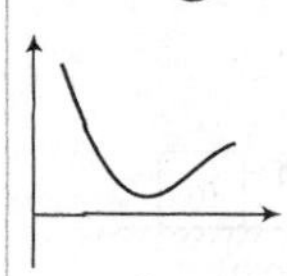
		s.	$-\frac{b^2}{4a}$
		t.	$\frac{b^2}{4a}$

Codes

	i	ii	iii
a.	p	q	t
b.	q	t	s
c.	s	p	q
d.	p	q	r

One Integer Value Correct Type

Working Space

***Direction** (Q. Nos. 24-29) This section contains 6 questions. When worked out will result in an integer from 0 to 9 (both inclusive).*

24 A point of the surface of a solid sphere of radius 5 cm is attached directly to a pivot on the ceiling. The sphere swings back and forth as a physical pendulum with small amplitude. What is the length of simple pendulum (in cm) that has the same period as this physical pendulum?

25 The angular frequency (in rad s^{-1}) of small oscillations of a uniform rod with length $l = 14.7$ cm, pivoted at one end is 2β. Find the value of β. (Given, $g = 9.8 \, ms^{-2}$)

26 In the figure is shown, a cord attached between a 2 kg block and a spring with force constant $k = 20 \, Nm^{-1}$. The other end of the spring is attached to the wall and the cord is placed over a pulley ($I = 0.6MR^2$) of mass 5 kg and radius 0.5 m. Assuming no slipping occurs, what is the angular frequency (in rad s^{-1}) of oscillations, when the body is set into motion?

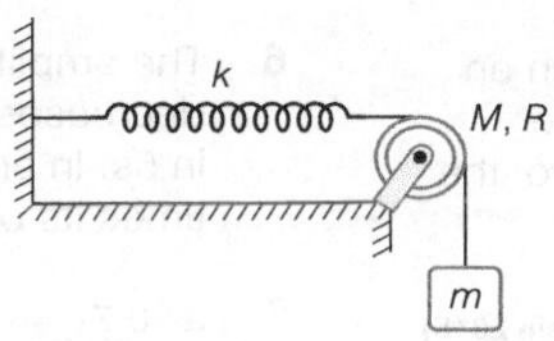

27 A heavy container containing an ideal gas is kept on a horizontal surface. A smooth piston of mass M is at rest as shown in the figure. The natural length of the spring is L_0. Now, the piston is given a small downward push. Assuming the temperature of the gas to be constant and there is vacuum over the piston, the time period of small oscillations is $\frac{2\pi}{\psi}$ s. Find the value of ψ. (Given, $L_0 = 2$, $M = 5$, $k = 295$, $g = 10$ all in SI units)

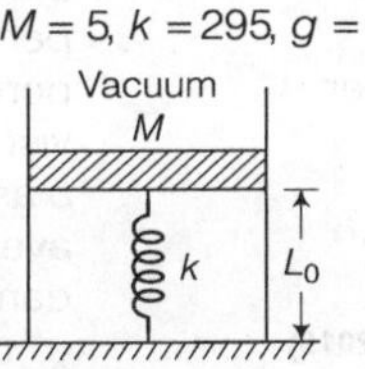

28 A simple pendulum is making small oscillations with its bob immersed in a liquid of density four times less than the density of the bob. The time period is $4\pi\sqrt{\frac{L}{\beta g}}$. Find the value of β, where L is length of the pendulum.

29 A particle moves according to the law $x = A\cos \pi t$. The distance covered by it in 2.5 s is $*A$. Find the value of *.

JEE Main & AIEEE Archive

(Compilation of Last 13 Years' Questions)

1 The period of oscillation of a simple pendulum is $T = 2\pi\sqrt{\frac{L}{g}}$. Measured value of L is 20.0 cm known to 1 mm accuracy and time for 100 oscillations of the pendulum is found to be 90 s using a wrist watch of 1s resolution. The accuracy in the determination of g is **(JEE Main 2015)**

a. 2% b. 3% c. 1% d. 5%

2 A pendulum made of a uniform wire of cross-sectional area A has time period T. When an additional mass M is added to its bob, the time period changes to T_M. If the Young's modulus of the material of the wire is Y, then $\frac{1}{Y}$ is equal to

(g = gravitational acceleration) **(JEE Main 2015)**

a. $\left[\left(\frac{T_M}{T}\right)^2 - 1\right]\frac{A}{Mg}$ b. $\left[\left(\frac{T_M}{T}\right)^2 - 1\right]\frac{Mg}{A}$

c. $\left[1 - \left(\frac{T_M}{T}\right)^2\right]\frac{A}{Mg}$ d. $\left[1 - \left(\frac{T}{T_M}\right)^2\right]\frac{A}{Mg}$

3 For a simple pendulum, a graph is plotted between its Kinetic Energy (KE) and Potential Energy (PE) against its displacement d. Which one of the following represents these correctly? (graphs are schematic and not drawn to scale) **(JEE Main 2015)**

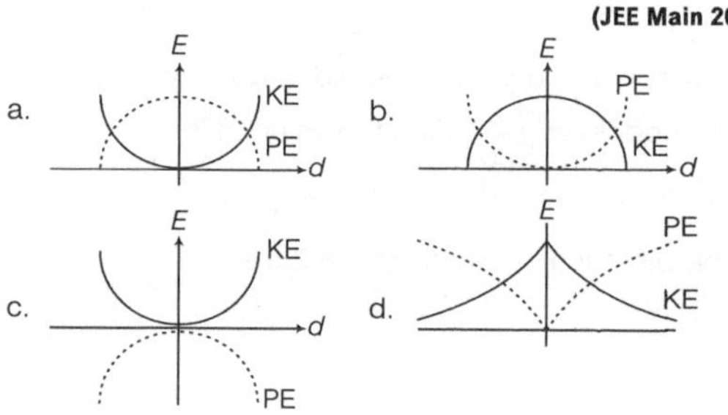

4 Two long current carrying thin wires, both with current I, are held by insulating threads of length L and are in equilibrium as shown in the figure, with threads making an angle θ with the vertical. If wires have mass λ per unit length then, the value of I is

(g = gravitational acceleration) **(JEE Main 2015)**

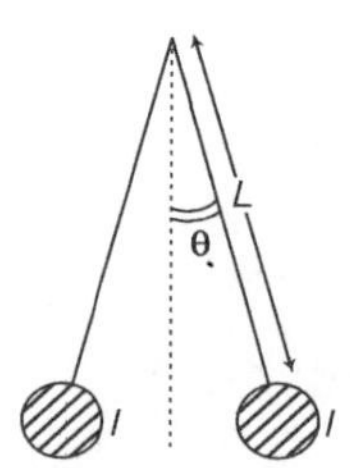

a. $\sin\theta\sqrt{\frac{\pi\lambda gL}{\mu_0 \cos\theta}}$ b. $2\sin\theta\sqrt{\frac{\pi\lambda gL}{\mu_0 \cos\theta}}$

c. $2\sqrt{\frac{\pi gL}{\mu_0}\tan\theta}$ d. $\sqrt{\frac{\pi\lambda gL}{\mu_0}\tan\theta}$

5 A particle moves with simple harmonic motion in a straight line. In first t second after starting from rest, it travels a distance a and in next t second it travels $2a$ in same direction, then **(JEE Main 2014)**

a. amplitude of motion is $3a$

b. time period of oscillations is 5π

c. amplitude of motion is $4a$

d. time period of oscillations is 6π

6 The amplitude of a damped oscillator decreases to 0.9 times its original magnitude in 5s. In another 10s, it will decrease to α times its original magnitude, where α equals **(JEE Main 2013)**

a. 0.7 b. 0.81

c. 0.729 d. 0.6

7 If a simple pendulum has significant amplitude (upto a factor of $1/e$ of original) only in the period between $t = 0$ s to $t = \tau$ s, then τ may be called the average life of the pendulum. When the spherical bob of the pendulum suffers a retardation (due to viscous drag) proportional to its velocity with b as the constant of proportionality, the average life time of the pendulum (assuming damping is small) (in seconds) is **(AIEEE 2012)**

a. $\frac{0.693}{b}$ b. b

c. $\frac{1}{b}$ d. $\frac{2}{b}$

8 A wooden cube (density of wood d) of side l floats in a liquid of density ρ with its upper and lower surfaces horizontal. If the cube is pushed slightly down and released, it performs simple harmonic motion of period T. Then, T is equal to **(AIEEE 2011)**

a. $2\pi\sqrt{\frac{l\rho}{(\rho - d)g}}$ b. $2\pi\sqrt{\frac{ld}{\rho g}}$

c. $2\pi\sqrt{\frac{l\rho}{dg}}$ d. $2\pi\sqrt{\frac{ld}{(\rho - d)g}}$

9 If a spring of stiffness k is cut into two parts A and B of length $l_A : l_B = 2 : 3$, then the stiffness of spring A is given by **(AIEEE 2011)**

a. $\frac{5}{2}k$

b. $\frac{3k}{5}$

c. $\frac{2k}{5}$

d. k

10 Two particles are executing simple harmonic motion of the same amplitude A and frequency ω along the X-axis. Their mean position is separated by distance X_0 ($X_0 > A$). If the maximum separation between them is $(X_0 + A)$, the phase difference between their motion is **(AIEEE 2011)**

a. $\frac{\pi}{3}$ b. $\frac{\pi}{4}$ c. $\frac{\pi}{6}$ d. $\frac{\pi}{2}$

11 A mass M, attached to a horizontal spring, executes SHM with amplitude A_1. When the mass M passes through its mean position, then a smaller mass m is placed over it and both of them move together with amplitude A_2. The ratio of $\left(\frac{A_1}{A_2}\right)$ is **(AIEEE 2011)**

a. $\frac{M+m}{M}$ b. $\left(\frac{M}{M+m}\right)^{1/2}$

c. $\left(\frac{M+m}{M}\right)^{1/2}$ d. $\frac{M}{M+m}$

12 The potential energy function for the force between two atoms in a diatomic molecule is approximately given by $U(x) = \frac{a}{x^{12}} - \frac{b}{x^6}$, where a and b are constants and x is the distance between the atoms. If the dissociation energy of the molecule is $D = [U(x=\infty) - U_{\text{at equilibrium}}]$, then D is **(AIEEE 2010)**

a. $\frac{b^2}{2a}$ b. $\frac{b^2}{12a}$ c. $\frac{b^2}{4a}$ d. $\frac{b^2}{6a}$

13 If x, v and a denote the displacement, the velocity and the acceleration of a particle executing simple harmonic motion of time period T, then which of the following does not change with time? **(AIEEE 2009)**

a. $a^2T^2 + 4\pi^2v^2$ b. $\frac{aT}{x}$

c. $aT + 2\pi v$ d. $\frac{aT}{v}$

14 A point mass oscillates along the X-axis according to the law $x = x_0 \cos(\omega t - \pi/4)$. If the acceleration of the particle is written as $a = A\cos(\omega t + \delta)$, then **(AIEEE 2007)**

a. $A = x_0$, $\delta = -\frac{\pi}{4}$ b. $A = x_0\omega^2$, $\delta = \frac{\pi}{4}$

c. $A = x_0\omega^2$, $\delta = -\frac{\pi}{4}$ d. $A = x_0\omega^2$, $\delta = \frac{3\pi}{4}$

15 Two springs of force constants k_1 and k_2, are connected to a mass m as shown. The frequency of oscillation of the mass is f. If both k_1 and k_2 are made four times their original values, the frequency of oscillation becomes **(AIEEE 2007)**

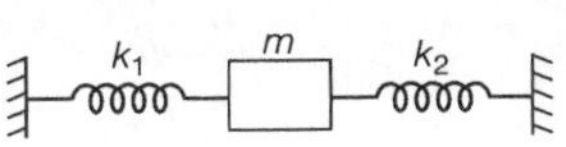

a. $\frac{f}{2}$ b. $\frac{f}{4}$ c. $4f$ d. $2f$

16 A particle of mass m executes simple harmonic motion with amplitude a and frequency ν. The average kinetic energy during its motion from the position of equilibrium to the end is **(AIEEE 2007)**

a. $\pi^2 ma^2\nu^2$ b. $\frac{1}{4}ma^2\nu^2$

c. $4\pi^2 ma^2\nu^2$ d. $2\pi^2 ma^2\nu^2$

17 The displacement of an object attached to a spring and executing simple harmonic motion is given by $x = 2 \times 10^{-2} \cos \pi t$ metre. The time at which the maximum speed first occurs is **(AIEEE 2007)**

a. 0.5 s b. 0.75 s

c. 0.125 s d. 0.25 s

18 The maximum velocity of a particle, executing simple harmonic motion with an amplitude 7 mm, is 4.4 ms^{-1}. The period of oscillation is **(AIEEE 2006)**

a. 0.01 s b. 10 s

c. 0.1 s d. 100 s

19 The function $\sin^2(\omega t)$ represents

a. a periodic but not simple harmonic motion with a period $2\pi/\omega$ **(AIEEE 2005)**

b. a periodic but not simple harmonic motion with a period π/ω

c. a simple harmonic motion with a period $2\pi/\omega$

d. a simple harmonic motion with a period π/ω

20 Two simple harmonic motions are represented by the equations $y_1 = 0.1\sin\left(100\pi t + \frac{\pi}{3}\right)$ and $y_2 = 0.1\cos \pi t$. The phase difference of the velocity of particle 1 with respect to the velocity of particle 2 is **(AIEEE 2005)**

a. $\frac{-\pi}{6}$ b. $\frac{\pi}{3}$

c. $\frac{-\pi}{3}$ d. $\frac{\pi}{6}$

21 If a simple harmonic motion is represented by $\frac{d^2x}{dt^2} + \alpha x = 0$, its time period is **(AIEEE 2005)**

a. $\frac{2\pi}{\alpha}$ b. $\frac{2\pi}{\sqrt{\alpha}}$

c. $2\pi\alpha$ d. $2\pi\sqrt{\alpha}$

22 The bob of a simple pendulum is a spherical hollow ball filled with water. A plugged hole near the bottom of the oscillating bob gets suddenly unplugged. During observation, till water is coming out, the time period of oscillation would **(AIEEE 2005)**

a. first increase and then decrease to the original value

b. first decrease and then increase to the original value

c. remain unchanged

d. increase towards a saturation value

23 The bob of a simple pendulum executes simple harmonic motion in water with a period t, while the period of oscillation of the bob is t_0 in air. Neglecting frictional force of water and given that the density of the bob is $(4/3)\times 1000\ \text{kgm}^{-3}$. What relationship between t and t_0 is true? **(AIEEE 2004)**

a. $t = t_0$
b. $t = \dfrac{t_0}{2}$
c. $t = 2t_0$
d. $t = 4t_0$

24 A particle at the end of a spring executes simple harmonic motion with a period t_1, while the corresponding period for another spring is t_2. If the period of oscillation with the two springs in series is T, then **(AIEEE 2004)**

a. $T = t_1 + t_2$
b. $T^2 = t_1^2 + t_2^2$
c. $T^{-1} = t_1^{-1} + t_2^{-1}$
d. $T^{-2} = t_1^{-2} + t_2^{-2}$

25 The total energy of a particle, executing simple harmonic motion is **(AIEEE 2004)**

a. $\propto x$
b. $\propto x^2$
c. independent of x
d. $\propto x^{1/2}$

where, x is the displacement from the mean position.

26 A particle of mass m is attached to a spring (of spring constant k) and has a natural angular frequency ω_0. An external force $F(t)$ proportional to $\cos \omega t (\omega \neq \omega_0)$ is applied to the oscillator. The time displacement of the oscillator will be proportional to **(AIEEE 2004)**

a. $\dfrac{m}{\omega_0^2 - \omega^2}$
b. $\dfrac{1}{m(\omega_0^2 - \omega^2)}$
c. $\dfrac{1}{m(\omega_0^2 + \omega^2)}$
d. $\dfrac{m}{\omega_0^2 + \omega^2}$

27 In forced oscillation of a particle, the amplitude is maximum for a frequency ω_1 of the force, while the energy is maximum for a frequency ω_2 of the force, then **(AIEEE 2004)**

a. $\omega_1 = \omega_2$
b. $\omega_1 > \omega_2$
c. $\omega_1 < \omega_2$, when damping is small and $\omega_1 > \omega_2$, when damping is large
d. $\omega_1 < \omega_2$

28 A mass M is suspended from a spring of negligible mass. The spring is pulled a little and then released so that the mass executes SHM of time period T. If the mass is increased by m, the time period becomes $\dfrac{5T}{3}$, then the ratio of $\dfrac{m}{M}$ is **(AIEEE 2003)**

a. $\dfrac{3}{5}$
b. $\dfrac{25}{9}$
c. $\dfrac{16}{9}$
d. $\dfrac{5}{3}$

29 Two particles A and B of equal masses are suspended from two massless springs of spring constants k_1 and k_2, respectively. If the maximum velocities, during oscillations are equal, the ratio of amplitudes of A and B is **(AIEEE 2003)**

a. $\sqrt{\dfrac{k_1}{k_2}}$
b. $\dfrac{k_1}{k_2}$
c. $\sqrt{\dfrac{k_2}{k_1}}$
d. $\dfrac{k_2}{k_1}$

30 The length of a simple pendulum executing simple harmonic motion is increased by 21%. The percentage increase in the time period of the pendulum of increased length is **(AIEEE 2003)**

a. 11%
b. 21%
c. 42%
d. 10%

31 A metal wire of linear mass density of 9.8 g / m is stretched with a tension of 10 kg-wt between two rigid supports 1 m apart. The wire passes at its middle point between the poles of a permanent magnet and it vibrates in resonance when carrying an alternating current of frequency n. The frequency n of the alternating source is **(AIEEE 2003)**

a. 50 Hz
b. 100 Hz
c. 200 Hz
d. 25 Hz

32 A body executes simple harmonic motion. The potential energy (PE), the kinetic energy (KE) and total energy (TE) are measured as function of displacement x. Which of the following statements is true? **(AIEEE 2003)**

a. KE is maximum when $x = 0$
b. TE is zero when $x = 0$
c. KE is maximum when x is maximum
d. PE is maximum when $x = 0$

33 A child swinging on a swing in sitting position, stands up, then the time period of the swing will **(AIEEE 2002)**

a. increase
b. decrease
c. remain same
d. increase, if the child is long and decrease, if the child is short

34 In a simple harmonic oscillator, at the mean position **(AIEEE 2002)**

a. kinetic energy is minimum, potential energy is maximum
b. both kinetic and potential energies are maximum
c. kinetic energy is maximum, potential energy is minimum
d. both kinetic and potential energies are minimum

35 If a spring has time period T and is cut into n equal parts, then the time period of each part will be **(AIEEE 2002)**

a. $T\sqrt{n}$
b. $\dfrac{T}{\sqrt{n}}$
c. nT
d. T

JEE Advanced & IIT JEE Archive

(Compilation of Last 10 Years' Questions)

1 Two independent harmonic oscillators of equal masses are oscillating about the origin with angular frequencies ω_1 and ω_2 and have total energies E_1 and E_2, respectively. The variations of their momenta P with position x are shown in the figures. If $\frac{a}{b} = n^2$ and $\frac{\alpha}{R} = n$, then the correct equation(s) is (are)

(2015 Adv. One or More than One Options Correct Type)

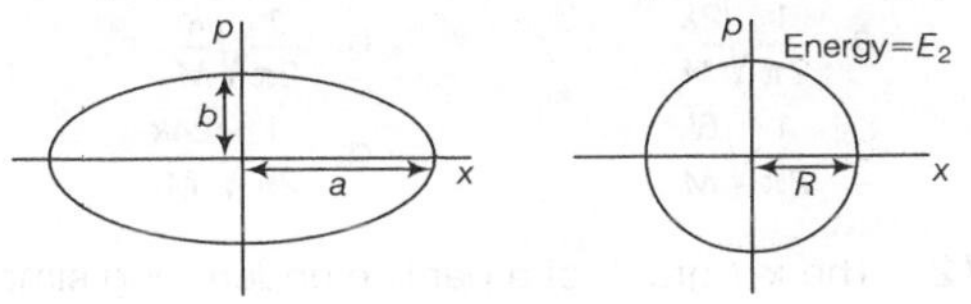

a. $E_1\omega_1 = E_2\omega_2$ b. $\frac{\omega_2}{\omega_1} = n^2$

c. $\omega_1\omega_2 = n^2$ d. $\frac{E_1}{\omega_1} = \frac{E_2}{\omega_2}$

2 Four harmonic waves of equal frequencies and equal intensities I_0 have phase angles 0, $\frac{\pi}{3}$, $\frac{2\pi}{3}$ and π. When they are superposed, the intensity of the resulting wave is nI_0. The value of n is

(2015 Adv. Integer Type)

3 A small block is connected to one end of a massless spring of unstretched length 4.9 m. The other end of the spring (see the figure) is fixed. The system lies on a horizontal frictionless surface. The block is stretched by 0.2 m and released from rest at $t = 0$. It then executes simple harmonic motion with angular frequency $\omega = \frac{\pi}{3}$ rad s^{-1}. Simultaneously at $t = 0$, a small pebble is projected with speed v from point P at an angle of 45° as shown in the figure. Point P is at a horizontal distance of 10 m from O. If the pebble hits the block at $t = 1$ s, the value of v is (Take, $g = 10$ ms^{2}) **(2012, Only One Option Correct Type)**

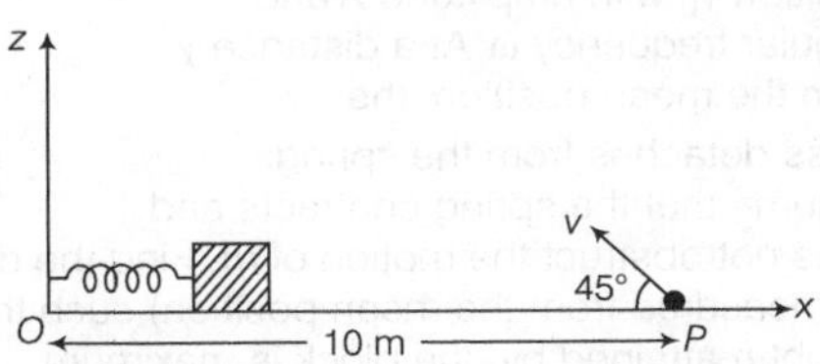

a. $\sqrt{50}$ ms^{-1}

b. $\sqrt{51}$ ms^{-1}

c. $\sqrt{52}$ ms^{-1}

d. $\sqrt{53}$ ms^{-1}

4 A wooden block performs SHM on a frictionless surface with frequency ν_0. The block carries a charge $+Q$ on its surface. If now a uniform electric field E is switched-on as shown, then the SHM of the block will be

(2011, Only One Option Correct Type)

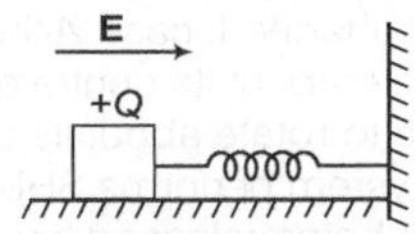

a. of the same frequency and with shifted mean position

b. of the same frequency and with the same mean position

c. of changed frequency and with shifted mean position

d. of changed frequency and with the same mean position

Passage for Q. Nos. (5-7)

When a particle of mass m moves on the x-axis in a potential of the form $V(x) = kx^2$, it performs simple harmonic motion. The corresponding time period is proportional to $\sqrt{\frac{m}{k}}$, as can be seen easily using dimensional analysis. However, the motion of a particle can be periodic even when its potential energy increases on both sides of $x = 0$ in a way different from kx^2 and its total energy is such that the particle does not escape to infinity. Consider a particle of mass m moving on the X-axis. Its potential energy is $V(x) = \alpha x^4 (\alpha > 0)$ for $|x|$ near the origin and becomes a constant equal to V_0 for $|x| \geq X_0$ (see figure).

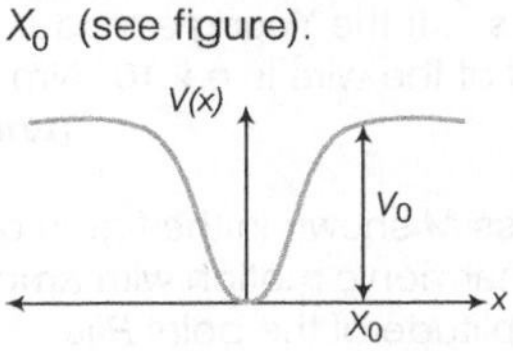

5 If the total energy of the particle is E, it will perform periodic motion only if

a. $E < 0$ b. $E > 0$

c. $V_0 > E > 0$ d. $E > V_0$

6 For periodic motion of small amplitude A, the time period T of this particle is proportional to

a. $A\sqrt{\frac{m}{\alpha}}$ b. $\frac{1}{A}\sqrt{\frac{m}{\alpha}}$

c. $A\sqrt{\frac{\alpha}{m}}$ d. $\frac{1}{A}\sqrt{\frac{\alpha}{m}}$

7 The acceleration of this particle for $|x| > X_0$ is

a. proportional to V_0

b. proportional to $\frac{V_0}{mX_0}$

c. proportional to $\sqrt{\frac{V_0}{mX_0}}$

d. zero

8 A metal rod of length L and mass m is pivoted at one end. A thin disk of mass M and radius $R (< L)$ is attached at its centre to the free end of the rod. Consider two ways the disc is attached. case A-the disc is not free to rotate about its centre and case B-the disc is free to rotate about its centre. The rod-disc system performs SHM in vertical plane after being released from the same displaced position. Which of the following statement(s) is/are true?

(2011, One or More than One Options Correct Type)

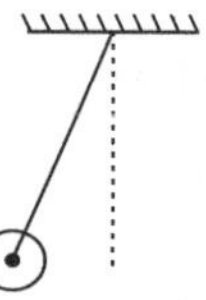

a. Restoring torque in case A = Restoring torque in case B

b. Restoring torque in case A < Restoring torque in case B

c. Angular frequency for case A > Angular frequency for case B

d. Angular frequency for case A < Angular frequency for case B

9 A 0.1 kg mass is suspended from a wire of negligible mass. The length of the wire is 1 m and its cross-sectional area is 4.9×10^{-7} m^2. If the mass is pulled a little in the vertically downward direction and released, it performs simple harmonic motion of angular frequency 140 rad s^{-1}. If the Young's modulus of the material of the wire is $n \times 10^9$ Nm^{-2}, the value of n is. **(2010, Integer Type)**

10 The mass M shown in the figure oscillates in simple harmonic motion with amplitude A. The amplitude of the point P is

(2009, Only One Option Correct Type)

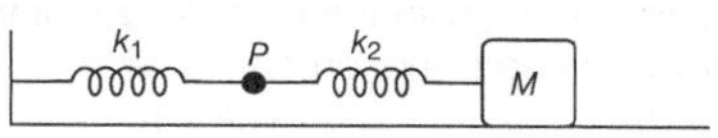

a. $\frac{k_1 A}{k_2}$

b. $\frac{k_2 A}{k_1}$

c. $\frac{k_1 A}{k_1 + k_2}$

d. $\frac{k_2 A}{k_1 + k_2}$

11 A uniform rod of length L and mass M is pivoted at the centre. Its two ends are attached to two springs of equal spring constants k. The springs are fixed to rigid supports as shown in the figure, and rod is free to oscillate in the horizontal plane. The rod is gently pushed through a small angle θ in one direction and released. The frequency of oscillation is

(2009, Only One Option Correct Type)

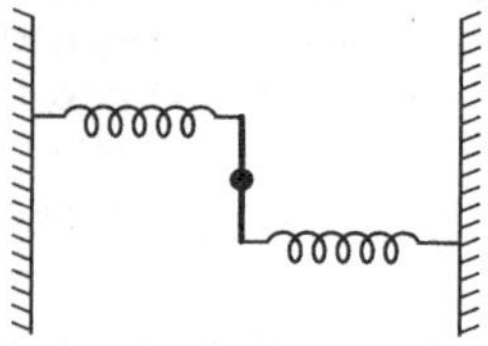

a. $\frac{1}{2\pi}\sqrt{\frac{2k}{M}}$

b. $\frac{1}{2\pi}\sqrt{\frac{k}{M}}$

c. $\frac{1}{2\pi}\sqrt{\frac{6k}{M}}$

d. $\frac{1}{2\pi}\sqrt{\frac{24k}{M}}$

12 The x-t graph of a particle undergoing simple harmonic motion is shown below. The acceleration of the particle at $t = 4/3$ s is

(2009, Only One Option Correct Type)

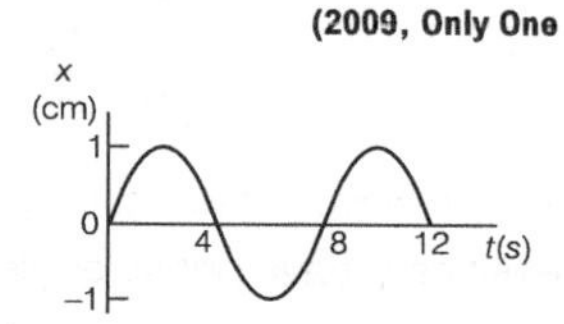

a. $\frac{\sqrt{3}}{32}\pi^2$ cms^{-2}

b. $\frac{-\pi^2}{32}$ cms^{-2}

c. $\frac{\pi^2}{32}$ cms^{-2}

d. $-\frac{\sqrt{3}}{32}\pi^2$ cms^{-2}

13 A simple pendulum has time period T_1. The point of suspension is now moved upward according to the relation $y = kt^2$, $(k = 1 \text{m/s}^2)$ where y is the vertical displacement. The time period now becomes T_2. The ratio of $\frac{T_1^2}{T_2^2}$ is

(Take, $g = 10$ ms^{-2}) **(2005, Only One Option Correct Type)**

a. 6/5

b. 5/6

c. 1

d. 4/5

14 A mass m is undergoing SHM in the vertical direction about the mean position y_0 with amplitude A and angular frequency ω. At a distance y from the mean position, the mass detaches from the spring. Assume that the spring contracts and does not obstruct the motion of m. Find the distance y. (measured from the mean position) such that the height h attained by the block is maximum. $(A\omega^2 > g)$

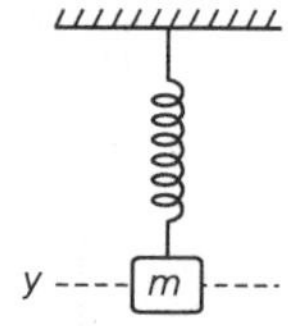

(2005, Subjective Type)

2. Progressive Waves

DPP

DAILY PRACTICE PROBLEMS

DPP-1 Basic Parameters and Wave Equation

Subjective Question

Working Space

Direction *(Q. No. 1) This question is subjective in nature, need to be solved completely on notebook.*

1 The displacement of a wave is represented by $y = 0.25 \times 10^{-3} \sin(500t - 0.25x)$, where y, t and x are in cm, second and metre respectively. Deduce

i. Amplitude
ii. Time period
iii. Angular frequency
iv. Wavelength

Also, deduce the amplitude of particle velocity and particle acceleration.

Only One Option Correct Type

Direction *(Q. Nos. 2-16) This section contains 15 multiple choice questions. Each question has four choices (a), (b), (c) and (d), out of which ONLY ONE is correct.*

2 A wave is expressed by the equation $y = 0.5\sin\pi(0.01x - 3t)$, where y and x are in metre and t in second. Its speed of propagation is

a. $200\,\text{ms}^{-1}$ b. $300\,\text{ms}^{-1}$
c. $400\,\text{ms}^{-1}$ d. $350\,\text{ms}^{-1}$

3 A wave propagates on a string in the positive X-direction with a velocity v. The shape of the string at $t = t_0$ is given by $g(x, t_0) = A\sin\left(\frac{x}{a}\right)$. The wave equation for a general time t is

a. $f(x, t) = A\sin\left(\frac{x - vt}{a}\right)$ b. $f(x, t) = A\sin\left(\frac{x - v(t - t_0)}{a}\right)$
c. $f(x, t) = A\sin\left(\frac{x + v(t - t_0)}{a}\right)$ d. $f(x, t) = A\sin\left(\frac{x + v(t + t_0)}{a}\right)$

4 A light pointer fixed to one prong of a tuning fork touches a vertical plate. The fork is set vibrating and the plate is allowed to fall freely. 8 complete oscillations are counted, when the plate falls through 10 cm. The frequency of the tuning fork is

a. 82 Hz b. 72 Hz
c. 66 Hz d. 56 Hz

5 The wave equation of a pulse is given by $y = \frac{3}{2+(x-4t)^2}$, where y is in metre and t in second. The wave velocity of the pulse is

a. $+4\,\text{ms}^{-1}$ b. $-4\,\text{ms}^{-1}$
c. $+2\,\text{ms}^{-1}$ d. $-2\,\text{ms}^{-1}$

Working Space

6 The wave has the shape shown in the figure, at time $t = 0$ s. Its speed is $v = 300\text{ms}^{-1}$.

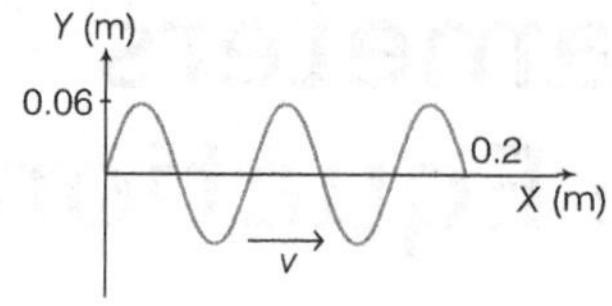

Then, choose the correct option(s).

a. $(0.06\text{m})\sin(7500\pi t - 25\pi x)$
b. $(0.06\text{m})\sin(25\pi x + 7500\pi t)$
c. $(0.06\text{m})\sin(25\pi x - 7500\pi t)$
d. $(0.06\text{m})\cos(25\pi x - 7500\pi t)$

7 A progressive simple harmonic wave is moving in air along the X-axis. The part of this wave at a given point $x = x_0$ from the source and at a certain instant $t = t_0$ has the waveform shown below in the displacement-time (y-t) graph and velocity-time (v-t) graph respectively. Velocity of the wave has value v_0 and its angular frequency is ω. Which of the following equations will correctly represent the complete wave at $x = x_0$, agreeing with above waveforms ?

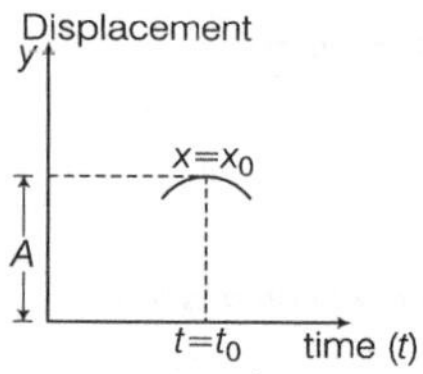

y-t graph

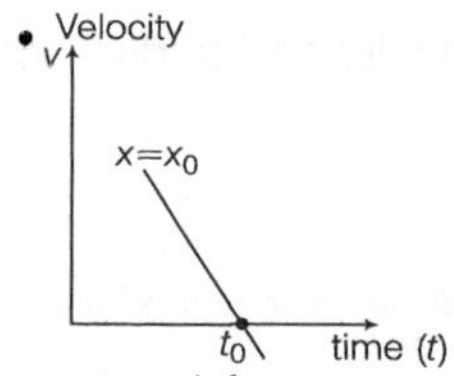

v-t graph for $x = x_0$

a. $y = -A\left[\cos\left\{\frac{2\pi}{T}(t - t_0) - \frac{\pi}{2}\right\}\right]$

b. $y = -A\left[\sin\left\{\frac{2\pi}{T}(t - t_0) + \frac{\pi}{2}\right\}\right]$

c. $y = A\left[\sin\left\{\frac{2\pi}{T}(t - t_0) + \frac{\pi}{2}\right\}\right]$

d. $y = A\left[\cos\left\{\frac{2\pi}{T}(t - t_0) - \frac{\pi}{2}\right\}\right]$

8 A sinusoidal wave travelling in the positive direction has an amplitude of 15 cm, wavelength 40 cm and frequency 8 Hz. The vertical displacement of the medium at $t = 0$ s and $x = 0$ is also 15 cm (see figure). What are the phase constant δ and general expression for wave function ?

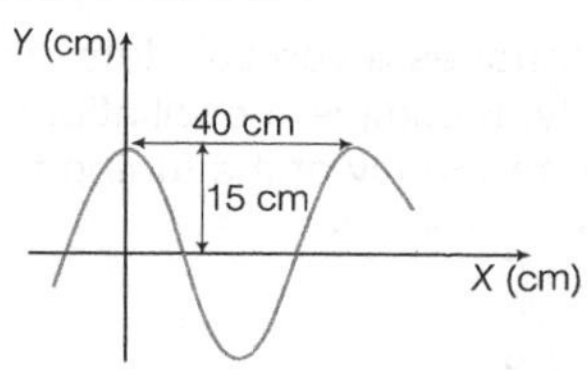

a. $\frac{\pi}{2}, \cos\left(16\pi t - \frac{\pi}{20}x\right)$

b. $0, \sin\left(16\pi t - \frac{\pi}{20}x\right)$

c. $\frac{\pi}{2}, \sin\left(16\pi t - \frac{\pi}{20}x\right)$

d. $\frac{\pi}{4}, \sin\left(16\pi t - \frac{\pi}{20}x + \frac{\pi}{4}\right)$

Working Space

9 One end of each of two identical springs, each of force constant 0.5 Nm^{-1} are attached on the opposite sides of a wooden block of mass 0.01 kg. The other ends of the springs are connected to separate rigid supports such that, the springs are unstretched and collinear in horizontal plane. To the wooden piece is fixed a pointer which touches a vertically moving plane of paper. The wooden piece kept on a smooth horizontal table is now displaced by 0.02 m along the line of spring and released. If the speed of the paper, perpendicular to the length of the springs, is 0.1 ms^{-1}, find the the distance between two consecutive maxima on the path.

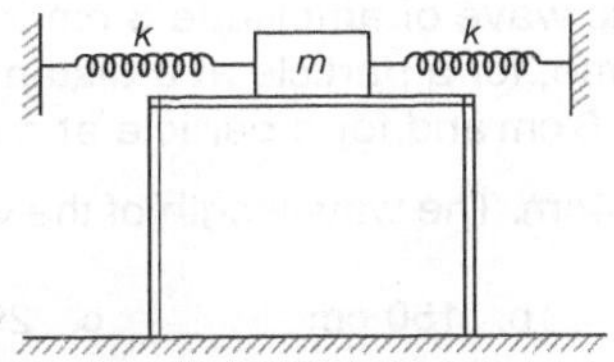

a. $\frac{\pi}{100}$ m b. $\frac{\pi}{50}$ m c. $\frac{\pi}{25}$ m d. $\frac{\pi}{10}$ m

10 In the above problem, find the equation of the path traced by the pointer on the paper.

a. $y = 0.02\cos(50x)$ b. $y = 0.02\cos(75x)$
c. $y = 0.02\cos(100x)$ d. $y = 0.02\sin(25x)$

11 Figure shows a snapshot of a sinusoidal travelling wave taken at $t = 0.3$ s. The wavelength is 7.5 cm and amplitude is 2 cm.

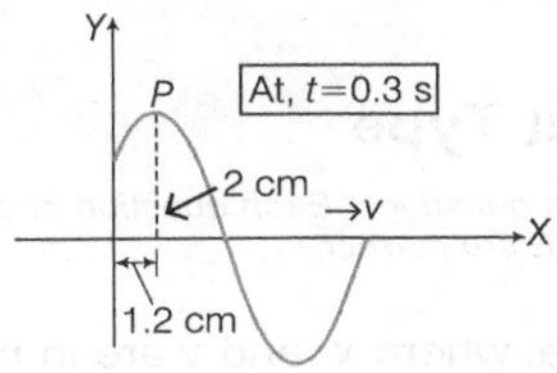

If the speed of wave is 4 cm s^{-1}, write the equation of travelling wave.

a. $y = 2\sin\left(\frac{4\pi}{15}x - \frac{16\pi}{15}t\right)$ b. $y = -2\cos\left(\frac{16\pi}{15}t - \frac{4\pi}{15}x\right)$
c. $y = 2\sin\left(\frac{16\pi}{15}t - \frac{4\pi}{15}x\right)$ d. $y = 2\cos\left(\frac{16\pi}{15}t - \frac{4\pi}{15}x\right)$

12 Figure shows the position of particles of medium at $t = 0$ s. Supposing a simple harmonic wave travelling either along or opposite to the positive X-axis, find the angle made by the tangent at point P with the axis.

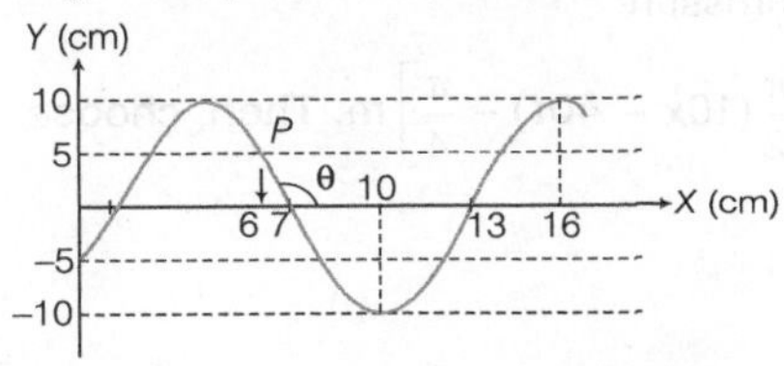

a. $\tan^{-1}\left(-\frac{7\pi}{\sqrt{3}}\right)$ b. $\tan^{-1}\left(\frac{-7\pi}{2\sqrt{3}}\right)$
c. $\tan^{-1}\left(\frac{-5\pi}{3\sqrt{3}}\right)$ d. $\tan^{-1}\left(-\frac{5\pi}{2\sqrt{3}}\right)$

13 Which of the following functions represents a travelling wave ? Here, a, b, c are constants.

I. $y = a\cos bx \sin ct$ II. $y = a\sin(bx + ct)$
III. $y = a\sin(bx + ct) + a\sin(bx - ct)$ IV. $y = a\sin(bx - ct)$

a. Both I and IV b. Both I and III
c. Both II and IV d. Both II and III

Working Space

14 A simple harmonic wave of amplitude 8 cm travels along positive X-axis. At a given instant of time, for a particle at a distance of 10 cm from the origin, the displacement is $+6$ cm and for a particle at a distance of 25 cm from the origin, the displacement is $+4$cm. The wavelength of the wave is (Given, $\sin^{-1}\left(\frac{3}{4}\right) = 0.8481\,\text{rad}$)

a. 100 cm
b. 150 cm
c. 290 cm
d. 200 cm

15 A wave pulse is described by $y(x, t) = ae^{-(bx-ct)^2}$, where a, b and c are positive constants. What is the speed of the wave ?

a. $\frac{c}{b}$
b. $\frac{b}{c}$
c. $\frac{a}{c}$
d. $\frac{c}{a}$

16 The displacement of a wave disturbance propagating in the positive x-direction is given by

$y = \frac{1}{1+x^2}$ at $t = 0$ and $y = \frac{1}{1+(x-1)^2}$ at $t = 2$ s, where x and y are in metre. The shape of the wave disturbance does not change during the propagation. What is the velocity of the wave ?

a. $0.75\,\text{ms}^{-1}$
b. $1\,\text{ms}^{-1}$
c. $0.5\,\text{ms}^{-1}$
d. $0.25\,\text{ms}^{-1}$

One or More than One Options Correct Type

Direction (Q. Nos. 17-20) This section contains 4 multiple choice questions. Each question has four choices (a), (b), (c) and (d), out of which ONE or MORE THAN ONE are correct.

17 $y(x, t) = \frac{0.8}{(4x+5t)^2+5}$ represents a moving pulse, where x and y are in metre and t in second. Then, choose the correct options.

a. Pulse is moving in positive x-direction
b. In 2 s, it will travel a distance of 2.5 m
c. Its maximum displacement is 0.16 m
d. It is a symmetric pulse

18 In a wave motion $y = A\sin(kx - \omega t)$, y can represent the

a. electric field
b. magnetic field
c. displacement
d. pressure

19 The equation of a wave is $y(x,t) = 0.05\sin\left[\frac{\pi}{2}(10x - 40t) - \frac{\pi}{4}\right]$ m. Then, choose the correct options.

a. The wavelength of the wave is 0.4 m
b. The wave velocity is $+4\,\text{ms}^{-1}$
c. The particle velocity at $x = 0.5$ m and $t = 0.05$ s is 2.22 ms^{-1}
d. The particle acceleration at $x = 0.5$m and $t = 0.05$ s is $140\,\text{ms}^{-2}$

20 The equation of a travelling wave is $y(x, t) = 0.02\sin\left[\frac{x}{0.05} + \frac{t}{0.01}\right]$m. Then, choose the correct options. (Given, $\cos\theta = -0.85$ when, $\theta = 34\,\text{rad}$)

a. Wave velocity is $-5\,\text{ms}^{-1}$
b. Wave velocity is $+5\,\text{ms}^{-1}$
c. The particle velocity at $x = 0.2$m and $t = 0.3$s is $-1.7\,\text{ms}^{-1}$
d. The particle velocity at $x = 0.2$m and $t = 0.3$s is $+1.7\,\text{ms}^{-1}$

Working Space

Matching List Type

Direction (Q. Nos. 21 and 22) *Choices for the correct combination of elements from Column I and Column II are given as options (a), (b), (c) and (d), out of which one is correct.*

21 Consider a sinusoidal wave shown in the figure. The wave velocity is $+40\text{ cms}^{-1}$. Its frequency is f. The phase difference between points 2.5 cm apart is $\Delta\phi$. It takes a time Δt for the phase, at a given position, to change by 60°. The velocity of the particle P at the instant shown is v_P. Match the Column I with Column II and mark the correct option from the codes given below.

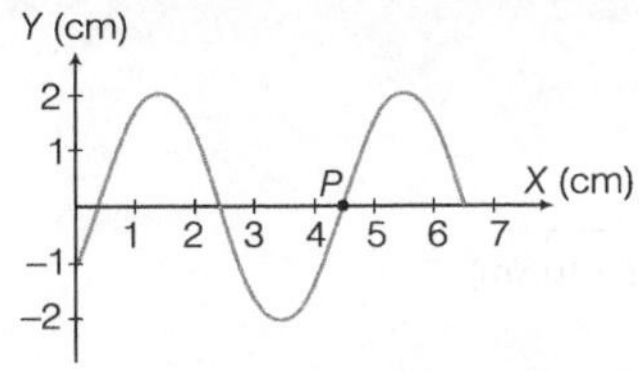

	Column I		Column II
i.	f (in Hz)	p.	3.925
ii.	$\Delta\phi$ (in rad)	q.	0.0167
iii.	Δt (in s)	r.	10
iv.	$\lvert v_P \rvert$ (in ms^{-1})	s.	1.214
		t.	2.12

Codes

	i	ii	iii	iv
a.	r	p	q	s
b.	p	q	r	s
c.	s	q	p	r
d.	q	p	s	r

22 Transverse waves on a string have wave speed 12 ms^{-1}, amplitude 0.05 m and wavelength 0.4 m. The waves travel in the $+x$-direction and at $t = 0$, the $x = 0$ end of the string has zero displacement and is moving upwards. A particle P is at a distance $x = 0.25$m from the left end i.e. $x = 0$. Match the Column I with Column II and mark the correct option from the codes given below.

	Column I		Column II
i.	Wave function of the motion is	p.	+3.54
ii.	The transverse displacement of P at $t = 0.15$s is (in cm)	q.	2.723
iii.	The time that must elapse from the instant in part (ii) until the particle P has zero displacement is (in ms)	r.	$y = 0.05 \sin(60\pi t - 5\pi x)$
		s.	$y = 0.05 \cos(60\pi t - 5\pi x)$
		t.	−3.54
		u.	4.167

Codes

	i	ii	iii
a.	u	t	p
b.	p	q	r
c.	r	t	u
d.	s	t	u

Comprehension Type

***Direction** (Q. Nos. 23-28) This section contains 2 paragraphs, each describing theory, experiments, data, etc. Six questions related to the paragraphs have been given. Each question has only one correct answer among the four given options (a), (b), (c) and (d).*

Passage I

Figure shows a snapshot of a vibrating string at $t = 0$ s. At this instant, the particle P is observed to be moving up with velocity $20\sqrt{3}$ cms^{-1}. The tangent at P makes an angle of 60° with the X-axis.

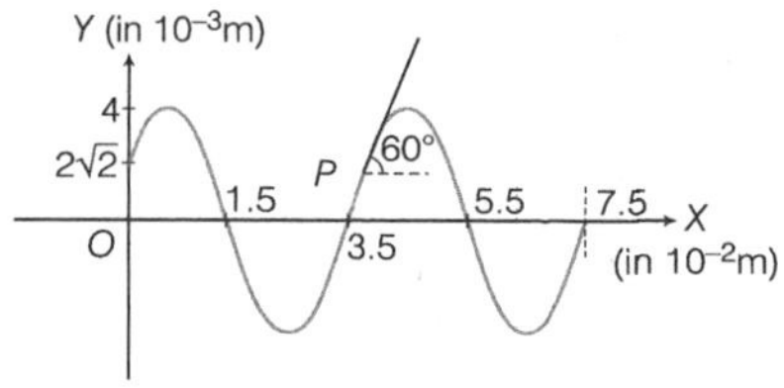

23 The velocity of the wave is

a. -20 cms^{-1} b. $+20$ cm s^{-1} c. -15 cms^{-1} d. $+15$ cms^{-1}

24 The equation of the wave is

a. $y = 0.4\sin\left[10\pi t + \frac{\pi}{2}x + \frac{\pi}{4}\right]$

b. $y = 0.4\sin\left[10\pi t - \frac{\pi}{2}x + \frac{\pi}{4}\right]$

c. $y = 0.4\cos\left[10\pi t + \frac{\pi}{2}x + \frac{3\pi}{4}\right]$

d. $y = 0.4\cos\left[10\pi t - \frac{\pi}{2}x + \frac{3\pi}{4}\right]$

Passage II

Figure shows two snapshots of a travelling wave on a string of mass per unit length $\mu = 0.25$ kgm^{-1}. The first snap is taken at $t = 0$ s and the second is taken at $t = 0.05$ s.

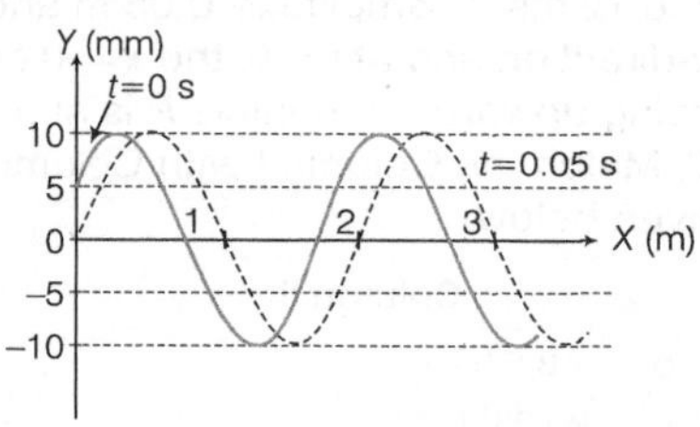

25 The speed of the wave is

a. 5ms^{-1} b. $\frac{20}{3}$ms^{-1} c. $\frac{5}{3}$ms^{-1} d. $\frac{10}{3}$ms^{-1}

26 The wavelength and frequency are respectively

a. 1m, $\frac{10}{3}$Hz b. 2 m, $\frac{10}{3}$Hz c. 2 m, $\frac{5}{3}$Hz d. 1 m, $\frac{5}{3}$Hz

27 The maximum speed of the particle is

a. $\frac{\pi}{20}$ms^{-1} b. $\frac{\pi}{30}$ms^{-1} c. $\frac{\pi}{10}$ms^{-1} d. $\frac{\pi}{5}$ms^{-1}

28 The equation of the wave is

a. $y = 10\times10^{-3}\sin\left(\frac{10\pi}{3}t - \pi x + \frac{5\pi}{6}\right)$

b. $y = 10\times10^{-3}\sin\left(\frac{10\pi}{3}t + \pi x - \frac{5\pi}{6}\right)$

c. $y = 10\times10^{-3}\sin\left(\frac{10\pi}{3}t - \pi x\right)$

d. $y = 10\times10^{-3}\cos\left(\frac{10\pi}{3}t - \pi x\right)$

DPP-2 Speed of Waves

Only One Option Correct Type

Working Space

Direction *(Q. Nos. 1-11) This section contains 11 multiple choice questions. Each question has four choices (a), (b), (c) and (d), out of which ONLY ONE is correct.*

1 A string has a mass 10 g and a length $L = 3$ m. Its two ends are tied to two walls a distance $d = 2$ m apart. Two blocks, each of mass $m = 2$kg, are suspended from the string as shown in the figure. In what time will a wave pulse sent from a point P_1 reach point P_2?

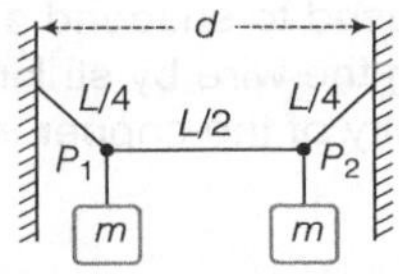

a. 52.9 ms
b. 42.9 ms
c. 32.9 ms
d. 22.9 ms

2 A uniform rope of length 12 m and mass 6 kg hangs vertically from a rigid support. A block of mass 2 kg is attached to the free end of the rope. A transverse pulse of wavelength 0.06 m is produced at the lower end of the rope. What is the wavelength of the pulse, when it reaches the top of the rope ?

a. 0.18 m
b. 0.03 m
c. 0.09 m
d. 0.12 m

3 A rope of mass m_0 and length L is suspended vertically. If a mass m is suspended from the bottom of the rope, find the time for a transverse wave to travel the length of the rope.

a. $2\sqrt{\frac{L}{m_0 g}}(\sqrt{m_0 + m} - m)$
b. $\sqrt{\frac{L}{m_0 g}}(\sqrt{m_0 + m} - \sqrt{m})$
c. $\sqrt{\frac{L}{2m_0 g}}(\sqrt{m_0 + m} - \sqrt{m})$
d. None of these

4 One end of 12 m long rubber tube with a total mass of 900 g is fastened to a fixed support. A cord attached to the other end passes over a pulley and supports an object of mass 5 kg. The tube is struck a transverse blow at one end. Find the time required for the pulse to reach the other end. (Take, $g = 9.8\text{ms}^{-2}$)

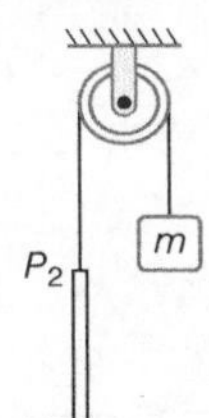

a. 0.47 s
b. 0.57 s
c. 0.67 s
d. 0.77 s

Working Space

5 Two wires of different densities but same area of cross-section are soldered together at one end and are stretched to a tension T. The velocity of a transverse wave in the first wire is double of that in the second wire. The ratio of density of the first wire to that of the second wire is

a. $\frac{1}{2}$ b. 2

c. $\frac{1}{4}$ d. 4

6 Two long strings P and Q, each having linear mass density $1.2 \times 10^{-2}\,\text{kgm}^{-1}$ are stretched by different tensions 4.8 N and 7.5 N respectively and are kept parallel to each other with their left ends at $x = 0$. Wave pulses are produced on the strings at the left ends at $t = 0$ s on string P and at $t = 20$ ms on string Q. When and where will the pulse on Q overtake that on P?

a. $t = 0.1\text{s}, x = 2\text{m}$ b. $t = 0.15\text{s}, x = 2.5\text{m}$

c. $t = 0.05\text{s}, x = 1.5\text{m}$ d. $t = 0.2\text{s}, x = 3\text{m}$

7 A copper wire 2.4 mm in diameter is 3 m long and is used to suspend a 2 kg mass from a beam. If a transverse disturbance is sent along the wire by striking with a pencil, how fast will the disturbance travel ? The density of the copper is $8920\,\text{kgm}^{-3}$.

a. 11ms^{-1} b. $22\,\text{ms}^{-1}$

c. $33\,\text{ms}^{-1}$ d. $30\,\text{ms}^{-1}$

8 A wire of variable mass per unit length $\mu = \mu_0 y$, is hanging from the ceiling as shown in the figure. The length of the wire is L_0. A small transverse disturbance is produced at its lower end. The time after which the disturbance will reach the other end is

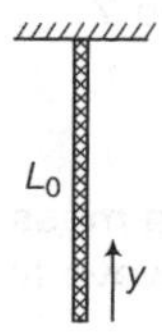

a. $\frac{2\sqrt{L_0}}{g}$ b. $\sqrt{\frac{8L_0}{g}}$

c. $\sqrt{\frac{L_0}{2g}}$ d. $\sqrt{\frac{12L_0}{g}}$

9 A block of mass m is attached to a vertical string of mass m_0 and length L_0 as shown in the figure. The whole system is placed on a planet whose mass is three times that of the earth and whose radius is 3 times that of the earth. Acceleration due to gravity on the earth is g. The time taken by a wave pulse to travel from one end A to other end B of the string is

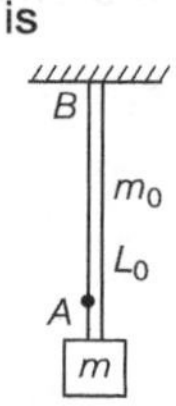

a. $\left[\frac{\sqrt{m+m_0}-\sqrt{m}}{\sqrt{m_0}}\right]\sqrt{\frac{2L_0}{g}}$ b. $\left[\frac{\sqrt{m+m_0}-\sqrt{m}}{\sqrt{m_0}}\right]\sqrt{\frac{3L_0}{g}}$

c. $\left[\frac{\sqrt{m+m_0}-\sqrt{m}}{\sqrt{m_0}}\right]\sqrt{\frac{L_0}{2g}}$ d. $2\left[\frac{\sqrt{m+m_0}-\sqrt{m}}{\sqrt{m_0}}\right]\sqrt{\frac{3L_0}{g}}$

Working Space

10 In the above problem, the ratio of maximum velocity to minimum velocity of the wave pulse is

a. $\sqrt{1+\frac{m_0}{m}}$ b. $\sqrt{\frac{m_0}{m}}$

c. $\sqrt{1-\frac{m_0}{m}}$ d. $\sqrt{\frac{m}{m_0}}$

11 A wave moves with speed 300 ms^{-1} on a wire, which is under a tension of 500 N. Find how much the tension must be changed, to increase the speed to 302 ms^{-1} ?

a. Tension should be increased by 16.67 N

b. Tension should be increased by 6.67 N

c. Tension should be increased by 10.67 N

d. Tension should be increased by 7.67 N

One or More than One Options Correct Type

Direction *(Q. Nos. 12 and 13) This section contains 2 multiple choice questions. Each question has four choices (a), (b), (c) and (d), out of which ONE or MORE THAN ONE are correct.*

12 Two blocks each having a mass of 3.2 kg are connected by a wire *CD* and the system is suspended from the ceiling by another wire *AB*. The linear mass density of the wire *AB* is 10×10^{-3} kgm^{-1} and that of *CD* is 8×10^{-3} kg m^{-1}. The speeds of transverse wave pulse produced in *AB* and *CD* respectively are v_1 and v_2. Then, (Take, $g = 10$ ms^{-2})

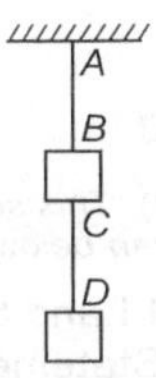

a. $v_1 = 80 ms^{-1}$ b. $v_2 = 63 ms^{-1}$

c. $v_1 = 70 ms^{-1}$ d. $v_2 = 50 ms^{-1}$

13 A particle on a stretched string supporting a travelling wave, takes 5 ms to move from its mean position to the extreme position. The distance between two consecutive particles, which are at their mean positions, is 2 cm. The frequency, the wavelength and wave speed respectively are f, λ and v. Then,

a. $f = 50$ Hz b. $\lambda = 4$ cm

c. $v = 2 ms^{-1}$ d. $v = 1.5 ms^{-1}$

Comprehension Type

Direction *(Q. Nos. 14-17) This section contains a paragraph, describing theory, experiments, data, etc. Four questions related to the paragraph have been given. Each question has only one correct answer among the four given options (a), (b), (c) and (d).*

Passage

The wave speed (v) in a uniform freely hanging rope varies with the distance y from the top. The tension T varies while the mass per unit length μ remains same. We want to design a rope in which v is independent of y. For such a rope μ cannot be constant.

14 Let μ_0 is the mass per unit length at the top of the rope ($y = 0$). If the rope is infinitely long and the velocity of wave is constant throughout it, then

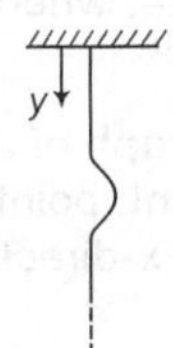

a. $\mu(y) = \mu_0 e^{\left(-\frac{g}{v^2} y\right)}$ b. $\mu(y) = \frac{\mu_0}{2} e^{\left(-\frac{g}{v^2} y\right)}$

c. $\mu(y) = \frac{\mu_0}{2} e^{\left(-\frac{2g}{v^2} y\right)}$ d. $\mu(y) = \mu_0 e^{\left(-\frac{g}{2v^2} y\right)}$

Working Space

15 If the total mass of the rope is m, then the wave speed is

a. $\sqrt{\frac{mg}{2\mu_0}}$ b. $\sqrt{\frac{mg}{\mu_0}}$

c. $\sqrt{\frac{2mg}{\mu_0}}$ d. $\frac{1}{2}\sqrt{\frac{mg}{\mu_0}}$

16 μ, in terms of μ_0, m and y only, is

a. $\mu_0 e^{-\mu_0 y/2m}$ b. $\frac{\mu_0}{2} e^{-\mu_0 y/m}$

c. $\mu_0 e^{-\mu_0 y/m}$ d. None of these

17 Suppose we cut the rope off at length L (from the top). We can still have constant wave speed on the finite rope, if we hang a mass m_0 on the end provided

a. $m_0 = \frac{m}{2} e^{-\mu_0 L/m}$ b. $m_0 = m e^{-\mu_0 L/2m}$

c. $m_0 = \frac{m}{2} e^{-\mu_0 L/2m}$ d. $m_0 = m e^{-\mu_0 L/m}$

Statement Type

Direction (Q. Nos. 18-22) *This section is based on Statement I and Statement II. Select the correct answer from the codes given below.*

a. Both Statement I and Statement II are correct and Statement II is the correct explanation of Statement I
b. Both Statement I and Statement II are correct but Statement II is not the correct explanation of Statement I
c. Statement I is correct but Statement II is incorrect
d. Statement II is correct but Statement I is incorrect

18 **Statement I** Transverse waves cannot travel in gaseous medium.

Statement II Gases do not possess modulus of rigidity.

19 **Statement I** Transverse waves can travel on the surface of water.

Statement II Surface tension property of water plays the role of modulus of rigidity.

20 **Statement I** Wave speed is given by $v = f\lambda$. If frequency f is doubled, v will become two times.

Statement II For given conditions of medium, wave speed remains constant.

21 **Statement I** Two strings shown in the figure, have same tension.

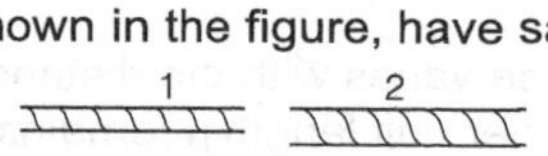

Speed of transverse wave in string 1 will be more.

Statement II $v \propto \frac{1}{\sqrt{\mu}}$, where μ is mass per unit length of the wire.

22 **Statement I** Y-X graph of a transverse wave on a string is as shown in the figure. At the given instant, point P is moving downwards. Hence, we can say that wave is moving along $+x$-direction.

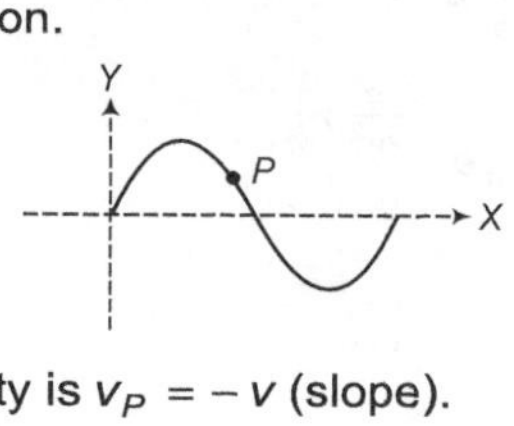

Statement II Particle velocity is $v_P = -v$ (slope).

Working Space

One Integer Value Correct Type

Direction (Q. No. 23 and 24) This section contains 2 questions. When worked out will result in an integer from 0 to 9 (both inclusive).

23 A copper wire is held at the two ends by rigid supports. At 30°, the wire is just taut with negligible tension. The speed of transverse waves in this wire at 10°C is $10\,D\,\text{ms}^{-1}$. Find the value of D. (Take, $Y_{cu} = 1.3 \times 10^{11}$ Pa, $\alpha_{cu} = 1.7 \times 10^{-5}\,°\text{C}^{-1}$, $\rho_{cu} = 9000\ \text{kg m}^{-3}$)

24 A string of length 0.3 m and mass 10^{-2} kg is clamped at both its ends. The tension in the string is 1.2 N. When a pulse travels along the string, the shape the string is found to be the same at times t and $t + \Delta t$. The minimum value of Δt is $\frac{1}{2k}$ s. Find the value of k.

Subjective Questions

Direction (Q. Nos. 25-27) These questions are subjective in nature, need to be solved completely on notebook.

25 A loop of rope is whirled at a high angular velocity ω, so that it becomes taut circle of radius R. A kink develops in the whirling rope.

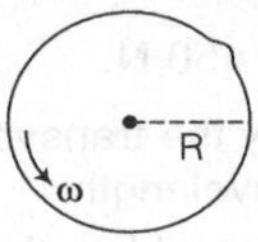

i. Show that the speed of the kink in the rope is $v = \omega R$.
ii. Under what conditions the kink remains stationary relative to an observer on the ground ?

26 A long string having a cross-sectional area $0.8\ \text{mm}^2$ and density $12.5\ \text{g cm}^{-3}$, is subjected to a tension of 64 N along the positive X-axis. One end of this string is attached to a vibrator at $x = 0$ moving in transverse direction at a frequency of 20 Hz. At $t = 0$, the source is at a maximum displacement $y = 1$ cm.

i. Find the speed of the wave travelling on the string.
ii. Write the equation for the wave.
iii. What is the displacement of the particle of the string at $x = 50$ cm at time $t = 0.05$ s?
iv. What is the velocity of this particle at this instant?

27 For the speed of a transverse wave on a string, we write the formula $v = \sqrt{\frac{F}{\mu}}$ and sometimes we also write $v = f\lambda$. Are these two same or different ? Explain.

DPP-3 Energy, Intensity and Power of Waves

Subjective Questions

Working Space

Direction (Q. Nos. 1-3) *These questions are subjective in nature, need to be solved completely on notebook.*

1 For a transverse wave travelling on a string, $y = A\sin(kx - \omega t)$, prove that the potential energy per unit volume is $\frac{1}{2}\rho A^2\omega^2 \cos^2(kx - \omega t)$.

2 A wave travels out in all directions from a point source. Justify the expression $y = \frac{a_0}{r}\sin k(r - vt)$, at a distance r from the source. Find the speed, periodicity and intensity of the wave. What are the dimensions of a_0?

3 A steel wire has a mass of 50 gm^{-1} and is under tension 450 N.

i. Find the maximum average power that can be carried by the transverse wave in the wire, if the amplitude is not to exceed 20% of the wavelength.

ii. The change in maximum average power, if the mass per unit length of the wire is doubled. (Take, $\pi^2 = 10$)

Only One Option Correct Type

Direction (Q. Nos. 4-13) *This section contains 10 multiple choice questions. Each question has four choices (a), (b), (c) and (d), out of which ONLY ONE is correct.*

4 An electric siren of 100 W sends out sound waves in air at a frequency of 1 kHz. Assuming that the sound energy travels equally in all directions in space, calculate the intensity of sound and amplitude of air particles, at a point 100 m away from the source. The velocity of sound is 350 ms^{-1} and density of air = 1.29 kg m^{-3}.

a. 9.96×10^{-4} W m^{-2}, 4.49×10^{-7} m

b. 8.96×10^{-4} W m^{-2}, 3.99×10^{-7} m

c. 7.96×10^{-4} Wm^{-2}, 2.99×10^{-7} m

d. 6.96×10^{-4} Wm^{-2}, 1.99×10^{-7} m

5 A thin string is held at one end and oscillates vertically so that, $y(x = 0, t) = 8\sin 4t$ (cm). Neglect the gravitational force. The string's linear mass density is 0.2 kg m^{-1} and its tension is 1 N. The string passes through both filled with 1 kg water. Due to friction, heat is transferred to the both. The heat transfer efficiency is 50%. Calculate how much time passes before the temperature of the bath rises by one degree kelvin.

a. 3.1 day
b. 4.2 day
c. 2.5 day
d. 1.1 day

Working Space

6 The figure shows a snapshot of a vibrating string at $t = 0$. The particle P is observed to be moving up with velocity $20\sqrt{3}$ cms^{-1}. The tangent at P makes an angle 60° with X-axis. Calculate the total energy carried by the wave per cycle of the string, assuming that mass per unit length of the string is 50 g m^{-1}.

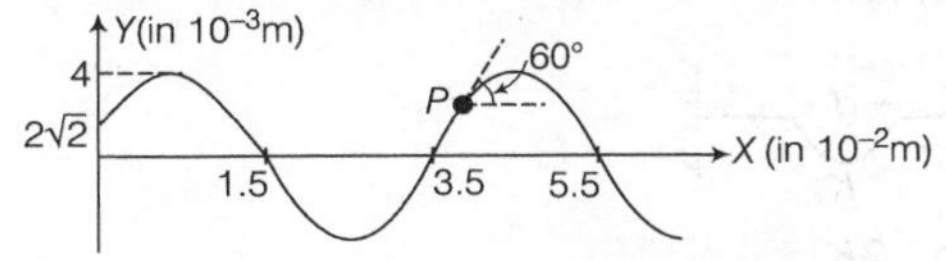

a. 3.2×10^{-4} J
b. 1.6×10^{-4} J
c. 3.2×10^{-3} J
d. 1.6×10^{-3} J

7 Two harmonic waves travelling in the same medium have frequency ratio 1 : 2 and intensity ratio 1 : 36. Their amplitude ratio is

a. 1 : 3
b. 1 : 6
c. 1 : 8
d. 1 : 72

8 A stretched rope having linear mass density 5×10^{-2} kgm^{-1} is under a tension of 80 N. The power that has to be supplied to the rope to generate harmonic waves at a frequency of 60 Hz and an amplitude of 6 cm is

a. 215 W
b. 251 W
c. 518 W
d. 538 W

9 Two pulses (of same dimensions) in a stretched string, whose centres are initially 8 cm apart, are moving towards each other as shown in the figure. The speed of each pulse is 2 cms^{-1}. After 2 second, the total energy of the pulse will be

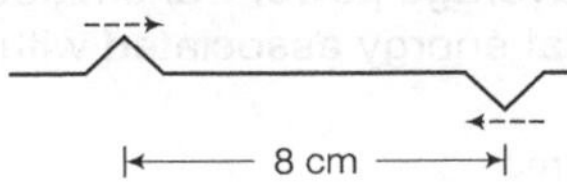

a. zero
b. purely kinetic
c. purely potential
d. partly kinetic and partly potential

10 A 10 W source of sound of frequency 1000 Hz sends out waves in air. The displacement-amplitude at a distance of 10 m from the source is (Given, speed of sound in air = 340 ms^{-1} and density of air = 1.29 kgm^{3}).

a. 0.62 μm
b. 1.91μm
c. 1.6μm
d. 0.96μm

11 A 2 m long rope has mass of 0.1 kg and tension of 60 N. A power source at one end sends a harmonic wave with an amplitude of 1 cm down the rope. The wave is extracted at the other end without any reflection. What is the frequency of the power source, if the power transmitted is 100 W ?

a. 180 Hz
b. 171 Hz
c. 160 Hz
d. 150 Hz

12 Spherical waves are emitted from a 1W source in an isotropic non-absorbing medium. The wave intensity 1 m from the source is

a. 8×10^{-2} Wm^{-2}
b. 16×10^{-2} Wm^{-2}
c. 12×10^{-2} Wm^{-2}
d. 4×10^{-2} Wm^{-2}

Working Space

13 Two identical pulses move in opposite directions, with same uniform speeds on a stretched string. The width and kinetic energy of each pulse are d and K respectively. At the instant, they completely overlap, the kinetic energy of the width d of the string (in the overlap region) is

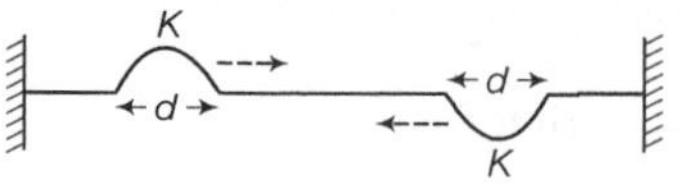

a. $8K$
b. $2K$
c. $4K$
d. zero

One or More than One Options Correct Type

Direction (Q. Nos. 14-19) *This section contains 6 multiple choice questions. Each question has four choices (a), (b), (c) and (d), out of which ONE or MORE THAN ONE are correct.*

14 As, a wave propagates

a. the wave intensity remains constant for a plane wave
b. the wave intensity decreases as the inverse of the distance from the source for a spherical wave
c. the wave intensity decreases as the inverse square of the distance from the source for a spherical wave
d. the wave intensity decreases as the inverse of the distance from a line source

15 A 200 Hz wave with an amplitude 1 mm travels on a long string of linear mass density 6 g m^{-1} kept under a tension of 60 N. The average power transmitted across a given point on the string is P_{av} and the total energy associated with the wave in a 2 m long portion of the string is E. Then,

a. $P_{av} = 0.47$ W
b. $E = 9.4$ mJ
c. $P_{av} = 0.94$ W
d. $E = 18.8$ mJ

16 A sinusoidal wave travels on a string. The string has a length 8 m and mass 6 g. The wave speed is 30 ms^{-1} and the wavelength is 0.2 m. If the wave is to have an average power of 50 W, the amplitude of the wave is A. For this same string, if the amplitude is A, wavelength is 0.2 m (i.e. same as earlier) and the tension is increased such that, wave speed is doubled then the power increases to x times initial power. Then,

a. $A = 9.08$ cm
b. $A = 7.08$ cm
c. $x = 8$
d. $x = 2$

17 For plane waves in the air of frequency 1000 Hz and the displacement amplitude 2×10^{-8} m, the velocity amplitude is v_{max} and the intensity is I. (Take, $\rho = 1.3$ kgm^{-3} and speed of wave = 340 ms^{-1}). Then,

a. $v_{max} = 1.3 \times 10^{-4}$ ms^{-1}
b. $I = 7.4 \times 10^{-6}$ Wm^{-2}
c. $v_{max} = 2.6 \times 10^{-4}$ ms^{-1}
d. $I = 3.7 \times 10^{-6}$ Wm^{-2}

18 A piano wire with mass 3 g and length 80 cm is stretched with a tension of 25 N. A wave with frequency 120 Hz and amplitude 1.6 mm travels along the wire. The average power carried by the wave is P. The average power becomes P', if the wave amplitude is halved. Then,

a. $P = 0.446$ W
b. $P = 0.223$ W
c. $P' = 0.056$ W
d. $P' = 0.223$ W

Working Space

19 Energy density u (energy per unit volume) of the medium at a distance r from a sound source varies according to the curve shown in the figure. Which of the following are possible?

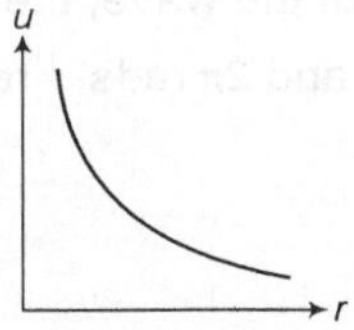

a. The source may be a point isotropic source
b. If the source is a plane source, then the medium particles have clamped oscillations
c. If the source is a plane source, then power of the source is decreasing with time
d. Density of the medium decreases with distance r from the source

Comprehension Type

Direction *(Q. Nos. 20-25) This section contains 2 paragraphs, each describing theory, experiments, data, etc. Six questions related to the paragraphs have been given. Each question has only one correct answer among the four given options (a), (b), (c) and (d).*

Passage I

Two identical strings 1 and 2 are stretched by the same force and waves are generated in the two strings travelling in positive X-direction. Shapes of 1 and 2 at some instant are shown in the figure. Based on this information, answer the following questions.

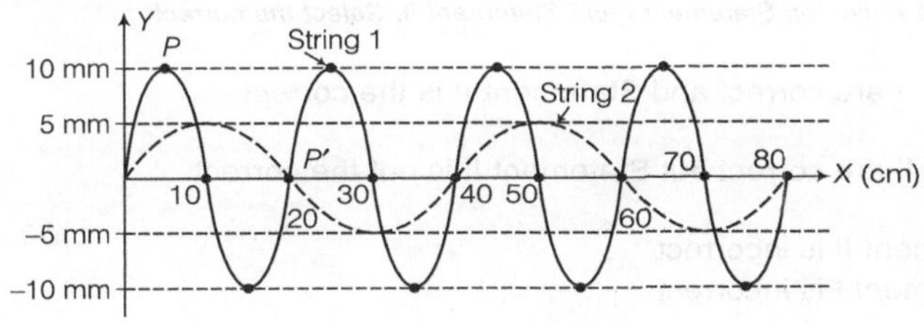

20 Ratio of intensities of waves in strings 1 and 2 is

a. 1 : 1 b. 1 : 4
c. 16 : 1 d. 4 : 1

21 Phase difference between point P (on the first string) and P' (on the second string) at the instant shown is

a. $\frac{\pi}{2}$ b. $\frac{2\pi}{3}$
c. π d. zero

22 Phase difference between particles of string 1 and 2 at $x = 20$ cm at the instant shown is

a. zero b. $\frac{\pi}{2}$
c. $\frac{2\pi}{3}$ d. π

Working Space

Passage II

A sinusoidal wave is propagating in negative x-direction in a string stretched along X-axis. A particle of string at $x = 2$ m is found at its mean position and it is moving in positive y-direction at $t = 1$s. The amplitude of the wave, the wavelength and the angular frequency of the wave are 0.1 m, $\frac{\pi}{2}$ m and 2π rads^{-1} respectively.

23 The equation of the wave is

a. $y = 0.1\sin[4\pi(t-1) + 8(x-2)]$
b. $y = 0.1\sin[2\pi(t-1) + 4(x-2)]$
c. $y = 0.1\sin[(t-1) - (x-2)]$
d. None of the above

24 The speed of particle at $x = 2$ m and $t = 1$s is

a. 0.6π ms^{-1}
b. 0.4π ms^{-1}
c. 0.2π ms^{-1}
d. zero

25 The instantaneous power transfer through $x = 2$ m and $t = 1.25$ s is

a. 10 Js^{-1}
b. $\frac{4\pi}{3}$ Js^{-1}
c. $\frac{2\pi}{3}$ Js^{-1}
d. zero

Statement Type

Direction (Q. Nos. 26 and 27) This section is based on Statement I and Statement II. Select the correct answer from the codes given below.

a. Both Statement I and Statement II are correct and Statement II is the correct explanation of Statement I
b. Both Statement I and Statement II are correct but Statement II is not the correct explanation of Statement I
c. Statement I is correct but Statement II is incorrect
d. Statement II is correct but Statement I is incorrect

26 **Statement I** In a progressive longitudinal wave, the amplitude of the wave will not be the same at all points of the medium along the direction of propagation of the wave.

Statement II There is a continuous change of the phase angle of the wave as we progress in the direction of motion.

27 **Statement I** The intensity of a plane progressive wave does not change with the distance from the source.

Statement II The wavefronts associated with a plane progressive wave are planar.

Matching List Type

Direction (Q. No. 28) Choices for the correct combination of elements from Column I and Column II are given as options (a), (b), (c) and (d), out of which one is correct.

28 A stretched string is forced to transmit transverse waves by means of an oscillator coupled to one end. The string has a diameter of 4 mm. The amplitude of oscillations is 10^{-4} m and the frequency is 10 Hz. Tension in the string is 100 N and mass density of the wire is 4200 kg m^{-3}.

Working Space

Match the Column I with Column II and mark the correct option from the codes given below.

	Column I		Column II (In SI units)
i.	The equation of wave along the string is	p.	$y = 10^{-4} \sin[1.44x - 62.83t]$
ii.	Energy density is	q.	$y = 10^{-4} \sin[2.88x - 125.66t]$
iii.	The average energy flow per unit time across any section of the string is	r.	7.25×10^{-5}
iv.	The power required to drive the oscillator is	s.	8.29×10^{-2}
		t.	4.53×10^{-5}

Codes

	i	ii	iii	iv
a.	r,t	p	q	t
b.	p	s	t	t
c.	q	r,t	p	s
d.	p,s	q,r	r	s,t

One Integer Value Correct Type

Direction *(Q. Nos. 29-32) This section contains 4 questions. When worked out will result in an integer from 0 to 9 (both inclusive).*

29 A taut string for which $\mu = 0.05$ kg m^{-1} is under a tension of 80 N. The power to be supplied to the string, to generate sinusoidal waves at a frequency of 60 Hz and an amplitude of 6 cm is approximately 171β watt. Find the value of β.

30 A certain 120 Hz wave on a string has an amplitude of 0.16 mm. The energy that exists in an 80 g length of the string is $(0.29)\,\beta$ mJ. Find the value of β.

31 A taut string for which $\mu = 0.05$ kgm^{-1} is under a tension of 80 N. The power that must be supplied to the string to generate sinusoidal waves at a frequency of 60 Hz and an amplitude of 6 cm is $64x$ watt. Find the value of x.

32 A transverse wave of amplitude 0.5×10^{-3} m and frequency 100 Hz is produced on a wire stretched to a tension of 100 N. The wave speed is 100 ms^{-1}. The average power that the source is transmitting to the wire is $(0.007)\,\alpha$ watt. Find the value of α.

DPP-4 Reflection and Transmission of Waves

Subjective Questions

Working Space

Direction (Q. Nos. 1-4) *These questions are subjective in nature, need to be solved completely on notebook.*

1 A harmonic wave is travelling on a string. At a junction with string 2, it is partly reflected and partly transmitted. The linear mass density of the second string is four times that of the first string and the boundary between the two strings is at $x = 0$. If the expression for the incident wave is $y_i = A_i \cos(k_1 x - \omega_1 t)$.

i. What are the expressions for the transmitted and the reflected waves in terms of A_i, k_1 and ω_1?

ii. Show that the average power carried by the incident wave is equal to the sum of the average power carried by the transmitted wave and the reflected wave.

2 A traingular pulse moving at 2 cms^{-1} on a rope approaches one end at which, it is free to slide on a vertical pole.

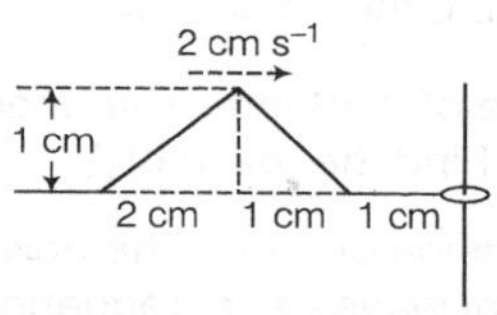

i. Draw the pulse at 0.5 s interval until it is completely reflected.

ii. What is the particle speed on the trailing edge at the instant depicted?

3 A wave pulse on a string has the dimensions, shown in the figure. The wave speed is $v = 1\,\text{cms}^{-1}$.

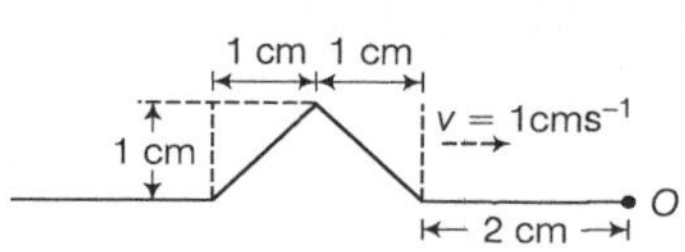

i. If point O is fixed, draw the resultant wave at $t = 3$ s and $t = 4$ s.

ii. Repeat part (i) for the case in which O is a free end.

4 A wave pulse on a string has the dimensions shown in the figure at $t = 0$ s. The wave speed is 5 ms^{-1}.

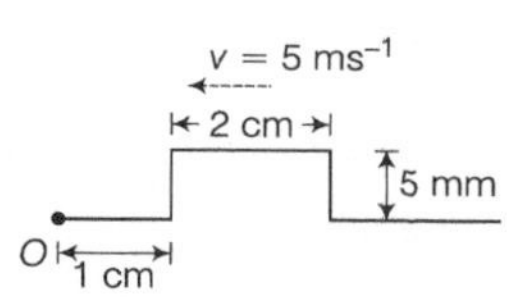

i. If point O is fixed end, draw the total wave on the string at $t = 1$ ms, 2 ms, 3 ms, 4 ms, 5 ms, 6 ms and 7 ms.

ii. Repeat part (i) for the case in which point O is a free end.

Only One Option Correct Type

Direction *(Q. Nos. 5-12) This section contains 8 multiple choice questions. Each question has four choices (a), (b), (c) and (d), out of which ONLY ONE is correct.*

5 The pulse in the figure shown has a speed of 0.1 ms^{-1}. The linear mass density of the right string is 0.25 that of the left string. At what speed does the transmitted wave travel?

a. 25 cm s^{-1} b. 20 cm s^{-1}
c. 15 cm s^{-1} d. 17.5 cm s^{-1}

6 In the above problem, the ratio of amplitude of transmitted wave to that of reflected wave is

a. -4 b. $-\frac{1}{4}$
c. $\frac{1}{4}$ d. 4

7 A wave $y_i = 0.3\cos(2x - 40t)$ is travelling along a string towards a boundary at $x = 0$. If at $x = 0$, the string has fixed end, the expression for reflected wave is (Assume SI units)

a. $y_r = 0.3\cos(2x + 40t)$ b. $y_r = 0.3\cos\left(2x + 40t + \frac{\pi}{2}\right)$
c. $y_r = 0.3\cos(2x + 40t + \pi)$ d. $y_r = 0.3\cos(2x - 40t + \pi)$

8 A wave $y_i = 0.3\cos(2x - 40t)$ is travelling along a string towards a boundary at $x = 0$. If at $x = 0$, the string has free end, the expression for reflected wave is (Assume SI units)

a. $y_r = 0.3\cos(2x + 40t)$ b. $y_r = 0.3\cos\left(2x + 40t + \frac{\pi}{2}\right)$
c. $y_r = 0.3\cos(2x + 40t + \pi)$ d. $y_r = 0.3\cos(2x - 40t + \pi)$

9 A harmonic wave $y_i = 0.002\cos(2\pi x - 50\pi t)$ travels along a string towards a boundary at $x = 0$, with a second string. The wave speed on the second string is 50 ms^{-1}. The expression for reflected wave is (Assume SI units)

a. $y_r = (6.67\times10^{-3})\cos(2\pi x + 50\pi t)$ b. $y_r = (6.67\times10^{-4})\cos(2\pi x + 50\pi t)$
c. $y_r = (6.67\times10^{-5})\cos(2\pi x + 50\pi t)$ d. $y_r = -6.67\times10^{-5}\cos(2\pi x + 50\pi t)$

10 A harmonic wave $y_i = 0.002\cos(2\pi x - 50\pi t)$ travels along a string towards a boundary at $x = 0$, with a second string. The wave speed on the second string is 50 ms^{-1}. The expression for transmitted wave is (Assume SI units)

a. $y_t = (2.67\times10^{-4})\cos(2\pi x - 50\pi t)$ b. $y_t = (2.67\times10^{-4})\cos(\pi x - 50\pi t)$
c. $y_t = (2.67\times10^{-3})\cos(2\pi x - 50\pi t)$ d. $y_t = (2.67\times10^{-3})\cos(\pi x - 50\pi t)$

11 Figure shows a square pulse and a triangular pulse approaching each other. The speed of each pulse is 0.5 cms^{-1}. The resultant pulse at $t = 2$ s will look as

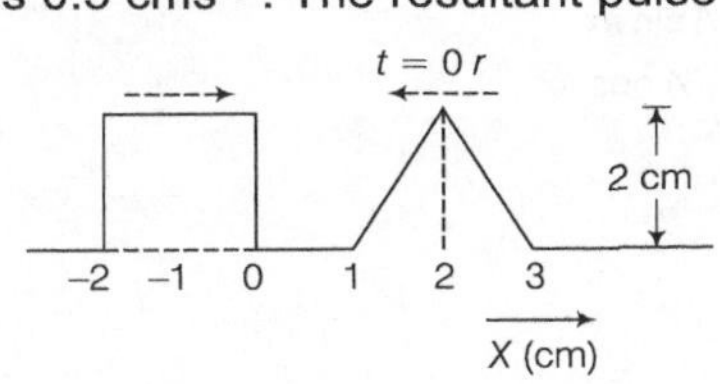

a. 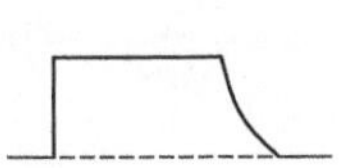 b. 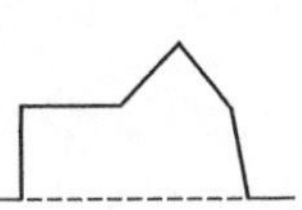 c. 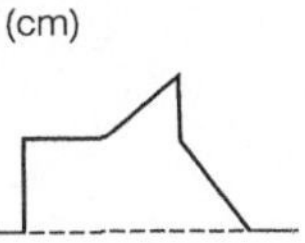 d. 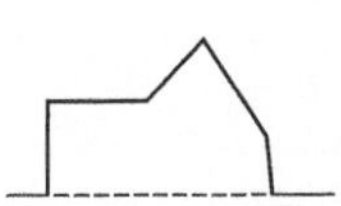

Working Space

12 A progressive wave gets reflected at a boundary such that, the ratio of amplitudes of the reflected wave and incident wave is 1 : 2. The percentage of energy transmitted per unit time is

a. 90% b. 75%
c. 50% d. 30%

One or More than One Options Correct Type

Direction (Q. Nos. 13 and 14) This section contains 2 multiple choice questions. Each question has four choices (a), (b), (c) and (d), out of which ONE or MORE THAN ONE are correct.

13 A non-uniform wire under constant tension has a variable linear mass density. Linear mass density is uniform for $-\infty \le x \le 0$. In this region, a transverse wave has the form $y = 0.003 \cos(25x - 50t)$, where x and y are in metre and t in second. From $x = 0$ to $x = 2$ m, the linear mass density decreases gradually from μ_1 to $\frac{\mu_1}{4}$. For $20 \le x \le \infty$, the linear mass density is $\frac{\mu_1}{4}$. Assume that variation in linear mass density is gradual so that, an incident wave is transmitted without reflection. Wave velocity for $-\infty \le x \le 0$ is v_1 and that for $20 \le x \le \infty$ is v_2. Amplitude of the wave and the wave equation for the region $20 \le x \le \infty$ are represented by A_2 and y_2 respectively. Then,

a. $v_1 = 2 \text{ ms}^{-1}$ b. $v_2 = 4 \text{ ms}^{-1}$
c. $A_2 = 0.0042$ m d. $y_2 = 0.0042 \cos(12.5x - 50t)$

14 A progressive wave travels in a medium M_1 and enters a medium M_2 in which its speed decreases to 75%. The amplitude of incident wave, reflected wave and transmitted wave are A_i, A_r and A_t respectively. Then,

a. $\left|\frac{A_r}{A_i}\right| = \frac{1}{7}$ b. $\left|\frac{A_r}{A_i}\right| = \frac{1}{5}$
c. $\left|\frac{A_t}{A_i}\right| = \frac{6}{7}$ d. $\left|\frac{A_t}{A_i}\right| = \frac{4}{5}$

Matching List Type

Direction (Q. Nos. 15-17) Choices for the correct combination of elements from Column I and Column II are given as options (a), (b), (c) and (d), out of which one is correct.

15 An incident wave train is given by the wave equation $y_i = A \sin(\omega t - kx)$ and is travelling on a stretched string whose right end is fixed. The wave is incident on the fixed end at $t = 0$. If T is the period match for time elapsed t in Column I with the function for reflected wave on Column II. Match the Column I with Column II and mark the correct option from the codes given below.

	Column I		Column II
i.	$t = T/4$	p.	$y_r = -A \sin kx$
ii.	$t = T/2$	q.	$y_r = A \cos kx$
iii.	$t = 3T/4$	r.	$y_r = A \sin kx$
iv.	$t = T$	s.	$y_r = -A \cos kx$

Codes

	i	ii	iii	iv
a.	s	r	q	p
b.	p	r	q	s
c.	s	q	p	r
d.	q	s	r	p

Working Space

16 An incident wave is given by the wave function $y_i = A\sin(\omega t - kx)$ and travels on a string that has a free boundary at its right end. T is the time period. Compare for time elapsed (t) in Column I with the wave function of reflected wave in Column II. Match the Column I with Column II and mark the correct option from the codes given below.

	Column I		Column II
i.	$t = T/4$	p.	$y_r = -A\cos kx$
ii.	$t = T/2$	q.	$y_r = -A\sin kx$
iii.	$t = 3T/4$	r.	$y_r = A\sin kx$
iv.	$t = T$	s.	$y_r = A\cos kx$

Codes

	i	ii	iii	iv		i	ii	iii	iv
a.	r	p	q	s	b.	p	q	r	s
c.	s	q	p	r	d.	s	p	q	r

17 Column I contains incident wave functions and the end from which it is reflected and Column II contains the reflected wave function. Incident wave is reflected from free end or rigid end (i.e. $x = 0$). Match the Column I with Column II and mark the correct option from the codes given below.

	Column I		Column II
i.	$y_i = A\sin(kx - \omega t)$, free end	p.	$y_r = -A\cos(kx + \omega t)$
ii.	$y_i = A\sin(kx - \omega t)$, rigid end	q.	$y_r = A\sin(kx + \omega t)$
iii.	$y_i = A\cos(kx - \omega t)$, free end	r.	$y_r = -A\sin(kx + \omega t)$
iv.	$y_i = A\cos(kx - \omega t)$, rigid end	s.	$y_r = A\cos(kx + \omega t)$

Codes

	i	ii	iii	iv		i	ii	iii	iv
a.	s	p	q	r	b.	r	q	s	p
c.	p	q	r	s	d.	r	p	q	s

Comprehension Type

***Direction** (Q. Nos. 18-22) This section contains 2 paragraphs, each describing theory, experiments, data, etc. Five questions related to the paragraphs have been given. Each question has only one correct answer among the four given options (a), (b), (c) and (d).*

Passage I

Consider two strings of equal lengths to be joined at B. The mass of the string BC is four times that of string AB. If a wave is generated in string AB, which travels towards boundary at B with speed v and the equation of incident wave is given by $y_i = A_i\sin(\omega t - kx)$. Now, answer the following questions.

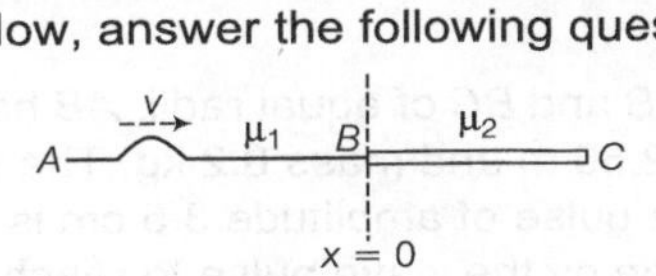

18 Amplitude of the wave reflected back after incidence on boundary at point B is

a. $-\dfrac{A_i}{3}$ b. $-\dfrac{2A_i}{3}$

c. $\dfrac{A_i}{3}$ d. $\dfrac{2A_i}{3}$

19 Speed of transmitted wave on string BC is

a. v b. $\dfrac{v}{2}$ c. $2v$ d. $\dfrac{v}{3}$

Working Space

20 Equations for reflected wave and transmitted wave respectively are

a. $y_r = -\frac{A_i}{3}\sin(\omega t + kx)$, $y_t = \frac{2A_i}{3}\sin(\omega t - 2kx)$

b. $y_r = \frac{A_i}{3}\sin(\omega t + kx)$, $y_t = \frac{2A_i}{3}\sin(\omega t + 2kx)$

c. $y_r = -\frac{A_i}{3}\sin(\omega t - kx)$, $y_t = \frac{2A_i}{3}\sin(\omega t - 2kx)$

d. $y_r = -\frac{A_i}{3}\sin(\omega t - kx)$, $y_t = \frac{2A_i}{3}\sin\left(\omega t + \frac{k}{2}x\right)$

Passage II

Consider the stretched strings 1 and 2, joined at a point (P) as shown in the figure. The linear mass densities of the strings are μ_1 and μ_2 respectively. Tension on both the strings is same. Now, answer the following questions :

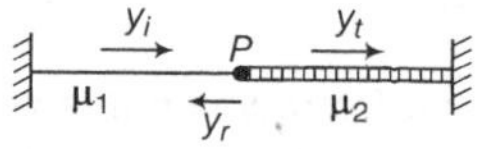

21 The amplitude of reflected wave is

a. $A_r = \frac{2\sqrt{\mu_1}}{(\sqrt{\mu_1}+\sqrt{\mu_2})}A_i$

b. $A_r = \left(\frac{\sqrt{\mu_1}-\sqrt{\mu_2}}{\sqrt{\mu_1}+\sqrt{\mu_2}}\right)A_i$

c. $A_r = \left(\frac{\sqrt{\mu_1}+\sqrt{\mu_2}}{\sqrt{\mu_1}-\sqrt{\mu_2}}\right)A_i$

d. None of these

22 The amplitude of transmitted wave is

a. $A_t = \frac{2\sqrt{\mu_1}}{(\sqrt{\mu_1}+\sqrt{\mu_2})}A_i$

b. $A_t = \left(\frac{\sqrt{\mu_1}-\sqrt{\mu_2}}{\sqrt{\mu_1}+\sqrt{\mu_2}}\right)A_i$

c. $A_t = \frac{\sqrt{\mu_1}}{\sqrt{\mu_1}+\sqrt{\mu_2}}A_i$

d. None of these

One Integer Value Correct Type

Direction *(Q. Nos. 23-25) This section contains 3 questions. When worked out will result in an integer from 0 to 9 (both inclusive).*

23 Two wires of different linear mass densities are soldered together end to end and are then stretched under tension. The wave speed in the first wire is twice in the second wire. Fraction of the incident power that is reflected at the junction is $\frac{1}{\beta}$. Find the value of β.

24 A long wire ABC is made by joining two wires AB and BC of equal radii. AB has length 4.8 m and mass 0.06 kg. BC has length 2.56 m and mass 0.2 kg. The wire ABC is under a tension 80 N. A sinusoidal wave pulse of amplitude 3.5 cm is sent along the wire AB from the end A. The time taken by the wave pulse to reach the other end C of the wire is $\frac{2}{100}\alpha$ second. Find the value of α.

25 Two wires of different linear mass densities are soldered together end to end then stretched under a tension T. The wave speed in the first wire is thrice that in the second. If a harmonic wave travelling in the first wire is incident on the junction of the wires and if the amplitude of incident wave is $\sqrt{13}$ cm, the amplitude of transmitted wave will be approximately $\Psi\left(\frac{2}{10}\right)$ cm. Find the value of Ψ.

SET - 1

Revisal Problems
for JEE Main

Only One Option Correct Type

Working Space

***Direction** (Q. Nos. 1-23) This section contains 23 multiple choice questions. Each question has four choices (a), (b), (c) and (d), out of which ONLY ONE option is correct.*

1 A transverse wave on a stretched string travelling along positive X-axis is shown in the figure. The mathematical form of shown wave is $y = (3\text{ cm})\sin\left(2\pi \times 0.1t - \frac{2\pi}{100}x\right)$, where t is in second and x in cm. Find the total distance trravelled by particle P is 10 min 15 s, measured from the instant shown in the figure and direction of its motion at the end of this time.

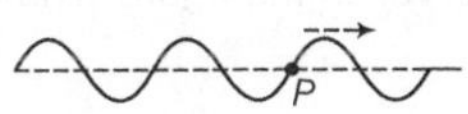

a. 732 cm in upward direction
b. 738 cm in upward direction
c. 6 cm in downward direction
d. 6 cm in upward direction

2 The figure shows three progressive waves A, B and C moving in the same direction. What can be concluded with respect to wave A from the figure?

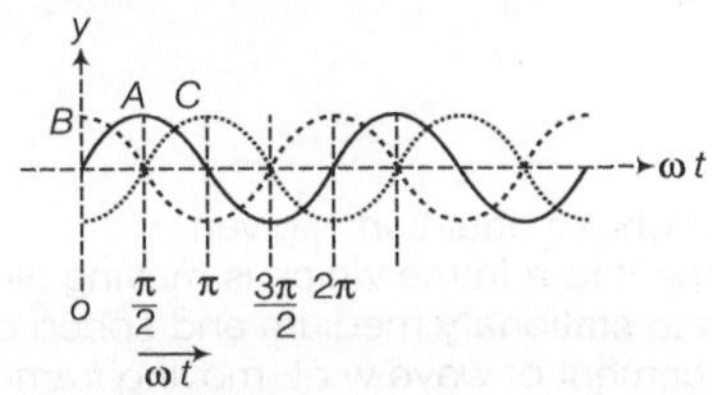

a. The wave C lags behind by a phase angle of $\pi/2$ and the wave B is ahead by a phase angle $\pi/2$
b. The wave C is ahead by a phase angle of $\pi/2$ and the wave B lags behind by a phase angle $\pi/2$
c. The wave C is ahded by a phase angle of π and the wave B lags behind by a phase angle π
d. The wave C lags behind by a phase angle π and the wave B is ahead by a phase angle π

3 The figure shows the snapshot of two waves w_1 and w_2 at a certain time t. The equation for wave w_1 is $y = A\sin(kx - \omega t + \phi)$ and for w_2, it is $y = A\sin(kx - \omega t)$. It is clearly shown in the figure, that wave w_1 is ahead by w_2 by a distance ϕ/k.

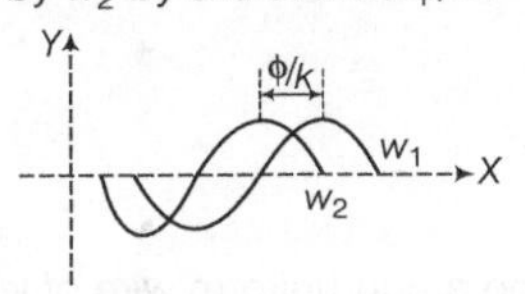

The motion of a single point in time i.e. *y versus t* for two waves is best represented by

a. 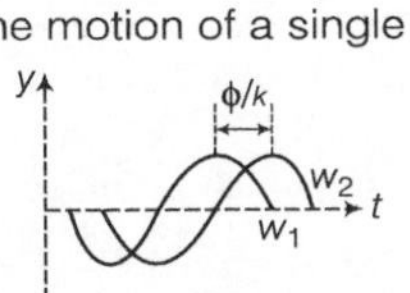

b. 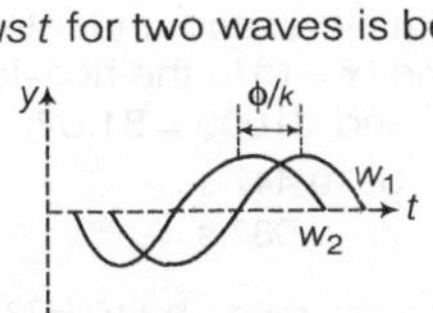

c. 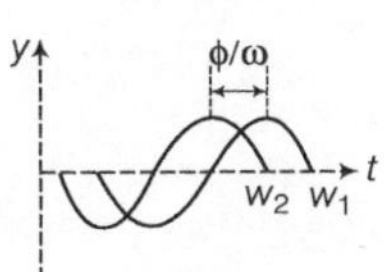

d. 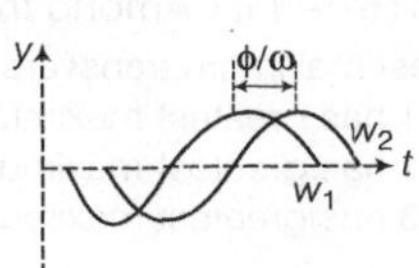

Working Space

4 A transverse sinusoidal wave is generated at one end of a long horizontal string by a bar, that moves the end up and down through a distacne by 2 cm. The motion of the bar is continuous and is repeated regularly 125 times per second. If the distance between adjacent wave crests is observed to be 15.6 cm and the wave is moving along positive x-direction and at $=0$, the element of the string at $x = 0$ is at mean position $y = 0$ and is moving downward, the equation of the wave is best described by

a. $y = (1\text{ cm})\sin[(40.3\text{ rad m}^{-1})x - (786\text{ rad s}^{-1})t]$

b. $y = (1\text{ cm})\cos[(40.3\text{ rad m}^{-1})x - (780\text{ rad s}^{-1})t]$

c. $y = (2\text{ cm})\sin[(40.3\text{ rad m}^{-1})x - (786\text{ rad s}^{-1})t]$

d. $= (2\text{ cm})\cos[(40.3\text{ rad m}^{-1}x - (786\text{ rad s}^{-1})t]$

5 If the maximum speed of a particle on a sinusoidal travelling wave is v_0, then find the speed of the particle, when the displacement is half the maximum value.

a. $\frac{v_0}{2}$
b. $\frac{\sqrt{3}}{4}v_0$
c. $\frac{\sqrt{3}}{2}v_0$
d. v_0

6 Two waves in the same medium are represented by y-t curves in the figure. The ratio of their average intensities is

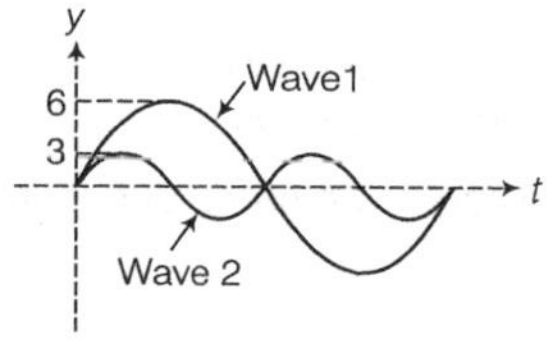

a. 4
b. 1/4
c. 9
d. 1

7 A harmonic wave is travelling in a stationary medium whose equation is given $y = A\sin(\omega t - kx)$. The equation of the wave with respect to a frame which is moving along negative X-axis with a constant speed v with respect to stationary medium and speed of the wave in moving frame respectively are (x' displacement of wave w.r.t. moving frame).

a. $y = A\sin\left[\left(\frac{\omega + kv}{2}\right)t - kx'\right], \frac{\omega + kv}{2k}$
b. $y = A\sin[(\omega + kv)t - kx'], \frac{\omega + kv}{k}$
c. $y = A\sin\left[\left(\frac{\omega}{2} + kv\right)t - kx'\right], \frac{\omega + kv}{2}$
d. None of these

8 A plane undamped harmonic wave propagates in a medium. Find the mean space density of total energy, if at any point, the space density of energy becomes equal to u_0 at an instant $t = t_0 + \frac{T}{6}$, where t_0 is the instant, when amplitude is maximum at this location and T is the time period of oscillation.

a. $\frac{1}{4}u_0$
b. $\frac{4}{3}u_0$
c. $\frac{2}{3}u_0$
d. $\frac{1}{2}u_0$

9 A wave pulse starts propagating in the + x-direction along a non-uniform wire of length 10 m with mass per unit length given by $\mu = \mu_0 + ax$ and under tension 100 N. Find the time taken by a pulse to travel from the lighter end ($x = 0$) to the heavier end.
(Given, $\mu_0 = 10^{-2}$ kg m^{-1}, $a = 9\times10^{-3}$ kg m^{-2} and $\sqrt{1000} = 31.62$)

a. 0.227 s
b. 0.447 s
c. 0.707 s
d. 0.087 s

10 The wave equations of three travelling waves are given by $y_1 = (2\text{ cm})\sin(3x - 6t)$, $y_2 = (3\text{ cm})\sin(4x - 12t)$ and $y_3 = (4\text{ cm})\sin(5x - 11t)$. Among these waves

a. wave 1 has greatest wave speed and greatest maximum transverse string speed
b. wave 2 has greatest wave speed and wave 1 has greatest maximum transverse string speed
c. wave 3 has greatest wave speed and wave 1 has greatest maximum transverse string speed
d. wave 2 has greatest wave speed and wave 3 has greatest maximum transverse string speed

Working Space

11 A progressive wave is given $y = 3\sin 2\pi\left[\frac{t}{0.04} - \frac{x}{0.01}\right]$, where x, y are in cm and t in s. The frequency of wave and maximum acceleration will be

a. 100 Hz, 470 ms^{-2} b. 50 Hz, 750 ms^{-2} c. 25 Hz, 470 ms^{-2} d. 25 Hz, 750 ms^{-2}

12 Sound waves travel in air with an amplitude 10^{-8} cm and frequency $f = 1000$ Hz. If density of the air at that temperature be 1.29 kg m^{-3} and speed of sound in air be 332 ms^{-1}, the energy density is

a. 3.55×10^{-13} Jm^{-3} b. 2.55×10^{-13} Jm^{-3}

c. 1.55×10^{-13} Jm^{-3} d. 0.55×10^{-13} Jm^{-3}

13 The equation of a travelling wave is given by $y = \pm \frac{b}{a}\sqrt{a^2 - (x - ct)^2}$, where $-a < x < a = 0$, otherwise. Now, the initial particle velocity at $x = \frac{a}{2}$ will be

a. $\sqrt{\frac{3}{2}}\frac{bc}{a}$ b. $\frac{bc}{\sqrt{2}a}$

c. $\frac{bc}{\sqrt{3}a}$ d. $\sqrt{\frac{2}{3}}\frac{bc}{a}$

14 In the figure is shown, a metal wire having a linear mass density 2.25×10^{-2} kg m^{-1}, with a block of mass 5 kg attached to the free end. Determine the speed of transverse wave in the string, if the block is immersed completely in a liquid, whose density bears a ratio of $\frac{17}{81}$ with that of the block.

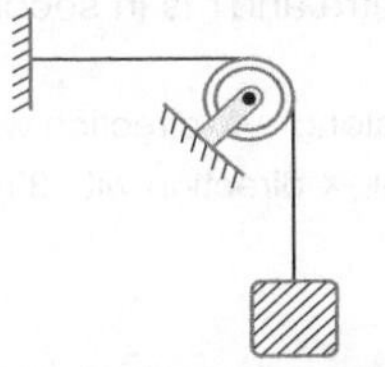

a. 46.7 ms^{-1} b. 41.5 ms^{-1}

c. 37.2 ms^{-1} d. 31.7 ms^{-1}

15 The form of a wave (of wavelength λ) $y = A\sin(kx - 23562t)$ at $t = 0$, is shown in the figure. Find the X-coordinate of point P in metre, if the wave speed is 300 ms^{-1}. (Assume SI units)

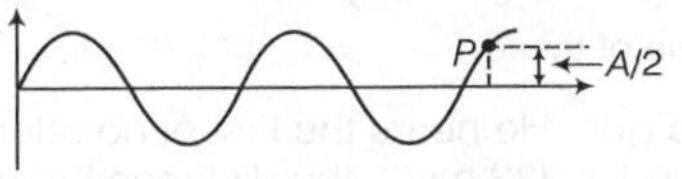

a. $\frac{25\lambda}{12}$ b. $\frac{27\lambda}{12}$

c. $\frac{29\lambda}{12}$ d. $\frac{31\lambda}{12}$

16 Two uniform ropes (1 and 2) of ratio of mass per unit length η are hanging from the rigid supports. Different masses are attached to the free ends of the ropes such that, tension at point A in the first rope and tension at B in the second rope are equal. Find the the ratio of wavelengths of the pulses reaching at points A and B, if ratio of frequencies of the pulses produced at lower ends is β.

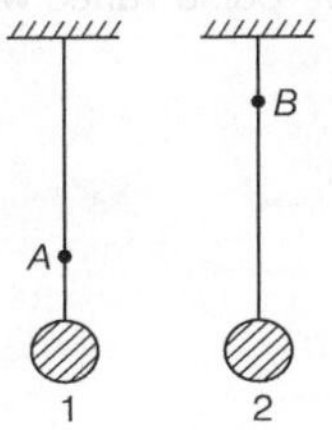

a. $\beta\sqrt{\eta}$ b. $\beta\eta$

c. $\frac{\beta}{\sqrt{\eta}}$ d. $\frac{1}{\beta\sqrt{\eta}}$

17 If a number of waves having slightly different wavelengths travel in the same medium, the medium is said to be dispersive and the energy is not transmitted in the medium with the wave velocity (or phase velocity), but it travels with the group velocity v_g given by

$$v_g = \frac{d\omega}{dk}$$

Waves passing through a certain medium have a dispersion relation $\omega(k) = \sqrt{\omega_0^2 + \alpha^2 k^2}$, where α and ω_0 are constants. The phase velocity (v) and group velocity (v_g) in this medium are respectively

a. $\frac{1}{k}\sqrt{\omega_0^2 + \alpha^2 k^2}, \frac{\alpha^2}{v}$
b. $\frac{\sqrt{\omega_0^2 + \alpha^2 k^2}}{2k}, \frac{\alpha^2}{\sqrt{2}v}$
c. $\frac{2\sqrt{\omega_0^2 + \alpha^2 k^2}}{k}, \frac{2\alpha^2}{v}$
d. None of these

18 In a sine wave, position of different particles at time, $t = 0$ is shown in the figure. If the wave is travelling in + X-direction, the equation of this wave is

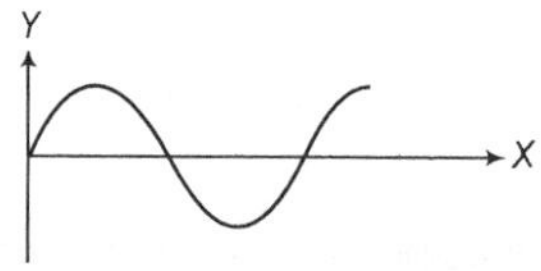

a. $y = A \sin(\omega t - kx)$
b. $y = A \cos(\omega t - kx)$
c. $y = A \sin(kx - \omega t)$
d. $y = A \cos(kx - \omega t)$

19 A motion is described by $y = 3e^x e^{-3t}$, where y, x are in metre and t is in second. Choose the correct statements.

a. This represents equation of progressive wave propagating along – X-direction with 3 ms^{-1}
b. This represents equation of progressive propagating along + X-direction with 3 ms^{-1}
c. This does not represent a progressive wave equation
d. Data is insufficient to arrive at any conclusion of this sort

20 Consider a wave propagating in the negative X-direction whose frequency is 100 Hz. At $t = 5$ s, the displacement associated with wave is given by $y = 0.5 \cos(0.1x)$, where x and y are measured in centimetre and t in seconds. Obtain the displacement as a function of x at $t = 10$ s.

a. $y = 0.5 \cos\left(\frac{\pi}{4} + 0.1x\right)$
b. $y = 0.5 \cos\left(\frac{\pi}{3} + 0.1x\right)$
c. $y = 0.5 \cos(0.1x)$
d. None of these

21 A person, standing between two parallel hills, fires a gun. He hears the first echo after 1.5 s and the second after 2.5 s. If the speed of sound is 332 ms^{-1}, the distance between the hills is

a. 964 m
b. 864 m
c. 764 m
d. 664 m

22 In a plane progressive harmonic wave, particle speed is always less than the wave speed, if (A is amplitude, λ is wavelength and μ is linear mass density).

a. $A < \frac{\lambda}{2\pi}$
b. $A > \frac{\lambda}{2\pi}$
c. $A < \lambda$
d. $A > \frac{\lambda}{\mu}$

23 A uniform rope having mass m hangs vertically from a rigid support. A transverse wave pulse is produced at the lower end. The speed (v) of the wave pulse varies with height (h) from the lower end as

a.
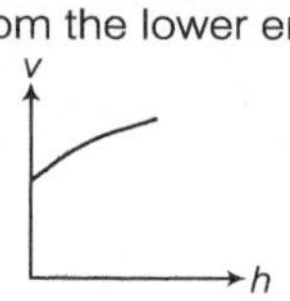

b.
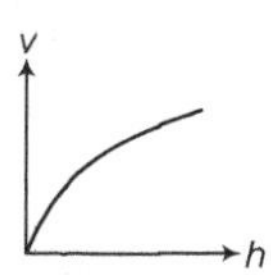

c.
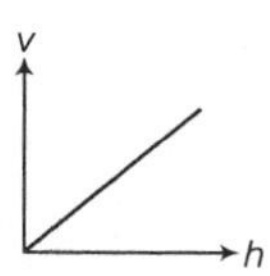

d.
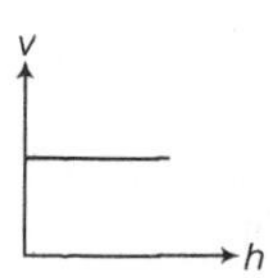

Working Space

SET - 2

Revisal Problems

for JEE Advanced

Only One Option Correct Type

Direction *(Q. Nos. 1-11) This section contains 11 multiple choice questions. Each question has four choices (a), (b), (c) and (d), out of which ONLY ONE is correct.*

Working Space

1 A man standing in front of a mountain beats a drum at regular intervals. The rate of drumming is increased and he finds that the echo is not heard distinctly, when the rate becomes 40 per minute. He then moves nearer to the mountain by 90 m and finds that echo is again not heard, when drumming rate becomes 60 per minute. The distance between the mountain and the initial position of man is

a. 205 m b. 300 m
c. 180 m d. 270 m

2 At $t = 0$, a transverse wave pulse travelling on the + x-direction with a speed of 2 ms^{-1} in a wire is described by the function $y = \frac{6}{x^2}$ given that, $x \neq 0$. Transverse velocity of a particle at $x = 2$ m and $t = 2$ s is

a. $-8\ ms^{-1}$ b. $8\ ms^{-1}$
c. $-3\ ms^{-1}$ d. $3\ ms^{-1}$

3 A pulse is incident on a yielding surface, then the possible form of reflected pulse is

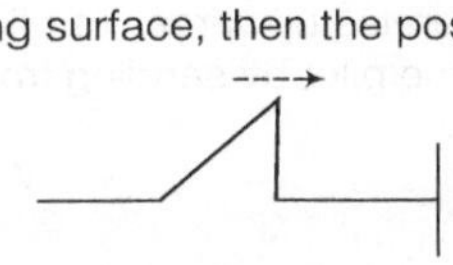

a. 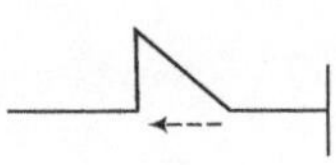
b. 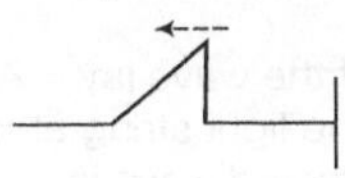

c. 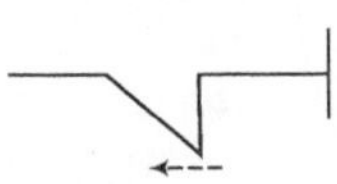
d. 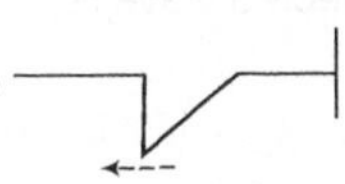

4 A transverse wave is described by the function $y = y_0 \sin 2\pi \left(ft - \frac{x}{\lambda}\right)$. The maximum particle velocity (v_p) is equal to four times the wave velocity if

a. $\lambda = \frac{\pi y_0}{4}$ b. $\lambda = \frac{\pi y_0}{2}$
c. $\lambda = \pi y_0$ d. $\lambda = 2\pi y_0$

5 In the shown arrangement, the horizontal surface is smooth and pulley is ideal. Mass per unit length of the string is μ and its mass is much less than the masses of blocks. At $t = 0$, the system is released from rest (the initial distance between block m_2 and pulley is L_0) and a transverse pulse is produced at point A. The time required by the pulse to reach pulley is (Given, $m_1 = 10$ kg, $m_2 = 10$ kg, $\mu = 20\ gm^{-1}$, $L_0 = 4$ m and $g = 10\ ms^{-2}$)

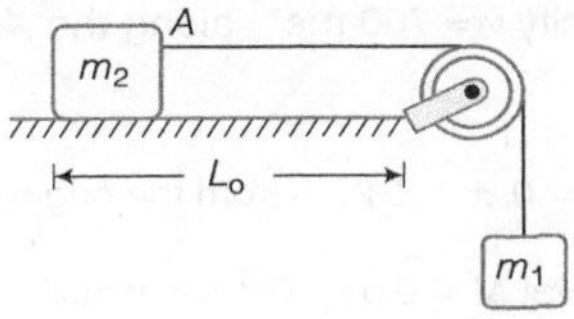

a. 0.01 s b. 0.06 s
c. 0.08 s d. 0.02 s

Working Space

6 Two canoes are 10 m apart on a lake. Each bobs up and down with a period of 4 s. When one canoe is at its highest point, the other canoe is at its lowest point. Both canoes are always within a single cycle of the waves. The speed of the wave is

a. $2.5\ ms^{-1}$ b. $5\ ms^{-1}$
c. $4.5\ cms^{-1}$ d. $4\ ms^{-1}$

7 Two particles of medium disturbed by the wave propagation are at $x_1 = 0$ and $x_2 = 1$ cm. The respective displacements (in cm) of the particles can be given by the equations

$$y_1 = 2 \sin 3\pi t,\ y_2 = 2 \sin\left(3\pi t - \frac{\pi}{8}\right)$$

The wave velocity is

a. $16\ cms^{-1}$ b. $24\ cms^{-1}$
c. $12\ cms^{-1}$ d. $8\ cms^{-1}$

8 A taut string having tension 100 N and linear mass density $0.25\ kg\ m^{-1}$ is used inside a cart to generate a wave pulse starting at the left end, as shown. What should be the velocity of the cart, so that pulse remains stationary w.r.t. ground.

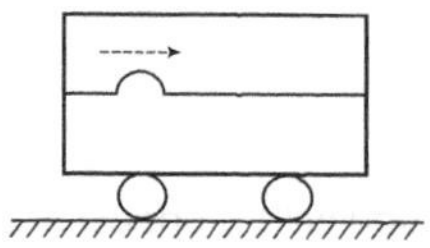

a. $20\ ms^{-1}$ towards left b. $10\ ms^{-1}$ towards left
c. $15\ ms^{-1}$ towards left d. $40\ ms^{-1}$ towards left

9 An 80 kg hiker is trapped on a mountain ledge following a storm. A helicopter rescues the hiker by hovering above him and lowering a cable to him. The mass of the cable is 8 kg and its length is 15 m. A sting of mass 70 kg is attached to the end of the cable. The hiker attaches himself to the sting and the helicopter then accelerates upwards. Terrified by hanging from the cable in mid-air, the hiker tries to signal the pilot by sending transverse pulses up the cable. The acceleration of the helicopter is

a. $1\ ms^{-2}$ b. $2\ ms^{-2}$
c. $3\ ms^{-2}$ d. $4\ ms^{-2}$

10 A wave travels on a light string. The equation of the wave is $y = A \sin(kx - \omega t + 30°)$. It is reflected from a heavy string tied at an end of the light string at $x = 0$. If 64% of the incident energy is relfected, then the equation of the relfected wave is

a. $y = 0.8A \sin(kx - \omega t + 30°)$
b. $y = 0.8A \sin(kx - \omega t - 30°)$
c. $y = 0.8A \sin(kx + \omega t + 30° + 180°)$
d. $y = 0.8A \sin(kx - \omega t + 30° + 180°)$

11 A transverse wave described by an equation $y = 0.02 \sin(x + 30t)$, where x and t are in metre and second, is travelling along a wire of cross-section $1\ mm^2$ and density $8000\ kg\ m^{-3}$. The tension in the string is

a. 20 N b. 7.2 N
c. 30 N d. 14.4 N

One or More than One Options Correct Type

Direction (Q. Nos. 12-15) *This section contains 4 multiple choice questions. Each question has four choices (a), (b), (c) and (d), out of which ONE or MORE THAN ONE are correct.*

12 A wave of frequency $f = 1000$ Hz, propagates at a velocity $v = 700\ ms^{-1}$ along the X-axis. Choose the correct statements.

a. The wavelength of the wave is 0.7 m
b. The points at a distance $(0.117 + 0.7n)$ metre, where $n = 0, \pm 1, \pm 2, \cdots$ from the origin differ in phase at the origin by $\pi/3$ rad
c. The phase gained at a given point x during as time interval $\Delta t = 0.5 \times 10^{-3}$ s is π rad
d. The phase gained at a given point x during a time interval $\Delta t = 10^{-4}$ s is 0.628 rad

Working Space

13 A harmonic wave has been set up on a very long string, which travels along the length of the string. The wave has a frequency of 50 Hz, amplitude 1 cm and wavelength 0.5 m. The time taken by the wave to travel a distance of 8 m along the length of string is t_1. The time taken by a point on the string to travel a distance of 8 m, once the the wave has reached the point and sets it into motion is t_2. Then,

a. $t_1 = 0.32$ s
b. $t_2 = 2$ s
c. $t_1 = 0.64$ s
d. $t_2 = 4$ s

14 A simple harmonic progressive wave in a gas has a particle displacement of $y = a$ at time $t = T/4$ at the origin of the wave and a particle velocity of y is v' at the same instant but at a distance $\frac{\lambda}{4}$ from the origin. Here, T and λ are time period and wavelength of the wave respectively. Then, choose the correct option(s).

(Wave speed is v and at $t = \frac{T}{4}$, the particle at the origin is momentarily at rest)

a. The amplitude of the wave is $2a$
b. The amplitude of the wave is a
c. The equation of the wave can be represented by $y = a \sin \frac{v'}{a}\left[t - \frac{x}{v}\right]$
d. The equation of the wave can be represented by $y = 2a \cos \frac{v'}{a}\left[t - \frac{x}{v}\right]$

15 Mark out the correct statement(s) for a transverse travelling mechanical wave on a string.

a. The wave speed is same for the entire wave, while particle velocity is different for different points at a particular instant
b. Wave speed depends upon property of the medium but not on the wave properties
c. Wave speed depends upon both the properties of the medium and on the properties of wave
d. Particle velocity depends upon properties of the wave and not on medium properties

Statement Type

Direction (Q. Nos. 16 and 17) ***This section is based on Statement I and Statement II. Select the correct answer from the codes given below.***

a. Both Statement I and Statement II are correct and Statement II is the correct explanation of Statement I
b. Both Statement I and Statement II are correct but Statement II is not the correct explanation of Statement I
c. Statement I is correct but Statement II is incorrect
d. Statement II is correct but Statement I is incorrect

16 **Statement I** For a travelling wave in a string, for small amplitudes, the instantaneous values of the kinetic energy and the potential energy of an element are equal.

Statement II $dU = \frac{1}{2} T dx \left(\frac{\partial y}{\partial x}\right)^2$ and $d(\text{KE}) = \frac{1}{2} \mu dx \left(\frac{\partial y}{\partial t}\right)^2$, where T is the tension and μ is mass per unit length of the string.

17 **Statement I** A plane progressive harmonic wave is propagating in a string. If tension in the string is made two times, then average power transmitted through the string becomes two times.

Statement II Average power transmitted in a string is given by

$$P = \frac{F\omega^2 A^2}{2v}$$

Matching List Type

Direction *(Q. Nos. 18-21) Choices for the correct combination of elements from Column I and Column II are given as options (a), (b), (c) and (d), out of which one is correct.*

18 In the figure are shown, two strings. Linear mass density of string 2 is nine times that of string 1. Tension in both the strings is same. A transverse wave in string 1 is incident their boundary. Part of wave is reflected into medium 1 (i.e. string 1) and part is transmitted. If A, v, I and P represent amplitude, speed of wave, intensity and power. Match the Column I with Column II and mark the correct option from the codes given below.

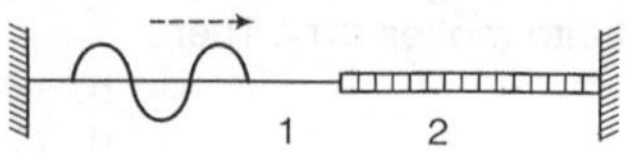

	Column I		Column II
i.	$\vert A_1/A_2 \vert$	p.	9
ii.	v_1/v_2	q.	1
iii.	I_1/I_2	r.	3
iv.	P_2/P_1	s.	Data insufficient

Codes

	i	ii	iii	iv		i	ii	iii	iv
a.	q	r	s	r	b.	s	q	r	p
c.	q	r	s	p	d.	p	s	q	r

19 For a transverse wave on a string. Match the Column I with Column II and mark the correct option from the codes given below.

	Column I		Column II
i.	If amplitude increases	p.	Average power decreases
ii.	If frequency increases	q.	Average power increases
iii.	If amplitude decreases	r.	Maximum instantaneous power decreases
iv.	If frequency decreases	s.	Maximum instantaneous power increases

Codes

	i	ii	iii	iv		i	ii	iii	iv
a.	q,s	q,s	p,r	p,r	b.	p,s	q,p	r,s	s,p
c.	s	r	p	q	d.	q	p	r	s

20 Three travelling sinusoidal waves are on identical strings having same tension. The mathematical forms of the waves are $y_1 = A\sin(3x - 6t)$, $y_2 = A\sin(4x - 8t)$, and $y_3 = A\sin(6x - 12t)$. Match the Column I with Column II and mark the correct option from the codes given below.

	Column I		Column II
i.	Speed of each wave is	p.	$2\ ms^{-1}$
ii.	y_1 is best represented by	q.	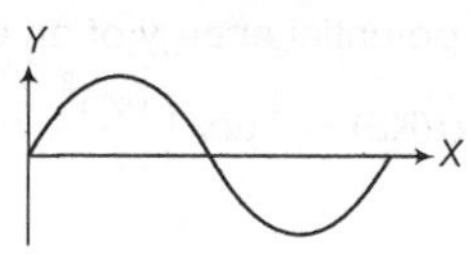
iii.	y_2 is best represented by	r.	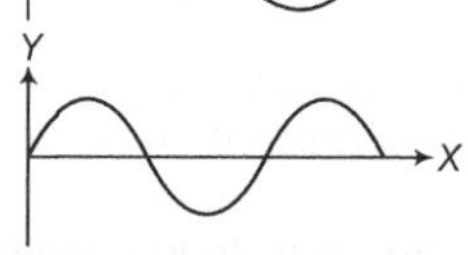
iv.	y_3 is best represented by	s.	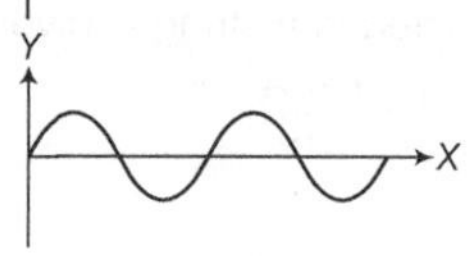
		t.	$4\ ms^{-1}$

Codes

	i	ii	iii	iv		i	ii	iii	iv
a.	p	q,p	r,p	s,p	b.	r	p	q,r	s,p
c.	s	q	r	p	d.	p	q	r	s

Working Space

Working Space

21 Match the Column I with Column II and mark the correct option from the codes given below. Symbols have their usual meanings.

	Column I		Column II
i.	String slope at $x = 0$ *versus* time t for $y = A \sin (kx + \omega t)$	p.	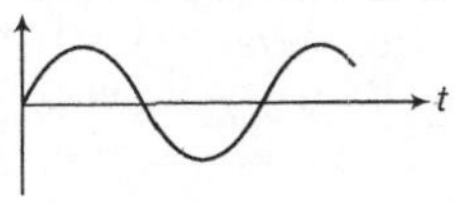
ii.	Instantaneous velocity of particle at $x = 0$ *versus* time t for $y = A\sin(kx - \omega t)$	q.	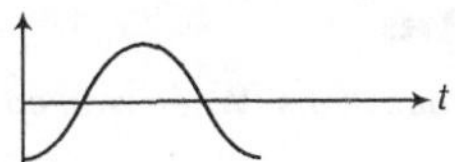
iii.	Displacement of particle at $x = 0$ *versus* t for $y = A\sin (\omega t - kx)$	r.	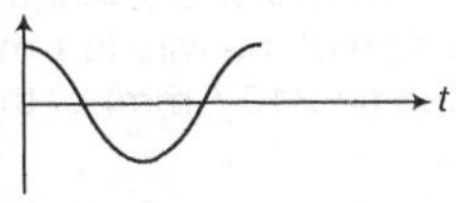

Codes

	i	ii	iii
a.	r	q	p
b.	q	r	p
c.	p	q	r
d.	r	p	q

Comprehension Type

Direction *(Q. Nos. 22-26) This section contains 2 paragraphs, each describing theory, experiments, data, etc. Five questions related to the paragraphs have been given. Each question has only one correct answer among the four given options (a), (b), (c) and (d).*

Passage I

Four pieces of string each of length L are joined end to end to make a long string $4L$. The linear mass density of the strings are μ, 4μ, 9μ and 16μ respectively. One end of the combined string is tied to a fixed support and a transverse wave has been generated at the other end having frequency f. Ignore any reflection and absorption. String has been stretched under a tension F.

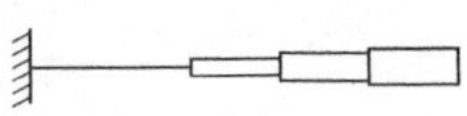

22 The time taken by wave to reach from the source end to fixed end is

a. $\frac{L}{\sqrt{F/\mu}}$ b. $\frac{4L}{\sqrt{F/\mu}}$ c. $\frac{10L}{\sqrt{F/\mu}}$ d. $\frac{25}{12}\frac{L}{\sqrt{F/\mu}}$

23 Find the ratio of wavelengths of the waves on four strings, starting from the right hand side.

a. 12 : 6 : 4 :3 b. 4 : 3 : 2 :1
c. 3 : 4 : 6 :12 d. 1:2 : 3 : 4

Passage II

A child playing with a long rope ties one end and holds the other. The rope is stretched taut along the horizontal. The child shakes the end he is holding, up and down, in a sinusoidal manner with amplitude 10 cm and frequency 3 Hz. Speed of the wave is 15 ms^{-1} and at $t = 0$, displacement at the child's end is maximum positive. Assuming that, there is no wave reflected from the fixed end, so that the waves in the rope are plane progressive waves, answer the following. Also assume that wave propagates in + x-direction.

24 A wave function that describes the wave in the given situation is

a. $y = (0.1 \text{ m}) \cos (2x - 12.5t)$ b. $y = (0.1 \text{ m}) \sin (1.5x - 10t)$
c. $y = (0.1 \text{ m}) \cos (18.8t - 1.26x)$ d. $y = (0.1 \text{ m}) \sin (1.5x - 4t)$

Working Space

25 Equation of displacement of a point 2.5 m from the child's end can be expressed as

a. $y = -$ (0.1 m) cos (18.8t)
b. $y =$ (0.1 m) cos (12.5t)
c. $y =$ (0.1 m) sin (4t)
d. $y = -$ (0.1 m) sin (10t)

26 Phase difference between the child's end and a point 2.5 m from the child end's will be

a. $\pi/2$
b. $3\pi/2$
c. $\pi/4$
d. π

One Integer Value Correct Type

***Direction** (Q. Nos. 27-31) This section contains 5 questions. When worked out will result in an integer from 0 to 9 (both inclusive).*

27 The speed of a transverse wave, going on a wire having a length 50 cm and mass 5 g is 80 ms^{-1}. The area of cross-section of the wire is 1 mm^2 and its Young's modulus is 8×10^{11} Nm^{-2}. Find the extension (in $\times 10^{-2}$ mm) of the wire over its natural length.

28 A plane progressive wave is given by $x =$ (40 cm) cos $(50\pi t - 0.02\pi y)$, where y is in cm and t in second. The particle velocity at $y = 25$ cm in time $t = 0.005$ s will be $10\pi\sqrt{\alpha}$ ms^{-1}. What is the value of α?

29 A 4 kg block is suspended from the ceiling of an elevator through a string having a linear mass density of 19.2×10^{-3} kg m^{-1}. The speed (w.r.t. the string) with which a wave pulse can proceed on the string (in ms^{-1}), if the elevator accelerates up at the rate of 2 ms^{-2} is 12.5 β. Find the value of β. (Take, $g = 10$ ms^{-2})

30 A travelling wave is given by $y = \dfrac{0.8}{(3x^2 + 24xt + 48t^2 + 4)}$, where x and y are in metre and t is in second. Find the velocity in ms^{-1}.

31 An ant with mass m is standing peacefully on top of a horizontal, stretched rope. The rope has mass per unit length μ and is under tension F. Without warning, a student starts a sinusoidal transverse wave of wavelength λ propagating along the rope. What minimum wave amplitude (in mm) will make the ant feel weightless momentarily? Assume that m is very small and that the presence of the ant has no effect on the propagation of the wave. (Given, $\lambda = 0.5$ m, $\mu = 0.1$ kg m^{-1}, $F = 3.125$ N, $g = \pi^2$).

JEE Main & AIEEE Archive

(Compilation of Last 13 Years' Questions)

1 A travelling wave represented by $y = A\sin(\omega t - kx)$ is superimposed on another wave represented by $y = A\sin(\omega t + kx)$. The resultant is **(AIEEE 2011)**

a. a standing wave having nodes at $x = \left(n + \frac{1}{2}\right)\frac{\lambda}{2}, n = 0, 1, 2$

b. a wave travelling along $+x$ direction

c. a wave travelling along $-x$ direction

d. a standing wave having nodes at $x = \frac{n\lambda}{2}; n = 0, 1, 2$

2 The equation of a wave on a string of linear mass density 0.04 kg m^{-1} is given by

$$y = 0.02\,(m)\sin\left[2p\left(\frac{t}{0.04\,(s)} - \frac{x}{0.50\,(m)}\right)\right]$$

The tension in the string is **(AIEEE 2010)**

a. 4.0 N
b. 12.5 N
c. 0.5 N
d. 6.25 N

3 A wave travelling along the X-axis is described by the equation $y(x,t) = 0.005\cos(\alpha x - \beta t)$. If the wavelength and the time period of the wave are 0.08 m and 2.0 s respectively, then α and β in appropriate units are **(AIEEE 2008)**

a. $\alpha = 25.00\pi, \beta = \pi$
b. $\alpha = \frac{0.08}{\pi}, \beta = \frac{2.0}{\pi}$
c. $\alpha = \frac{0.04}{\pi}, \beta = \frac{1.0}{\pi}$
d. $\alpha = 12.50\pi, \beta = \frac{\pi}{2.0}$

4 The displacement y of a particle in a medium can be expressed as

$$y = 10^{-6}\sin\left(100t + 20x + \frac{\pi}{4}\right)\text{m}$$

where, t is in second and x in metre. The speed of the wave is **(AIEEE 2004)**

a. 2000 ms^{-1}
b. 5 ms^{-1}
c. 20 ms^{-1}
d. 5π ms^{-1}

5 The displacement y of a wave travelling in the x-direction is given by

$$y = 10^{-4}\sin\left(600t - 2x + \frac{\pi}{3}\right)\text{m}$$

where, x is expressed in metres and t in seconds. The speed of the wave motion, in ms^{-1} is **(AIEEE 2003)**

a. 300
b. 600
c. 1200
d. 200

6 The displacement of a particle varies according to the relation $x = 4(\cos \pi t + \sin \pi t)$. The amplitude of the particle is **(AIEEE 2003)**

a. -4
b. 4
c. $4\sqrt{2}$
d. 8

JEE Advanced & IIT JEE Archive

(Compilation of Last 10 Years' Questions)

1 A point mass is subjected to two simultaneous sinusoidal displacements in x-direction, $x_1(t) = A\sin\omega t$ and $x_2(t) = A\sin\left(\omega t + \frac{2\pi}{3}\right)$. Adding a third sinusoidal displacement $x_3(t) = B\sin(\omega t + \phi)$ brings the mass to a complete rest. The values of B and ϕ are

(2011, Only One Option Correct Type)

a. $\sqrt{2}A, \frac{3\pi}{4}$ b. $A, \frac{4\pi}{3}$

c. $\sqrt{3}A, \frac{5\pi}{6}$ d. $A, \frac{\pi}{3}$

2 A massless rod BD is suspended by two identical massless strings AB and CD of equal lengths. A block of mass m is suspended from point P such that BP is equal to x. If the fundamental frequency of the left wire is twice the fundamental frequency of right wire, then the value of x is **(2006, Only One Option Correct Type)**

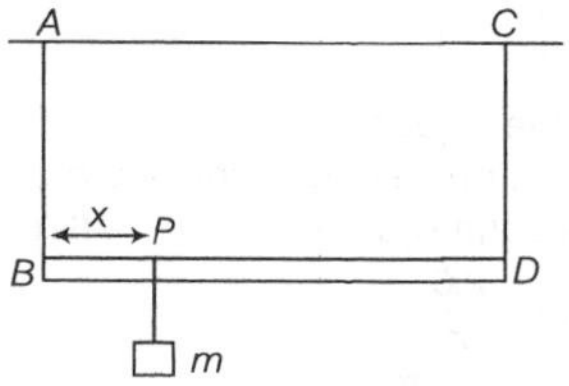

a. $l/5$ b. $l/4$

c. $4l/5$ d. $3l/4$

3 A transverse sinusoidal wave moves along a string in the positive x-direction at a speed of 10 cms^{-1}. The wavelength of the wave is 0.5 m and its amplitude is 10 cm. At a particular time t, the snap-shot of the wave is shown in the figure. The velocity of point P, when its displacement is 5 cm is **(2008, Only One Option Correct Type)**

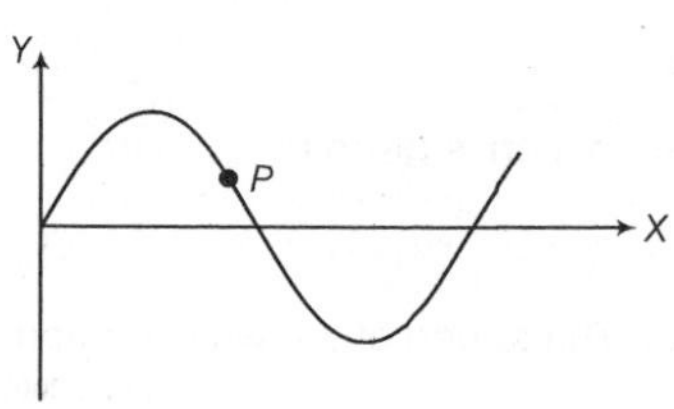

a. $\frac{\sqrt{3}\pi}{50}\hat{j}$ ms^{-1} b. $-\frac{\sqrt{3}\pi}{50}\hat{j}$ ms^{-1}

c. $\frac{\sqrt{3}\pi}{50}\hat{i}$ ms^{-1} d. $-\frac{\sqrt{3}\pi}{50}\hat{i}$ ms^{-1}

3. Sound

DPP
DAILY PRACTICE PROBLEMS

DPP-1 Basic Parameters of Sound Waves

Subjective Questions

Working Space

Direction (Q. Nos. 1-4) *These questions are subjective in nature, need to be solved completely on notebook.*

1 i. What experimental evidence can be cited to show that the speed of sound is the same for all wavelengths?

ii. Explain, if an observer places his ear to one end of a long iron pipeline, he can distinctly here two sounds, when a work man hammers the other end of the pipe line.

iii. A man stands on the ground at a fixed distance from a siren, which emits a clear sound during night than during day. Explain why?

2 The equation of a travelling plane sound wave has the form $y = 50 \cos(1800t - 5.3x)$, where y is expressed in micrometer, t in second and x in metre. Find

i. the ratio of the displacement amplitude, with which the particles of medium vibrate, to the wavelength.

ii. the velocity oscillation amplitude of particles of the medium and its ratio to the wave propagation velocity.

3 A sound wave of frequency 100 Hz is travelling in air. The speed of sound in air is 350 ms^{-1}.

i. By how much is the phase changed at a given point in 2.5 ms?

ii. What is the phase difference at a given instant between two points separated by a distance of 10 cm along the direction of propagation?

4 The equation of a sound wave in air is given by $\Delta p = 0.02 \sin(3000t - 9x)$, where all variables are in SI units.

i. Find the frequency, wavelength and the speed of sound wave in air.

ii. If the equilibrium pressure of air is 1.01×10^5 Nm^{-2}, what are the maximum pressure and minimum pressure at a point as the wave passes through that point?

Only One Option Correct Type

Direction (Q. Nos. 5-13) *This section contains 9 multiple choice questions. Each question has four choices (a), (b), (c) and (d), out of which ONLY ONE is correct.*

5 An aeroplane is going towards East at a speed of 510 kmh^{-1} at a height of 2000 m. At a certain instant, the sound of the plane heard by a ground observer appears to be coming from a point vertically above him. What is the horizontal distance between the observer and the plane? Speed of sound in air is 340 ms^{-1}.

a. 833 m　　b. 1000 m
c. 933 m　　d. 733 m

Working Space

6 A road runs midway between two parallel rows of buildings. A motorist moving with a speed of 36 km h^{-1} sounds the horn. He hears the first echo one second after he has sounded the horn. Speed of sound in air is 330 ms^{-1}. Find the distance between the rows of buildings and the time after which (from the time of sounding the horn) he hears the second echo.

a. 350 m, 2.5 s
b. 329.8 m, 2 s
c. 360 m, 2.3 s
d. 300.2m, 1.5 s

7 An engine approaches a hill with a constant speed. When it is at a distance of a 0.9 km, it blows a whistle, whose echo is heard by the driver after 5 s. If the speed of sound in air is 330 ms^{-1}, the speed of the engine is

a. 120 km h^{-1}
b. 90 km h^{-1}
c. 108 km h^{-1}
d. 88 km h^{-1}

8 From a cloud at an angle of 30° to the horizontal, we hear the thunder clap 8 s after seeing the lightning flash. The speed of sound in air is 330 ms^{-1}. The height of the cloud above the ground is

a. 2.32 km
b. 1.16 km
c. 0.98 km
d. 1.32 km

9 The pressure variation in a sound wave in air is given by

$\Delta p = 12 \sin\left(8.18x - 2700t + \frac{\pi}{4}\right)$ Nm^{-2}. If density of air is 1.29 kg m^{-3}, the displacement amplitude is

a. 0.089×10^{-5} m
b. 1.044×10^{-5} m
c. 2.011×10^{-5} m
d. 2.732×10^{-5} m

10 Speed of sound in water is 1450 ms^{-1}. The minimum wavelength and maximum wavelength of sound in water, for audible range of an average human ear, respectively are

a. 72.5 cm, 72.5 m
b. 7.25 mm, 7.25 m
c. 72.5 mm, 72.5 m
d. 7.25 m, 72.5 m

11 Calculate the bulk modulus (B) of air from the following data for a sound wave of wavelength 35 cm travelling in air. The pressure at a point varies between $(10^5 \pm 14)$ Pa and the particles of the air vibrate in SHM with amplitude 5.5×10^{-6} m.

a. 2.12×10^5 Pa
b. 1.22×10^5 Pa
c. 1.32×10^5 Pa
d. 1.42×10^5 Pa

12 For a sound wave travelling in + x-direction, sinusoidal longitudinal displacement y at a certain time is given as a function of x as shown in the figure. If bulk modulus (B) of air is 5×10^5 Pa, the variation of excess pressure is best represented by

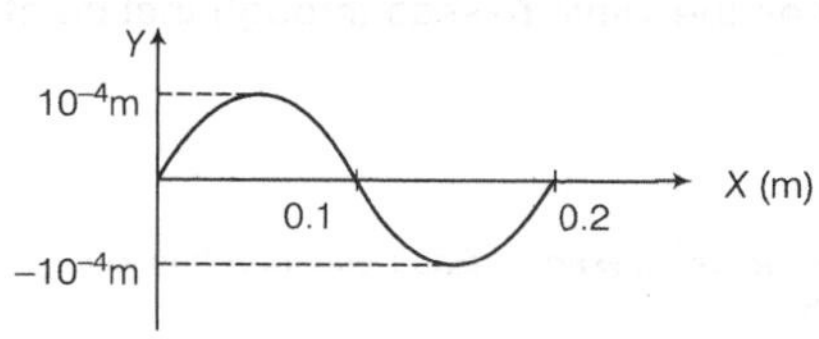

a.
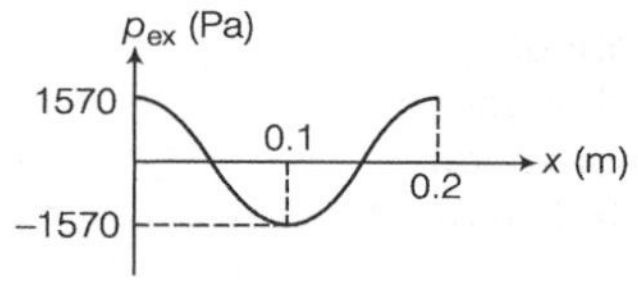

b.
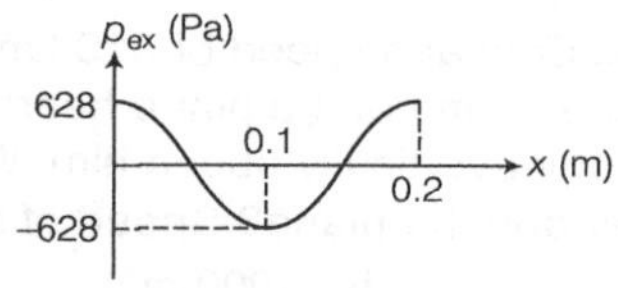

c.
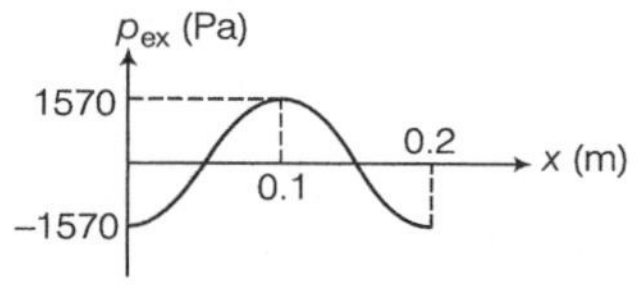

d.
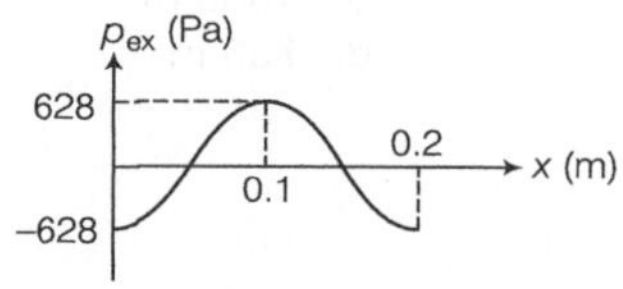

Working Space

13 A sound wave is propagating through a gas medium whose displacement (s) *versus* x graph is shown in the figure. Its pressure (Δp) *versus* x graph is

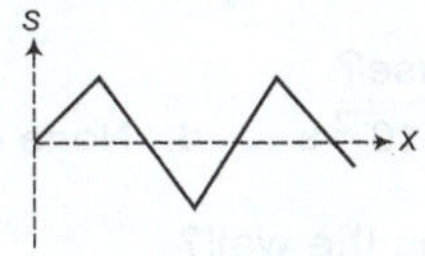

a. 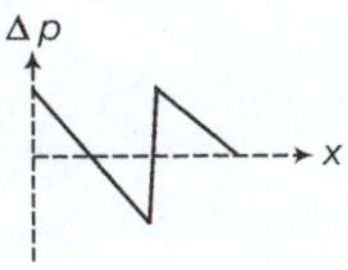 b. 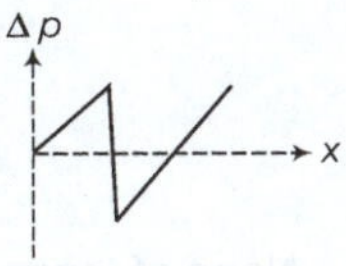 c. 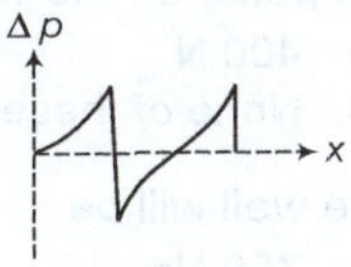 d. 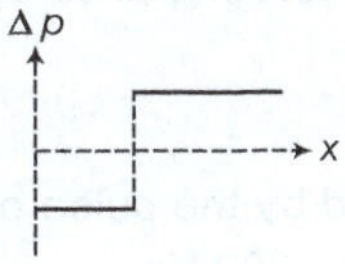

One or More than One Options Correct Type

Direction *(Q. Nos. 14-16) This section contains 3 multiple choice questions. Each question has four choices (a), (b), (c) and (d), out of which ONE or MORE THAN ONE are correct.*

14 Speed of sound in air is 340 ms^{-1} and density of air is 1.29 kg m^{-3}. The pressure amplitude for a sound wave having a frequency of 100 Hz is 10 Pa and its displacement amplitude is A_1. The displacement amplitude for a sound wave having a frequency of 300 Hz is 10^{-7} m and its pressure amplitude is $(\Delta p)_{m2}$. Then,

a. $A_1 = 3.63 \times 10^{-5}$ m b. $A_1 = 4.63 \times 10^{-5}$ m

c. $(\Delta p)_{m2} = 8.26 \times 10^{-2}$ Nm^{-2} d. $(\Delta p)_{m2} = 9.26 \times 10^{-2}$ Nm^{-2}

15 A typical sound wave has a frequency 1000 Hz and has a pressure amplitude 10 Pa. Density of air is 1.29 kg m^{-3} and speed of sound in air is 340 ms^{-1}. The maximum value of displacement at any time and place is A. At $t = 0$, pressure is maximum at some point x_1 and the displacement is y_1. Then,

a. $A = 2.63$ μm b. $A = 3.63$ μm

c. $y_1 = 1.63$ μm d. $y_1 =$ zero

16 In the figure is shown, an instantaneous displacement *versus* position graph of a sound wave travelling in air along $+x$-direction. Choose the correct options.

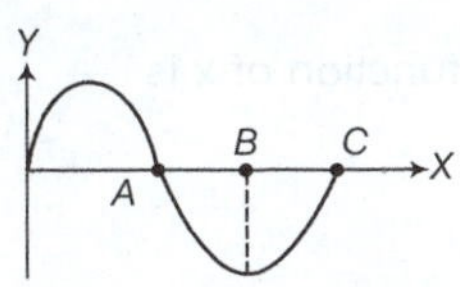

a. At A, pressure is maximum

b. At B, pressure is minimum

c. At C, pressure is minimum

d. At B, pressure is atmospheric

Comprehension Type

Direction *(Q. Nos. 17-24) This section contains 2 paragraphs, each describing theory, experiments, data, etc. Eight questions related to the paragraphs have been given. Each question has only one correct answer among the four given options (a), (b), (c) and (d).*

Passage I

A plane pressure pulse triangular in shape approaches a rigid wall along normal at a speed 400 ms^{-1}. The situation at $t = 0$s is shown in the figure. The peak pressure is 100 Pa. By the wall pulse gets reflected and pressure near the wall gets doubled. Height of the wall is 2 m and width is also 2 m. A detector on the wall records a minimum excess pressure of 16 Pa.

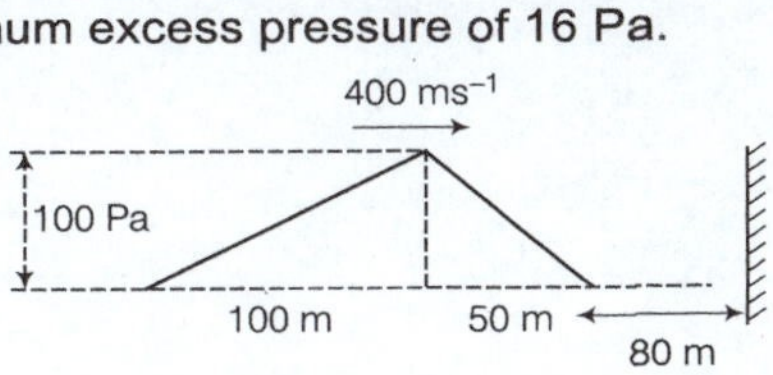

Working Space

17 When for the first time, detector will record the pressure pulse?

a. 31.5×10^{-2} s b. 22×10^{-2} s c. 21×10^{-2} s d. None of these

18 For how much time the detector will record the pulse?

a. 34.5×10^{-2} s b. 37.5×10^{-2} s c. 31.5×10^{-2} s d. None of these

19 What is the maximum force applied by the pulse on the wall?

a. 200 N b. 400 N

c. 800 N d. None of these

20 Total impulse imparted by the pulse on the wall will be

a. 150 Ns b. 300 Ns c. 750 Ns d. None of these

Passage II

The speed of longitudinal waves in a rod given by $v=\sqrt{\dfrac{Y}{\rho}}$, where Y is the Young's modulus and ρ its density. Longitudinal waves are set up in a rod of variable density, by striking at one of the ends (LL') of the rod. The density varies as $\rho=\rho_0(1+\beta x)$, where x is the distance from the struck end and β is a constant. The vibrations set up by a simple harmonic source have angular frequency ω and the amplitude at the end LL' is A_0. Assume that, the Young's modulus of elasticity Y is constant throughout the rod. (The cross-section at a distance x is MM' and there is no loss of energy)

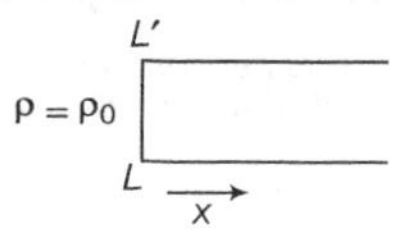

21 The amplitude of vibrations, as a function of x, along the rod is

a. $\dfrac{A_0}{2}\sqrt{\dfrac{1}{1+\beta x}}$ b. $A_0\sqrt{\dfrac{2}{1+\beta x}}$

c. $A_0\sqrt{\dfrac{1}{2(1+\beta x)}}$ d. $A_0\sqrt{\dfrac{1}{1+\beta x}}$

22 Wavelength as a function of x is

a. $\dfrac{\pi}{\omega}\sqrt{\dfrac{Y}{\rho_0(1+\beta x)}}$ b. $\dfrac{2\pi}{\omega}\sqrt{\dfrac{Y}{2\rho_0(1+\beta x)}}$

c. $\dfrac{2\pi}{\omega}\sqrt{\dfrac{Y}{\rho_0(1+\beta x)}}$ d. $\dfrac{2\pi}{\omega}\sqrt{\dfrac{2Y}{\rho_0(1+\beta x)}}$

23 Time taken to travel a distance x is

a. $\dfrac{4}{3\beta}\sqrt{\dfrac{\rho_0}{Y}}(1+\beta x)^{3/2}$ b. $\dfrac{2}{3\beta}\sqrt{\dfrac{\rho_0}{Y}}(1+\beta x)^{3/2}$

c. $\dfrac{3}{2\beta}\sqrt{\dfrac{\rho_0}{Y}}(1+\beta x)^{3/2}$ d. None of these

24 Equation of longitudinal wave set up in the rod, at cross-section MM', is

a. $y=A_0\sqrt{\dfrac{1}{1+\beta x}}\sin\omega\left[t-\sqrt{\dfrac{\rho_0(1+\beta x)}{Y}}\cdot x\right]$

b. $y=A_0\sqrt{\dfrac{1}{1+\beta x}}\sin\omega\left[t-\sqrt{\dfrac{2\rho_0(1+\beta x)}{Y}}\cdot x\right]$

c. $y=A_0\sqrt{\dfrac{2}{1+\beta x}}\sin\omega\left[t-\sqrt{\dfrac{\rho_0(1+\beta x)}{2Y}}\cdot x\right]$

d. $y=A_0\sqrt{\dfrac{1}{2(1+\beta x)}}\sin\omega\left[t-\sqrt{\dfrac{\rho_0(1+\beta x)}{2Y}}\cdot x\right]$

DPP-2 Speed of Sound

Only One Option Correct Type

Working Space

Direction (Q. Nos. 1-11) *This section contains 11 multiple choice questions. Each question has four choices (a), (b), (c) and (d), out of which ONLY ONE is correct.*

1 Calculate the (approximate) increase in velocity of sound for 1°C rise of temperature, if the velocity of sound at 0°C is $332 ms^{-1}$.

a. $0.41 ms^{-1}$ b. $0.51 ms^{-1}$ c. $0.61 ms^{-1}$ d. $0.71 ms^{-1}$

2 At what temperature will the speed of sound in air becomes double of its value at 0°C?

a. 1092°C b. 819°C c. 919°C d. 719°C

3 The planet Jupiter has an atmosphere mainly of methane at a temperature −130°C. Calculate the velocity of sound on this planet assuming γ for the mixture to be 1.3. (Given, gas constant, $R = 8.3 \text{ Jmol}^{-1} \text{°C}^{-1}$)

a. $311 ms^{-1}$ b. $292 ms^{-1}$ c. $275 ms^{-1}$ d. $332 ms^{-1}$

4 How long will it take for sound waves to travel a distance L_0 between the points I and F, if the air temperature between them varies linearly from T_1 to T_2? The velocity of sound propagation in air is equal to $v = \alpha\sqrt{T}$, where α is a constant.

a. $\frac{L_0}{\alpha(\sqrt{T_2} + \sqrt{T_1})}$ b. $\frac{L_0}{2\alpha(\sqrt{T_2} + \sqrt{T_1})}$ c. $\frac{\sqrt{2} L_0}{\alpha(\sqrt{T_2} + \sqrt{T_1})}$ d. $\frac{2L_0}{\alpha(\sqrt{T_2} + \sqrt{T_1})}$

5 The speed of sound as measured by a student in the laboratory on a winter day is $340 ms^{-1}$, when the room temperature is 17°C. What speed will be measured by another student repeating the experiment on a day, when the room temperature is 32°C?

a. $349 ms^{-1}$ b. $344 ms^{-1}$ c. $354 ms^{-1}$ d. $352 ms^{-1}$

6 At what temperature will the speed of sound in hydrogen be the same as in oxygen at 100°C? Molar masses of oxygen and hydrogen are in the ratio 16 : 1.

a. −219.7°C b. −249.7°C c. −229.7°C d. −239.7°C

7 A mixture of two diatomic gases exists in a closed cylinder. The volumes of two gases are V_1 and V_2 (the volume of the cylinder is $V_1 + V_2$). The speeds of sound in these gases are v_1 and v_2 respectively. The speed of sound in the mixture is

a. $v_1 v_2 \sqrt{\frac{V_1 + V_2}{V_1 v_2^2 + V_2 v_1^2}}$ b. $\frac{v_1 v_2}{2} \sqrt{\frac{V_1 + V_2}{V_1 v_2^2 + V_2 v_1^2}}$

c. $\frac{v_1 v_2}{\sqrt{2}} \sqrt{\frac{V_1 + V_2}{V_1 v_2^2 + V_2 v_1^2}}$ d. None of these

8 A mixture of two diatomic gases are obtained by mixing m_1 mass of first gas and m_2 mass of second gas. The speeds of sound in two gases are v_1 and v_2 respectively. The speed of sound in the given mixture is

a. $\sqrt{\frac{m_1 v_1^2 + m_2 v_2^2}{2(m_1 + m_2)}}$ b. $\sqrt{\frac{2(m_1 v_1^2 + m_2 v_2^2)}{m_1 + m_2}}$ c. $2\sqrt{\frac{m_1 v_1^2 + m_2 v_2^2}{m_1 + m_2}}$ d. $\sqrt{\frac{m_1 v_1^2 + m_2 v_2^2}{m_1 + m_2}}$

Working Space

9 In a liquid with density 900 kg m^{-3}, longitudinal waves with frequency 250 Hz are found to have wavelength 8 m. The bulk modulus of the liquid is

a. 3.6×10^8 Pa
b. 1.8×10^9 Pa
c. 3.6×10^9 Pa
d. 1.8×10^8 Pa

10 A metal bar with a length of 1.5 m has a density 6400 kgm^{-3}. Longitudinal sound waves take 3.9×10^{-4} s to travel from one end of the bar to the other. Young's modulus for this metal is

a. 9.47×10^{12} Pa
b. 9.47×10^9 Pa
c. 9.47×10^{11} Pa
d. 9.47×10^{10} Pa

11 There are cases, when an explosion at point P_1 will be heard at point P_2, that is far away from point P_1, while in a certain region located much closer to P_1 than to P_2, the explosion is not heard due to obstacle. This will be possible

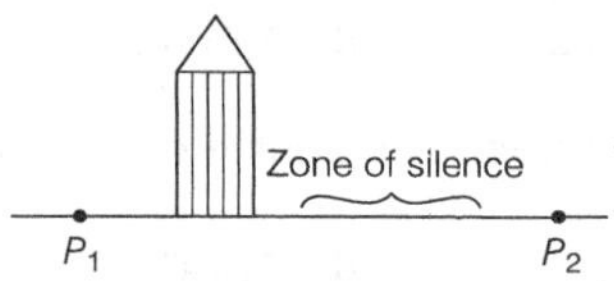

a. if air temperature increases with altitude
b. if air temperature decreases with altitude
c. if air is blowing from P_2 to P_1
d. if air is blowing from P_1 to P_2

One or More than One Options Correct Type

Direction *(Q. No. 12) This section contains 1 multiple choice question. Each question has four choices (a), (b), (c) and (d), out of which ONE or MORE THAN ONE are correct.*

12 The speeds of longitudinal waves in the following gases at 0°C and 1 atm ($= 10^5$ Pa) respectively are v_1 and v_2

I. Oxygen for which bulk modulus is 1.41×10^5 Pa and density is 1.43 kgm^{-3}
II. Helium for which bulk modulus is 1.7×10^5 Pa and density is 0.18 kg m^{-3}

a. 314 ms^{-1}
b. 972 ms^{-1}
c. 332 ms^{-1}
d. 872 ms^{-1}

Matching List Type

Direction *(Q. Nos. 13 and 14) Choices for the correct combination of elements from Column I and Column II are given as options (a), (b), (c) and (d), out of which one is correct.*

13 Regarding speed of sound in a gas, match the following columns and mark the correct option from the codes given below.

	Column I		Column II
i.	Temperature of the gas is made 4 times and pressure 2 times	p.	Speed becomes $2\sqrt{2}$ times
ii.	Only pressure is made 4 times without change in temperature	q.	Speed becomes 2 times
iii.	Only temperature is changed to 4 times	r.	Speed remains unchanged
iv.	Molecular mass of the gas is made 4 times	s.	Speed becomes half

Codes

	i	ii	iii	iv		i	ii	iii	iv
a.	r	p	q	s	b.	p	q	r	s
c.	q	r	q	s	d.	s	r	q	p

Working Space

14 Match the items in Column I with respective properties in Column II and mark the correct option from the codes given below.

	Column I		Column II
i.	In refraction	p.	Speed of wave does not change
ii.	In reflection	q.	Wavelength is decreased
iii.	In refraction from rarer to denser medium	r.	Frequency does not change
iv.	In reflection from a denser medium	s.	Phase change of π takes place

Codes

	i	ii	iii	iv
a.	p	q	r	s
b.	r	p,r	q,r	p,r,s
c.	q	r	s	p
d.	s	p	q,r	p,s

One Integer Value Correct Type

Direction *(Q. Nos. 15-19) This section contains 5 questions. When worked out will result in an integer from 0 to 9 (both inclusive).*

15 The stress $\left(\frac{F}{A}\right)$ in a stretched wire of a material (whose Young's modulus is Y), for the speed of longitudinal waves to be equal to 30 times speed of transverse waves, is $\frac{Y}{100x}$. Find the value of x.

16 A plane sound wave passes from medium 1 to medium 2. The speed of sound in medium 1 is 200 ms^{-1} and in medium 2 is 100 ms^{-1}. If the ratio of amplitude of the transmitted wave to that of incident wave is $\frac{2}{k}$, then find the value of k.

17 The change in volume of 6 L of alcohol, if pressure is decreased from 200 cm of Hg to 70 cm of Hg, is 15β cc. Find the value of β. (Given, velocity of sound in alcohol = 1280 ms^{-1}, density of alcohol = 0.81 g cm^{-3}, density of Hg = 13.6 g cm^{-3} and g = 9.81 ms^{-2})

18 A point source of sound is located somewhere along the X-axis. Experiments show that the same wavefront simultaneously reaches listeners at $x = -8$ m and $x = +2$ m. A third listener is positioned on the positive Y-axis. What is his y-coordinate (in m), if the same wavefront reaches him at the same instant as it does the first two listeners ?

19 The speed of sound in hydrogen at 0°C is 1200 ms^{-1}. When some amount of oxygen is mixed with hydrogen, the speed decreases to 500 ms^{-1}. If density of oxygen is 16 times that of hydrogen, the ratio of hydrogen to oxygen by volume in this mixture is $\frac{11}{10}\alpha$. Find the value of α.

DPP-3 Intensity of Sound

Only One Option Correct Type

Working Space

Direction *(Q. Nos. 1-10) This section contains 10 multiple choice questions. Each question has four choices (a), (b), (c) and (d), out of which ONLY ONE is correct.*

1 Consider that 10 identical sitars are playing a note, each at 70 dB. If they all are in chorus, what will be the sound level?

a. 80 dB　　b. 90 dB
c. 85 dB　　d. 95 dB

2 Two loudspeakers are being compared and one is perceived to be 32 times louder than the other. What will be the difference in intensity levels betweeen the two, when measured in decibel? (Take, $\log_{10} 32 \simeq 1.5$)

a. 15 dB　　b. 7.5 dB
c. 22.5 dB　　d. 30 dB

3 A point isotropic source is located on the perpendicular to the plane of a ring, drawn through the centre O of the ring. The distance between the point O and the source is $l = 1$ m and the radius of the ring is $r = 0.5$ m. Find the mean energy flow across the area enclosed by the ring, if at the point O, the intensity of sound is equal to $I_0 = 30\mu \text{Wm}^{-2}$.

a. $10 \mu\text{Js}^{-1}$　　b. $15 \mu\text{Js}^{-1}$
c. $20 \mu\text{Js}^{-1}$　　d. $25 \mu\text{Js}^{-1}$

4 A point isotropic source with sonic power $P = 0.1$W is located at the centre of a round hollow cylinder with radius $R = 1$m and height $h = 2$m. Assuming the sound to be absorbed by the walls (both curved and plane) of the cylinder, find the mean energy of flow reaching the lateral curved surface of the cylinder.

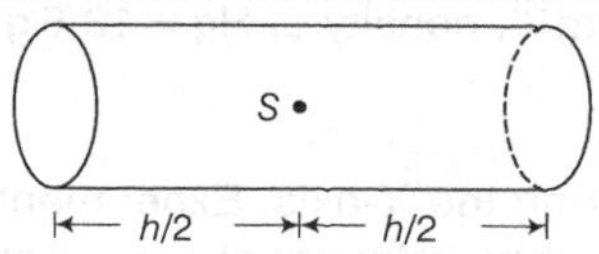

a. 0.08 W　　b. 0.07 W
c. 0.06 W　　d. 0.05 W

5 A dog while barking delivers about 1 mW of power. *Assume* this power is uniformly distributed over a hemispherical area. Now, 5 dogs start barking at the same time from the same point, each delivering 1 mW of power. What would be the sound level at a distance of 5 m ? (Take, $\log_{10}(6.37) = 0.8$ and $\log_{10} 5 = 0.7$)

a. 70 dB　　b. 75 dB
c. 72 dB　　d. 80 dB

6 What is the maximum possible sound level (in dB) of sound waves in air ? Speed of sound in air is 332 ms^{-1} and density of air is 1.3 kg m^{-3}. The atmospheric pressure $p_{atm} = 1.01 \times 10^5 \text{ Nm}^{-2}$.

a. 210 dB　　b. 180 dB
c. 200 dB　　d. 190 dB

Working Space

7 A window whose area is $2 m^2$ opens to a street where street noise results in an intensity level of 60 dB at the window. If an acoustic absorber is fitted to the window, how much energy from street will it cover in 10 hours ?

a. 36 mJ b. 72 mJ c. 108 mJ d. 18 mJ

8 In a good FM radio receiver, the radio signal detected may be as much as, 65 dB greater than the noise signal. What is the ratio of signal intensity to noise intensity?

a. 1.82×10^6 b. 2.39×10^6 c. 0.91×10^6 d. 3.16×10^6

9 A family ice show is held at an enclosed arena. The skaters perform to music with level of 80 dB. This level is too loud for your baby. So, she yells at 75 dB. What total sound intensity (in $\times 10^{-5}$ Wm^{-2}) engulfs you ?

a. $\sqrt{10}$ b. 10 c. 14.92 d. 13.2

10 In the above problem, what is the combined sound level? (Given, $\log_{10}(13.2) = 1.12$)

a. 82.1dB b. 81.2 dB c. 82.2 dB d. 83.0dB

One or More than One Options Correct Type

Direction (Q. Nos. 11-15) *This section contains 5 multiple choice questions. Each question has four choices (a), (b), (c) and (d), out of which ONE or MORE THAN ONE are correct.*

11 The speaker of a public address system emits 20 W power. Considering it a point source, sound level at a point 4 m away from it is β. And the sound level at a distance x from the speaker is $\frac{\beta}{2}$. Then,

a. $\beta = 120$ dB b. $\beta = 110$ dB c. $x = 1.24$ km d. $x = 2.24$ km

12 Suppose the sound level in Bengaluru has been increasing by about 1 dB annually. This corresponds to x percentage increase in intensity. In about y number of years, the intensity of sound is doubled, if it increased at 1 dB annually. Then, (Take, $\ln 2 = 0.693$, $\ln 10 = 2.3026$)

a. $x = 23\%$ b. $y = 3$ yr c. $x = 33\%$ d. $y = 25$ yr

13 A point sound source is situated in a medium having bulk modulus 1.6×10^5 Nm^{-2}. An observer standing at a distance 10 m from the source, writes down the equation for the wave as $y = A\sin(15\pi x - 6000\pi t)$, where y and x are in metre and t in second. The maximum pressure amplitude tolerable to the observer's ear is 24π Nm^{-2}. Then,

a. density of the medium is 1 kgm^{-3}

b. displacement amplitude (A) of the wave received by the observer is 10^{-5} m

c. displacement amplitude (A) of the wave received by the observer is 10^{-6} m

d. maximum power of the sound source is 8.93×10^3 W

14 A plane longitudinal wave having angular frequency 500 s^{-1} is travelling in positive x-direction in a medium of density $\rho = 1$ kgm^{-3} and bulk modulus 4×10^4 Nm^{-2}. The loudness at a point in the medium is observed to be 20 dB. Assume at $x = 0$, initial phase of the medium particles is zero. Maximum pressure change in the medium is $(\Delta p)_{max}$. Then,

a. $(\Delta p)_{max} = 10^{-4}$ Nm^{-2}

b. $(\Delta p)_{max} = 2 \times 10^{-4}$ Nm^{-2}

c. equation of wave is $y = 2 \times 10^{-9} \sin\left(500t - \frac{7}{2}x\right)$

d. equation of wave is $y = 2 \times 10^{-9} \sin\left(500t - \frac{5}{2}x\right)$

Working Space

15 A point Q is located at a distance $r = 1.5$ m from a point source of sound of frequency 600 Hz. The power of the source is 0.8 W. Speed of sound in air is $340 ms^{-1}$ and density of air is $1.29 kgm^{-3}$. At the point Q, the pressure oscillation amplitude is $(\Delta p)_{max}$ and displacement oscillation amplitude is A. Then,

a. $(\Delta p)_{max} = 4.98 Nm^{-2}$

b. $(\Delta p)_{max} = 6.98 Nm^{-2}$

c. $A = 2 \times 10^{-6}$ m

d. $A = 3 \times 10^{-6}$ m

Matching List Type

Direction (Q. No. 16) *Choices for the correct combination of elements from Column I and Column II are given as options (a), (b), (c) and (d), out of which one is correct.*

16 A loudspeaker diaphragm 0.2 m in diameter is vibrating at 1000 Hz with an amplitude of 0.01×10^{-3} m. Assume that the air molecules in the vicinity have the same amplitude of vibration. Density of air is $1.29 kgm^{-3}$. Speed of sound is $340 ms^{-1}$. Then, match the Column I with Column II and mark the correct option from the codes given below.

	Column I		Column II
i.	Pressure amplitude immediately in front of diaphragm (in Nm^{-2})	p.	2.7×10^{-2}
ii.	Sound intensity in front of the diaphragm (in Wm^{-2})	q.	2.15×10^{-5}
iii.	The acoustic power radiated (in W)	r.	27.55
iv.	Intensity at 10 m from the loudspeaker (in Wm^{-2})	s.	0.865

Codes

	i	ii	iii	iv
a.	p	q	r	s
b.	s	p	q	r
c.	r	s	p	q
d.	q	r	p	s

Statement Type

Direction (Q. Nos. 17 and 18) *This section is based on Statement I and Statement II. Select the correct answer from the codes given below.*

a. Both Statement I and Statement II are correct and Statement II is the correct explanation of Statement I

b. Both Statement I and Statement II are correct but Statement II is not the correct explanation of Statement I

c. Statement I is correct but Statement II is incorrect

d. Statement II is correct but Statement I is incorrect

17 **Statement I** If two people talk simultaneously and each creates an intensity level of 60 dB at a point Q, then total intensity level at point Q is 120 dB.

Statement II Sound level is defined on a non-linear scale.

18 **Statement I** In case of a plane progressive wave, intensity remains same. (in case of no damping).

Statement II The energy crossing unit area in unit time is same.

Comprehension Type

Working Space

Direction *(Q. Nos. 19-22) This section contains a paragraph, describing theory, experiments, data, etc. Four questions related to the paragraph have been given. Each question has only one correct answer among the four given options (a), (b), (c) and (d).*

Passage

When a sound wave enters a ear, it sets the eardrum into oscillation, which in turn causes of oscillation of three tiny bones in the middle ear called 'ossicles'. This oscillation is finally transmitted to the fluid filled in inner portion of the ear termed as inner ear. The motion of the fluid disturbs hair cells within the inner ear, which transmit nerve impulses to the brain, with the information that a sound is present. The three bones present in the middle ear are named as hammer, anvil and stirrup. Out of these the stirrup is the smallest one and this only connects the middle ear to inner ear as shown in the figure. The area of stirrup and its extent of connection with the inner ear limits the sensitivity of the human ear.

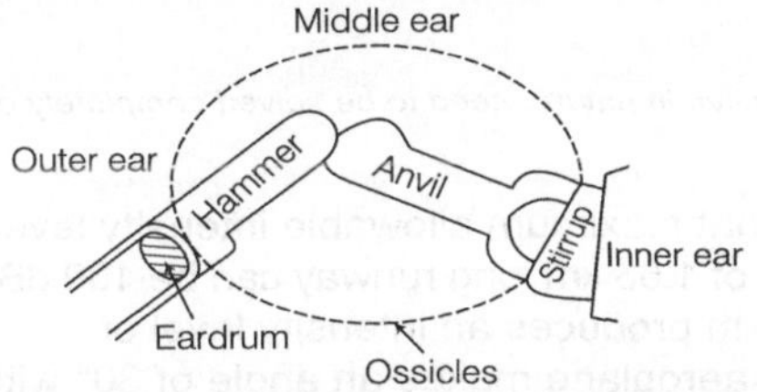

Consider a person's ear whose moving part of the eardrum has an area of about 50 mm^2 and the area of stirrup is about 5 mm^2. As a result, force exerted by sound wave in air on eardrum and ossicles is same as the force exerted by ossicles on the inner ear. This is because ossicles have negligible mass. Consider a sound wave having maximum pressure fluctuation of 4×10^{-2} Pa from its normal equilibrium pressure value, which is equal of 10^5 Pa. Frequency of sound wave is 1200 Hz.

19 Speeds of sound wave in air and in fluid (present in inner ear) are 332 ms^{-1} and 1500ms^{-1} respectively. Bulk modulus of air is 1.42×10^5 Pa and bulk modulus of fluid is 2.18×10^9 Pa. The pressure amplitude of given sound wave in the fluid of inner ear is

a. 0.03 Pa b. 0.3 Pa
c. 0.04 Pa d. 0.4 Pa

20 The displacement amplitude of given sound wave in the fluid of inner ear is

a. 4.42×10^{-11} m b. 7.81×10^{-11} m
c. 8.21×10^{-12} m d. 3.65×10^{-11} m

21 If the person is using a hearing aid, which increase the sound intensity level by 30 dB, then by what factor the intensity of given sound wave change as perceived by inner ear ?

a. 1000 b. 100
c. 10000 d. None of these

22 This person (without hearing aid machine) is sitting inside a busy restaurant, where average sound intensity is 3.2×10^{-5} Wm^{-2}. How much energy in the form of sound is taken up by the person in his meal time of 1 h?

a. 1.2×10^{-5} J b. 1.8×10^{-4} J
c. 2.4×10^{-5} J d. 3.6×10^{-4} J

One Integer Value Correct Type

Working Space

Direction (Q. Nos. 23-25) *This section contains 3 questions. When worked out will result in an integer from 0 to 9 (both inclusive).*

23 The intensity of sound from a point source is $1 \times 10^{-8}\ \text{Wm}^{-2}$ at a distance of 5 m from the source. What will be the intensity (in $10^{-10}\ \text{Wm}^{-2}$) at a distance of 25 m from the source?

24 The resultant loudness at a point A is ndB higher than the loudness of S_1, which is one of the two identical sound sources S_1 and S_2, from which sound waves are reaching at point A in phase. Find the value of n. (Take, $\log_{10} 4 = 0.6$)

25 Loudness of sound from an isotropic point source at a distance of 70 cm is 20 dB. What is the distance (in metre) at which it is not heard?

Subjective Questions

Direction (Q. Nos. 26 and 27) *These question are subjective in nature, need to be solved completely on notebook.*

26 Airport authority has made the regulations that maximum allowable intensity level detected by a microphone situated at the end of 1.63 km long runway can be 100 dB. An aeroplane when flying at a height of 200 m produces an intensity level of 100 dB on the ground. While taking off, this aeroplane makes an angle of 30° with horizontal. Find the maximum distance this aeroplane can cover on the runway, so that the regulations are not violated (assume no reflection).

27
i. The power of sound from the speaker of a radio is 20 mW. By turning the knob of volume control the power of sound is increased to 400 mW. What is the power increase (in dB) as compared to original power?
ii. How much more intense is an 80 dB sound than 20 dB whisper?

DPP-4 Doppler Effect-I

Only One Option Correct Type

Working Space

Direction *(Q. Nos. 1-11) This section contains 11 multiple choice questions. Each question has four choices (a), (b), (c) and (d), out of which ONLY ONE is correct.*

1 A policeman on duty detects a drop of 10% of the frequency of the horn of a moving car as it crosses him. If the speed of sound is 330 ms^{-1}, the speed of the car is

a. 36.67 ms^{-1} b. 17.4 ms^{-1}
c. 23.32 ms^{-1} d. 19.97 ms^{-1}

2 A locomotive whistling with a frequency 256 Hz, is moving towards a stationary observer with a velocity $\frac{1}{20}$ th of that of sound. What will be the frequencies of the notes heard by the observer before and after the engine passes him?

a. 269.5 Hz, 243.8 Hz b. 243.8 Hz, 269.5 Hz
c. 267.5 Hz, 245.8 Hz d. 245.8 Hz, 267.5 Hz

3 A source of sonic oscillations with frequency f_0 = 1700 Hz and a receiver R are located on the same normal to a wall. Both the source and receiver are stationary and the wall recedes from the source with a velocity 6cms^{-1}. The speed of sound is equal to 340 ms^{-1}. What is the apparent frequency (of the reflected sound from the wall) heard by the receiver ?

a. 1698.2 Hz b. 1699.4 Hz
c. 1699 Hz d. 1698.6 Hz

4 A vibrating tuning fork tied to the end of a string 1.988 m long is whirled round a circle. If it makes the revolutions in a second, calculate the ratio of the frequencies of the highest and the lowest notes heard by a stationary observer situated in the plane of rotation of tuning fork at a large distance from the tuning fork. Speed of sound is 350ms^{-1}. (In both, the observer is assumed to be along the line of velocity sound).

a. 1.732 b. 1.154
c. 1.278 d. 1.000

5 A source of sonic oscillations with frequency f_0 = 1700Hz and a receiver are located at the same point. At the moment t = 0, the source starts receding from the receiver with constant acceleration a = 10 ms^{-2}. Assuming the speed of sound to be equal to v = 340 ms^{-1}, find the frequency registered by the stationary receiver at t = 10s after the start of the motion.

a. 1650 Hz b. 1414 Hz
c. 1314 Hz d. 1350 Hz

6 A locomotive approaching a crossing at a speed of 80 mile h^{-1}, sounds a whistle of frequency 400Hz, when 1 mile from the crossing. There is no wind and the speed of sound in air is 0.2 mile s^{-1}. What frequency is heard by an observer 0.6 mile from the crossing on the straight road, which crosses the rail road at right angles ?

a. 450 Hz b. 442 Hz
c. 420 Hz d. 432 Hz

Working Space

7 A source emits a frequency of 261.6 Hz. An observer is moving away from the source with a velocity 20 ms^{-1} and the source is moving away from the observer with a velocity 10ms^{-1}. The wind is blowing in a direction from the observer to the source with a velocity 5ms^{-1}. The apparent frequency heard by the observer is (Take, velocity of sound = 343 ms^{-1})

a. 244 Hz
b. 242 Hz
c. 239 Hz
d. 237 Hz

8 Naval vessels use sonar (an acronym for sound navigation and ranging) to detect submarines. A sonar has a sound source and a sensitive microphone, both placed below the water line. Short bursts of sound are emitted from the source and any reflected sound is detected by a microphone. The time interval between emission and detection can be used to locate position of the object, whereas Doppler effect can be used to determine the speed of the reflector.

Suppose a warship is moving due West at 9.5 ms^{-1} w.r.t. the floor of the ocean. An ocean current moves due to East at 1.3 ms^{-1}. A submarine is headed directly towards the warship at 14ms^{-1}. What is the value of the apparent frequency heard by the microphone ? Source frequency is 8000Hz and speed of sound in sea water is 1470 ms^{-1}.

a. 8131 Hz
b. 8262 Hz
c. 8293 Hz
d. 8111 Hz

9 A fighter plane is moving in a vertical loop of radius R with a constant speed. The centre of the loop is at a height h directly overhead of an observer standing on the ground. The observer receives maximum frequency of the sound produced by the plane, when it is nearest to him. Find the speed of fighter plane. Velocity of sound in air is v. (Given, h = 200 units, R = 100 units and v = 340 units) (all units are consistent)

a. 205 units
b. 180 units
c. 195 units
d. 215 units

10 Two listeners are standing on the same line (which is perpendicular to a straight railway line) at different distances from the railway line. When a train passes them, each hears how the frequency of the train whistle changes. The change occurring along curve 1 is for observer 1 and that occurring along curve 2 is for observer 2. Which of the two observers is standing closer to the railway line?

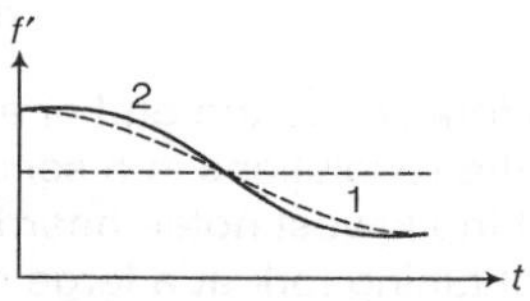

a. 1
b. 2
c. Both 1 and 2 at the same distance from railway track
d. Data is insufficient be predict any thing

11 A detector is released from rest over a source of sound of frequency f_0 = 1000 Hz. The frequency observed by the detector at time t is plotted in the graph. The speed of sound in air is (Take, g = 10ms^{-2})

a. 350 ms^{-1}
b. 330 ms^{-1}
c. 310 ms^{-1}
d. 300 ms^{-1}

Working Space

One or More than One Options Correct Type

Direction *(Q. Nos. 12-14) This section contains 3 multiple choice questions. Each question has four choices (a), (b), (c) and (d), out of which ONE or MORE THAN ONE are correct.*

12 A car travelling at 10 ms^{-1} sounds its horn which has a frequency of 500 Hz and a second car is travelling behind the first car in the same direction, with a velocity 20 ms^{-1}. Sound directly coming from the first car appears to have an apparent frequency f_1 when heard by the driver of second car. Sound is also heard in the second car by reflection from a wall ahead. This has an apparent frequency f_2. Speed of sound in air is 340 ms^{-1}. Then,

a. $f_1 = 515.2$ Hz
b. $f_2 = 512.2$ Hz
c. $f_1 = 514.3$ Hz
d. $f_2 = 515.5$ Hz

13 A source of sound with natural frequency $f_0 = 1800$ Hz moves uniformly along a straight line separated from a stationary observer by a distance $L_0 = 250$ m. The velocity of source is equal to 0.8 v, where v is the speed of sound. The frequency of sound received by the observer at the moment, when the source gets closest to him is f'. The distance between the source and the observer at the moment, when the observer receives a frequency $f_{ap} = f_0$ is d. Then

a. $f' = 4000$ Hz
b. $f' = 5000$ Hz
c. $d = 250$ m
d. $d = 320$ m

14 A driver in a stationary car blows a horn which produces monochromatic sound waves of frequency 1000 Hz normally towards a reflecting wall. The wall approaches the car with a speed 3.3 ms^{-1}. Speed of sound is 340 ms^{-1}. Then, choose the correct option(s).

a. The frequency of sound reflected from wall and heard by the driver is 1020 Hz
b. The frequency of sound reflected from wall and heard by the driver is 980 Hz
c. The percentage increase in frequency (as heard by the driver) of sound after reflection from wall is 2%
d. The percentage decrease in frequency (as heard by the driver) of sound after reflection from wall is 2%

Matching List Type

Direction *(Q. No. 15) Choices for the correct combination of elements from Column I and Column II are given as options (a), (b), (c) and (d), out of which one is correct.*

15 The horn of a car gives a sound of frequency 800 Hz. The listener is on a bicycle. Match the Column I with Column II and mark the correct option from the codes given below. Speed of sound is 343 ms^{-1} and there is no wind.

	Column I		Column II
i.	The frequency that the listener hears when he moves towards the stationary car with a velocity 15 ms^{-1}.	p.	873 Hz
ii.	The frequency that the listener hears when the car moves towards the stationary listener, with a velocity 15 ms^{-1}.	q.	733 Hz
iii.	The frequency that the listener hears, when the car and the listener move towards each other, each with a velocity 15 ms^{-1}.	r.	835 Hz
iv.	The frequency that the listener hears when the car and the listener recede from each other, each with a velocity 15 ms^{-1}.	s.	836.6 Hz

Codes

	i	ii	iii	iv		i	ii	iii	iv
a.	p	q	r	s	b.	r	s	p	q
c.	q	s	r	p	d.	s	q	r	p

Working Space

Comprehension Type

Direction *(Q. Nos. 16-22) This section contains 2 paragraphs, each describing theory, experiments, data, etc. Seven questions related to the paragraphs have been given. Each question has only one correct answer among the four given options (a), (b), (c) and (d).*

Passage I

A source(S), emitting a sound wave at a certain frequency moves with a constant velocity along the X-axis as shown in the figure (i). The source moves directly towards a stationary detector D_1 and directly away from another stationary detector D_2. The superimposed three plots shown in the figure (ii), indicate pressure fluctuations $p(x)$ of the sound wave as measured by detector D_1, detector D_2 and by someone (D_3) in the rest frame of source(S).

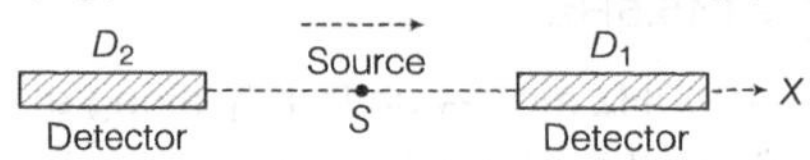

Fig (i)

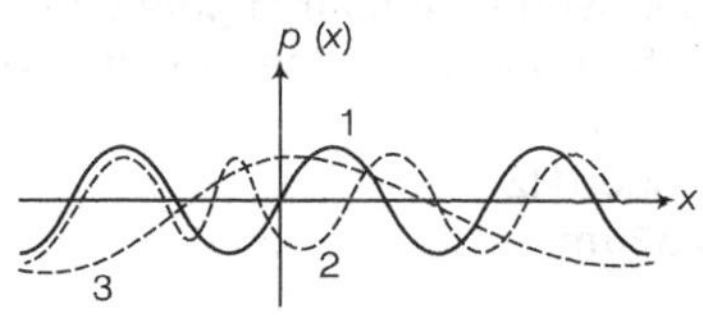

Fig (ii)

16 Which of the plots may correspond to the measurement done by detector D_1?

a. 1
b. 2
c. 3
d. These plots are not possible

17 Which of the plots may correspond to the measurement done by detector D_2?

a. 1
b. 2
c. 3
d. These plots are not possible

18 Which of the plots may correspond to the measurement done by detector D_3?

a. 1
b. 2
c. 3
d. These plots are not possible

19 Now, the source stops and begins to move along Y-axis with same speed, the plot which may be correspond to the measurement of D_2 now is

a. 1
b. 2
c. 3
d. None of these

Passage II

A man of mass 50 kg runs on a plank of mass 150 kg with a speed of 8 ms^{-1} relative to the plank as shown in the figure. Both were initially at rest. The velocity of man with respect to ground remains constant. Plank is placed on a smooth horizontal surface. The man while running whistles with frequency 500 Hz. A detector (D) fixed to the plank detects the frequency. The man jumps off with same velocity (w.r.t ground) from point D and slides on the smooth horizontal surface. The speed of sound in air is 330ms^{-1}. Answer the following questions. [Assume that the whistle and detector are on the same horizontal line]

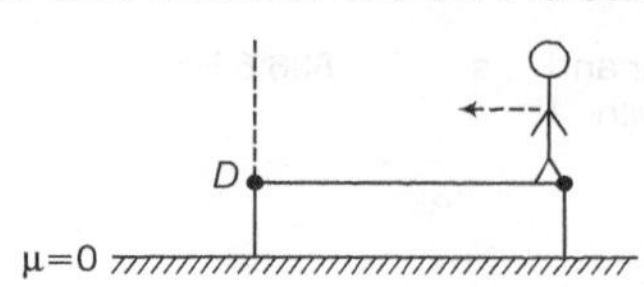

Working Space

20 The frequency detected by detector *D*, before the man jumps off the plank is

a. 516.35 Hz b. 512.35 Hz
c. 508.35 Hz d. 504.35 Hz

21 The frequency detected by detector *D*, after the man jumps off the plank is

a. 484.1 Hz b. 486.1 Hz
c. 488.1 Hz d. 489.1 Hz

22 Choose the best plot between the frequency (f') detected by detector *D versus* position of the man relative to detector.

a. 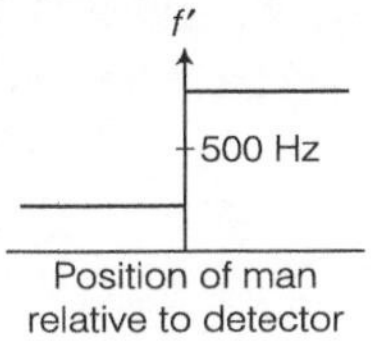

b. 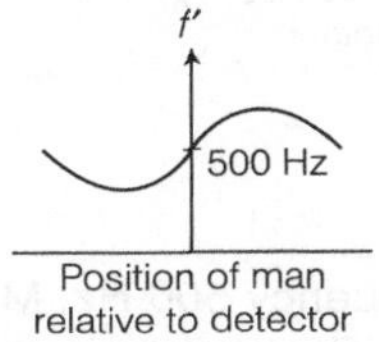

c. 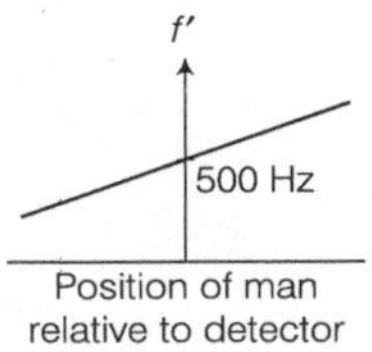

d. 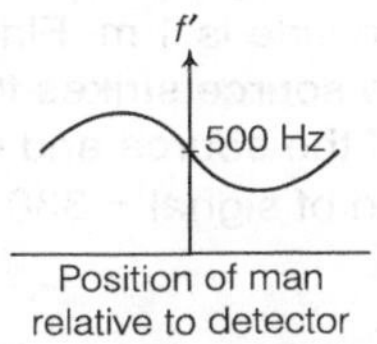

Subjective Questions

Direction (Q. Nos. 23-26) *These questions are subjective in nature, need to be solved completely on notebook.*

23 A boat is travelling in a river with a speed 10 ms^{-1} along the stream flowing with a speed 2ms^{-1}. From the boat, a sound transmitter is lowered into the river through a rigid support. The wavelength of the sound emitted from the transmitter inside the water is 14.45 mm. Assume that attenuation of sound in water and air is negligible. The frequency detected by a receiver kept inside the river downstream is f_1'.

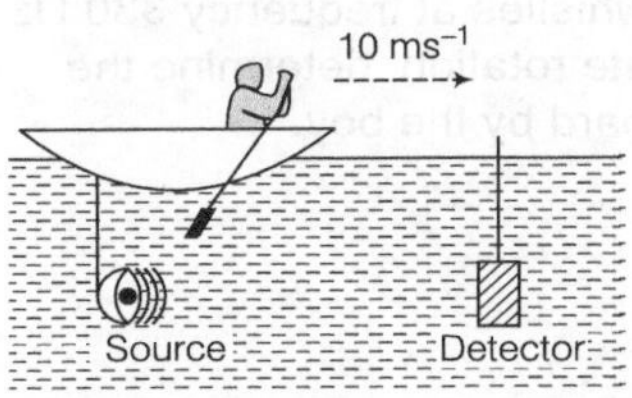

The transmitter and the receiver are now pulled up into the air. The air is blowing with a speed 5ms^{-1} in the direction opposite the river stream. The frequency detected by the receiver (i.e. detector) now is f_2'.

(Given,

Temperature of air and water = 20° C, Density of river water = 1000 kgm^{-3},

Bulk modulus of water = 2.088×10^9 Pa, Gas constant = 8.31 Jmol^{-1} K^{-1},

Mean molar mass of air = 28.8×10^{-3} kg - mol^{-1}, $\dfrac{C_p}{C_V}$ for air = 1.4)

Find the values of f_1' and f_2'.

24 The figure shows the location of a source and detector at time $t = 0$. The source and detector are moving with velocities $\overline{v}_S = 5\hat{i}\ ms^{-1}$ and $\overline{v}_{O'} = 10\hat{j}\ ms^{-1}$ respectively. Frequency of the source is 100 Hz. Determine the frequency of signals received by the detector (O') at the moment, when the source crosses origin. Speed of sound is $330\ ms^{-1}$.

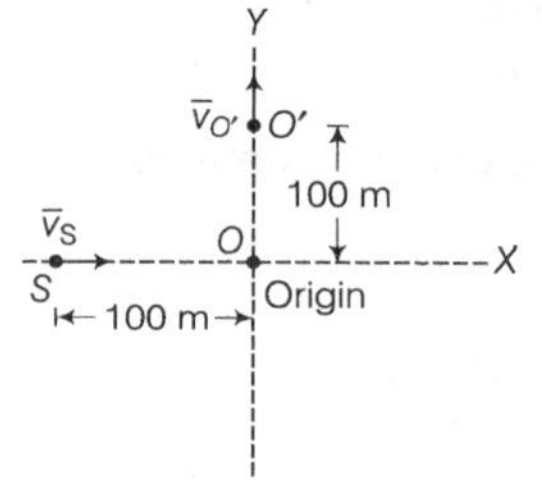

25 In the figure shown, the source has frequency 500 Hz. Mass of the source and detector are 2 kg and 8 kg respectively. The system is released from rest at time $t = 0$, when initial length of string on each side is 1 m. Find the frequency registered by the detector just before the source strikes the pulley. Assume the pulley to be very small, so that motion of the source and detector is along the same line. (Take, $g = 10\ ms^{-2}$ and speed of signal $= 330\ ms^{-1}$)

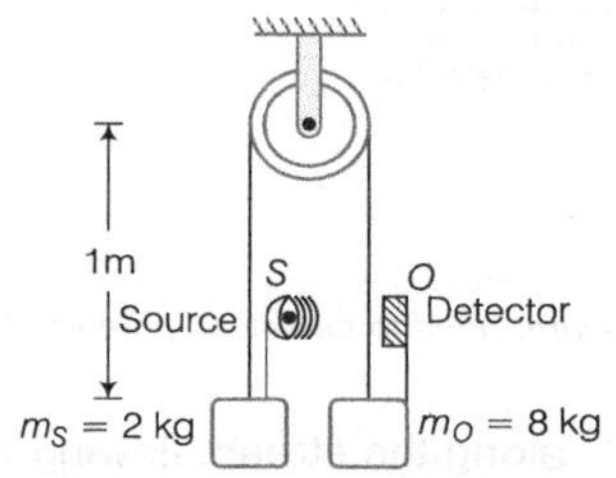

26 A boy is moving along a circular track in anticlockwise sense. A toy train moves along a square path with centres of circular track and square both coinciding, as shown in the figure. The train as well as the boy start from points P_2 and P_1 such that points O, P_1 and P_2 always lie on the same radial line. The speed of listener (i.e. the boy) is $11\ ms^{-1}$. The train continuously whistles at frequency 330 Hz. Speed of sound is $330\ ms^{-1}$. During one complete rotation, determine the maximum frequency and minimum frequency heard by the boy.

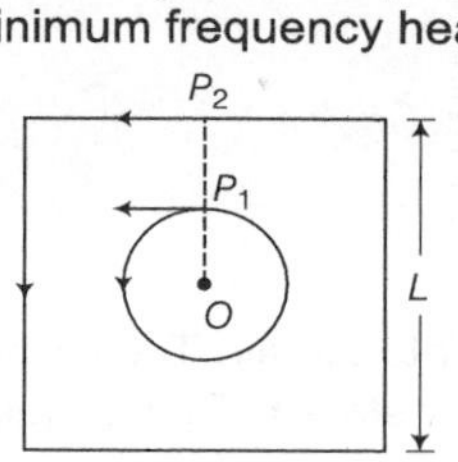

Working Space

DPP-5 Doppler Effect-II

Only One Option Correct Type

Working Space

Direction *(Q. Nos. 1-8) This section contains 8 multiple choice questions. Each question has four choices (a), (b), (c) and (d), out of which ONLY ONE is correct.*

1 A source of sound S starts falling from a height of 180 m from horizontal surface under gravity and emits 5500 Hz sound continuously. The distance between the source and observer at an instant, when the frequency heard by the observer is also 5500 Hz, is (speed of sound in air is 330 ms^{-1} and collision of source with surface is perfectly inelastic)

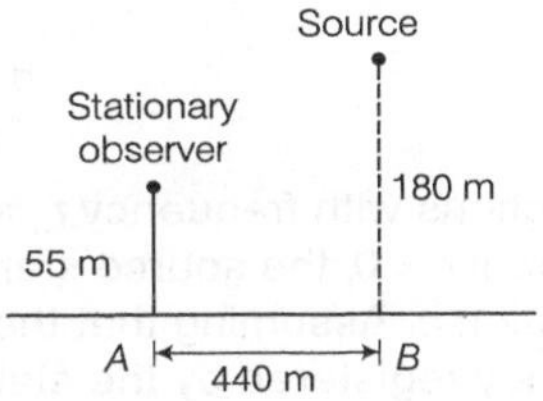

a. $55\sqrt{80}$ m
b. $65\sqrt{55}$ m
c. $55\sqrt{65}$ m
d. $55\sqrt{75}$ m

2 A source of sound S and observers are placed as shown in the figure. S is always at rest and O_1 and O_2 start moving with speed v_0 each at $t = 0$ as shown. At $t = 5$s, f_1' and f_2' represent apparent frequencies of sound received by O_1 and O_2 respectively. Then $\frac{f_1'}{f_2'}$ is (SO_1 and v_0 are not along same line)

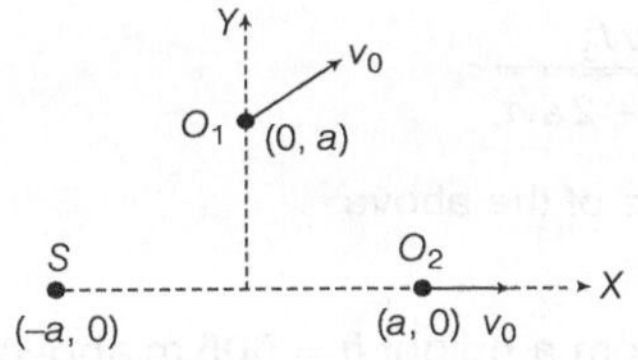

a. zero
b. $0 < \frac{f_1'}{f_2'} < 1$
c. 1
d. > 1

3 A source of sound is moving with velocity $\frac{u}{2}$ and two observers R_1 and R_2 are moving with velocity u as shown. If the velocity of sound is 10 u, then the ratio of wavelengths received by R_1 and R_2 is

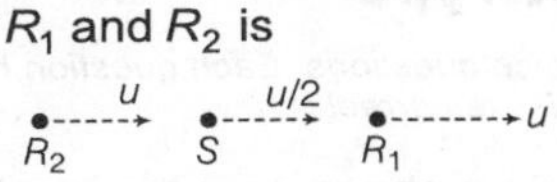

a. $\frac{19}{21}$
b. $\frac{21}{23}$
c. $\frac{17}{23}$
d. $\frac{17}{21}$

4 Two cars C_1 and C_2 depart simultaneously from the same position and in the same direction on a straight road. C_1 starts with initial velocity 2 ms^{-1} and acceleration 2 ms^{-2}, while C_2 starts with constant velocity 2 ms^{-1}. The driver of car C_1 hears a sound of frequency 352 Hz emitted by car C_2 after 10s, after the start. Find the actual frequency of the sound emitted by C_2. Velocity of sound is 330 ms^{-1}.

a. 370.12 Hz
b. 372.23 Hz
c. 377.45 Hz
d. 374.86 Hz

Working Space

5 A train approaching a hill at a speed of 40 km h^{-1} sounds a whistle of frequency 580 Hz, when it is at a distance of 1 km from a hill. A wind with a speed of 40 km h^{-1} is blowing in the direction of motion of the train. Velocity of sound in air is $v = 1200$ km h^{-1}. Find the frequency of the whistle as heard by an observer on the hill.

a. 604 Hz b. 599 Hz c. 592 Hz d. 595 Hz

6 A source of sonic oscillations with frequency f_0 is placed at a large distance from a detector. The source starts moving towards the detector with a constant acceleration a. Find the frequency corresponding to the wave emitted first after the source starts. The speed of sound in the medium is v.

a. $\frac{2vf_0^2}{2vf_0 - a}$ b. $\frac{vf_0^2}{2vf_0 - a}$

c. $\frac{2vf_0^2}{vf_0 - a}$ d. None of these

7 A source of oscillations with frequency f_0 and a detector are located at the same point. At the moment $t = 0$, the source starts receding from the receiver with constant acceleration a. Assuming that the velocity of sound is v, find the oscillation frequency registered by the stationary observer after the start of motion.

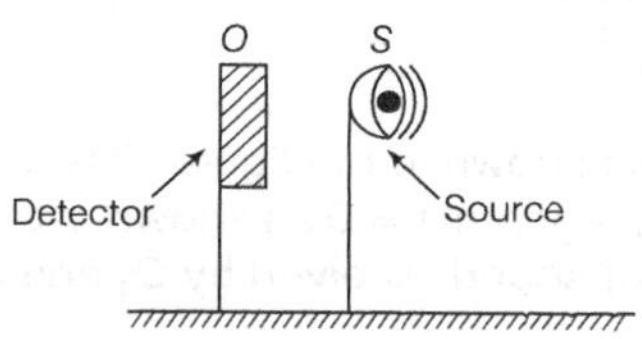

At $t = 0$, both are at the same point

a. $\frac{2vf_0}{\sqrt{v^2 + 2avt}}$ b. $\frac{vf_0}{\sqrt{v^2 + 2avt}}$

c. $\frac{2vf_0}{\sqrt{v^2 + avt}}$ d. None of the above

8 A whistle of frequency $f_0 = 1300$ Hz is dropped from a height $h = 505$ m above the ground. At the same time, a detector is projected upwards with velocity $= 50\ ms^{-1}$ along the same line. If the velocity of sound is $v \approx 300\ ms^{-1}$, then find the frequency detected by the detector after $t = 5$ s. (Take, $g = 10\ ms^{-2}$)

a. 1450 Hz b. 1600 Hz c. 1400 Hz d. 1500 Hz

One or More than One Options Correct Type

Direction (Q. Nos. 9 and 10) *This section contains 2 multiple choice questions. Each question has four choices (a), (b), (c) and (d), out of which ONE or MORE THAN ONE are correct.*

9 A source S of sound wave of fixed frequency f_0 and an observer O are located in air initially at the space points P_1 and P_2, a fixed distance apart. State in which of the following cases, the observer will not see any Doppler effect and will receive same frequency f_0 as produced by the source.

a. Both S and O remain stationary but a wind blows with a constant speed in arbitrary direction
b. The observer O remains stationary but the source S moves parallel to and in the same direction and with the same velocity as the wind
c. The source S remains stationary but the observer and the wind have the same speed away from the source
d. The source and the observer move directly against the wind but both with the same speed

Working Space

10 A stationary observer receives a sound of frequency $f_0 = 2000$ Hz. The source is moving on a straight line on which observer is not present. The apparent frequency f' varies with time as shown in the figure. Speed of sound is $300\ ms^{-1}$. Choose the correct alternative (s).

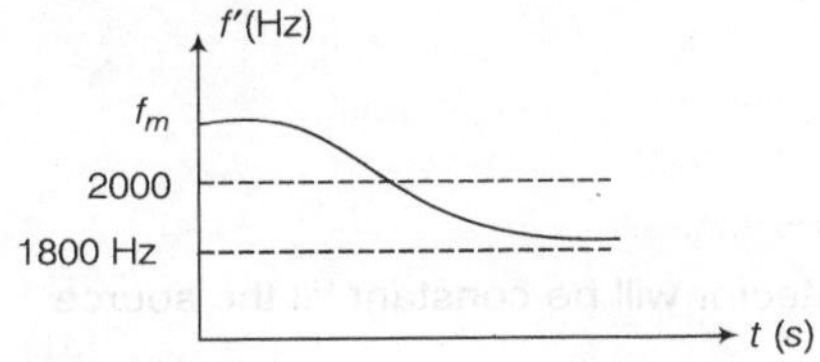

a. Speed of source is $66.7\ ms^{-1}$

b. f_m shown in the figure cannot be greater than 2500 Hz

c. Speed of source is $33.33\ ms^{-1}$

d. f_m shown in the figure cannot be greater than 2250 Hz

Comprehension Type

Direction (Q. Nos. 11-16) *This section contains 2 paragraphs, each describing theory, experiments, data, etc. Six questions related to the paragraphs have been given. Each question has only one correct answer among the four given options (a), (b), (c) and (d).*

Passage I

A source (S) emitting sound of frequency $f_0 = 960$ Hz is projected with velocity $10\sqrt{2}\ ms^{-1}$ at an angle $\theta = 45°$ with the horizontal. Simultaneously, a detector (O) is projected vertically upwards with a speed $10\ ms^{-1}$. There is no wind and speed of sound in air is $310\ ms^{-1}$. After the source and detector fall to the ground, they come to rest. (Take, $g = 10\ ms^{-2}$)

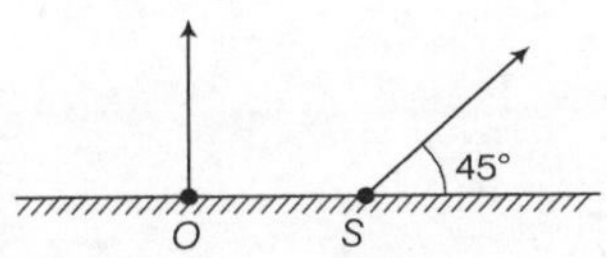

11 The maximum apparent frequency received by the detector is equal to

a. 992 Hz
b. 952 Hz
c. 682 Hz
d. 930 Hz

12 The graph of apparent frequency (f') *versus* time (t) is best represented by (for time $0 \le t \le 2$s)

a.

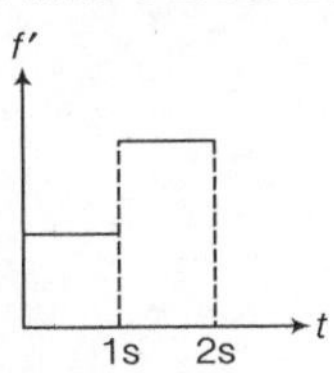

b.

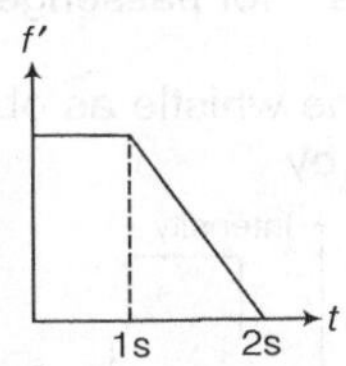

c.

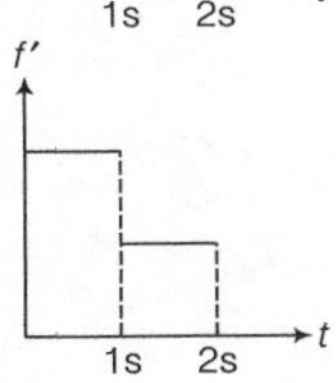

d.

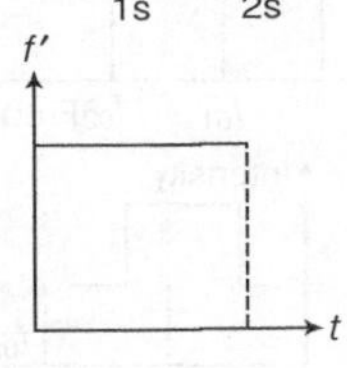

Working Space

13 Suppose the source and detector are projected horizontally in opposite direction from the top of a tower as shown in the figure. Then, choose the correct options.

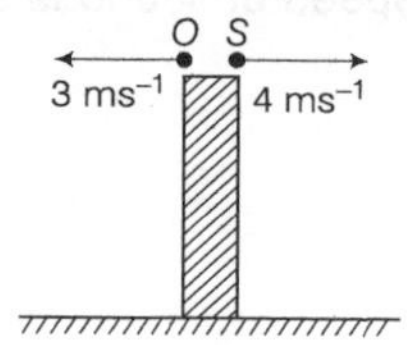

a. The apparent frequency received by the detector will be constant till the source and the detector fall on to the ground
b. The apparent frequency received by detector will continuously decrease till the sound and detector fall on to the ground
c. The apparent frequency must be same as the frequency emitted by the source at least once during the complete journey
d. The apparent frequency received by the detector will continuously increase till the source and detector fall on to the ground

Passage II

Two trains T_1 and T_2 are moving with velocities 20 ms^{-1} and 30 ms^{-1} respectively in the same direction on the same straight track. T_2 is a head of T_1. The engines are at the front ends. The engine of train T_1 blows a long whistle. Assume that the sound of the whistle is composed of components varying in frequency from f_{01} = 800 Hz and f_{02} = 1120 Hz, as shown in the figure. The spread in the frequency (highest frequency – lowest frequency) is thus 320 Hz. The speed of sound in still air is 340 ms^{-1} and there is no wind.

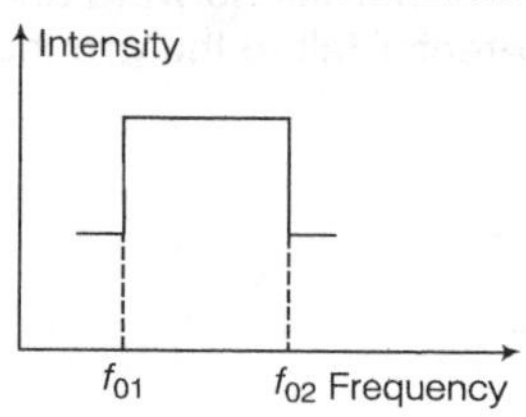

14 The speed of sound of the whistle is

a. 340 ms^{-1} for passengers in T_1 and 310 ms^{-1} for passengers in T_2
b. 320 ms^{-1} for passengers in T_1 and 310 ms^{-1} for passengers in T_2
c. 360 ms^{-1} for passengers in T_1 and 310 ms^{-1} for passengers in T_2
d. 310 ms^{-1} for passengers in T_1 and 360 ms^{-1} for passengers in T_2

15 The distribution of the sound intensity of the whistle as observed by the passengers in train T_1 is best represented by

a.
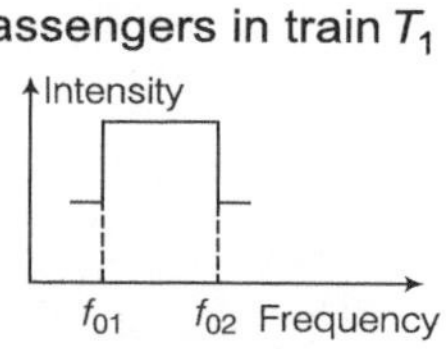

b.
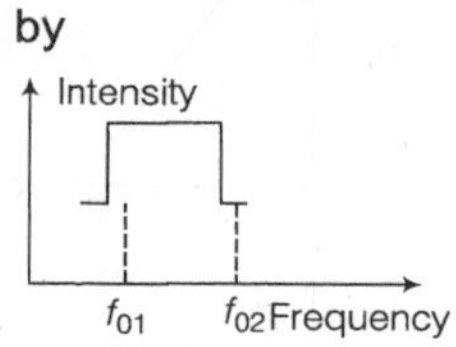

c.
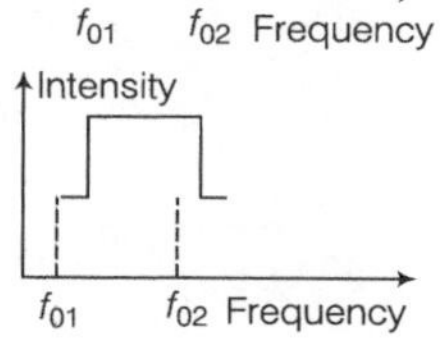

d.
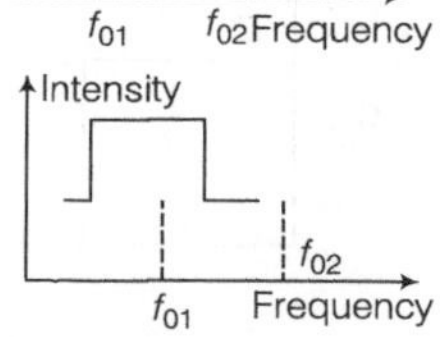

16 The spreed of frequency as observed by the passengers in train T_2 is

a. 330 Hz
b. 310 Hz
c. 350 Hz
d. 290 Hz

Working Space

Subjective Questions

Direction *(Q. Nos. 17-20) These questions are subjective in nature, need to be solved completely on notebook.*

17 Bats emit sounds from their nostrils, then listen the frequency of the sound reflected from their prey to determine the prey's speed. A bat flying at a speed v_{bat} emits sound of frequency f_{bat}. The sound it hears is reflected from an insect flying towards it and has a higher frequency f_{refl}. Show that the speed of the insect is $v_{insect} = \left[\frac{f_{refl}(v - v_{bat}) - f_{bat}(v + v_{bat})}{f_{refl}(v - v_{bat}) + f_{bat}(v + v_{bat})}\right] v$, where v is the speed of sound.

18 A source of oscillations S is fixed to the river bed of a river with stream velocity v. Two receivers R_1 and R_2 also are fixed to the river bed. If the source generates a frequency f_0, what frequencies are received by receivers R_1 and R_2?

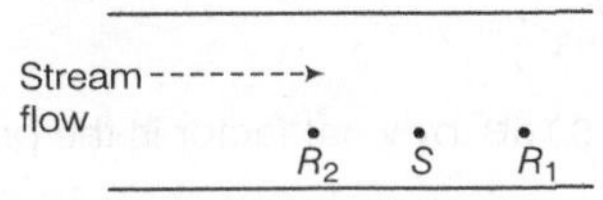

19 A source of sound is moving along a circular orbit of radius 3 m with an angular velocity of 10 rad s^{-1}. A sound detector located far away from the source is executing linear simple harmonic motion PQ with amplitude $PM = MQ = 6$ m. The frequency of oscillation of the detector is $\frac{5}{\pi}$ per second. The source is at point N, when the detector is at point P. If the source emits a continuous sound wave of frequency 340 Hz, find the maximum frequency and the minimum frequency recorded by the detector. Speed of sound is 330 ms^{-1}.

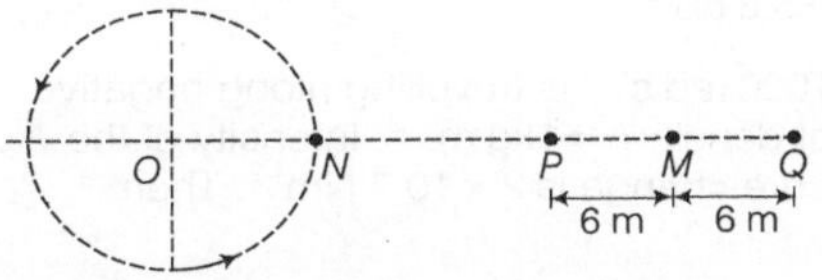

Assume that the time taken by the sound to reach the detector from the source is integral multiple of time period of rotation.

20 By observing an alternating Doppler shift of a spectral line, the astronomers can deduce the existence of a binary star system, which otherwise cannot be resolved visually. Suppose that an astronomical observation shows that the source of light is eclipsed once every 18 h. The wavelength of the spectral line changes for a maximum of 563 nm to a minimum of 539 nm. Assume that, the binary system consists of two identical stars. Determine the separation between the objects and the mass of each object.

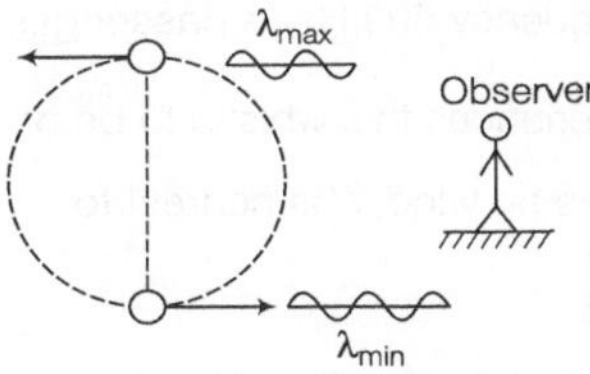

SET - 1

Revisal Problems
for JEE Main

Only One Option Correct Type

Working Space

Direction (*Q. Nos. 1-21*) *This section contains 21 multiple choice questions. Each question has four choices (a), (b), (c) and (d), out of which ONLY ONE option is correct.*

1 Seven grams of nitrogen is mixed with 12 g of oxygen in a tube and then sealed. The velocity of sound through the tube at 27° C is

a. $338.6\,ms^{-1}$ b. $328.6\,ms^{-1}$
c. $334.6\,ms^{-1}$ d. $330.6\,ms^{-1}$

2 If the sound level in a room is increased from 50 dB to 60 dB, by what factor is the pressure amplitude increased ?

a. 50 b. $\sqrt{10}$
c. 100 d. 150

3 A fire work charge is detonated some metres above the ground. At a distance of 400 m from the explosion, the acoustic pressure reaches a maximum of $10\,Nm^{-2}$. Assume that the speed of sound is constant at $343\,ms^{-1}$ throughout the atmosphere over the region considered. Also, assume the ground absorbs all the sound falling on it and the air absorbs sound energy at the rate of $7\,dB\,km^{-1}$. The sound level (in dB) at 4 km from the explosion is ($\rho_{air} = 1.2\,kg\,m^{-3}$)

a. 45.6 dB b. 55.6 dB
c. 75.6 dB d. 65.6 dB

4 A plane longitudinal wave of angular frequency $1000\,rad\,s^{-1}$ is travelling along negative x-direction in a homogenous gaseous medium of density $\rho = 1\,kg\,m^{-3}$. Intensity of the wave is $I = 10^{-10}\,Wm^{-2}$ and the maximum pressure change is $2 \times 10^{-4}\,Nm^{-2}$. Then displacement equation is

a. $y = 10^{-9} \sin\left(1000t - 5x + \frac{\pi}{2}\right)$
b. $y = 10^{-9} \cos(1000t + 5x)$
c. $y = 10^{-9} \tan(1000t - 5x)$
d. $y = 10^{-9} \cos(1000t - 5x)$

5 When a person wears a hearing aid, the second intensity level increases by 30 dB. The sound intensity increases by

a. e^3 b. 30
c. 1000 d. 100

6 An engine running at speed $\frac{v}{10}$ sounds a whistle of frequency 600 Hz. A passenger in a train coming from the opposite side at a speed $\frac{v}{15}$ experiences this whistle to be of frequency f'. If v is the speed of sound in air and there is no wind, f' is nearest to

a. 711 Hz b. 630 Hz
c. 580 Hz d. 610 Hz

7 A source of sound produces waves of wavelength 60 cm, when it is stationary. If the speed of sound in air is $320\,ms^{-1}$ and source moves with a speed $20\,ms^{-1}$, the wavelength of sound in the forward direction will be nearest to

a. 56 cm b. 60 cm
c. 64 cm d. 65 cm

Working Space

8 Your clock radio awakens you with a steady and irritating sound of frequency 600 Hz. On a morning, it malfunctions and cannot be turned off. In frustration, you drop the clock radio out of your fourth storey door window 15 m above the ground. As, you listen to the falling clock radio, what frequency do you hear, just before you hear it striking the ground ? (Speed of sound = 343 ms^{-1})

a. 591 Hz b. 581 Hz c. 571 Hz d. 578 Hz

9 A source S emitting sound of 300 Hz is fixed on block A which is attached to the free end of spring 1 as shown. Th detector D fixed on block B attached to the free end of spring 2 detects this sound. The block A and B are simultaneously displaced towards each other through a distance of 1 m and then left to vibrate. If the vibrational frequency of each block is 2 Hz and speed of sound is 340 ms^{-1}, the maximum and minimum frequencies of sound detected by D, respectively are

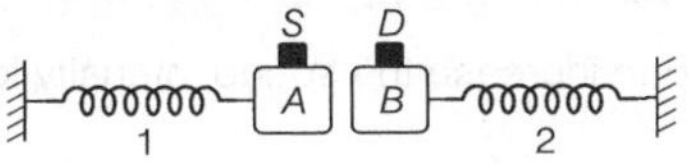

a. 320 Hz, 275.6 Hz b. 323 Hz, 278.6 Hz
c. 326 Hz, 281.6 Hz d. 317 Hz, 272.6 Hz

10 The apparent frequency of a whistle of an engine changes in the ratio 6 : 5 as the engine passes a stationary observer. If the velocity of sound is 330 ms^{-1}, then the velocity of the engine is

a. 66 ms^{-1} b. 15 ms^{-1}
c. 30 ms^{-1} d. 0.33 ms^{-1}

11 A train is moving in an elliptical orbit in anticlockwise sense with a speed of 110 ms^{-1}. Guard is also moving in the given direction with same speed as that of train. The ratio of the lengths of major axis to minor axis is 4/3. Driver blows a whistle of 1900 Hz at P, which is received by guard at S. The frequency received by guard is (velocity of sound = 330 ms^{-1})

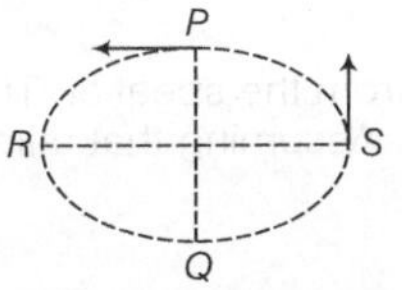

a. 1800 Hz b. 1900 Hz
c. 2000 Hz d. 1500 Hz

12 In a car race, sound signals emitted by two cars are detected by the detector on the straight track at the end point of the race. Frequencies observed are 330 Hz and 360 Hz and the original frequency is 300 Hz for both the cars. Assume that both cars move with constant velocity and velocity of sound is 330 ms^{-1}. Race ends with the separation of 100 m between the cars. The time taken by winning car is

a. 10 s b. 8 s
c. 4 s d. 12 s

13 The intensity of sound wave gets reduced by 20% on passing through a slab. The reduction in intensity on passing through two such consecutive slabs is

a. 40% b. 36%
c. 30% d. 50%

14 In a sports meet the timing of a 200 m straight dash is recorded at the finish point by starting an accurate stopwatch on hearing the sound of starting gun fired at the starting point. The time recorded will be more accurate

a. in winter b. in summer
c. in all seasons d. None of these

15 Two radio stations broadcast their programmes at the same amplitude A_0 and intensity I_0. Their frequency difference is $f_1 - f_2$ (= 1000 Hz). A detector receives from the two stations simultaneously. It can detect signals of intensity $\geq 2\, I_0$. Find the time for which the detector remains idle in each cycle of the intensity of the signal.

a. 333.33 μs b. 250 μs
c. 750 μs d. 500 μs

Working Space

16 A point source of sound emits a constant power with intensity inversely proportional to the square of the distance from the source. By how many decibles does the sound intensity level drop, when you move from point P_1 to P_2 ? Distance of P_2 from the source is two times the distance of source from P_1.

a. 6 dB
b. 4 dB
c. 8 dB
d. 10 dB

17 A car approaching a crossing at a speed of 20 ms^{-1}, sounds a horn of frequency 500 Hz, when 80 m from the crossing. Speed of sound in air is 330 ms^{-1}. What frequency is heard by an observer 60 m from the crossing on the straight road, which crosses car road at right angles ?

a. 530 Hz
b. 535.5 Hz
c. 525.5 Hz
d. 520 Hz

18 By what factor must be sound intensity be increased to increase the sound intensity level by 13 dB ? (Given, $(10)^{0.3} = 2$)

a. 12
b. 13
c. 20
d. 22

19 If it were possible to generate a sinusoidal 300 Hz sound wave in air that has a displacement amplitude of 0.2 mm. What would be the sound level of the wave? (Assume $v = 330\ ms^{-1}$ and $\rho_{air} = 1.29\ kgm^{-1}$). (Given $\log_{10} 3 = 0.447$)

a. 145.7 dB
b. 140.7 dB
c. 124.7 dB
d. 134.7 dB

20 A swimming duck paddles the water with its feet once every 1.6 s, producing surface waves with this period. The duck is moving at constant speed in a pond, where the speed of surface waves is 0.32 ms^{-1} and the crests of the waves ahead of the duck are spaced 0.12 m apart. The speed of the duck is

a. 0.245 ms^{-1}
b. 0.300 ms^{-1}
c. 0.255 ms^{-1}
d. 0.275 ms^{-1}

21 A siren creates a sound level of 60 dB at a location 500 m from the speaker. The siren is powered by a battery that delivers a total energy of 1000 J. Assuming that efficiency of siren is 30%, the total time the siren can sound is

a. 105 s
b. 95 s
c. 90 s
d. 110 s

Statement Type

Direction *(Q. Nos. 21 and 22) This section is based on Statement I and Statement II. Select the correct answer from the codes given below.*

a. Both Statement I and Statement II are correct and Statement II is the correct explanation of Statement I
b. Both Statement I and Statement II are correct but Statement II is not the correct explanation of Statement I
c. Statement I is correct but Statement II is incorrect
d. Statement II is correct but Statement I is incorrect

22 **Statement I** An observer is approaching a stationary point source of sound. The apparent frequency heard by him is more than the real frequency because of decrease of wavelength of the sound wave.

Statement II If there is no relative motion between source and observer, apparent frequency is equal to actual frequency.

23 **Statement I** In a longitudinal wave, pressure is maximum where displacement is zero.

Statement II There is a phase difference of $\frac{\pi}{2}$ between displacement wave and pressure wave in case of longitudinal wave.

Revisal Problems
for JEE Advanced

Only One Option Correct Type

Working Space

Direction (Q. Nos. 1-9) *This section contains 9 multiple choice questions. Each question has four choices (a), (b), (c) and (d), out of which ONLY ONE option is correct.*

1 A source of sound is travelling with a velocity of 30 ms^{-1} towards a stationary observer. If actual frequency of source is 1000 Hz and the wind is blowing with a velocity 20 ms^{-1} in a direction at 60° with the direction of motion of source, then the apparent frequency heard by observer is (speed of sound is $v = 340\ ms^{-1}$)

a. 1094 Hz
b. 1110 Hz
c. 1085 Hz
d. 1082 Hz

2 Spherical sound waves are emitted uniformly in all directions from a point source. The variation in sound level (β) as a function of distance r from the source can be written as (k_1 and k_2 are constants)

a. $\beta = -k_2 \log r^k$
b. $\beta = k_1 - k_2 (\log r)^2$
c. $\beta = k_1 - \frac{k_2}{r^2}$
d. $\beta = k_1 - k_2 \log r$

3 The difference between the apparent frequencies of a source of sound as perceived by a stationary observer during its approach and recession is 2% of the actual frequency of the source. If the speed of sound is 300 ms^{-1}, speed of source is

a. 3 ms^{-1}
b. 4 ms^{-1}
c. 6 ms^{-1}
d. 1.5 ms^{-1}

4 The driver of a car approaching a vertical wall notices that the frequency of the horn of his car changes from 400 Hz to 450 Hz on receiving after reflection from the wall. Speed of sound is 340 ms^{-1}. The speed of approach of the car towards the wall is

a. 10 ms^{-1}
b. 20 ms^{-1}
c. 40 ms^{-1}
d. 30 ms^{-1}

5 When source and observer are stationary and the wind is blowing at a speed v_w (from source to observer), the apparent wavelength λ' is related to the actual wavelength λ_0 by (v is speed of sound in air)

a. $\lambda' = \lambda_0$
b. $\lambda' = \frac{v_w}{v} \lambda_0$
c. $\lambda' = \frac{v_w + v}{v} \lambda_0$
d. $\lambda' = \frac{v}{v - v_w} \lambda_0$

6 In the figure shown, the displacement Y *versus* distance X along the direction of propagation of a longitudinal wave. The pressure is maximum at position marked.

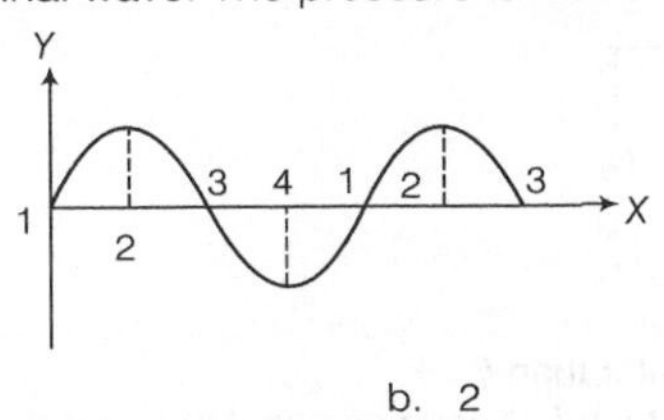

a. 1
b. 2
c. 3
d. 4

Working Space

7 Consider a source of sound S and an observer O. The source emits a sound wave of frequency f_0. The frequency observed by O is found to be

i. f_1, if O approaches S and S is stationary
ii. f_2, if S approaches O and O is stationary
iii. f_3, if both S and O approach each other with the same speed
In all the cases, relative speed of S w.r.t. O is the same. Choose the incorrect option.

a. $f_1 \neq f_2 \neq f_3$
b. $f_1 < f_2$
c. $f_3 < f_0$
d. $f_1 < f_3 < f_2$

8 A source S and a detector O are initially at a distance of 1 km from each other. Both start moving towards each other with same acceleration, $a_0 = 10\ ms^{-2}$. Frequency of source is $f_0 = 2000$ Hz and speed of sound in air is $300\ ms^{-1}$. The frequency observed by the detector at time $t = 4$ s, is

a. 2350 Hz
b. 2354 Hz
c. 2324 Hz
d. 2341 Hz

9 The frequency of a radar is 780 MHz. When it is reflected from an approaching aeroplane, the apparent frequency is more than the actual frequency by 2.6 kHz. The speed of the aeroplane is

a. $250\ ms^{-1}$
b. $500\ ms^{-1}$
c. $750\ ms^{-1}$
d. $1000\ ms^{-1}$

One or More than One Options Correct Type

Direction (Q. Nos. 10-15) *This section contains 6 multiple choice questions. Each question has four choices (a), (b), (c) and (d), out of which ONE or MORE THAN ONE are correct.*

10 The figure represents a longitudinal wave (displacement) travelling in + x-direction. Then,

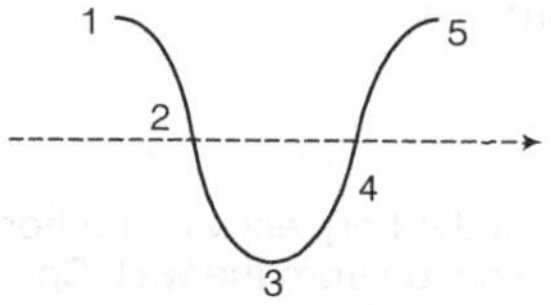

a. part 123 represents compression
b. part 123 represents rarefaction
c. part 345 represents compression
d. part 345 represents rarefaction

11 Which of the following statements is/are incorrect?

a. Wave pulses on strings are transverse waves
b. Sound waves in air are transverse waves of compressions and rarefactions
c. The speed of sound in air at 20 °C is twice that at 5°C
d. A 60 dB sound has twice the intensity of a 30 dB sound

12 A source of sound and detector are moving as shown in the figure, at $t = 0$. The speed of sound waves is v. For this situation choose the correct option(s).

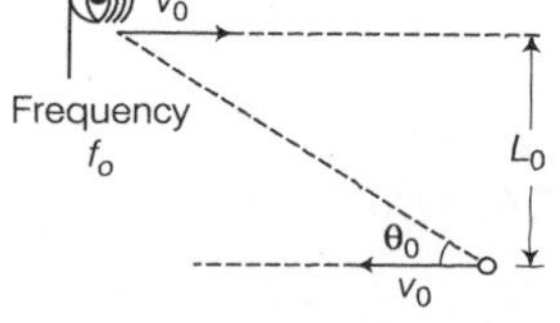

a. The frequency received by the detector is always greater than f_0
b. Initially, frequency received by the detector is greater than f_0, becomes equal to f_0 and then decreases with time
c. Frequency received by the detector is equal to f_0 at $t_0 = \dfrac{L_0 \cot \theta_0}{2v_0}$
d. None of the above

Working Space

13 A sonic source, located in a uniform medium, emits waves of frequency f_0. If intensity, energy density (energy per unit volume) and maximum speed of oscillations of medium particles are, respectively I, u and v_{max} at a point, then which of the following graphs are correct ?

a. 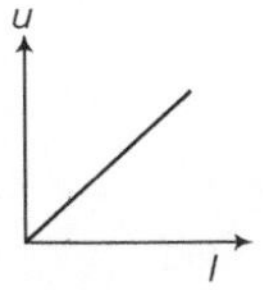

b. 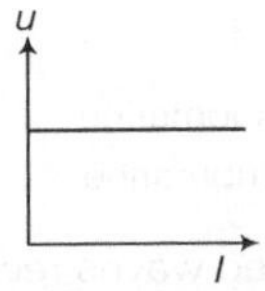

c. 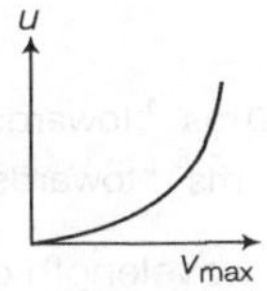

d. 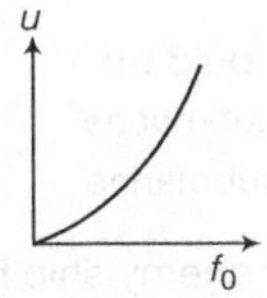

14 Among the following, choose correct statement(s).

a. When a sound wave strikes a wall, the compression pulse is reflected as compression pulse
b. When a sound wave strikes a wall, the compression pulse is reflected as a rarefaction pulse
c. When a sound wave is coming out after passing through a narrow pipe, then refection would be at the open end
d. When a sound wave is coming out after passing through a narrow pipe, then compression pulse is reflected as a rarefaction pulse

15 Which of the following statement(s) is/are correct ?

a. Change in air temperature have no effect on the speed of sound
b. Change in air pressure have no effect on the speed of sound
c. The speed of sound in water is higher than in air
d. The speed of sound in water is lower than in air

Statement Type

Direction (Q. Nos. 16 - 18) ***This section is based on Statement I and Statement II. Select the correct answer from the codes given below.***

a. Both Statement I and Statement II are correct and Statement II is the correct explanation of Statement I
b. Both Statement I and Statement II are correct but Statement II is not the correct explanation of Statement I
c. Statement I is correct but Statement II is incorrect
d. Statement II is correct but Statement I is incorrect

16 **Statement I** Sound travels faster in solids than is gases.

Statement II Solid possess greater density than gases.

17 **Statement I** The speed of sound in solids is maximum though their density is large.

Statement II This is because their coefficient of elasticity is large.

18 **Statement I** The bells are made of metals and not of wood.

Statement II Wood offers high damping on the sound waves.

Comprehension Type

Direction (Q. Nos. 19-25) ***This section contains 2 paragraphs, each describing theory, experiments, data, etc. Seven questions related to the paragraphs have been given. Each question has only one correct answer among the four given options (a), (b), (c) and (d).***

Passage I

An Indian submarine is moving in the Arabian sea with a constant velocity. To detect enemy it sends out sonar waves which travel with velocity 1050 ms^{-1} in water. Intially, the waves are getting reflected from a fixed island and the reflected waves are coming back to submarine. The frequency of reflected waves are detected by the submarine and found to be 10% greater than that of sent waves. Now, an enemy ship comes in front, due to which the frequency of reflected waves detected by submarine becomes 21% greater than that of sent waves.

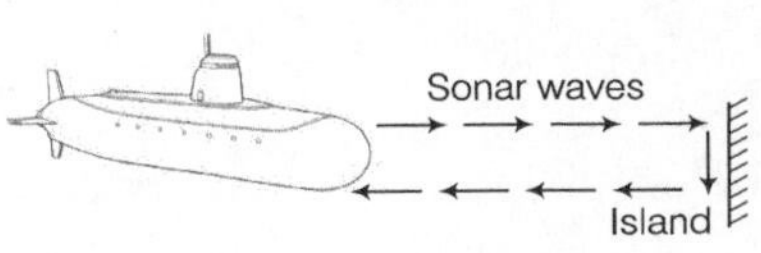

Working Space

19 The speed of Indian submarine is

a. 20 ms^{-1}
b. 100 ms^{-1}
c. 50 ms^{-1}
d. 10 ms^{-1}

20 The velocity of enemy ship should be

a. 100 ms^{-1} away from Indian submarine
b. 100 ms^{-1} towards Indian submarine
c. 50 ms^{-1} away from Indian submarine
d. 50 ms^{-1} towards Indian submarine

21 If the wavelength received by enemy ship is λ' and wavelength of reflected waves received by submarine is λ'', then $\frac{\lambda'}{\lambda''}$ is equal to

a. 2
b. 1.2
c. 1.1
d. 1

22 Bulk modulus of sea water approximately is ($\rho_{water} = 1000 \text{ kg m}^{-3}$)

a. 10^{11} Nm^{-2}
b. 10^{10} Nm^{-2}
c. 10^{5} Nm^{-2}
d. 10^{9} Nm^{-2}

Passage II

A conveyor belt moves to the right with speed $v = 300 \text{ m min}^{-1}$. A very fast pieman puts pies on the belt at a rate of 20 per minute and they are received at the other end by a pieeater.

23 If the pieman is stationary, find the spacing x between the pies and the frequency with which they are received by the stationary pieeater.

a. $15 \text{ m}, 20 \text{ min}^{-1}$
b. $30 \text{ m}, 40 \text{ min}^{-1}$
c. $30 \text{ m}, 10 \text{ min}^{-1}$
d. $15 \text{ m}, 40 \text{ min}^{-1}$

24 The pieman now walks with speed 30 m min^{-1} towards the receiver while continuing to put pies on the belt at 20 per minute. What is the spacing between the pies now ?

a. 15 m
b. 13.5 m
c. 14.5 m
d. 12.5 m

25 In the previous problem, what is the frequency with which the pies are received by the stationary pieeater ?

a. 20 min^{-1}
b. 25.3 min^{-1}
c. 22.2 min^{-1}
d. 27.4 min^{-1}

One Integer Value Correct Type

Direction *(Q. No. 26) This section contains 1 question. When worked out will result in an integer from 0 to 9 (both inclusive).*

26 The average power transmitted across a cross-section by two sound waves moving in the same direction are equal. The wavelengths of two sound waves are in the ratio 1 : 2, then find the ratio of their pressure amplitudes.

JEE Main & AIEEE Archive

(Compilation of Last 13 Years' Questions)

1 A train is moving on a straight track with speed $20 ms^{-1}$. It is blowing its whistle at the frequency of 1000 Hz. The percentage change in the frequency heard by a person standing near the track as the train passes him is close to (speed of sound = $320 ms^{-1}$) **(JEE Main 2015)**

a. 6% b. 12%
c. 18% d. 24%

2 A motor cycle starts from rest and accelerates along a straight path at $2 ms^{-2}$. At the starting point of the motor cycle, there is a stationary electric siren. How far has the motor cycle gone when the driver hears the frequency of the siren at 94% of its value when the motor cycle was at rest? (Speed of sound = $330 ms^{-1}$) **(AIEEE 2009)**

a. 49 m b. 98 m
c. 147 m d. 196 m

3 Three sound waves of equal amplitudes have frequencies $(\nu - 1), \nu, (\nu + 1)$. They superpose to give beat. The number of beats produced per second will be **(AIEEE 2009)**

a. 4 b. 3
c. 2 d. 1

4 The speed of sound in oxygen (O_2) at a certain temperature is $460 ms^{-1}$. The speed of sound in helium (He) at the same temperature will be (assume both gases to be ideal) **(AIEEE 2008)**

a. $460 ms^{-1}$ b. $500 ms^{-1}$
c. $650 ms^{-1}$ d. $1420 ms^{-1}$

5 While measuring the speed of sound by performing a resonance column experiment, a student gets the first resonance condition at a column length of 18 cm during winter. Repeating the same experiment during summer, she measures the column length to be x cm for the second resonance. Then, **(AIEEE 2008)**

a. $18 > x$ b. $x > 54$
c. $54 > x > 36$ d. $36 > x > 18$

6 A sound absorber attenuates the sound level by 20 DB. The intensity decreases by a factor of **(AIEEE 2007)**

a. 1000 b. 10000
c. 10 d. 100

7 A whistle producing sound waves of frequencies 9500 Hz and above is approaching a stationary person with speed $v\ ms^{-1}$. The velocity of sound in air is $300 ms^{-1}$. If the person can hear frequencies upto a maximum of 10000 Hz, the maximum value of v upto which he can hear the whistle is **(AIEEE 2006)**

a. $15\sqrt{2}\ ms^{-1}$ b. $15/\sqrt{2}\ ms^{-1}$
c. $15\ ms^{-1}$ d. $30\ ms^{-1}$

8 When two tuning forks (fork 1 and fork 2) are sounded simultaneously, 4 beats per second are heard. Now, some tape is attached on the prong of the fork 2. When the tuning forks are sounded again, 6 beats per second are heard. If the frequency of fork 1 is 200 Hz, then what will be the original frequency of fork 2? **(AIEEE 2005)**

a. 200 Hz b. 202 Hz
c. 196 Hz d. 204 Hz

9 An observer moves towards a stationary source of sound, with a velocity one-fifth of the velocity of sound. What is the percentage increase in the apparent frequency? **(AIEEE 2005)**

a. Zero
b. 0.5%
c. 5%
d. 20%

10 A tuning fork of known frequency 256 Hz makes 5 beats per second with the vibrating string of a piano. The beat frequency decreases to 2 beats per second when the tension in the piano string is slightly increased. The frequency of the piano string before increasing the tension was **(AIEEE 2003)**

a. (256 + 2) Hz
b. (256 – 2) Hz
c. (256 – 5) Hz
d. (256 + 5) Hz

11 When temperature increases, the frequency of a tuning fork **(AIEEE 2002)**

a. increases
b. decreases
c. remains same
d. increases or decreases depending on the material

12 A tuning fork arrangement (pair) produces 4 beats per second with one fork of frequency 288 cps. A little wax is placed on the unknown fork and then it produces 2 beats per second. The frequency of the unknown fork is **(AIEEE 2002)**

a. 286 cps
b. 292 cps
c. 294 cps
d. 288 cps

JEE Advanced & IIT JEE Archive

(Compilation of Last 13 Years' Questions)

1 Two vehicles, each moving with speed u on the same horizontal straight road, are approaching each other. Wind blows along the road with velocity w. One of these vehicles blows a whistle of frequency f_1. An observer in the other vehicle hears the frequency of the whistle to be f_2. The speed of sound in still air is v. The correct statement(s) is (are) **(2013 Adv., One or More than One Options Correct Type)**

a. If the wind blows from the observer to the source, $f_2 > f_1$.

b. If the wind blows from the source to the observer, $f_2 > f_1$

c. If the wind blows from the observer to the source, $f_2 < f_1$

d. If the wind blows from the source to the observer, $f_2 < f_1$

2 A person blows into open-end of a long pipe. As a result, a high-pressure pulse of air travels down the pipe. When this pulse reaches the other end of the pipe **(2012, One or More than One Options Correct Type)**

a. a high-pressure pulse starts travelling up the pipe, if the other end of the pipe is open

b. a low-pressure pulse starts travelling up the pipe, if the other end of the pipe is open

c. a low-pressure pulse starts travelling up the pipe, if the other end of the pipe is closed

d. a high-pressure pulse starts travelling up the pipe, if the other end of the pipe is closed

3 A student is performing the experiment of resonance column. The diameter of the column tube is 4 cm. The frequency of the tuning fork is 512 Hz. The air temperature is 38° C in which the speed of sound is 336 ms^{-1}. The zero of the meter scale coincides with the top end of the resonance column tube. When the first resonance occurs, the reading of the water level in the column is **(2012, Only One Option Correct Type)**

a. 14.0 cm b. 15.2 cm c. 16.4 cm d. 17.6 cm

4 A police car with a siren of frequency 8 kHz is moving with uniform velocity 36 kmh^{-1} towards a tall building which reflects the sound waves. The speed of sound in air is 320 ms^{-1}. The frequency of the siren heard by the car driver is **(2011, Only One Option Correct Type)**

a. 8.50 kHz b. 8.25 kHz

c. 7.75 kHz d. 7.50 kHz

5 A student performed the experiment to measure the speed of sound in air using resonance air-column method. Two resonances in the air-column were obtained by lowering the water level. The resonance with the shorter air-column is the first resonance and that with the longer air column is the second resonance. Then, **(2009, One or More than One Options Correct Type)**

a. the intensity of the sound heard at the first resonance was more than that at the second resonance

b. the prongs of the tuning fork were kept in a horizontal plane above the resonance tube

c. the amplitude of vibration of the ends of the prongs is typically around 1 cm

d. the length of the air-column at the first resonance was somewhat shorter than 1/4th of the wavelength of the sound in air

Passage for Q. Nos. (6-8)

Two trains A and B are moving with speeds 20 ms^{-1} and 30 ms^{-1} respectively in the same direction on the same straight track, with B ahead of A. The engines are at the front ends. The engine of train A blows a long whistle.

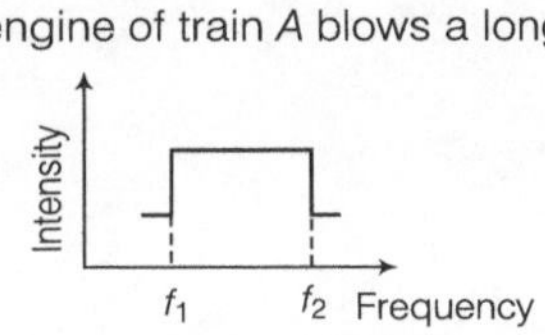

Assume that the sound of the whistle is composed of components varying in frequency from $f_1 = 800$ Hz to $f_2 = 1120$ Hz, as shown in the figure. The spread in the frequency (highest frequency–lowest frequency) is thus 320 Hz. The speed of sound in air is 340 ms^{-1}. **(2007, Comprehension Type)**

6 The speed of sound of the whistle is

a. 340 ms^{-1} for passengers in A and 310 ms^{-1} for passengers in B

b. 360 ms^{-1} for passengers in A and 310 ms^{-1} for passengers in B

c. 310 ms^{-1} for passengers in A and 360 ms^{-1} for passengers in B

d. 340 ms^{-1} for passengers in both the trains

7 The distribution of the sound intensity of the whistle as observed by the passengers in train *A* is best represented by

a. 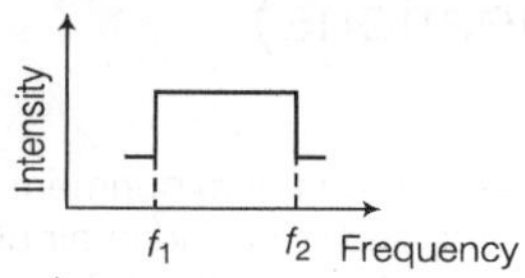

b. 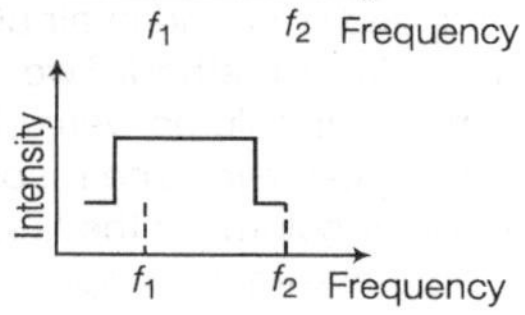

c. 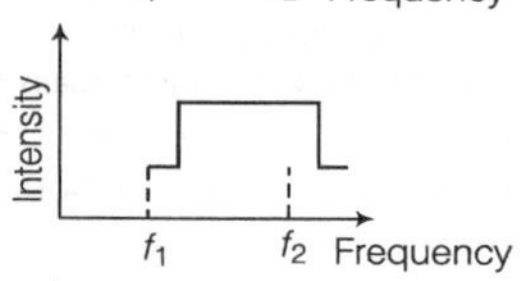

d. 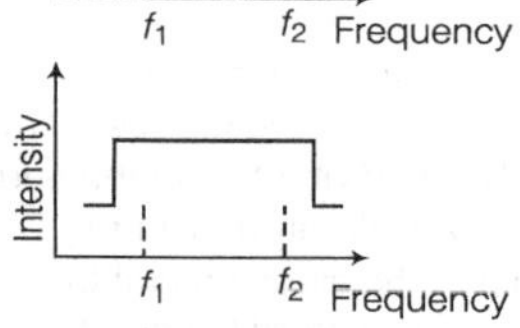

8 The spread of frequency as observed by the passengers in train *B* is

a. 310 Hz
b. 330 Hz
c. 350 Hz
d. 290 Hz

Passage for Q. Nos. (9-11)

Two plane harmonic sound waves are expressed by the equations.

$$y_1(x,t) = A\cos(0.5\pi x - 100\pi t)$$
$$y_2(x,t) = A\cos(0.46\pi x - 92\pi t)$$

(All parameters are in MKS)

(2006, Comprehension Type)

9 How many times does an observer hear maximum intensity in one second?

a. 4 b. 10
c. 6 d. 8

10 What is the speed of the sound?

a. 200 ms^{-1} b. 180 ms^{-1}
c. 192 ms^{-1} d. 96 ms^{-1}

11 At $x = 0$ how many times the amplitude of $y_1 + y_2$ is zero in one second?

a. 192 b. 48
c. 100 d. 96

12 A tuning fork of 512 Hz is used to produce resonance in a resonance tube experiment. The level of water at first resonance is 30.7 cm and at second resonance is 63.2 cm. The error in calculating velocity of sound is

(2005, Only One Option Correct Type)

a. 204.1 $cm s^{-1}$ b. 110 $cm s^{-1}$
c. 58 $cm s^{-1}$ d. 280 $cm s^{-1}$

4. Superposition and Standing Waves

DPP
DAILY PRACTICE PROBLEMS

DPP-1 Superposition and Interference

Only One Option Correct Type

Working Space

Direction *(Q. Nos. 1-14) This section contains 14 multiple choice questions. Each question has four choices (a), (b), (c) and (d), out of which ONLY ONE is correct.*

1 Two waves of equal frequencies have their amplitudes in the ratio of 3 : 5. They are superimposed on each other. The ratio of minimum intensity to maximum intensity of the resultant wave is

a. $\frac{1}{16}$ b. $\frac{9}{25}$

c. $\frac{3}{5}$ d. $\frac{1}{4}$

2 Sound waves from a tuning fork T reach a point Q by two separate paths TPQ and TRQ. When TRQ is greater than TPQ by 11.5 cm, there is a silence at Q. When the difference is 23 cm, the sound becomes loudest at Q and when 34.5 cm there is silence at Q and so on. Speed of sound is 331.2 ms^{-1}. The minimum frequency of the fork is

a. 1550 Hz b. 1220 Hz

c. 1440 Hz d. 1330 Hz

3 A wave is represented by $y_1 = 10\cos(5x + 25t)$, where x is measured in metre and t in second. A second wave given by $y_2 = 20\cos\left(5x + 25t + \frac{\pi}{3}\right)$, interferes with the first wave. The resultant wave is

a. $y = 10\sqrt{5}\cos[5x + 25t + \tan^{-1}(\sqrt{2})]$ b. $y = 10\sqrt{7}\cos\left[5x + 25t + \tan^{-1}\left(\frac{\sqrt{3}}{2}\right)\right]$

c. $y = 10\sqrt{7}\cos[5x + 25t + \tan^{-1}(\sqrt{2})]$ d. $y = 10\sqrt{5}\cos\left[5x + 25t + \tan^{-1}\left(\frac{\sqrt{3}}{2}\right)\right]$

4 Find the resultant amplitude at a point at which n sinusoidal waves interfere. All the waves have same amplitude A and their phase increases in arithmetic progression of common difference ϕ. The phase difference between 2nd wave and Ist wave is ϕ. That between 3rd wave and 2nd wave is ϕ, that between 4th wave and 3rd wave is ϕ, etc.

a. $A\left[\frac{\sin\left(\frac{n\phi}{2}\right)\cos\left(\frac{n\phi}{2}\right)}{\sin\left(\frac{\phi}{2}\right)}\right]$ b. $A\left[\frac{\sin\left(\frac{n\phi}{2}\right)}{\sin\left(\frac{\phi}{2}\right)}\right]$

c. $A\left[\frac{\cos\left(\frac{n\phi}{2}\right)}{\sin\left(\frac{\phi}{2}\right)}\right]$ d. $A\left[\frac{\sin\left(\frac{n\phi}{2}\right)}{\cos\left(\frac{\phi}{2}\right)}\right]$

Working Space

5 Two speakers connected to the same source of fixed frequency are placed 2 m apart in a box. A sensitive microphone placed at a distance of 4 m from their mid-point along the perpendicular bisector shows maximum response. The box is slowly rotated till the speakers are in line with the microphone. The distance between the mid-point of the speakers and microphone remains unchanged. In doing this exactly 5 maximum responses are observed in the microphone. The wavelength of the sound wave is

a. 0.8 m b. 0.6 m c. 0.4 m d. 0.2 m

6 Three component sinusoidal waves progressing in the same direction along the same path have the same period, but their amplitudes are A, $\frac{A}{2}$ and $\frac{A}{3}$. The phases of vibration at position x on their paths at time $t = 0$ s are 0, $-\frac{\pi}{2}$ and $-\pi$ respectively. The amplitude and phase of resulting wave respectively, are

a. $\frac{6}{5}A, \tan^{-1}\left(\frac{3}{4}\right)$ b. $\frac{6}{5}A, -\tan^{-1}\left(\frac{3}{4}\right)$ c. $\frac{5}{6}A, \tan^{-1}\left(\frac{3}{4}\right)$ d. $\frac{5}{6}A, -\tan^{-1}\left(\frac{3}{4}\right)$

7 Two coherent narrow slits emitting sound of wavelength λ_0, in same phase, are placed parallel to each other at a small separation of $2\lambda_0$. The sound is detected by a moving detector on the screen at a distance $D(>> \lambda_0)$ from the slit S_1 as shown in the figure. Find the distance y such that the intensity at P is equal to intensity at O.

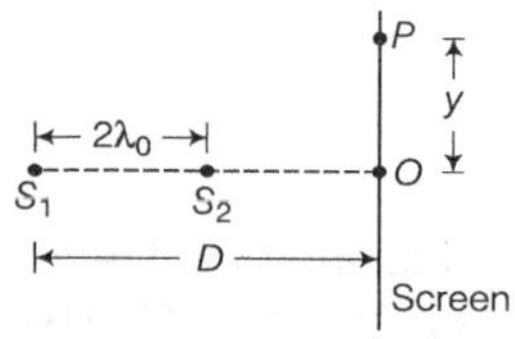

a. $2\sqrt{3}D$ b. $2D$ c. $\frac{\sqrt{3}}{2}D$ d. $\sqrt{3}D$

8 A source emitting sound of frequency 180 Hz is placed in front of an obstacle at a distance 2 m from it. A detector is also placed in front of the obstacle at the same distance from it. Find the minimum distance between the source and the detector for which the detector detects a maximum sound. (Take, speed of sound = 360 ms^{-1})

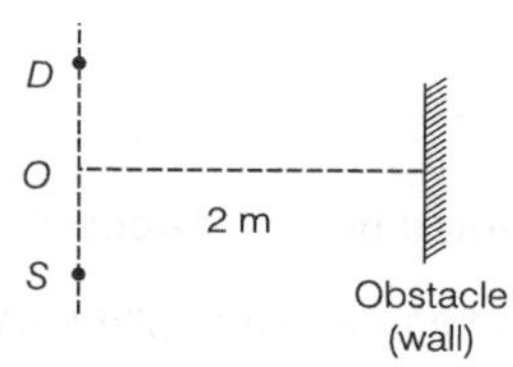

a. 6 m b. 7.5 m c. 10 m d. 9.5 m

9 Two identical speakers 10 m apart are driven by the same oscillator with a frequency of $f = 21.44$ Hz (see figure). If the receiver has to be moved in the plane of the speakers, what path should it take so that, the intensity remains at a minimum? That is determine the relationship between x and y (the coordinates of the receiver), that causes the receiver to record a minimum sound intensity. Speed of the sound is 343 ms^{-1}. (In the options, n = 1, 2, 3, ⋯)

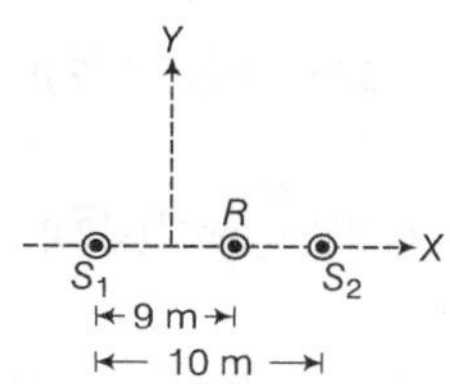

a. $\sqrt{(x+5)^2+y^2}+\sqrt{(x-5)^2+y^2}=(8n-1)$

b. $\sqrt{(x-5)^2-y^2}+\sqrt{(x+5)^2+y^2}=(8n-4)$

c. $\sqrt{(x+5)^2+y^2}-\sqrt{(x-5)^2+y^2}=(16n-8)$

d. None of the above

10 Two loudspeakers L_1 and L_2 driven by a common oscillator and amplifier, are arranged as shown in the figure. The frequency of the oscillator is gradually increased from zero and the detector D records a series of maxima and minima. If the speed of sound is 330 ms^{-1}, then the frequency at which the first maximum is observed, is

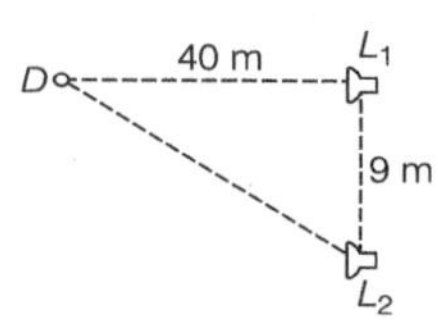

a. 165 Hz b. 660 Hz c. 495 Hz d. 330 Hz

Working Space

11 Two identical loudspeakers placed 3 m apart are driven by the same oscillator. A listeners is originally at point O, located 8 m from the centre of the line connecting the speakers as shown in the figure. The listener then moves to point P, which is at a perpendicular distance 0.35 m from point O and he experiences the first minimum sound intensity. Speed of sound is 343 ms^{-1}. What is the frequency of the oscillator?

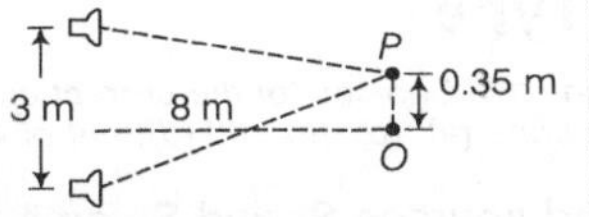

a. 1320 Hz
b. 1450 Hz
c. 1395 Hz
d. 1425 Hz

12 A source and a detector D of high frequency waves are a distance d apart. Maximum signal is received by D, when the reflecting layer is at a height H. When the layer rises by a distance h, no signal is detected at D. Absorption by atmosphere is negligible. The relation between d, h, H and wavelength (λ) of the waves is

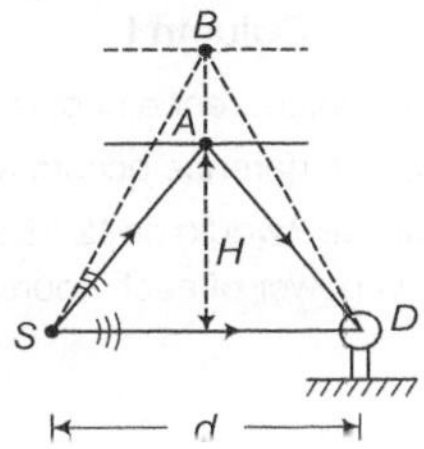

a. $\lambda = \sqrt{4(H+h)^2 + d^2} - \sqrt{4H^2 - d^2}$

b. $\lambda = 2\sqrt{4(H+h)^2 + d^2} - 2\sqrt{4H^2 + d^2}$

c. $\lambda = \frac{3}{2}\sqrt{4(H+h)^2 + d^2} - \frac{3}{2}\sqrt{4H^2 + d^2}$

d. None of the above

13 Figure shows a tube structure in which signal is sent from one end and is received at the other end. The frequency of the sound source can be varied electronically between 2000 Hz and 5000 Hz. Find the frequency at which maximum intensity is detected, if speed of sound in air is 343ms^{-1}.

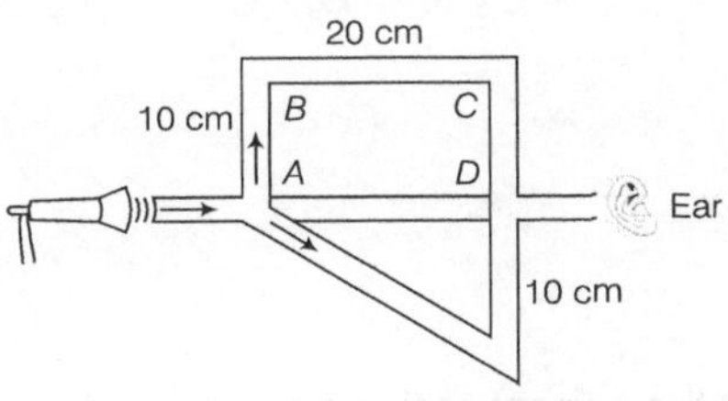

a. 4280 Hz
b. 4950 Hz
c. 4490 Hz
d. 4000 Hz

14 A short wave, radio receiver receives simultaneously two signals from a transmitter 500 km away, one by a path along the surface of the earth, and one by reflection from a portion of the ionospheric layer situated at a height of 200 km. The layer acts as a perfect (horizontal) reflector. When the frequency of the transmitted wave is 10 MHz, it is observed that the combined signal strength varies from maximum to minimum and back to maximum 8 times per minute. With what slow vertical speed is the ionospheric layer moving? Ignore atmospheric disturbances and assume the earth is flat.

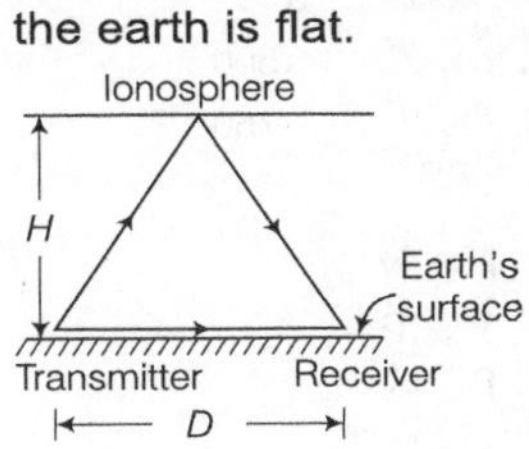

a. 4.2ms^{-1}
b. 2.0ms^{-1}
c. 1.2ms^{-1}
d. 3.2ms^{-1}

Matching List Type

Working Space

Direction *(Q. Nos. 15 and 16) Choices for the correct combination of elements from Column I and Column II are given as options (a), (b), (c) and (d), out of which one is correct.*

15 Two similar sound sources S_1 and S_2 emit pure sinusoidal waves in phase. Speed of sound is 350 ms^{-1}. Match for the condition in Column I with the frequency of the source (f) in Column II. Match the Column I with Column II and mark the correct option from the codes given below. (Given, $\cos^{-1}(0.7827) = 38.5°$)

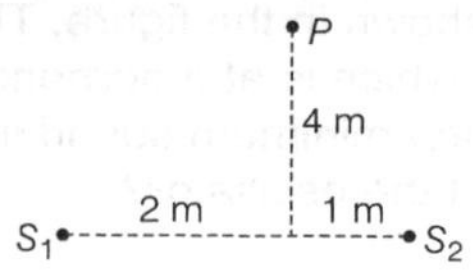

	Column I		Column II
i.	Constructive interference occurs at P	p.	$f = 1000$ Hz
ii.	Destructive interference occurs at P	q.	$f = 1500$ Hz
iii.	The intensity at P would be $12.98 \times 10^{-3} P_0$ (where P_0 is power of each source).	r.	$f = 2000$ Hz
		s.	$f \simeq 167$ Hz
		t.	$f \simeq 1267$ Hz

Codes

	i	ii	iii			i	ii	iii
a.	p,r	q	s		b.	q	r	s
c.	p	q	s		d.	r	p	q

16 Three identical radio sources each of intensity I_0, vibrating in phase, are positioned along Y-axis. The frequency of each source is 10^8 Hz and separation between two adjacent sources is $d = 1$ km. A detector moves parallel to Y-axis at a perpendicular distance $D (>> d)$ from the sources. The separation between two consecutive maxima registered is 10 km. Imagine loss of intensity during propagation is negligible.

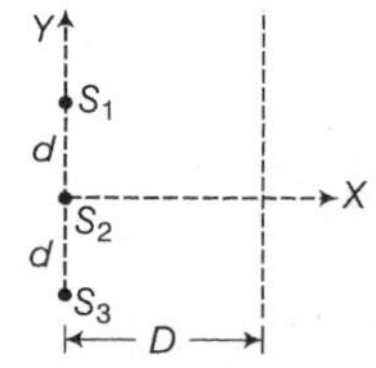

Match the Column I with Column II and mark the correct option from the codes given below.

	Column I		Column II (in SI units)
i.	The radiation wavelength is	p.	9
ii.	D is equal to	q.	3
iii.	The ratio of resultant intensities at maxima and minima is	r.	$\frac{10^7}{3}$
iv.	If the source S_2 stops radiating, the new separation between the remaining two sources, so that positions of maxima are unchanged, is	s.	$\frac{2}{3} \times 10^7$
		t.	500
		u.	750

Codes

	i	ii	iii	iv			i	ii	iii	iv
a.	q	r	p	t		b.	s	q	r	p
c.	p	q	r	s		d.	s	q	p	t

Working Space

Comprehension Type

Direction (Q. Nos. 17-22) *This section contains 2 paragraphs, each describing theory, experiments, data, etc. Six questions related to the paragraph have been given. Each question has only one correct answer among the four given options (a), (b), (c) and (d).*

Passage I

A narrow tube is bent in the form of circle of radius R as shown. Two small holes S and D are made in the tube at the positions right angles to each other. A source placed at S generates a wave of intensity I_0, which is equally divided into two parts. One part travels along the longer path, while the other travels along the shorter path. These two part waves meet at point D, where a detector is placed.

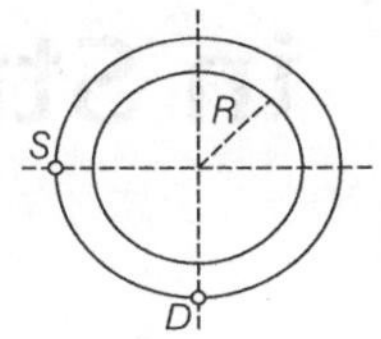

17 Maximum intensity produced at D is given by

a. $4I_0$ b. $3I_0$ c. $2I_0$ d. I_0

18 The maximum value of wavelength λ to produce a maximum at D is given by

a. $\frac{3}{2}\pi R$ b. $\frac{\pi R}{2}$ c. $2\pi R$ d. πR

19 The maximum value of wavelength λ to produce a minimum at D is given by

a. $\frac{3}{2}\pi R$ b. $\frac{\pi R}{2}$ c. $2\pi R$ d. πR

Passage II

A sinusoidal pulse is generated at $x = 0$ and $t = 0$ on a string lying along X-axis. The pulse is travelling with a speed $1\,\text{ms}^{-1}$. A rectangular pulse is generated at $t = 0$ and is travelling in opposite direction with speed $1\,\text{ms}^{-1}$ as shown in the figure.

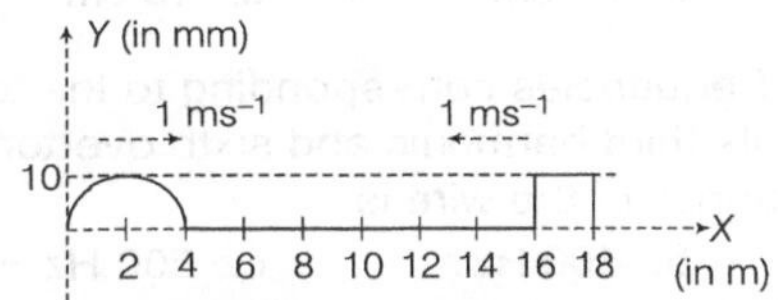

At t=0

20 The velocity of particle at $x = 4$ m and $t = 0$ s is

a. zero b. $\frac{\pi}{6}$ cms^{-1} c. $\frac{\pi}{2}$ cms^{-1} d. $\frac{\pi}{4}$ cms^{-1}

21 The displacement of the particle at $x = 9$ m and $t = 8$ s is

a. $\sqrt{2}$ cm b. $\frac{3}{2}$ cm c. $(\sqrt{2}+1)$ cm d. $\left(\frac{\sqrt{2}+1}{\sqrt{2}}\right)$ cm

22 The velocity of particle at $x = 9$ m and $t = 8$s is

a. $\frac{\pi}{2\sqrt{2}}$ cms^{-1} in $-y$-direction

b. $\frac{\pi}{4\sqrt{2}}$ cms^{-1} in $-y$-direction

c. $\frac{\pi}{4}$ cms^{-1} in $+y$-direction

d. $\frac{\pi}{4}$ cms^{-1} in $-y$-direction

One Integer Value Correct Type

Direction (Q. No. 23) *This section contains 1 question. When worked out will result in an integer from 0 to 9 (both inclusive).*

23 Two loudspeakers S_1 and S_2 vibrating in the same phase, each emits sounds of frequency 220 Hz uniformly in all directions. S_1 has an acoustic output of 1.2 mW and S_2 has 1.8 mW. Consider a point Q such that, $S_1Q = 0.75$ m and $S_2Q = 3$ m. Speed of sound in air is $330\,\text{ms}^{-1}$. The resultant power at Q is $\frac{k}{10}$ mW. Find the value of k.

DPP-2 Standing Waves in Strings

Only One Option Correct Type

Working Space

Direction (Q. Nos. 1-10) *This section contains 10 multiple choice questions. Each question has four choices (a), (b), (c) and (d), out of which ONLY ONE is correct.*

1 A uniformly stretched string is made to vibrate in its fundamental mode. Maintaining the same fundamental frequency, the length (L) of the string is adjusted according to the tension (F). The two are plotted on a log-log scale, adopting length on the Y-axis. The slope of this line will be

a. 1/4 b. 2 c. 1 d. 1/2

2 A stretched wire of length 1 m, vibrates in its 4th overtone. What is the wavelength of the waves producing stationary waves?

a. 20 cm b. 25 cm c. 15 cm d. 22.5 cm

3 The difference in frequencies corresponding to the tones emitted by stretched string vibrating in its third harmonic and sixth overtone is 1600 Hz. The fundamental frequency of the wire is

a. 300 Hz b. 400 Hz c. 500 Hz d. 350 Hz

4 A stretched string of length 50 cm vibrating in its third overtone emits notes of frequency 2000 Hz. The speed of transverse waves in the string is

a. 400 ms^{-1}
b. 450 ms^{-1}
c. 500 ms^{-1}
d. 550 ms^{-1}

5 A certain string fixed to one end on a rigid support, carries a load at its other end passing through a pulley. When a length L of the wire is made to vibrate, it is found to resonate with a certain tuning fork. Upon immersing the load completely in water and kerosene separately, the corresponding lengths are found to be L_w and L_k. The specific gravity of the load (solid) is

a. $\left[1+\left(\frac{L_w}{L}\right)^2\right]^{-1}$ b. $\frac{L_w}{L}+1$ c. $1+\frac{L}{L_w}$ d. $\left[1-\left(\frac{L_w}{L}\right)^2\right]^{-1}$

6 In the above problem, the specific gravity of kerosene is

a. $\frac{L^2+L_k^2}{L^2+L_w^2}$
b. $\frac{L^2-L_k^2}{L^2-L_w^2}$
c. $\frac{L^2+L_k^2}{L^2-L_w^2}$
d. $\frac{L^2-L_k^2}{L^2+L_w^2}$

7 The fundamental frequency of a stretched string increases by 15 Hz, when its tension is increased by 144%. Maintaining the initial tension in the wire, if the length is increased by 40%, what will be the change in the fundamental frequency?

a. 19.4 Hz
b. 20.4 Hz
c. 21.4 Hz
d. 22.4 Hz

Working Space

8 Two uniform wires A and B with lengths in the ratio 4 :1and linear mass densities in the ratio 1: 4 are joined end to end, and stretched between two fixed points. The ratio of the number of loops in A and B supporting standing waves is

a. 8 : 1 b. 1 : 1 c. 1 : 2 d. 2 : 1

9 A stretched metal wire sustaining standing waves has a fundamental frequency f. The wire is stretched uniformly to double its length and is again subjected to the same initial tension. The fundamental frequency will now be

a. $\frac{f}{\sqrt{2}}$ b. $\sqrt{2}f$ c. $f/2$ d. f

10 A steel wire is rigidly fixed at both ends. Its length, mass and cross-sectional area are 1 m, 0.1 kg and 10^{-6} m^2, respectively. Then, the temperature of the wire is lowered by 20° C. If transverse waves are set up by plucking the wire at 0.25 m from one end, find the frequency of oscillations (assuming that, the wire vibrates with minimum number of loops possible for such a case). (Given, coefficient of linear expansion for steel = 1.21×10^{-5} °C^{-1} and its Young's modulus = 2×10^{11} Pa)

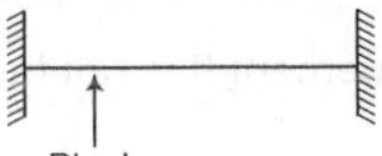

a. 44 Hz b. 33 Hz
c. 22 Hz d. 30 Hz

One or More than One Options Correct Type

Direction *(Q. Nos. 11-14) This section contains 4 multiple choice questions. Each question has four choices (a), (b), (c) and (d), out of which ONE or MORE THAN ONE are correct.*

11 A string 120 cm in length (fixed at both ends) sustains a standing wave with the points of the string at which the displacement amplitude is eqal to 3.5 mm being separated by 15 cm. Then, choose the correct options.

a. The maximum displacement amplitude is approximately 5 mm
b. The maximum displacement amplitude is approximately 5.5 mm
c. These oscillations correspond to fifth overtone
d. These oscillations correspond to third overtone

12 Two waves given by $y_1 = (10 \text{ cm}) \sin (100\pi t - \pi x)$ and $y_2 = (-10 \text{ cm}) \sin (100\pi t + \pi x)$, where t is in second and x is in metre are superposed. Now, choose the correct statement(s).

a. The equation of the standing wave obtained is $y = (20 \text{ cm}) \sin[\pi(x+1)] \cos (100\pi t)$
b. For positive values of x, position of nodes is $x = n - 1$, where $n = 1, 2, 3, 4, \cdots$
c. For positive values of x, position of antinodes is $x = \frac{2n-1}{2}$, where $n = 1, 2, 3, 4, \cdots$
d. The amplitude of particle at $x = \frac{1}{4}$ m is $10\sqrt{2}$ cm

13 The vibrations of a string of length 60 cm fixed at both ends are given by the equation $y = 5 \sin\left(\frac{\pi x}{20}\right) \cos (100\pi t)$, where x and y are in cm and t in second.

Then,

a. maximum displacement at $x = 10$ cm is 5 cm
b. positions of antinodes are 10 cm, 30 cm and 50 cm
c. positions of antinodes are 15 cm, 30 cm, 45 cm
d. the equations of constituent waves are
$y_1 = \frac{5}{2} \sin 2\pi\left(\frac{x}{10} - 50t\right)$ and $y_2 = \frac{5}{2} \sin 2\pi\left(\frac{x}{10} + 50t\right)$

Working Space

14 $y_1 = 8\sin(\omega t - kx)$ and $y_2 = 6\sin(\omega t + kx)$, are two waves travelling in a string of area of cross-section s and density ρ. These two waves are superimposed. Which of the following statement(s) is/are correct?

a. The energy of the wave between the consecutive nodes is $\frac{50\pi}{k}\rho\omega^2 s$

b. The energy of the wave between the consecutive nodes is $\frac{32\pi}{k}\rho\omega^2 s$

c. The total energy crossing through a node per second is zero

d. The total energy crossing through a node per second is $\frac{2\rho\omega^3 s}{k}$

Statement Type

Direction *(Q. Nos. 15-16) This section is based on Statement I and Statement II. Select the correct answer from the codes given below.*

a. Both Statement I and Statement II are correct and Statement II is the correct explanation of Statement I
b. Both Statement I and Statement II are correct but Statement II is not the correct explanation of Statement I
c. Statement I is correct but Statement II is incorrect
d. Statement II is correct but Statement I is incorrect

15 **Statement I** In a stationary wave, there is no transfer of energy across nodes.

Statement II The ratio of average kinetic energy to average potential energy is a constant, independent of the position.

16 **Statement I** In stationary waves, velocity of particles while crossing mean position, varies from maximum at antinodes at zero at nodes.

Statement II Amplitude of vibration at antinodes is maximum and at nodes, the amplitude is zero and all particles between the successive nodes cross the mean position together.

Matching List Type

Direction *(Q. No. 17) Choices for the correct combination of elements from Column I and Column II are given as options (a), (b), (c) and (d), out of which one is correct.*

17 In case of mechanical waves, a particle oscillates and during oscillations its kinetic energy and potential energy change. Match for the situations in Column I with the value of KE/PE in Column II. Match the Column I with Column II and mark the correct option from the codes given below.

	Column I		Column II
i.	When particle of travelling wave is passing through mean position.	p.	Kinetic energy is minimum
ii.	When particle of travelling wave is at its extreme position.	q.	Elastic potential energy is maximum
iii.	When particle between node and antinode in standing wave is passing through mean position.	r.	Kinetic energy is maximum
iv.	When particle between node and anitnode in standing wave is at extreme position.	s.	Elastic potential energy is minimum

Codes

	i	ii	iii	iv		i	ii	iii	iv
a.	q,r	p,s	r,s	p,q	b.	r,q	s,p	r	q
c.	p	q	r	s	d.	s	q	r	s

Comprehension Type

Working Space

Direction (Q. Nos. 18-21) *This section contains a paragraph, describing theory, experiments, data, etc. Four questions related to the paragraph have been given. Each question has only one correct answer among the four given options (a), (b), (c) and (d).*

Passage I

The vibrations of a string of length 60 cm fixed at both ends are represented by the equation $y = 4 \sin\left(\frac{\pi x}{15}\right) \cos(96\pi t)$, where x and y are in cm and t in second.

18 Maximum displacement of a point at $x = 5$ cm is

a. 4 cm
b. 2 cm
c. $2\sqrt{3}$ cm
d. 3 cm

19 The nodes are located along the string at the values of x given by

a. 0, 5 cm, 10 cm, 15 cm,
b. 0, 0.75 cm, 15 cm, 22.5 cm
c. 0, 30 cm, 60 cm, 90 cm,
d. 0, 15 cm, 30 cm, 45 cm,

20 The velocity of the particle at $x = 7.5$ cm at $t = 0.25$ s is

a. zero
b. 384 cms^{-1}
c. $192\sqrt{2}$ cms^{-1}
d. None of these

21 The given standing wave, i.e. $y = 4 \sin\left(\frac{\pi x}{15}\right) \cos(96\pi t)$ is obtained by the superposition of the two waves, y_1 and y_2. Then,

a. $y_1 = 2\sin\left(\frac{\pi x}{15} - 96\pi t\right)$, $y_2 = 2\sin\left(\frac{\pi x}{15} + 96\pi t\right)$

b. $y_1 = 2\sin\left(\frac{\pi x}{15} + 96\pi t\right)$, $y_2 = 2\sin\left(\frac{\pi x}{15} + 96\pi t\right)$

c. $y_1 = 2\sin\left(\frac{\pi x}{15} - 96\pi t\right)$, $y_2 = 2\sin\left(\frac{\pi x}{15} - 96\pi t\right)$

d. $y_1 = 4\sin\left(\frac{\pi x}{15} - 96\pi t\right)$, $y_2 = 4\sin\left(\frac{\pi x}{15} + 96\pi t\right)$

Subjective Questions

Direction (Q. Nos. 22-25) *These questions are subjective in nature, need to be solved completely on notebook.*

22 An aluminium wire of cross-sectional area 10^{-6} m^2 is joined to a steel wire of the same cross-sectional area. This compound wire is stretched on a sonometer pulled by a weight of 10 kg. The total length of compound wire is 1.5 m of which the aluminium wire is 0.6 m and the rest is steel wire. Transverse vibrations are set up in the wire by using an external source of variable frequency. Find the lowest frequency of excitation for which the standing waves are formed such that the joint in the wire is a node. What is the total number of nodes at this frequency? The density of aluminium is 2600 kg m^{-3} and that of steel is 10400 kg m^{-3}. (Take, $g = 10$ ms^{-2})

Working Space

23 The equation of a standing wave in a homogeneous elastic medium is given by $y = A \cos kx \cos \omega t$. Plot

i. y and $\partial y/\partial x$ as functions of x at the time $t = 0$ and $t = T/2$, where T is the oscillation period.

ii. the distribution of density $\rho(x)$ of the medium at the time $t = 0$ and $t = T/2$ in the case of the longitudinal vibrations and

iii. the velocity distribution of particles of the medium at the moment $t = T/4$.

24 A longitudinal standing wave $y = A \cos kx \cos \omega t$ (y and x are along same line) is maintained in a homeogenous medium of density ρ.

i. Find the expressions for space density of potential energy $U(x, t)$ and kinetic energy $K(x, t)$.

ii. Plot the space density distribution of the total energy E in the space between the displacement nodes at times $t = 0$ and $t = T/4$, where T is the oscillation period.

25 The following equations represent transverse waves,

$$Z_1 = A \cos (kx - \omega t)$$

$$Z_2 = A \cos (kx + \omega t)$$

and

$$Z_3 = A \cos (ky - \omega t)$$

Identify the combination of the waves, which will produce

i. a standing wave

ii. a wave travelling in the direction making an angle of 45° with the positive X-axis and positive Y-axis.

In each case find the positions, at which the resultant intensity is always zero.

DPP-3 Resonating Air Columns

Subjective Questions

Working Space

Direction *(Q. Nos. 1 and 2) These questions are subjective in nature, need to be solved completely on notebook.*

1 A 3.6 m long pipe resonates with a source of frequency 212.5 Hz, when water level is at certain heights in the pipe. Find the heights of water level (from the bottom of the pipe) at which resonances occur. Neglect end correction. Now, the pipe is filled to a height H (= 3.6 m). A small hole is drilled very close to its bottom and water is allowed to leak. Obtain an expression for the rate of fall of water level in the pipe as a function of H. If the radii of the pipe and the hole are 2×10^{-2} m and 1×10^{-3} m respectively, calculate the time interval between the occurrence of first two resonances. (Given that, the speed of sound in air is 340 ms^{-1} and $g = 10\ ms^{-2}$)

2 Two narrow cylindrical pipes P and Q have same length. Pipe P is open at both ends and is filled with a monatomic gas of molar mass M_1. Pipe Q is open at one end and closed at the other end and is filled with a diatomic gas of molar mass M_2. Both gases are at the same temperature.

I. If the frequency of second harmonic of the fundamental mode in pipe P is equal to the frequency of third harmonic of the fundamental mode in pipe Q, determine the value of $\frac{M_1}{M_2}$.

II. Now, the open end of pipe Q is also closed (so that, the pipe is closed at both ends). Find the ratio of fundamental frequency in pipe P to that in pipe Q.

Only One Option Correct Type

Direction *(Q. Nos. 3-13) This section contains 11 multiple choice questions. Each question has four choices (a), (b), (c) and (d), out of which ONLY ONE is correct.*

3 An open pipe of sufficient length is dipping in water with a speed u vertically. If at any instant, L is the length of tube above water, then the rate at which fundamental frequency of pipe changes, is (Take, speed of sound is v)

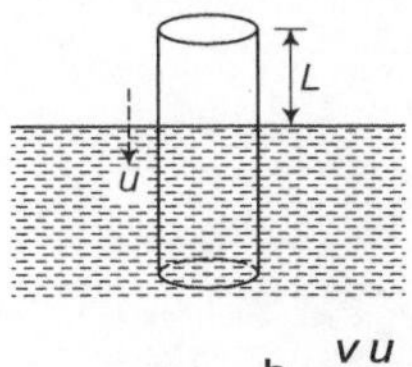

a. $\frac{vu}{2L^2}$

b. $\frac{vu}{L^2}$

c. $\frac{vu}{4L^2}$

d. $\frac{3vu}{4L^2}$

Working Space

4 First overtone frequency of a closed organ pipe is equal to the first overtone frequency of an open organ pipe. Further nth harmonic of closed organ pipe is also equal to the mth harmonic of open pipe, where n and m are

a. 7, 3
b. 9, 6
c. 7, 5
d. 5, 4

5 If L_1 and L_2 are the lengths of air column for the first and second resonance, when a tuning fork of frequency f is sounded on a resonance tube, taking into account the end correction, the distance of the antinode from the top end of the resonance tube is

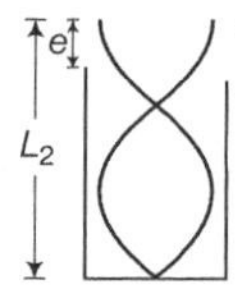

a. $\frac{L_2 - L_1}{2}$
b. $\frac{2L_1 - L_2}{2}$
c. $\frac{L_2 - 3L_1}{2}$
d. $2(L_2 - L_1)$

6 A pipe of length 1.5 m, closed at one end, is filled with a gas and it resonates in its fundamental with a tuning fork. Another pipe of same length put open at both ends, if filled with air, resonates in its fundamental with the same tuning fork. Calculate the speed of sound at 0°C in the gas, given that speed of sound in air is 360 ms^{-1} at 30°C, where the experiment is performed.

a. 683 ms^{-1}
b. 583 ms^{-1}
c. 783 ms^{-1}
d. 576 ms^{-1}

7 To determine the sound propagation velocity in air by acoustic resonance technique one can use a pipe with a piston and a sonic membrane closing one of its ends. Find the velocity of sound, if the distance between the adjacent positions of the piston at which resonance is observed at a frequency $f = 2000$ Hz is equal to $l = 8.5$ cm.

a. 330 ms^{-1}
b. 340 ms^{-1}
c. 350 ms^{-1}
d. 360 ms^{-1}

8 A U-tube having uniform cross-section, but unequal arm lengths L_1 and $L_2 (< L_1)$ has a liquid of density ρ_1 filled in it upto a height h as shown in the figure. Another liquid of density $\rho_2 \left(= \frac{\rho_1}{2}\right)$ is poured in arm A. Both liquids are immiscible. What length of the second liquid should be poured in A, so that first overtone of A is in unison with fundamental tone of B.

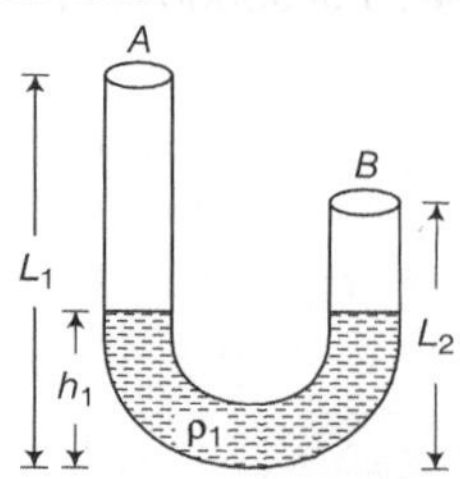

a. $\frac{L_1 - 3L_2 + 2h}{3}$
b. $\frac{2(L_1 - 3L_2 + 2h)}{3}$
c. $\frac{L_1 - 3L_2 + 2h}{2}$
d. $\frac{3(L_1 - 3L_2 + 2h)}{4}$

Working Space

9 Find the temperature (T_0) at which the fundamental frequency of an organ pipe is independent of small variation in temperature in terms of the coefficient of linear expansion (α) of the material of the tube.

a. $T_0 = \frac{3}{2\alpha}$ b. $T_0 = \frac{3}{\alpha}$

c. $T_0 = \frac{2}{3\alpha}$ d. $T_0 = \frac{1}{2\alpha}$

10 In the resonance tube experiment, the first resonance is heard, when length of air column is L_1 and second resonance is heard when length of air column is L_2. What should be the minimum length of the tube so that third resonance can also be heard?

a. $2L_2 - L_1$ b. $2L_1$

c. $7L_1$ d. $6L_1$

11 Four open organ pipes of different lengths and different gases at same temperature are shown in figure. Let f_1, f_2, f_3 and f_4 be their fundamental frequencies, then (Take, $\gamma_{CO_2} = 7/5$)

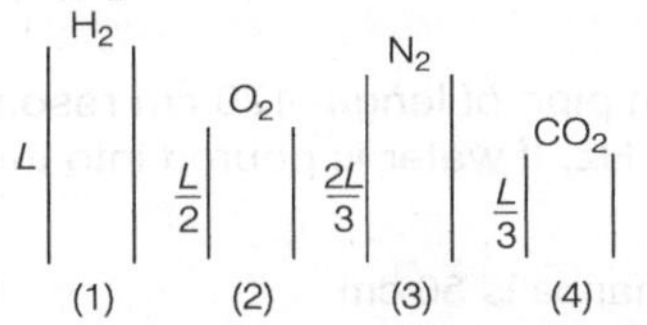

a. $f_4/f_1 = \sqrt{76/11}$ b. $f_2/f_3 = \sqrt{72/38}$

c. $f_3/f_4 = \sqrt{11/28}$ d. $f_1/f_2 = \sqrt{2}$

12 A closed organ pipe is sounded near a guitar causing one of the strings to vibrate with large amplitude. We vary the tension in the string until we find the maximum amplitude. The string is 80% as long as the closed pipe. If both the pipe and the string vibrate at their fundamental frequency, the ratio of the wave speed on the string to the speed of sound in air is

a. 0.4 b. 0.8 c. 2 d. 0.5

13 In a resonance tube experiment to determine the speed of sound in air, a pipe of diameter 5 cm is used. The column in pipe resonates with a tuning fork of frequency 480 Hz, when the minimum length of the air column is 16 cm. The speed of sound in air is

a. 330 ms^{-1} b. 307 ms^{-1}

c. 336 ms^{-1} d. 346 ms^{-1}

One or More than One Options Correct Type

Direction (Q. Nos. 14-18) *This section contains 5 multiple choice questions. Each question has four choices (a), (b), (c) and (d), out of which ONE or MORE THAN ONE are correct.*

14 A 3 m long organ pipe opened at both ends is driven to third harmonic standing wave. If the maximum amplitude of pressure oscillations is 1% of mean atmospheric pressure ($p_0 = 10^5 \text{ Nm}^{-2}$), the maximum amplitude of particle displacement is A and that of density oscillations is $(\Delta\rho)_{max}$. Speed of sound is $v = 332 \text{ ms}^{-1}$ and density of air $\rho = 1.03 \text{ kgm}^{-3}$. Choose the correct options.

a. $A = 0.28 \text{cm}$ b. $(\Delta\rho)_{max} = 9 \times 10^{-3} \text{ kgm}^{-3}$

c. $A = 0.56 \text{cm}$ d. $(\Delta\rho)_{max} = 1.8 \times 10^{-4} \text{ kgm}^{-3}$

15 The air column in a pipe closed at one end is made to vibrate in its second overtone by tuning fork of frequency 440 Hz. The speed of sound in air is 330 ms^{-1}. End corrections may be neglected. Let p_0 is the mean pressure at any point in the pipe and $(\Delta p)_{max}$ is the maximum amplitude of pressure variation. If length of the air column is L and the amplitude of pressure variation at the middle of the column is Δp, then

a. $L = \frac{16}{15}$ m b. $\Delta p = \frac{(\Delta p)_{max}}{\sqrt{3}}$

c. $L = \frac{15}{16}$ m d. $\Delta p = \frac{(\Delta p)_{max}}{\sqrt{2}}$

16 For a certain organ pipe, three successive resonance frequencies are observed at 425 Hz, 595 Hz and 765 Hz, respectively. The speed of sound in air is 340 ms^{-1}. Choose the correct options.

a. The pipe is open at both the ends
b. The pipe is closed at one end
c. Fundamental frequency of pipe is 85 cm
d. Length of the pipe is 1m

17 In a resonance tube experiment, a closed organ pipe of length 120 cm resonates, when tuned with a tuning fork of frequency 340 Hz. If water is poured into the pipe, then (Speed of sound in air = 340 ms^{-1})

a. minimum length of water column to have resonance is 50 cm
b. the distance between two successive nodes is 50 cm
c. the maximum length of water column to create resonance is 100 cm
d. None of the above

18 Figure shows a stretched string of length L and pipes of length L, $2L$, $\frac{L}{2}$ and $\frac{L}{2}$ in options (a), (b), (c) and (d) respectively. The tension in the string is adjusted until the speed of waves on the string equals the speed of sound waves in air. The fundamental mode of oscillation is then set up on the string. In which pipe(s) will the sound produced by the string causes resonance?

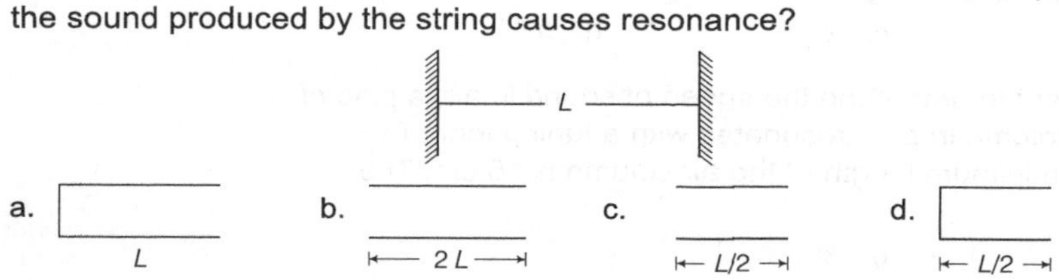

Comprehension Type

Direction *(Q. Nos. 19-23) This section contains 2 paragraphs, each describing theory, experiments, data, etc. Five questions related to the paragraphs have been given. Each question has only one correct answer among the four given options (a), (b), (c) and (d).*

Passage I

A 60 cm long flute can be considered as a pipe open at both the ends. The speed of sound in air is 340 ms^{-1}.

19 The fundamental frequency, when all the holes are covered is

a. 283 Hz b. 293 Hz
c. 303 Hz d. 273 Hz

20 How far from the mouth piece should a hole be uncovered for the fundamental frequency to be 330 Hz?

a. 34.5 cm b. 41.5 cm c. 44.5 cm d. 51.5 cm

Working Space

Passage II

A tube closed at one end has a vibrating diaphragm at the other end, which may be 'assumed' to be a displacement node. It is found that, when the frequency of the diaphragm is 2000 Hz, a stationary wave pattern is set up in which the distance between adjacent nodes is 8 cm. When the frequency is gradually reduced, the stationary wave pattern disappears, but another stationary wave pattern reappears at a frequency of 1600 Hz.

21 The speed of sound in air is

a. $320 ms^{-1}$ b. $330 ms^{-1}$

c. $340 ms^{-1}$ d. $350 ms^{-1}$

22 The distance between adjacent nodes at the frequency of 1600 Hz is

a. 14 cm b. 12 cm

c. 10 cm d. 11 cm

23 The distance between diaphragm and the closed end is

a. 30 cm b. 40 cm

c. 50 cm d. 60 cm

Matching List Type

***Direction** (Q. No. 24) Choices for the correct combination of elements from Column I and Column II are given as options (a), (b), (c) and (d), out of which one is correct.*

24 Equation of longitudinal stationary wave in second overtone mode in a closed organ pipe is $y = (4\text{mm})\sin \pi x \cos \pi t$, where x is in metre and t in second. Match the Column I with Column II and mark the correct option from the codes given below.

	Column I		Column II
i.	Length of pipe	p.	1m
ii.	Wavelength	q.	1.5 m
iii.	Distance of displacement node from closed end	r.	2 m
iv.	Distance of pressure node from closed end	s.	None

Codes

	i	ii	iii	iv
a.	r	p	q	s
b.	s	p,r	q	r
c.	s	q	p,r	q
d.	p	q	r	s

One Integer Value Correct Type

***Direction** (Q. No. 25) This section contains 1 question. When worked out will result in an integer from 0 to 9 (both inclusive).*

25 *PQ* is a cylinder of length 1m filled with a thin flexible diaphragm *R* at the middle and two other thin flexible diaphragms *P* and *Q* at the ends. The portions *PR* and *QR* contain hydrogen and oxygen gases respectively. The diaphragms *P* and *Q* are set into vibrations of same frequency. The minimum frequency of these vibrations for which the diaphragm *R* is a node is 550*p* Hz. Find out the value of *p*. (Under the conditions of the experiment, the velocity of sound in hydrogen is $1100 ms^{-1}$ and in oxygen is $300 ms^{-1}$)

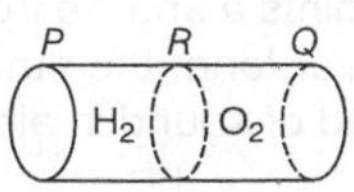

DPP-4 Standing Waves in Rods

Subjective Questions

Working Space

Direction *(Q. Nos. 1-3) These questions are subjective in nature, need to be solved completely on notebook.*

1 Discuss standing waves in rods.

2 A metallic rod of length 1m is rigidly clamped at its mid-point. Longitudinal stationary waves are set up in such a way that, there are two nodes on either side of the mid-point. The amplitude of an antinode is 2×10^{-6}m. Write the equation of motion at a point 2cm from the mid-point and those constituent waves in the rod. (Take, $Y = 2 \times 10^{11}\,\text{Nm}^{-2}$ and $\rho = 8000\,\text{kgm}^{-3}$)

3 A wave is given by the equation

$$y = 10\sin 2\pi(100t - 0.02x) + 10\sin 2\pi(100t + 0.02x).$$

Find the loop length, frequency, velocity and maximum amplitude of the stationary wave produced.

Only One Option Correct Type

Direction *(Q. Nos. 4-12) This section contains 9 multiple choice questions. Each question has four choices (a), (b), (c) and (d), out of which ONLY ONE is correct.*

4 A metal rod of length $L = 100$cm is clamped at two points P and Q as shown in the figure. Distance of each clamp from the nearer end is $d = 30$cm. Density and Young's modulus of elasticity of rod material are $\rho = 9000\,\text{kgm}^{-3}$ and $Y = 144$GPa respectively. Calculate the minimum frequency of natural longitudinal vibrations of the rod.

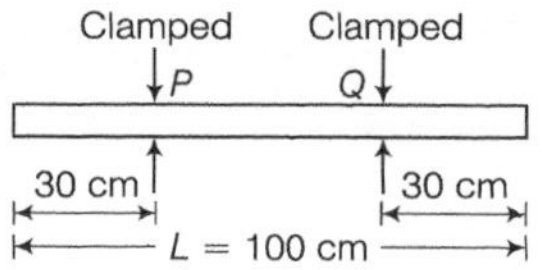

a. 14 kHz b. 8 kHz c. 12 kHz d. 10 kHz

5 In the above problem, what is the next higher frequency ?

a. 20 kHz b. 30 kHz c. 24 kHz d. 36 kHz

6 A closed organ pipe of length L_0 is resonating in 5th harmonic mode with a rod clamped at two points a and $3a$ from one end. If the length of the rod is $4a$ and it is vibrating in first over tone, then the length of the rod in terms of L_0, v_s, Y and ρ is (where, v_s = speed of sound in air, Y is Young's modulus and ρ density of the rod)

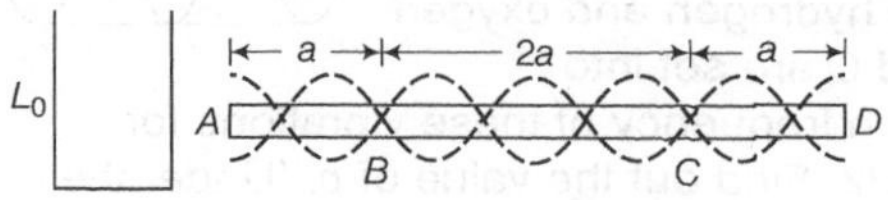

a. $\frac{12}{5}\frac{L_0}{v_s}\sqrt{\frac{Y}{\rho}}$ b. $\frac{7}{5}\frac{L_0}{v_s}\sqrt{\frac{Y}{\rho}}$ c. $\frac{17}{5}\frac{L_0}{v_s}\sqrt{\frac{Y}{\rho}}$ d. $\frac{14}{5}\frac{L_0}{v_s}\sqrt{\frac{Y}{\rho}}$

Working Space

7 The following equation represents a standing wave set up in a medium, $y = 4\cos\frac{\pi x}{3}\sin 40\pi t$, where x and y are in cm and t in second. The velocity of a medium particle at $x = 3$ cm at time $t = \frac{1}{8}$ s is, (x and y are in cm and t in second).

a. $16\pi\ \text{cms}^{-1}$
b. $160\pi\ \text{cms}^{-1}$
c. $18\pi\ \text{cms}^{-1}$
d. $200\pi\ \text{cms}^{-1}$

8 Six antinodes are observed in the air column, when a standing wave forms in a closed (at both ends) tube. A steel bar (in its fundamental mode) of length 1m and clamped at the middle is in unison with the air column. The speed of sound in steel is $5250\ \text{ms}^{-1}$ and in air, it is $343\ \text{ms}^{-1}$. Length of the air column is

a. 79 cm
b. 49 cm
c. 39 cm
d. 29 cm

9 A steel rod 100 cm long is clamped at its middle. The fundamental frequency of longitudinal vibrations of the rod is given to be 2.53 kHz. Speed of sound in steel is

a. $4.41\ \text{kms}^{-1}$
b. $5.23\ \text{kms}^{-1}$
c. $4.92\ \text{kms}^{-1}$
d. $5.06\ \text{kms}^{-1}$

10 A metal rod of length 1m is clamped at two points as shown in the figure. Distance of the clamps from two ends are 5 cm and 15 cm respectively. Find the minimum frequency of natural longitudinal oscillations of the rod. Given that, Young's modulus and density of the material of the rod are $Y = 1.6 \times 10^{11}$ Pa and $\rho = 2500\ \text{kgm}^{-3}$ respectively.

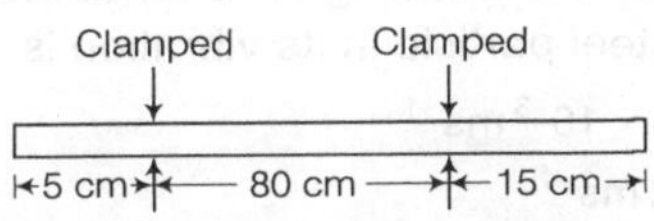

a. 30 kHz
b. 40 kHz
c. 50 kHz
d. 25 kHz

11 In the above problem, find out next higher frequency.

a. 80 kHz
b. 120 kHz
c. 100 kHz
d. 160 kHz

12 A uniform rod of linear mass density μ supports a stationary wave given by the equation $y = A\sin kx\cos\omega t$. Determine the total mechanical energy confined between two adjacent nodes.

a. $\frac{\mu\omega^2A^2\pi}{2k}$
b. $\frac{\mu\omega^2A^2\pi}{3k}$
c. $\frac{\mu\omega^2A^2\pi}{4k}$
d. $\frac{\mu\omega^2A^2\pi}{6k}$

One or More than One Options Correct Type

Direction (Q. Nos. 13 and 14) *This section contains 2 multiple choice questions. Each question has four choices (a), (b), (c) and (d), out of which ONE or MORE THAN ONE are correct.*

13 An aluminium rod 1.6 m long is clamped at its centre. It is stroked with a rosin-coated cloth to set up a longitudinal vibration. The speed of sound in thin rod of aluminum is $5100\ \text{ms}^{-1}$. The fundamental frequency of the waves established in the rod is f_0. If the rod were copper (in which speed of sound is $3560\ \text{ms}^{-1}$), the fundamental frequency would have been f_0'. Then,

a. $f_0 = 1590$ Hz
b. $f_0' = 1110$ Hz
c. $f_0 = 1690$ Hz
d. $f_0' = 1210$ Hz

Working Space

14 A metallic rod of length 1 m is rigidly clamped at one of its ends. The other end of the rod is free. Longitudinal stationary waves are set up in the rod in such a way that, there are total six antinodes observed along the rod. The amplitude of an antinode is 4×10^{-6} m. (Given, $Y = 6.4 \times 10^{10}$ Nm^{-2} and $\rho = 5000$ kgm^{-3}), choose the correct option(s).

a. Wavelength of constituent waves is approximately 0.36 m
b. Frequency of constituent waves is approximately 9939 Hz
c. Equation of motion at the mid-point of the rod is $y = 2.57 \times 10^{-6} \sin(62400t + \delta)$
d. Wavelength of constituent waves is 0.48 m

Comprehension Type

Direction (Q. Nos. 15-17) *This section contains a paragraph, describing theory, experiments, data, etc. Three questions related to the paragraph have been given. Each question has only one correct answer among the four given options (a), (b), (c) and (d).*

Passage I

A steel rod 2.5 m long is rigidly clamped at its centre *C* and longitudinal waves are setup on both sides *C*. Young's modulus of steel = 2×10^{11} Pa and density of steel = 8000 kgm^{-3}.

15 If two antinodes are observed on either side of *C*, the frequency of the mode in which the rod is vibrating will be

a. 1500 Hz
b. 7000 Hz
c. 3000 Hz
d. 1000 Hz

16 If the amplitude of the wave at its antinode, when it is vibrating in its fundamental mode is 2×10^{-6}m, the maximum velocity of a steel particle in its vibration is

a. 1.25×10^{-2} ms^{-1}
b. 1.25×10^{-3} ms^{-1}
c. 1ms^{-1}
d. 0.12 ms^{-1}

17 If the clamp of the rod be shifted to its end *A* and totally four antinodes are observed in the rod when, the longitudinal waves are set up in it, the frequency of vibration of the rod in this mode is

a. 500 Hz
b. 2500 Hz
c. 3500 Hz
d. 1500 Hz

DPP-5 Interference in Time-Beats

Subjective Question

Direction (Q. No. 1) *This question is subjective in nature, need to be solved completely on notebook.*

Working Space

1 i. Why one can call the phenomenon of beats 'interference in time'?

ii. Does the phenomenon of beats happen only when the frequency difference between sources is less than 10 Hz?

Only One Option Correct Type

Direction (Q. Nos. 2-17) *This section contains 16 multiple choice questions. Each question has four choices (a), (b), (c) and (d), out of which ONLY ONE is correct.*

2 A tuning fork T_1 produces 4 beats with tuning fork T_2 of frequency 256 Hz. When T_1 is filed, beats are found to occur at shorter intervals. What was its original frequency?

a. 260 Hz b. 252 Hz
c. 254 Hz d. 258 Hz

3 A tuning fork T_1, when sounded with another tuning fork T_2 of frequency 256 Hz gives 4 beats per second and when T_1 is loaded with a certain amount of wax, it is found to give again 4 beats s^{-1} with T_2. The frequency of T_1 is

a. 260 Hz b. 252 Hz
c. 254 Hz d. 258 Hz

4 Two identical sonometer wires have a fundamental frequency of 500 Hz, when kept under same tension. What fractional increase in the tension of one wire would cause an occurence of 5 beats s^{-1}, when both wires vibrate together?

a. 4% b. 3%
c. 2% d. 1%

5 Three sound sources S_1, S_2 and S_3 have frequencies 400 Hz, 401 Hz and 402 Hz. respectively. Calculate the number of beats noted per second.

a. 4 b. 3 c. 2 d. 1

6 Two vibrating tuning forks with natural frequencies 340 Hz each, move relative to a stationary observer. One fork moves away from the observer and the other moves towards him at the same speed. The observer hears beats of frequency 3 Hz. If velocity of sound in air is 340 ms^{-1}, the speed of each tuning fork is

a. 2.5 ms^{-1} b. 2 ms^{-1}
c. 1.5 ms^{-1} d. 1 ms^{-1}

7 A wave received by a detector is expressed by the wave function $y = \left(\cos^2 \pi t - \frac{1}{4}\right) \sin (500\,\pi t)$. What is the beat frequency recorded by the detector?

a. 1s^{-1} b. 2s^{-1} c. 3s^{-1} d. 4s^{-1}

Working Space

8 A tuning fork and an organ pipe at temperature 88°C produce 5 beats per second. When the temperature of the air column is decreased to 51°C, the two produce 1 beat s^{-1}. The frequency of the tuning fork is

a. 81 Hz
b. 71 Hz
c. 91 Hz
d. 61 Hz

9 A tuning fork T_1 of unknown frequency gives 7 beats in 2 second with another tuning fork T_2. When T_2 moves towards a wall with a speed of 5 ms^{-1}, it gives 5 beats per second with its echo. On loading with wax, T_1 gives 5 beats per second with T_2. Speed of sound is 332 ms^{-1}. The freqency of T_1 is

a. 200 Hz
b. 205 Hz
c. 177 Hz
d. 160 Hz

10 Two wires are fixed on a sonometer. Their tensions are in the ratio 8 : 1, the lengths in the ratio 36 : 35, the diameters in the ratio 4 : 1 and the densities in the ratio 1 : 2. Find the frequency of the beats produced, if the note of the higher pitch has frequency of 360 Hz.

a. 5 Hz
b. 10 Hz
c. 15 Hz
d. 3 Hz

11 A tuning fork of frequency 300 Hz resonates with an air column closed at one end at 27° C. How many beats will be heard in the vibrations of the fork and the air column at 0° C? End correction is negligible and air column is in its fundamental mode.

a. 10 Hz
b. 12 Hz
c. 14 Hz
d. 8 Hz

12 When beats are produced by two waves of nearly the same frequency,

a. the beat frequency depends on the position, where the beats are heard
b. the beat frequency decreases as time passes
c. the particles vibrate simple harmonically with a frequency equal to the difference of the two frequencies
d. the amplitude of vibration at any point changes simple harmonically with a frequency equal to the difference of the two frequencies

13 Two sounding bodies producing progressive waves given by $y_1 = 4 \sin(404\,\pi t)$ and $y_2 = 3 \sin(404\,\pi t)$, where t is in seconds, are situated near the ears of a person. The person will hear

a. 2 beats per second with intensity ratio $\frac{4}{3}$ between maxima and minima
b. 2 beats per second with intensity ratio 49 between maxima and minima
c. 4 beats per second with intensity ratio 7 between maxima and minima
d. 4 beats per second with intensity ratio $\frac{4}{3}$ between maxima and minima

14 A police car moving at 22 ms^{-1} chases a motorcyclist. The policeman sounds his horn at 176 Hz, while both of them move towards a stationary siren of frequency 165 Hz. It is given that, motorcyclist does not hear any beats. The speed of motorcyclist is (Given, speed of sound in air = 330 ms^{-1})

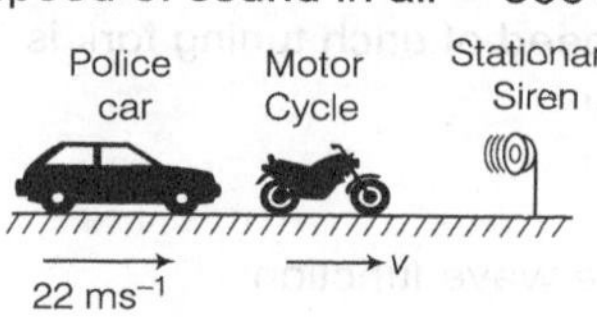

a. zero
b. 11ms^{-1}
c. 22ms^{-1}
d. 33ms^{-1}

Working Space

15 A string under a tension of 129.6 N produces 10 beats per second, when it is vibrated along with a tuning fork. When the tension in the string is increased to 160 N, it sounds in unison with the same tuning fork. Calculate the frequency of tuning fork.

a. 100 Hz
b. 90 Hz
c. 110 Hz
d. 120 Hz

16 The vibrating portion of a wire which is stretched with a weight of 6.48 kg, weighs 0.5 g. When sounding in fundamental note, it is found to give 20 beats in 5 seconds, with a vibrating tuning fork of frequency 256 Hz. If the length of the wire is slightly decreased, the note emitted by it is observed to be in unison with that of the fork. Calculate the original length of the wire.

a. 50 cm
b. 60 cm
c. 70 cm
d. 40 cm

17 A column of air at 51° C and a tuning fork produce 4 beats per second when sounded together. As the temperature of air column is decreased, the number of beats per second tends to decrease and when the temperature is 16° C, the two produce one beat per second. Find the frequency of tuning fork. End correction is negligible.

a. 69.8 Hz
b. 79.8 Hz
c. 59.8 Hz
d. 49.8 Hz

One or More than One Options Correct Type

Direction *(Q. Nos. 18 and 19) This section contains 3 multiple choice questions. Each question has four choices (a), (b), (c) and (d), out of which ONE or MORE THAN ONE are correct.*

18 A metal wire of diameter 1 mm is held on two knife edges separated by a distance of 50 cm. The tension in the wire is 100 N. The wire vibrating with its fundamental frequency and a vibrating tuning fork together produce 5 beats s^{-1}. The tension in the wire is then reduced to 81 N. When the two are excited again, beats are heard again at the same rate. The frequency of the tuning fork is f and density of the material of the wire is ρ. Then,

a. $f = 95$ Hz
b. $\rho = 1.274 \times 10^4$ kg m^{-3}
c. $f = 100$ Hz
d. $\rho = 1.724 \times 10^4$ kg m^{-3}

19 The first overtone of an open organ pipe beats with the first overtone of a closed organ pipe with a beat frequency of 2.2 Hz. The fundamental frequency of the closed organ pipe is 110 Hz. The lengths of open organ pipe and closed organ pipe, respectively, are l_O and l_c. Then, (Given, speed of sound is 330 ms^{-1})

a. $l_O = 0.8937$ m
b. $l_O = 0.9934$ m
c. $l_O = 1.0067$ m
d. $l_c = 0.75$ m

Comprehension Type

Direction *(Q. Nos. 20 and 21) This section contains a paragraph, describing theory, experiments, data, etc. Two questions related to the paragraph have been given. Each question has only one correct answer among the four given options (a), (b), (c) and (d).*

Passage I

A closed air column 32 cm long is in resonance with a tuning fork. Another open air column of length 66 cm is in resonance with another tuning fork. If two forks, when sounded together, produce 8 beats s^{-1}, find

20 The speed of sound in air

a. 40980 cms^{-1}
b. 31890 cms^{-1}
c. 35790 cms^{-1}
d. 33792 cms^{-1}

Working Space

21 The frequencies of the forks

a. 150 Hz, 158 Hz b. 250 Hz, 258 Hz
c. 231Hz, 239 Hz d. 264 Hz, 256 Hz

One Integer Value Correct Type

***Direction** (Q. Nos. 22-25) This section contains 4 questions. When worked out will result in an integer from 0 to 9 (both inclusive).*

22 Two identical wires are stretched by the same tension of 100 N and each emits a note of frequency 200 Hz. The tension in one wire is increased by 1 N. Calculate the number of beats heard per second, when the wires are plucked.

23 An organ pipe open at both ends sounds in unison with a tuning fork at 20° C. When the fork and the pipe are sounded together at 30° C, 5 beats per second are heard. If the frequency of the fork is approximately 37β, find the value of β. Assume that frequency of fork is not affected by the temperature change.

24 When 0.98 m long metallic wire is stressed, an extension of 0.02 m is produced. An organ pipe 0.5 m long and open at both ends, when sounded with this stressed metallic wire, produces 8 beats in its fundamental mode. By decreasing the stress in the wire, the number of beats is found to decrease. The density of metallic wire is 10^4 kg m^{-3} and sound velocity in air is 292 ms^{-1}. If the Young's modulus of the wire is $(0.588\,\alpha) \times 10^{11}$ Nm^{-2}. Find the value of α.

25 A set of 25 tuning forks is arranged in a series of decreasing frequencies. Each fork gives 3 beats with the succeeding one. The first fork is an octave higher of the last. If the frequency of the 16th tuning fork is 11 y Hz, what is the value of y?

SET - 1

Revisal Problems
for JEE Main

Only One Option Correct Type

Working Space

Direction (Q. Nos. 1-27) ***This section contains 27 multiple choice questions. Each question has four choices (a), (b), (c) and (d), out of which ONLY ONE is correct.***

1 The fundamental frequency of a sonometer wire increases by 6 Hz, if the tension is increased by 44%, keeping the length constant. Find the change in the fundamental frequency of the sonometer wire when the length of the wire is increased by 20%, keeping the original tension in the wire constant.

a. 25 Hz b. 5 Hz
c. 15 Hz d. 8 Hz

2 The length of a sonometer wire between two fixed ends is 1.1 m. Where should the two bridges be placed (from one end), to divide the wire into three segments, whose fundamental frequencies are in the ratio of 1 : 2 : 3 ?

a. 0.6 m, 0.9 m b. 0.5 m, 0.8 m
c. 0.4 m, 0.8 m d. 0.2 m, 0.9 m

3 One end of a horizontal string is attached to a vibrating blade and the other end passes over a pulley as shown in the figure. A sphere of mass 2 kg hangs at the end of the string. The string is vibrating in its second harmonic. A container of water is raised under the sphere so that the sphere is completely submerged. In this configuration, the string vibrates in its fifth harmonic. What is the radius of the sphere?

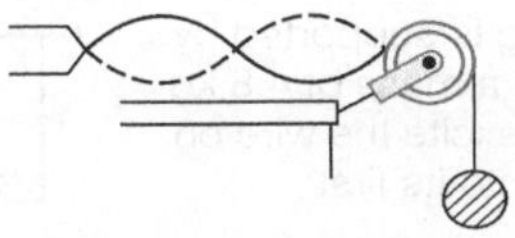

a. 5.38 cm b. 6.38 cm
c. 7.38 cm d. 4.38 cm

4 A section of drainage culvert 1.23 m in length makes a howling noise, when the wind blows across its open ends. Speed of sound in air is 343 ms^{-1}. The frequency of the first harmonic of the culvert, if it is cylindrical in shape and open at both ends, is

a. 239 Hz b. 189 Hz
c. 200 Hz d. 139 Hz

5 A tuning fork of frequency 340 Hz is vibrated just above a cylindrical tube. The length of the tube is $L = 120$ cm. Water is slowly poured into the tube. Determine the minimum height of water required for resonance. Speed of sound in air is 340 ms^{-1}.

a. 45 cm b. 35 cm
c. 95 cm d. 75 cm

6 An open pipe 40 cm long and a closed pipe 31 cm long, both having same diameter, are producing their first overtone and these are in unison. The end correction of these pipes is

a. 1.5 cm b. 2 cm
c. 1.2 cm d. 2.2 cm

7 Wavelengths of two notes in air is $\frac{90}{175}$ m and $\frac{90}{173}$ m, respectively. Each of these notes produces 4 beats per second with a third note of a fixed frequency. Calculate the velocity of sound in air.

a. 330 ms^{-1} b. 340 ms^{-1}
c. 360 ms^{-1} d. 370 ms^{-1}

8 In certain ranges of a piano keyboard, more than one string is tuned to the same note to provide extra loudness. For example, the note at 110 Hz has two strings at this frequency. If one string ships from its normal tension of 600 N to 540 N, what beat frequency is heard, when the hammer strikes the two strings simultaneously?

a. 6.6 Hz
b. 7.6 Hz
c. 4.6 Hz
d. 5.6 Hz

9 Two wires are welded together end to end. The wires are made of the same material, but the diameter of one is twice that of the other. They are subjected to a tension of 4.6 N. The thin wire has a length of 40 cm and a linear mass density of 2 gm^{-1}. The combination is fixed at both ends and vibrated in such a way that two antinodes are present, with the node between them being precisely at the weld. The length of the thick wire is

a. 30 cm
b. 25 cm
c. 20 cm
d. 22 cm

10 Sound waves from two coherent sources S_1 and S_2 are sent in phase and detected at point P equidistant from both the sources. Speed of sound in normal air is v_a but in same part in the path of S_1, there is a zone of hot air having temperature 4 times, the normal temperature and width L. What should be the minimum frequency of sound, so that minima can be found at P?

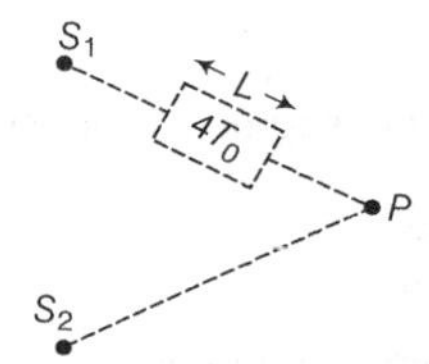

a. $\frac{v_a}{L}$
b. $\frac{v_a}{2L}$
c. $\frac{2v_a}{L}$
d. None of these

11 A uniform horizontal rod of length 0.4 m and mass 1.2 kg is supported by two identical wires as shown in the figure. Where should a mass of 4.8 kg be placed on the rod, so that the same tuning fork may excite the wire on the left into its fundamental vibrations and that on right into its first overtone? (Take, $g = 10\ ms^{-2}$)

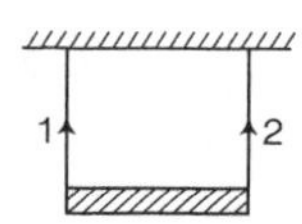

a. 15 cm from left end
b. 20 cm from left end
c. 10 cm from left end
d. 5 cm from left end

12 A sonometer wire under a tension of 128 N vibrates in resonance with a turning fork. The vibrating portion of sonometer wire has a length of 20 cm and mass 1 g. The vibrating tuning fork is now moved away from the vibrating wire at a constant speed of 0.75 ms^{-1} and an observer standing at the sonometer hears 1 beat s^{-1}. The speed of sound in air is

a. 300 ms^{-1}
b. 320 ms^{-1}
c. 330 ms^{-1}
d. 340 ms^{-1}

13 An audio oscillator capable of producing notes of frequencies ranging from 500 Hz to 1500 Hz is placed near a 25 cm long uniform stretched wire under a constant tension F. The linear mass density of the wire is 0.75 gm^{-1}. It is observed that by varying the frequency of the oscillator over the given range the sonometer wire sets into vibration at frequencies 840 Hz and 1120 Hz. The tension in the string is

a. 24.7 N
b. 14.7 N
c. 20 N
d. 18 N

14 Speed of sound in air is 320 ms^{-1}. The resonance pipe shown in the figure, cannot vibrate with a sound of frequency

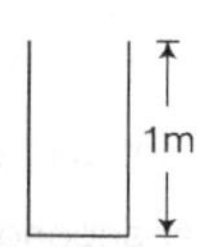

a. 80 Hz
b. 240 Hz
c. 320 Hz
d. None of these

15 A sufficiently long closed organ pipe has a small hole at its bottom. Initially, the pipe is empty. Water is poured into the pipe at a constant rate. The fundamental frequency of the air column in the pipe

a. continuously decreases
b. continuously increases
c. first decreases and then becomes constant
d. first increases and then becomes constant

Working Space

Working Space

16 A metal rod 40 cm long is dropped onto a wooden floor and rebounds into air. Compressional waves of many frequencies are thereby set up in the rod. If the speed of compressional waves in the rod is 5500 ms^{-1}, what is the lowest frequency of compressional waves to which the rod resonates as it rebounds?

a. 6875 Hz b. 16875 Hz
c. 675 Hz d. 0.11 Hz

17 A long cylindrical tube carries a highly polished piston and has a side opening. A tuning fork of frequency f is sounded at the open end of the tube. The intensity of the sound heard by the listener changes, if the piston is moved in or out. At a particular position of the piston, he hears a maximum sound. When the piston is moved through a distance of 9 cm, the intensity of sound becomes minimum. If the speed of sound is 360 ms^{-1}, the value of f is

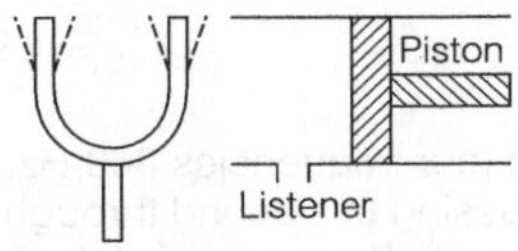

a. 2000 Hz b. 129.6 Hz
c. 500.6 Hz d. 1000 Hz

18 A sound wave of wavelength 0.4 m enters a thin tube at s (see figure). The smallest radius R of the circular segment to hear minimum sound at detector D must be,

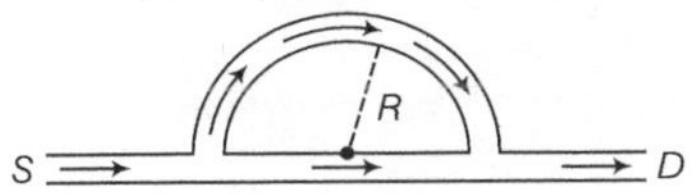

a. 9.3 m b. 0.93 m
c. 0.175 m d. 1.75 m

19 A stone is hung in air from a wire which is stretched over a sonometer. The bridges of the sonometer are 40 cm apart, when the wire is in unison with a tuning fork of frequency 256 Hz. When the stone is completely immersed in water, the length between the bridges is 22 cm for re-establishing unison. The specific gravity of the material of the stone is

a. $256 \times \frac{22}{40}$ b. $256 \times \frac{40}{22}$
c. $\frac{(40)^2}{(40)^2 + (22)^2}$ d. $\frac{(40)^2}{(40)^2 - (22)^2}$

20 An open organ pipe of length L is sounded together with another organ pipe of length $L + x$, both being in their fundamental modes. Speed of sound in air is v. The beat frequency heard will be ($x << L$),

a. $\frac{\sqrt{3}}{2}\frac{vx}{L^2}$ b. $\frac{vx}{L^2}$
c. $\frac{vx}{2L^2}$ d. $\frac{2vx}{L^2}$

21 One end of a 2.4 m long string is held fixed and the other end is attached to a weightless ring that can slide along a frictionless rod as shown in the figure. The three longest possible wavelengths for standing waves in this string respectively are

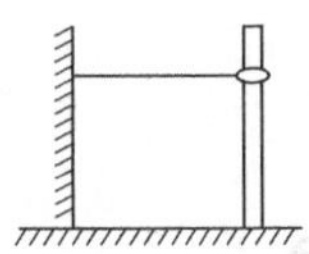

a. 9.6 m, 3.2 m, 1.92 m b. 4.8 m, 1.6 m, 0.96 m
c. 2.4 m, 0.8 m, 0.48 m d. 1.2 m, 0.4 m, 0.24 m

22 Which of the following travelling waves will produce standing waves, with node at $x = 0$, when superimposed on $y = A \sin(\omega t - kx)$.

a. $A \sin(\omega t + kx)$ b. $A \sin(\omega t + kx + \pi)$
c. $A \cos(\omega t + kx)$ d. $A \cos(\omega t + kx + \pi)$

23 A 100 m long rod of density 10^5 kg m^{-3} and Young's modulus $Y = 10^{11}$ Pa is clamped at one end. It is hammered at the other free end. The longitudinal pulse goes to the left end, gets reflected and again returns to the right end. How much time the pulse would take to go back to the initial point?

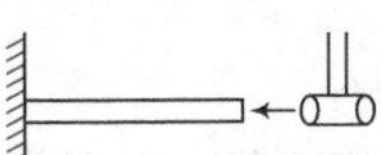

a. 0.3 s b. 0.2 s
c. 0.1 s d. 1 s

Working Space

24 Radiowaves coming at an angle θ to the vertical are received by a ladder after reflection from a near by water surface and also directly. What can be the height of antenna from water surface, so that it records maxima? (λ is wavelength of radiowaves)

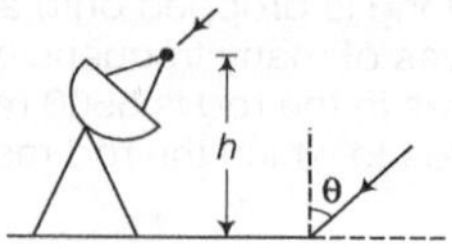

a. $\frac{\lambda}{4\cos\theta}$
b. $\frac{\lambda}{4\sin\theta}$
c. $\frac{\lambda}{2\cos\theta}$
d. $\frac{\lambda}{2\sin\theta}$

25 A sound consists of four frequencies 300 Hz, 600 Hz, 1200 Hz and 2400 Hz. A sound 'filter' is made by passing this sound through a bifurcate pipe as shown. The speed of sound in air is 300 ms^{-1}. Then, which of the following frequencies will be almost completely muffled or 'silenced' at the outlet?

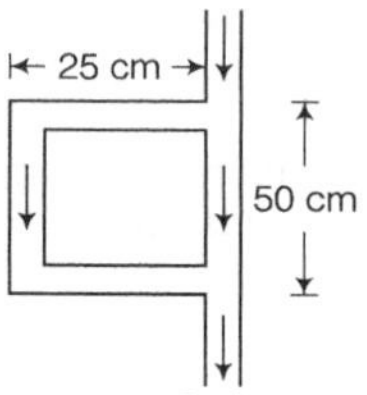

a. 300 Hz
b. 600 Hz
c. 1200 Hz
d. 2400 Hz

26 A wave is represented by $r = A\sin\left[\alpha\left(\frac{x-y}{2}\right)\right]\cos\left[\omega t - \alpha\left(\frac{x+y}{2}\right)\right]$, where x and y are in metre and t is second. Then,

a. the wave is a stationary wave
b. the wave is a progressive wave propagating along + X-axis
c. the wave is a travelling wave propagating at a right angles to X-axis
d. all points lying on line $y = x + \frac{4\pi}{\alpha}$ are always at rest

27 S_1 and S_2 are two coherent sources of radiations separated by a distance 100.25 λ, where λ is the wavelength of radiations. S_1 leads S_2 in phase by $\frac{\pi}{2}$. P_1 and P_2 are two points on the line joining S_1 and S_2. The ratio of amplitudes of sources S_1 and S_2 is in the ratio 1 : 2. The ratio of intensity at P_1 to that at P_2 (I_1/I_2) is

a. 9
b. 0
c. $\frac{1}{9}$
d. 4

Statement Type

Direction **(Q. Nos. 28 and 29)** ***This section is based on Statement I and Statement II. Select the correct answer from the codes given below.***

a. Both Statement I and Statement II are correct and Statement II is the correct explanation of Statement I
b. Both Statement I and Statement II are correct but Statement II is not the correct explanation of Statement I
c. Statement I is correct but Statement II is incorrect
d. Statement II is correct but Statement I is incorrect

28 **Statement I** A standing wave pattern is formed in a string. The power transfer through all the points is zero.

Statement II At anitnode tension is perpendicular to velocity.

29 **Statement I** When a closed organ pipe vibrates, the pressure of the gas at the closed end remains constant.

Statement II In a stationary wave system, displacement nodes are pressure anitnodes, and displacement antinodes are pressure nodes.

SET - 2

Revisal Problems
for JEE Advanced

Only One Option Correct Type

Working Space

Direction *(Q. Nos. 1-10) This section contains 10 multiple choice questions. Each question has four choices (a), (b), (c) and (d), out of which ONLY ONE is correct.*

1 Find the ratio of the fundamental tone frequencies of two identical strings, after one of them was stretched by $\varepsilon_1 = 2\%$ and the other by $\varepsilon_2 = 4\%$. The tension is assumed to be proportional to the elongation.

a. 1.2 b. 0.91 c. 0.71 d. 1.5

2 In a stationary wave pattern that forms as a result of reflection of waves from an obstacle, the ratio of the amplitude at an antinode and a node is 1.5. What percentage of the energy passes across the obstacle ?

a. 96% b. 86% c. 92% d. 82%

3 Two pulses travelling on the same string are described by

$$y_1 = \frac{5}{(3x-4t)^2+2} \quad \text{and} \quad y_2 = \frac{-5}{(3x+4t-6)^2+2}.$$

At what instant do the two cancel everywhere ?

a. 0.5 s b. 0.75 s
c. 1s d. 0.25 s

4 In a large room, a person receives direct sound waves from a source 120 m away from him. He also receives waves from the same source which reach him, being reflected from the 25 m high ceiling at a point halfway between them. Among the following, for which wavelength will these sound waves interfere destructively ?

a. 10 m b. $\frac{20}{3}$ m
c. 2 m d. None of these

5 In the figure are shown two coherent sources S_1 and S_2 which emit sound of wavelength λ in phase. The separation between the sources is 3λ. A circular wire of large radius (R) is placed in such a way that S_1 and S_2 lie in its plane and the middle point of S_1S_2 is at the centre of the wire. A man starts walking from point A moves along the circumference and reaches point A again. During his journey, how many times does he observe maximum sound ?

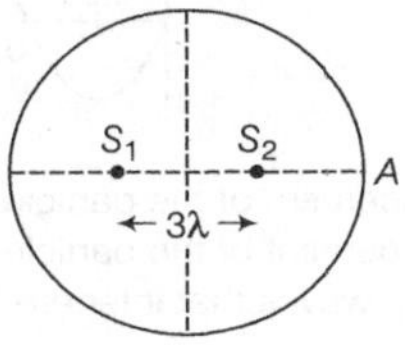

a. $\frac{R}{2\lambda}$ b. $\frac{R}{3\lambda}$
c. $\frac{R}{\lambda}$ d. None of these

6 In the above problem, how many number of times, he hears minimum sound ?

a. $\frac{2R}{(2\lambda+1)}$ b. $\frac{2R}{\lambda+1}$ c. $\frac{R}{\lambda+2}$ d. None of these

7 Three sources of sound S_1, S_2 and S_3 of equal intensity (I_0) are placed in a straight line with $S_1S_2 = S_2S_3$. At a point P, far away from the sources, the wave coming from S_2 is 120° ahead in phase of that from S_1. Also, the wave coming from S_3 is 120° ahead of that from S_2. What would be the resultant intensity of sound at P ?

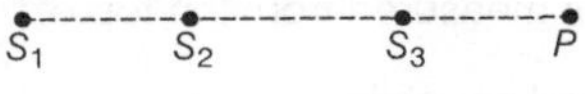

a. $\frac{\sqrt{3}}{2}I_0$ b. zero
c. $\frac{I_0}{2}$ d. $\frac{I_0}{\sqrt{2}}$

8 Two parts of a sonometer wire divided by a movable knife edge, differ in length by 2 mm and produce 1 beat s^{-1}, when sounded together. Find their frequencies, if the whole length of the wire is 1 m.

Working Space

a. 249 Hz, 250 Hz
b. 249.5 Hz, 248.5 Hz
c. 249.5 Hz, 250.5 Hz
d. 248 Hz, 249 Hz

9 Three successive frequencies for a string are 75 Hz, 125 Hz, 175 Hz. If the speed of transverse wave on the string is 400 ms^{-1}, the length of the string is

a. 6 m b. 5 m c. 4 m d. 3 m

10 A number of sonic sources emitting notes of same frequency f_0 are located at points $y = \pm na$ on Y-axis, where $n = 0, 1, 2$ and 3. If the phases of sources at $t = 0$ be 0 for S_0; $\left(\frac{2a^2\pi}{b\pi}\right)$ for S_1 and S_1', $\frac{4a^2\pi}{b\lambda}$ for S_2 and S_2', $\frac{6a^2\pi}{b\lambda}$ for S_3 and S_3', find the resultant intensity of sound at a point on X-axis at a distance b from O ($b >> a$). Intensity due to single source is I_0.

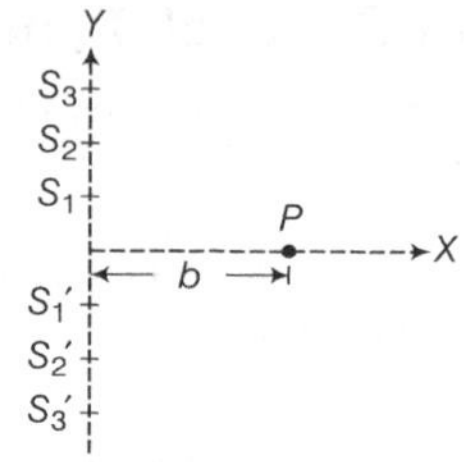

a. 49 I_0
b. 46 I_0
c. 36 I_0
d. 25 I_0

One or More than One Options Correct Type

Direction (Q. Nos. 11 and 12) This section contains 2 multiple choice questions. Each question has four choices (a), (b), (c) and (d), out of which ONE or MORE THAN ONE are correct.

11 Two adjacent natural frequencies of an organ pipe are found to be 550 Hz and 650 Hz. Speed of sound = 340 ms^{-1}. If the fundamental frequency is f_0 and length of the pipe is L, then

a. $f_0 = 50$ Hz b. $f_0 = 55$ Hz c. $L = 1.55$ m d. $L = 1.7$ m

12 For a certain transverse standing wave on a long string, an antinode is formed at $x = 0$ and next to it, a node is formed at $x = 0.1$m. The displacement $y(t)$ of the string particle at $x = 0$ is shown in the figure. Choose the correct statements.

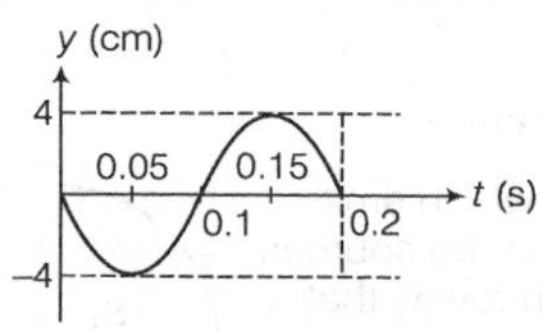

a. Transverse displacement of the particle at $x = 0.04$ m at $t = 0.025$ s is $-2\sqrt{2}$ cm
b. Transverse displacement of the particle at $x = 0.05$ m at $t = 0.05$ s is $-2\sqrt{2}$ cm
c. Speed of travelling waves that interfere to produce this standing wave is 2ms^{-1}
d. The transverse velocity of the string particle at $x = \frac{1}{15}$ m and $t = 0.1$ s is 20π cms^{-1}

Comprehension Type

Direction (Q. Nos. 13-19) This section contains 2 paragraphs, each describing theory, experiments, data, etc. Seven questions related to the paragraphs have been given. Each question has only one correct answer among the four given options (a), (b), (c) and (d).

Passage I

In an organ pipe (may be closed or open) of length 99 cm, standing wave is set up, whose equation is given by longitudinal displacement.

$$\varepsilon = (0.1\text{mm})\cos\left[\frac{2\pi}{0.8}(y + 1\text{cm})\right]\cos(400t),$$ where y is measured from the top of the tube in metre and t in seconds. Here, 1 cm is the end correction.

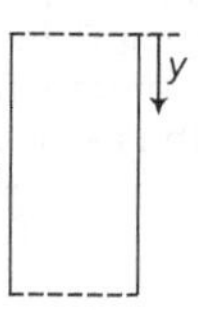

13 The upper end and the lower end of the tube, respectively, are

a. open-closed
b. closed-open
c. open-open
d. closed-closed

14 The air column is vibrating in

a. first overtone
b. second overtone
c. third overtone
d. fundamental mode

15 Equation of standing wave in terms of excess pressure is (Take, bulk modulus $= 5\times 10^5\ Nm^{-2}$).

a. $P_{ex} = (225\,\pi\ Nm^{-2})\cos\left(\frac{2\pi}{0.8}(y + 1\,cm)\right)\sin(200t)$

b. $P_{ex} = (225\,\pi\ Nm^{-2})\sin\left[\frac{2\pi}{0.8}(y + 1\,cm)\right]\cos(200t)$

c. $P_{ex} = (125\,\pi\ Nm^{-2})\cos\left[\frac{2\pi}{0.8}(y + 1\,cm)\right]\sin(400t)$

d. $P_{ex} = (125\,\pi\ Nm^{-2})\sin\left[\frac{2\pi}{0.8}(y + 1\,cm)\right]\cos(400t)$

16 Assume end correction approximately equals (0.3 × diameter of tube) and estimate the moles of air present inside the tube. [Also, assume tube is at NTP]

a. $\frac{10\,\pi}{60\times 22.4}$
b. $\frac{10\,\pi}{72\times 22.4}$
c. $\frac{10\,\pi}{18\times 22.4}$
d. $\frac{10\,\pi}{36\times 22.4}$

Passage II

In the arrangement shown, a mass can be hung from a string with a linear mass density of $2\times 10^{-3}\ kg\ m^{-1}$, that passes over a light pulley. The string is connected to a vibrator of frequency 700 Hz and the length of the string between the vibrator and the pulley is 1 m.

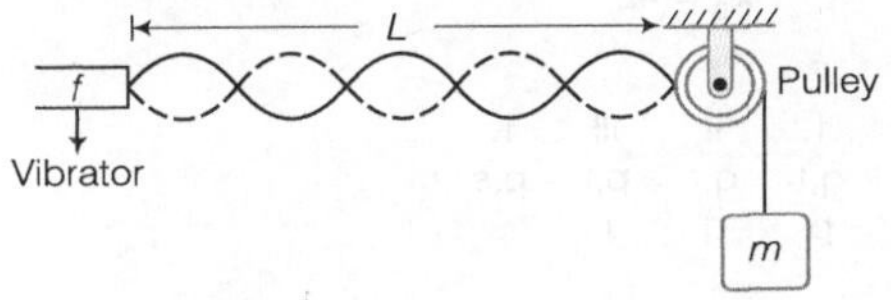

17 If the standing waves are observed, the largest mass to be hung is

a. 32 kg
b. 25 kg
c. 16 kg
d. 400 kg

18 If the mass suspended is 16 kg, then the number of loops formed in the string is

a. 8
b. 5
c. 3
d. 1

19 The string is set into vibrations and represented by the equation $y = 6\sin\left(\frac{\pi x}{10}\right)\cos(14000\,\pi t)$, where x and y are in cm and t in second. The maximum displacement at $x = 5$ m from the vibrator (in cm)

a. 2 cm
b. 3 cm
c. 6 cm
d. 5 cm

Matching List Type

***Direction** (Q. Nos. 20 and 21) Choices for the correct combination of elements from Column I and Column II are given as options (a), (b), (c) and (d), out of which one is correct.*

20 Match the standing waves in Column II due to plane progressive waves and also the conditions in Column I and mark the correct option from the codes given below.

	Column I		Column II
i.	Incident wave is $y_i = A\sin(kx - \omega t)$	p.	$y = 2A\cos kx \sin\omega t$
ii.	Incident wave is $y_i = A\cos(kx - \omega t)$	q.	$y = 2A\sin kx \sin\omega t$
iii.	$x = 0$ is rigid support	r.	$y = 2A\sin kx \cos\omega t$
iv.	$x = 0$ is free end	s.	$y = 2A\cos kx \cos\omega t$

Codes

	i	ii	iii	iv
a.	p	q	r	s
b.	q,r	p,s	s	p
c.	p,r	q,s	q,r	p,s
d.	q	p	s	r

21 Column I contains figures of systems, each of length L, for producing standing waves. The wave length in fundamental mode is represented by λ_0. Match the following columns and mark the correct option from the codes given below.

	Column I		Column II
i.	Pipe closed at one end 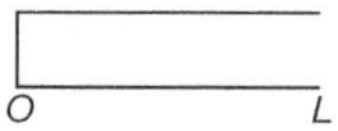	p.	Longitudinal waves
ii.	Pipe open at both ends 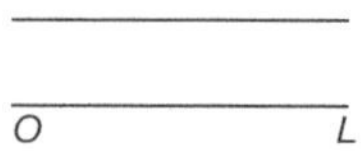	q.	Transverse waves
iii.	Stretched wire clamped at both ends 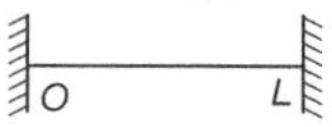	r.	$\lambda_0 = L$
iv.	Stretched wire clamped at both ends and the middle 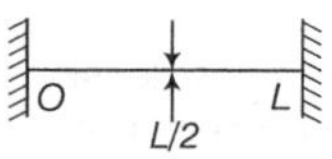	s.	$\lambda_0 = 2L$
		t.	$\lambda_0 = 4L$

Codes

	i	ii	iii	iv		i	ii	iii	iv
a.	p,t	p,s	q,s	q,r	b.	q,t	q,r	p,t	p,s
c.	p,t	p,r	q	r	d.	p	q	r	s

One Integer Value Correct Type

Direction *(Q. Nos. 22 and 23) This section contains 2 questions. When worked out will result in an integer from 0 to 9 (both inclusive).*

22 A 20 cm long string, having a mass of 1g is fixed at both the ends. The tension in the string is 0.5 N. The string is set into vibrations using an external vibrator of frequency 100 Hz. Find the separation (in cm) between the successive nodes in the string.

23 An aluminium wire of length $L = 60$ cm and of cross-sectional area 0.01 cm^2 is connected to a steel wire of the same cross-sectional area. The compound wire is loaded with a block of mass 10 kg as shown in the figure. The distance L_2 from the joint to the supporting pulley is 86.6 cm. Transverse waves are set up in the composite wire at its lowest frequency, such that the joint in the wire is a node. What is the total number of nodes observed at this frequency excluding the two at the ends of the wire ? Density of aluminium is 2.6g cm^{-3} and that of steel is 7.8g cm^{-3}.

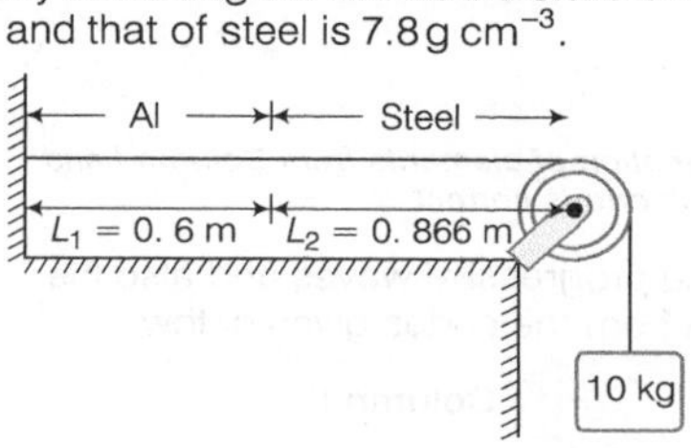

Working Space

JEE Main & AIEEE Archive

(Compilation of Last 13 Years' Questions)

1 A pipe of length 85 cm is closed from one end. Find the number of possible natural oscillations of air column in the pipe whose frequencies lie below 1250 Hz. The velocity of sound in air is 340 ms^{-1}. **(JEE Main 2014)**

a. 12 b. 8 c. 6 d. 4

2 A cylindrical tube open at both ends, has a fundamental frequency f, in air. The tube is dipped vertically in water so that half of it is in water. The fundamental frequency of the air-column is now **(AIEEE 2012)**

a. f b. $\frac{f}{2}$

c. $\frac{3f}{4}$ d. $2f$

3 **Statement I** Two longitudinal waves given by equations—$y_1(x,t) = 2a\sin(\omega t - kx)$ and $y_2(x,t) = a\sin(2\omega t - 2kx)$ will have equal intensity.

Statement II Intensity of waves of given frequency in same medium is proportional to the square of amplitude only. **(AIEEE 2011)**

a. Both Statement I and Statement II are correct and Statement II is the correct explanation of Statement I
b. Both Statement I and Statement II are correct but Statement II is not the correct explanation of Statement I
c. Statement I is correct but Statement II is incorrect
d. Statement II is correct but Statement I is incorrect

4 The transverse displacement $y(x,t)$ of a wave on a string is given by $y(x,t) = e^{-(ax^2 + bt^2 + 2\sqrt{ab}\,xt)}$.

This represents a **(AIEEE 2011)**

a. wave moving in $-x$ direction with speed $\sqrt{\frac{b}{a}}$
b. standing wave of frequency $\sqrt{b}$
c. standing wave of frequency $\frac{1}{\sqrt{b}}$
d. wave moving in $+x$ direction with speed $\sqrt{\frac{a}{b}}$

5 A string is stretched between fixed points separated by 75.0 cm. It is observed to have resonant frequencies of 420 Hz and 315 Hz. There are no other resonant frequencies between these two. Then, the lowest resonant frequency for this string is **(AIEEE 2006)**

a. 105 Hz b. 1.05 Hz
c. 1050 Hz d. 10.5 Hz

6 Length of a string tied to two rigid supports is 40 cm. Maximum length (wavelength in cm) of a stationary wave produced on it, is **(AIEEE 2002)**

a. 20 b. 80
c. 40 d. 120

JEE Archive

JEE Advanced & IIT JEE Archive

(Compilation of Last 13 Years' Questions)

1 One end of a taut string of length 3 m along the x-axis is fixed at $x = 0$. The speed of the waves in the string is $100\ \text{ms}^{-1}$. The other end of the string is vibrating in the y-direction so that stationary waves are set up in the string. The possible waveform(s) of these stationary wave is (are) **(2014 Adv. One or More than one Options)**

a. $y(t) = A\sin\frac{\pi x}{6}\cos\frac{50\pi t}{3}$ b. $y(t) = A\sin\frac{\pi x}{3}\cos\frac{100\pi t}{3}$

c. $y(t) = A\sin\frac{5\pi x}{6}\cos\frac{250\pi t}{3}$ d. $y(t) = A\sin\frac{5\pi x}{2}\cos 250\pi t$

2 A horizontal stretched string, fixed at two ends, is vibrating in its fifth harmonic according to the equation, $y(x,t) = (0.01\,\text{m})\,[\sin(62.8\,\text{m}^{-1})x]\cos[(628\,\text{s}^{-1})t]$. Assuming $\pi = 3.14$, the correct statement(s) is (are) **(2012 One or More than One Options)**

a. the number of nodes is 5

b. the length of the string is 0.25 m

c. the maximum displacement of the mid-point of the string from its equilibrium position is 0.01 m

d. the fundamental frequency is 100 Hz

3 Column I shows four systems, each of the same length L, for producing standing waves. The lowest possible natural frequency of a system is called its fundamental frequency, whose wavelength is denoted as λ_f. Match each system with statements given in Column II describing the nature and wavelength of the standing waves. **(2011, Matching Type)**

	Column I		Column II
i.	Pipe closed at one end 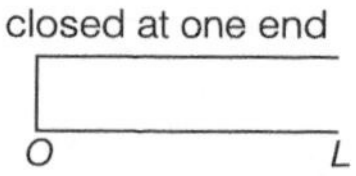	p.	Longitudinal waves
ii.	Pipe open at both ends 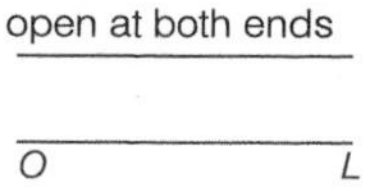	q.	Transverse waves
iii.	Stretched wire clamped at both ends 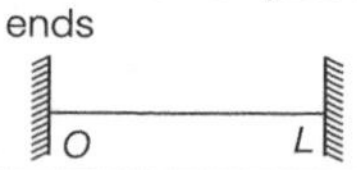	r.	$\lambda_f = L$
iv.	Stretched wire clamped at both ends and at mid-point 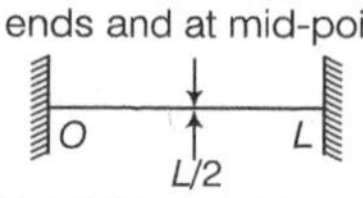	s.	$\lambda_f = 2L$
		t.	$\lambda_f = 4L$

Codes

	i	ii	iii	iv
a	p,s	q,t	p,r,s	q,s
b.	p,t	p,s	q,s	q,r
c.	p	q,t	q,r,s	p,q
d.	r,s	p,t	q,t	q,r

4 A hollow pipe of length 0.8 m is closed at one end. At its open end a 0.5 m long uniform string is vibrating in its second harmonic and it resonates with the fundamental frequency of the pipe. If the tension in the wire is 50 N and the speed of sound is $320\ \text{ms}^{-1}$, the mass of the string is **(2010, Only One Option Correct Type)**

a. 5 g

b. 10 g

c. 20 g

d. 40 g

5 A vibrating string of certain length l under a tension T resonates with a mode corresponding to the first overtone (third harmonic) of an air column of length 75 cm inside a tube closed at one end. The string also generates 4 beats s^{-1} when excited along with a tuning fork of frequency n. Now when the tension of the string is slightly increased the number of beats reduces to 2 per second. Assuming the velocity of sound in air to be $340\ \text{ms}^{-1}$, the frequency n of the tuning fork in Hz is **(2008,Only One Option Correct Type)**

a. 344 b. 336

c. 117.3 d. 109.3

6 An open pipe is in resonance in 2nd harmonic with frequency f_1. Now one end of the tube is closed and frequency is increased to f_2 such that the resonance again occurs in nth harmonic. Choose the correct option. **(2005, Only One Option Correct Type)**

a. $n = 3,\ f_2 = (3/4)\,f_1$

b. $n = 3,\ f_2 = (5/4)\,f_1$

c. $n = 5,\ f_2 = \frac{5}{4}f_1$

d. $n = 5,\ f_2 = \frac{3}{4}f_1$

Answers *with* Explanations

Answers

1. Oscillations

DPP-1 Kinematics of SHM

2. (i) Oscillatory (ii) Rectilinear (iii) SHM 3. *a* 4. *d* 5. *d* 6. *b* 7. *a*

8. *c* 9. *a* 10. *b* 11. *c* 12. *d* 13. *d* 14. *c* 15. *b* 16. *a*

17. *a,b,d* 18. *a,d* 19. *a,b,c,d* 20. *a,b,c,d* 21. *c,d* 22. *a,b,d* 23. *d* 24. *d*

25. *a* 26. *b* 27. *a* 28. *d* 29. *c* 30. *b*

DPP-2 Energy in SHM

1. *b* 2. *a* 3. *c* 4. *a* 5. *c* 6. *b* 7. *d* 8. *a* 9. *c* 10. *b*

11. *d* 12. *c* 13. *b* 14. *d* 15. *a* 16. *b,c* 17. *a,b,c* 18. *b* 19. *d* 20. *d*

21. *a* 22. *c* 23. *d* 24. *c* 25. *b* 26. *a* 27. 2 28. 5 29. 8

DPP-3 Simple Pendulum and Spring-Block System

1. *a* 2. *b* 3. *c* 4. *d* 5. *a* 6. *c* 7. *d* 8. *b* 9. *a* 10. *b*

11. *c* 12. *a* 13. *c* 14. *b* 15. *c* 16. *a,b,c* 17. *d* 18. *c,d* 19. *d* 20. *c*

21. *c* 22. *a* 23. *c* 24. *c* 25. *c* 26. *a* 27. *a* 28. 1 29. 2 30. 7

DPP-4 Time Period in SHM

1. *c* 2. *a* 3. *b* 4. *b* 5. *a* 6. *d* 7. *a* 8. *d* 9. *d* 10. *c*

11. *a* 12. *b* 13. *b,d* 14. *a,b,c* 15. *a,b,d* 16. *a* 17. *d* 18. *b* 19. *a* 20. *b*

21. *a* 22. *c* 23. *a* 24. *b* 25. *c* 26. 9 27. 4 28. 2 29. 6

DPP-5 Time Period : Physical Systems-I

1. *c* 2. *b* 3. *a* 4. *a* 5. *d* 6. *b* 7. *c* 8. *c* 9. *b* 10. *c*

11. *d* 12. *a* 13. *a,b* 14. *a,c* 15. *a,d* 16. *a* 17. *c* 18. *c* 19. *a* 20. *d*

21. *a* 22. 6 23. 4 24. 9 25. 7

DPP-6 Time Period : Physical Systems-II

1. *b* 2. *d* 3. *a* 4. *a* 5. *a* 6. *c* 7. *b* 8. *c* 9. *a* 10. *d*

11. *a* 12. *c* 13. *b* 14. *c* 15. *d* 16. *b,c* 17. *b,d* 18. *a,b,c* 19. *c* 20. *b*

21. *d* 22. *c* 23. *a* 24. 2 25. 8 26. 5 27. 4 28. 2 29. 5

DPP-7 Angular SHM

1. *b* 2. *d* 3. *b* 4. *a* 5. *c* 6. *c* 7. *d* 8. *b* 9. *a* 10. *c*

11. *a* 12. *a,c* 13. *c,d* 14. *b,c* 15. *b* 16. *a* 17. *b* 18. *c* 19. *a* 20. *a*

21. *b* 22. 5 23. 4 24. 8 25. 7 26. 6

DPP-8 Superposition of SHM

1. *c* 2. *b* 3. *a* 4. *d* 5. *d* 6. *c* 7. *c* 8. *b* 9. *a* 10. *d*

11. *a,b,c* 12. *a,c,d* 13. *a,b,c,d* 14. *a,b* 15. *a* 16. *d* 17. *c* 18. *d* 19. 3

DPP-9 Damped Oscillations and Resonance

4. *b* 5. *d* 6. *b* 7. *c* 8. *b* 9. *b* 10. *b* 11. *a* 12. *a* 13. *d*

14. *b* 15. *b* 16. *c* 17. *c* 18. *a* 19. *b* 20. *c* 21. *d*

Revisal Problems (JEE Main)

1. *a* 2. *c* 3. *b* 4. *d* 5. *b* 6. *a* 7. *d* 8. *b* 9. *b* 10. *c*

11. *c* 12. *a* 13. *d* 14. *d* 15. *c* 16. *b* 17. *d* 18. *a* 19. *c* 20. *b*

21. *a* 22. *d* 23. *a* 24. *b* 25. *c* 26. *d* 27. *b* 28. *b* 29. *a* 30. *b*

31. *d* 32. *d* 33. *b* 34. *a* 35. *c*

Revisal Problems(JEE Advanced)

1. *d* 2. *a* 3. *d* 4. *a* 5. *c* 6. *c* 7. *a,c* 8. *a,c* 9. *b,d* 10. *b,c*
11. *b,c,d* 12. *d* 13. *c* 14. *b* 15. *d* 16. *b* 17. *a* 18. *c* 19. *d* 20. *a*
21. *a* 22. *a* 23. *a* 24. 7 25. 5 26. 2 27. 8 28. 3 29. 5

JEE Main & AIEEE Archive

1. *b* 2. *a* 3. *b* 4. *b* 5. *d* 6. *c* 7. *d* 8. *b* 9. *a* 10. *a*
11. *c* 12. *c* 13. *b* 14. *d* 15. *d* 16. *a* 17. *a* 18. *a* 19. *b* 20. *a*
21. *b* 22. *a* 23. *c* 24. *b* 25. *c* 26. *b* 27. *a* 28. *c* 29. *c* 30. *d*
31. *a* 32. *a* 33. *b* 34. *c* 35. *b*

JEE Advanced & IIT JEE Archive

1. *b,d* 2. 3 3. *a* 4. *a* 5. *b* 6. *b* 7. *d* 8. *a,d* 9. 4 10. *d*
11. *c* 12. *d* 13. *a*

2. Progressive Waves

DPP-1 Basic Parameters and Wave Equation

1. (i) 0.25×10^{-3} cm (ii) $\frac{2\pi}{500}$ s (iii) 500 rad s^{-1} (iv) 8π m, $a_p = 62.5 \text{ cms}^{-2}$ $v_p = 0.125 \text{ cms}^{-1}$

2. *b* 3. *b* 4. *d* 5. *a* 6. *c* 7. *c* 8. *a* 9. *b* 10. *c* 11. *d*
12. *d* 13. *c* 14. *c* 15. *a* 16. *c* 17. *b,c,d* 18. *a,b,c,d* 19. *a,b,c,d*
20. *a,c* 21. *a* 22. *c* 23. *a* 24. *a* 25. *d* 26. *c* 27. *b* 28. *a*

DPP-2 Speed of Waves

1. *c* 2. *d* 3. *a* 4. *a* 5. *c* 6. *a* 7. *b* 8. *b* 9. *d* 10. *a*
11. *b* 12. *a,b* 13. *a,b,c* 14. *a* 15. *b* 16. *c* 17. *d* 18. *a* 19. *a* 20. *d*
21. *a* 22. *d* 23. 7 24. 5

26. (i) 80 ms^{-1} (ii) $y = (1\text{cm})\cos\left(40\pi t - \frac{\pi}{\Delta}x\right)$ (iii) $\frac{1}{\sqrt{2}}$ cm (iv) 89 cm s^{-1}

DPP-3 Energy, Intensity and Power of Waves

2. $v, \frac{2\pi}{kv}, \frac{1}{2}\rho A_0^2 k^2 v^3 \frac{v^3}{x^2}$ 3. (i) $1080\sqrt{10}$ W (ii) $\sqrt{2}$ times 4. *c* 5. *b* 6. *d*

7. *a* 8. *a* 9. *b* 10. *d* 11. *b* 12. *a* 13. *c* 14. *a,c,d* 15. *a,b* 16. *b,c*
17. *a,d* 18. *b,c* 19. *a,b,d* 20. *c* 21. *a* 22. *d* 23. *b* 24. *c* 25. *d* 26. *d*
27. *a* 28. *b* 29. 3 30. 2 31. 8 32. 7

DPP-4 Reflection and Transmission of Waves

1. (i) $y_t = \frac{2}{3} Ai \cos(2k_1 x - \omega_1 t)$, $y_r = \frac{Ai}{3}\cos(k_1 x + \omega_1 t + \pi)$ (ii) $p_i = p_t + p_r$ 2. (ii) 1 cms^{-1}

5. *b* 6. *d* 7. *c* 8. *a* 9. *b* 10. *d* 11. *c* 12. *b* 13. *a,b,c,d*
14. *a,c* 15. *a* 16. *c* 17. *b* 18. *a* 19. *b* 20. *a* 21. *b* 22. *a* 23. 9
24. 7 25. 2

Revisal Problems (JEE Main)

1. *b* 2. *a* 3. *d* 4. *a* 5. *c* 6. *d* 7. *b* 8. *c* 9. *a* 10. *d*
11. *d* 12. *b* 13. *c* 14. *b* 15. *a* 16. *d* 17. *a* 18. *c* 19. *c* 20. *c*
21. *d* 22. *a* 23. *b*

Revisal Problems(JEE Advanced)

1. *d* 2. *c* 3. *b* 4. *b* 5. *c* 6. *b* 7. *b* 8. *a* 9. *c* 10. *c*
11. *b* 12. *a,b,c,d* 13. *a,d* 14. *b,c* 15. *a,b,d* 16. *a* 17. *d* 18. *a* 19. *a*
20. *a* 21. *a* 22. *c* 23. *c* 24. *c* 25. *a* 26. *d* 27. 4 28. 2 29. 4
30. 4 31. 2

JEE Main & AIEEE Archive

1. *a* 2. *d* 3. *a* 4. *b* 5. *a* 6. *c*

JEE Advanced & IIT JEE Archive

1. *b* 2. *a* 3. *a*

3. Sound

DPP-1 Basic Parameters of Sound Waves

2. (i) 4.22×10^{-5} (ii) $339.6\ \text{ms}^{-1}, 2.65\times10^{-4}$ 3. (i) $\frac{\pi}{2}$ rad (ii) $\frac{2\pi}{35}$ rad

4. $\frac{1500}{\pi}$ Hz, $\frac{2\pi}{9}$ m, $\frac{1000}{3}$ ms^{-1} (ii) $(1.01\times10^{5}+0.02)$ Pa, $(1.01\times10^{5}-0.02)$ Pa 5. *a* 6. *b* 7. *c*

8. *d* 9. *b* 10. *c* 11. *d* 12. *c* 13. *d* 14. *a,c* 15. *b,d* 16. *a,c,d* 17. *c*
18. *a* 19. *c* 20. *a* 21. *d* 22. *c* 23. *b* 24. *a*

DPP-2 Speed of Sound

1. *c* 2. *b* 3. *a* 4. *d* 5. *a* 6. *b* 7. *a* 8. *d* 9. *c* 10. *d*
11. *a* 12. *a,b* 13. *c* 14. *b* 15. 9 16. 3 17. 5 18. 4 19. 2

DPP-3 Intensity of Sound

1. *a* 2. *a* 3. *c* 4. *b* 5. *b* 6. *d* 7. *b* 8. *d* 9. *d* 10. *b*
11. *b,d* 12. *a,b* 13. *a,b,d* 14. *b,d* 15. *a,d* 16. *c* 17. *d* 18. *a* 19. *d* 20. *d*
21. *a* 22. *a* 23. 4 24. 6 25. 7 26. 1230 m 27. (i) 13 dB (ii) $10^{6}\,I_1$

DPP-4 Doppler Effect-I

1. *a* 2. *a* 3. *b* 4. *b* 5. *d* 6. *b* 7. *c* 8. *b* 9. *a* 10. *b*
11. *d* 12. *c,d* 13. *b,d* 14. *a,c* 15. *b* 16. *b* 17. *c* 18. *a* 19. *d* 20. *b*
21. *c* 22. *a* 23. 1.00696×10^{5} Hz, 1.0304×10^{5} Hz 24. 97 Hz 25. 489.6 Hz
26. 301.6 Hz, 364.4 Hz

DPP-5 Doppler Effect-II

1. *c* 2. *d* 3. *a* 4. *d* 5. *b* 6. *a* 7. *b* 8. *d* 9. *a,d* 10. *c,d*
11. *d* 12. *d* 13. *a* 14. *c* 15. *a* 16. *b* 19. 442 Hz, 255 Hz
20. $d=1.35\times10^{11}$ m, $m=3.46\times10^{35}$ kg

Revisal Problems (JEE Main)

1. *a* 2. *b* 3. *d* 4. *b* 5. *c* 6. *a* 7. *a* 8. *c* 9. *b* 10. *c*
11. *a* 12. *c* 13. *b* 14. *b* 15. *d* 16. *a* 17. *c* 18. *c* 19. *d* 20. *a*
21. *b* 22. *d* 23. *d*

Revisal Problems(JEE Advanced)

1. *a* 2. *d* 3. *a* 4. *b* 5. *c* 6. *c* 7. *c* 8. *d* 9. *b* 10. *a,d*
11. *b,c,d* 12. *d* 13. *a,c* 14. *a,c,d* 15. *b,c* 16. *b* 17. *a* 18. *a* 19. *c* 20. *d*
21. *c* 22. *d* 23. *a* 24. *b* 25. *c* 26. 1

JEE Main & AIEEE Archive

1. *b* 2. *b* 3. *c* 4. *d* 5. *b* 6. *d* 7. *c* 8. *b* 9. *d* 10. *c*
11. *b* 12. *b*

JEE Advanced & IIT JEE Archive

1. *a,b* 2. *b,d* 3. *b* 4. *a* 5. *a,d* 6. *b* 7. *a* 8. *a* 9. *a* 10. *a*
11. *c* 12. *d*

4. Superposition and Standing Waves

DPP-1 Superposition and Interference

1. *a* 2. *c* 3. *b* 4. *b* 5. *c* 6. *d* 7. *d* 8. *b* 9. *c* 10. *d*
11. *a* 12. *b* 13. *c* 14. *b* 15. *a* 16. *a* 17. *c* 18. *d* 19. *c* 20. *d*
21. *d* 22. *b* 23. 6

DPP-2 Standing Waves in Strings

1. *d* 2. *a* 3. *b* 4. *c* 5. *d* 6. *b* 7. *c* 8. *d* 9. *a* 10. *c*
11. *a,d* 12. *a,b,c,d* 13. *a,b,d* 14. *a,d* 15. *b* 16. *a* 17. *a* 18. *c* 19. *d* 20. *a*
21. *a* 22. 163.4 Hz

DPP-3 Resonating Air Columns

1. $t = 43$ s 2. (i) $\frac{400}{189}$ (ii) $\frac{3}{4}$
3. *c* 4. *b* 5. *c* 6. *a* 7. *b* 8. *b* 9. *d* 10. *a*
11. *c* 12. *a* 13. *c* 14. *a,b* 15. *c,d* 16. *b,c,d* 17. *a,b,c* 18. *b,d* 19. *a* 20. *d*
21. *a* 22. *c* 23. *b* 24. *c* 25. 3

DPP-4 Standing Waves in Rods

2. $y = 10^{-6}\sin(25000\pi t + 5\pi x) + 10^{-6}\sin(25000\pi t - 5\pi x)$
3. Loop length $= \frac{\lambda}{2} = 25$ units, frequency $= 100$ units, maximum amplitude $= 20$ units, velocity $= 5000$ units
4. *d* 5. *b* 6. *a* 7. *b* 8. *c* 9. *d* 10. *b* 11. *b* 12. *c* 13. *a,b*
14. *a,b,c* 15. *c* 16. *a* 17. *c*

DPP-5 Interference in Time-Beats

2. *a* 3. *a* 4. *c* 5. *d* 6. *c* 7. *b* 8. *b* 9. *d* 10. *b*
11. *c* 12. *d* 13. *b* 14. *c* 15. *a* 16. *a* 17. *d* 18. *a,b* 19. *b,c,d* 20. *d*
21. *d* 22. 1 23. 8 24. 3 25. 9

Revisal Problems (JEE Main)

1. *b* 2. *a* 3. *c* 4. *d* 5. *a* 6. *b* 7. *c* 8. *d* 9. *c* 10. *a*
11. *d* 12. *a* 13. *b* 14. *c* 15. *d* 16. *a* 17. *d* 18. *c* 19. *d* 20. *c*
21. *a* 22. *b* 23. *b* 24. *a* 25. *a* 26. *d* 27. *c* 28. *d* 29. *d*

Revisal Problems(JEE Advanced)

1. *c* 2. *a* 3. *b* 4. *a* 5. *d* 6. *d* 7. *b* 8. *c* 9. *c* 10. *a*
11. *a,d* 12. *b,c,d* 13. *a* 14. *b* 15. *d* 16. *d* 17. *d* 18. *b* 19. *c* 20. *c*
21. *a* 22. 5 23. 6

JEE Main & AIEEE Archive

1. *c* 2. *d* 3. *d* 4. *a* 5. *a* 6. *b*

JEE Advanced & IIT JEE Archive

1. *a,c,d* 2. *b,c* 3. *b* 4. *b* 5. *a* 6. *c*

1. Oscillations

DPP-1 Kinematics of SHM

1. No, in the limit of small amplitudes, only those oscillatory motions can be treated as SHM for which the restoring force (or torque) becomes linear (i.e. $F \propto -x$ or $\tau \propto -\theta$).

For example the oscillatory motions of a simple pendulum and motion of atoms in a molecule become simple harmonic in the limit of small amplitude as restoring force(or torque) becomes linear. In case of oscillatory motion of a ball between two inclined planes or two perfectly elastic walls, the motion does not become simple harmonic even for a very small amplitude.

2. (i) $F = -2(x-3)^3$

$F = 0$ at $x = 3$. Force is along negative X-direction for $x > 3$ and it is along positive X-direction for $x < 3$. Thus, the motion of the particle is oscillatory (but not simple harmonic) about $x = 3$.

(ii) $F = -2(x-3)^2$

$F = 0$ for $x = 3$. But force is always along negative X-direction (except at $x = 3$). Thus, motion is rectilinear.

(iii) Let $x - 3 = X$. Then, the given equation becomes $F = -2X$.

Thus, the motion is SHM about $X = 0$ or $x = 3$.

3. (a) $x = A\sin(\omega t + \delta)$

If the particle is at mean position and moving towards positive extreme, $\delta = 0$

$\therefore \quad x = A\sin\omega t$

$\Rightarrow \quad \dfrac{A}{2} = A\sin\omega t$

$\therefore \quad \omega t = \dfrac{\pi}{6}$

$\Rightarrow \quad t = \dfrac{\pi}{6}\cdot\dfrac{T}{2\pi} = \dfrac{T}{12} = \dfrac{4}{12} = \dfrac{1}{3}\text{ s}$

4. (d) $v = A\omega\cos(\omega t + \delta)$

$a = -A\omega^2\sin(\omega t + \delta)$

$\therefore \quad v^2 = A^2\omega^2\cos^2(\omega t + \delta)$

and $\quad a^2 = A^2\omega^4\sin^2(\omega t + \delta)$

or $\quad a^2 = A^2\omega^4[1 - \cos^2(\omega t + \delta)]$

$= A^2\omega^4 - A^2\omega^4\cos^2(\omega t + \delta)$

or $\quad a^2 = A^2\omega^4 - \omega^2[A^2\omega^2\cos^2(\omega t + \delta)]$

$\Rightarrow \quad a^2 = A^2\omega^4 - \omega^2 v^2$

$\omega^2 v^2 = A^2\omega^4 - a^2$

$\Rightarrow \quad v^2 = A^2\omega^2 - \dfrac{1}{\omega^2}a^2$

$\therefore$ Best graph is (d)

5. (d) $v_{max} = A\omega = 0.04\text{ ms}^{-1}$

$x = A\sin\omega t$

$\Rightarrow \quad a = -A\omega^2\sin\omega t$

$\therefore \quad a = -\omega^2 x$

At $x = 0.02$ m, $\quad |a| = 0.06\text{ ms}^{-2}$ (given)

$\therefore \quad \omega^2(0.02) = 0.06$

$\Rightarrow \quad \omega = \sqrt{3}\text{ rads}^{-1}$

$\therefore \quad T = \dfrac{2\pi}{\omega} = \dfrac{2\pi}{\sqrt{3}} = 3.63\text{ s}$

$A = \dfrac{0.04}{\omega} = \dfrac{0.04}{\sqrt{3}} = 2.31 \times 10^{-2}\text{ m}$

6. (b) At the highest point, the situation is shown below.

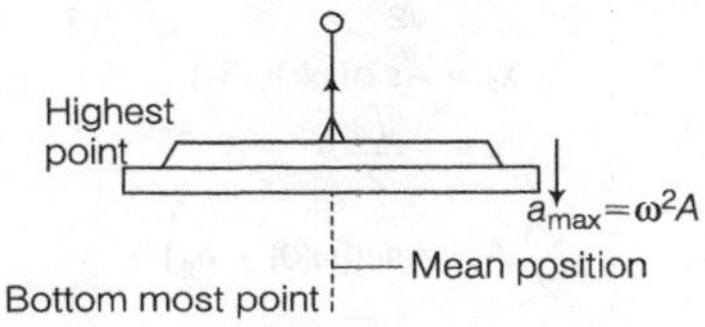

The man will feel weightlessness at the highest point if $\omega^2 A = g$.

According to the problem, $A = 0.5$ m and we know $\omega = 2\pi f$

$\therefore \quad 4\pi^2 f^2(0.5) = g$

$\Rightarrow \quad f^2 = \dfrac{g}{2\pi^2}$ or $f = \sqrt{\dfrac{g}{2}}\cdot\dfrac{1}{\pi}$

7. (a) The slope of given graph is

$-\tan 30° = -\dfrac{1}{\sqrt{3}}$

$\therefore \quad \dfrac{x}{a} = -\dfrac{1}{\sqrt{3}}$

We know in SHM, $\quad a = -\omega^2 x \Rightarrow \dfrac{x}{a} = -\dfrac{1}{\omega^2}$

$\therefore \quad \dfrac{1}{\omega^2} = \dfrac{1}{\sqrt{3}} \quad \Rightarrow \quad \omega = 3^{1/4}$

$\therefore \quad T = \dfrac{2\pi}{3^{1/4}}$

8. (c) Average speed is $v_{av} = \dfrac{\text{Total distance}}{\text{Total time}} = \dfrac{4A}{T}$ ...(i)

Maximum speed is v_0.

$\therefore \quad v_0 = A\omega = \dfrac{A\cdot 2\pi}{T}$

So, $\quad \dfrac{A}{T} = \dfrac{v_0}{2\pi}$

Using this in Eq. (i), we get

$v_{av} = \dfrac{2v_0}{\pi}$

9. (a) In 9 s, the distance travelled $= 4A \times 3 = 12A$

Imagine, now it is at $x = +\dfrac{A\sqrt{3}}{2}$ and $t = 0$.

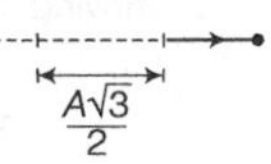

$x = A\sin(\omega t + \delta) \Rightarrow \dfrac{A\sqrt{3}}{2} = A\sin(0 + \delta)$

$\therefore \quad \sin\delta = \dfrac{\sqrt{3}}{2} \quad \Rightarrow \quad \delta = 60°$

$\therefore \quad x = A\sin(\omega t + 60°)$

$\omega = \dfrac{2\pi}{3}$

$\therefore \quad x = A\sin\left(\dfrac{2\pi}{3}t + \dfrac{\pi}{3}\right)$

After a time of 1 s, $x = A\sin\left(\dfrac{2\pi}{3}\cdot 1 + \dfrac{\pi}{3}\right)$ or $x = 0$

i.e. it is at mean position.

$\therefore$ Distance travelled in last 1 s $= \left(A - \dfrac{A\sqrt{3}}{2}\right) + A = 2A - \dfrac{A\sqrt{3}}{2}$

So, total distance travelled in 10 s

$= 12A + \left(2A - \dfrac{A\sqrt{3}}{2}\right) = 14A - \dfrac{A\sqrt{3}}{2} = A\left(14 - \dfrac{\sqrt{3}}{2}\right)$

$= \left(\dfrac{28 - \sqrt{3}}{2}\right)A$

10. (b) $x_1 = A\sin(\omega t + \delta_1)$

At $t = 0$, $\quad x = \dfrac{A}{\sqrt{2}}$

$$\Rightarrow \quad \frac{A}{\sqrt{2}} = A\sin[\omega(0) + \delta_1]$$

$$\sin\delta_1 = \frac{1}{\sqrt{2}} \quad \Rightarrow \quad \delta_1 = \frac{\pi}{4}$$

$$x_2 = A\sin(\omega t + \delta_2)$$

At $t = 0$, $\quad x = -\dfrac{\sqrt{3}}{2}A$

$$\Rightarrow \quad -\frac{\sqrt{3}}{2}A = A\sin[\omega(0) + \delta_2]$$

$$\sin\delta_2 = -\frac{\sqrt{3}}{2}$$

$$\Rightarrow \quad \delta_2 = \pi + \frac{\pi}{3} = \frac{4\pi}{3}$$

$$\delta_2 - \delta_1 = \frac{4\pi}{3} - \frac{\pi}{4} = \left(\frac{16-3}{12}\right)\pi = \frac{13}{12}\pi$$

11. (c) If this instant is taken as $t = 0$. Then,

initial phase for particle (1) is $\delta_1 = \dfrac{\pi}{2}$

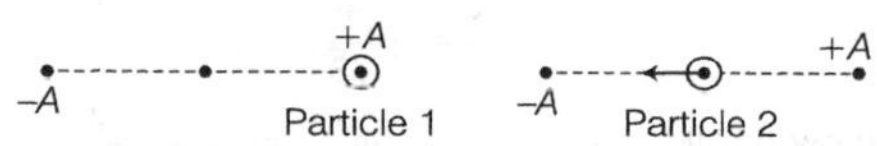

and initial phase for particle (2) is $\delta_2 = \pi$.

$$\therefore \quad x_1 = A\sin\left(\omega t + \frac{\pi}{2}\right)$$

and $\quad x_2 = A\sin(\omega t + \pi)$

When both of them have same displacement,

$$x_1 = x_2$$

$$\therefore \quad A\sin\left(\omega t + \frac{\pi}{2}\right) = A\sin(\omega t + \pi)$$

$$\Rightarrow \quad \cos\omega t = -\sin\omega t$$

$$\Rightarrow \quad \tan\omega t = -1$$

$$\Rightarrow \quad \frac{3\pi}{4} = \omega t \quad \text{or} \quad \frac{2\pi}{T}t = \frac{3\pi}{4}$$

$$\Rightarrow \quad t = \frac{3T}{8}$$

12. (d) Let particle (1) is moving towards right and particle (2) is moving towards left at this instant, $t = 0$.

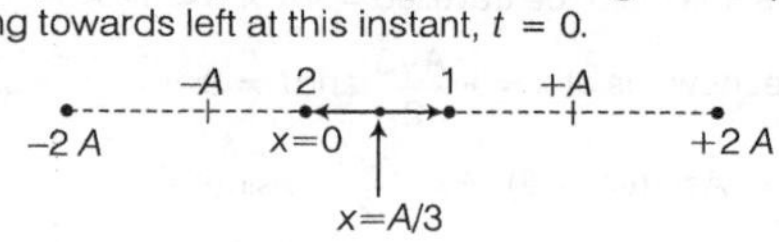

$$x_1 = A\sin(\omega t + \delta_1)$$

and $\quad x_2 = 2A\sin(\omega t + \delta_2)$

So, $\quad v_1 = A\omega\cos(\omega t + \delta_1)$

and $\quad v_2 = 2A\omega\cos(\omega t + \delta_2)$

At $t = 0$, $\quad x_1 = \dfrac{A}{3}$ and v_1 is positive

$$\therefore \quad \frac{A}{3} = A\sin[\omega(0) + \delta_1] \quad \Rightarrow \sin\delta_1 = \frac{1}{3}$$

$$\therefore \quad \delta_1 = \sin^{-1}\left(\frac{1}{3}\right)$$

At $t = 0$, $\quad x_2 = \dfrac{A}{3}$ and v_2 is negative.

$$\therefore \quad \frac{A}{3} = 2A\sin[\omega(0) + \delta_2]$$

$$\Rightarrow \quad \sin\delta_2 = \frac{1}{6}$$

As v_2 is negative at this instant, $\sin(\pi - \delta_2) = \dfrac{1}{6}$

$$\therefore \quad \pi - \delta_2 = \sin^{-1}\left(\frac{1}{6}\right)$$

$$\delta_2 = \pi - \sin^{-1}\left(\frac{1}{6}\right)$$

$\therefore$ Phase difference is

$$\delta_2 - \delta_1 = \pi - \sin^{-1}\left(\frac{1}{6}\right) - \sin^{-1}\left(\frac{1}{3}\right)$$

13. (d) We know, $\quad v = \omega\sqrt{A^2 - x^2}$

$$\therefore \quad v_0^2 = \frac{k}{m}(A^2 - d^2)$$

$$\Rightarrow \quad A^2 - d^2 = \frac{mv_0^2}{k}$$

$$\therefore \quad A = \sqrt{\frac{mv_0^2}{k} + d^2}$$

$$T = \frac{2\pi}{\omega} = 2\pi\sqrt{\frac{m}{k}}$$

$$y = A\sin(\omega t + \delta)$$

At $t = 0$, $\quad y = d$

$$\Rightarrow \quad d = A\sin(0 + \delta)$$

$$\sin\delta = \frac{d}{A}$$

$$\Rightarrow \quad \delta = \sin^{-1}\left(\frac{d}{\sqrt{\dfrac{mv_0^2}{k} + d^2}}\right)$$

When it reaches mean position, $y = 0$

$$\therefore \quad 0 = A\sin(\omega t + \delta)$$

$$\Rightarrow \quad \omega t + \delta = \pi$$

$$\omega t = \pi - \sin^{-1}\left(\frac{d}{\sqrt{\dfrac{mv_0^2}{k} + d^2}}\right)$$

$$\Rightarrow \quad t = \sqrt{\frac{m}{k}}\left[\pi - \sin^{-1}\left(\frac{d}{\sqrt{\dfrac{mv_0^2}{k} + d^2}}\right)\right]$$

14. (c) In equilibrium position, $mg = kx_0$

$$\Rightarrow \quad x_0 = \frac{mg}{k}$$

When the bullet gets embedded, the new equilibrium position changes.

Let it is a distance x from initial equilibrium position.

$$\therefore \quad \left(m + \frac{m}{2}\right)g = k(x_0 + x)$$

$$\Rightarrow \quad \frac{3}{2}mg = k\left(\frac{mg}{k} + x\right)$$

$$x = \frac{3}{2}\frac{mg}{k} - \frac{mg}{k}$$

$$\Rightarrow \quad x = \frac{mg}{2k}$$

Applying conservation of linear momentum,

$$\frac{m}{2}u + 0 = \left(m + \frac{m}{2}\right)v$$

$$v = \frac{m}{2} \times \frac{2}{3m}u = \frac{u}{3}$$

So, at $t = 0$, mass $\frac{3m}{2}$ is at a distance $\frac{mg}{2k}$ from its mean position and moving up with velocity $\frac{u}{3}$.

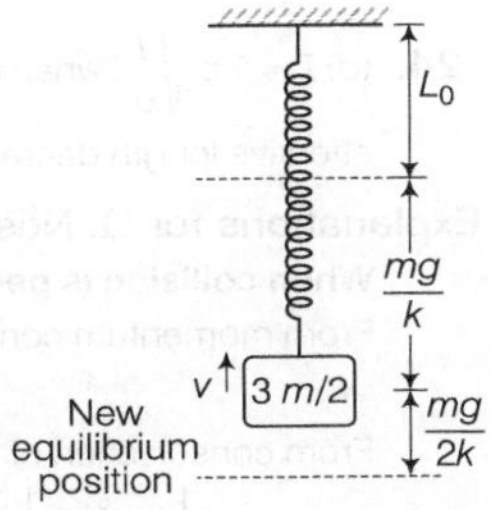

$$\omega = \sqrt{\frac{k}{(3m/2)}} = \sqrt{\frac{2k}{3m}}$$

$$v = \omega\sqrt{A^2 - x^2}$$

$$\Rightarrow \quad \frac{u}{3} = \sqrt{\frac{2k}{3m}\left[A^2 - \left(\frac{mg}{2k}\right)^2\right]}$$

$$\frac{u^2}{9} = \frac{2k}{3m}\left[A^2 - \left(\frac{mg}{2k}\right)^2\right]$$

$$A^2 - \left(\frac{mg}{2k}\right)^2 = \frac{mu^2}{6k}$$

$$\Rightarrow \quad A = \sqrt{\frac{mu^2}{6k} + \left(\frac{mg}{2k}\right)^2}$$

15. (b) $y = A\sin(\omega t + \delta)$; At $t = 0$,

$$y = \frac{mg}{2k}$$

$$\therefore \quad \frac{mg}{2k} = A\sin[\omega(0) + \delta] \quad \Rightarrow \quad \sin\delta = \frac{mg}{2kA}$$

$$\delta = \sin^{-1}\left[\frac{mg}{2k\sqrt{\frac{mu^2}{6k} + \left(\frac{mg}{2k}\right)^2}}\right]$$

At maximum extreme at the top, $y = A$

$\therefore \quad \sin(\omega t + \delta) = 1$

$$\Rightarrow \quad \omega t + \delta = \frac{\pi}{2}$$

$$\omega t = \frac{\pi}{2} - \sin^{-1}\left[\frac{mg}{2k\sqrt{\frac{mu^2}{6k} + \left(\frac{mg}{2k}\right)^2}}\right]$$

$$t = \sqrt{\frac{3m}{2k}}\left[\frac{\pi}{2} - \sin^{-1}\left\{\frac{mg}{2k\sqrt{\frac{mu^2}{6k} + \left(\frac{mg}{2k}\right)^2}}\right\}\right]$$

$$= \sqrt{\frac{3m}{2k}}\cos^{-1}\left\{\frac{mg}{2k\sqrt{\frac{mu^2}{6k} + \left(\frac{mg}{2k}\right)^2}}\right\}$$

16. (a) $x = A\sin(\omega t + \delta)$

When $x = \frac{A}{2}$, $\quad \frac{A}{2} = A\sin(\omega t + \delta)$

$$\Rightarrow \quad \sin(\omega t + \delta) = \frac{1}{2}$$

$\omega t + \delta$ is phase, let $\omega t + \delta = \phi$, then

$$\sin\phi = \frac{1}{2} \quad \Rightarrow \quad \phi_1 = \frac{\pi}{6}$$

and $\quad \phi_2 = \pi - \frac{\pi}{6} = \frac{5\pi}{6}$

Phase difference is $\quad \phi_2 - \phi_1 = \frac{5\pi}{6} - \frac{\pi}{6} = \frac{4\pi}{6} = \frac{2\pi}{3}$

17. (a,b,d) $x = A\sin^2\left(\omega t - \frac{\pi}{4}\right)$

$$= A\left[\frac{1 - \cos 2\left(\omega t - \frac{\pi}{4}\right)}{2}\right] = \frac{A}{2} - \frac{A}{2}\cos\left(2\omega t - \frac{\pi}{2}\right)$$

or $\quad x = \frac{A}{2} - \frac{A}{2}\sin 2\omega t \quad \ldots$(i)

$\therefore$ Amplitude $= \frac{A}{2}$, angular frequency $= 2\omega$

$$v = \frac{dx}{dt} = -\frac{A}{2}2\omega\cos 2\omega t$$

or $\quad v = -A\omega\cos 2\omega t$

From Eq. (i), $\quad \frac{A}{2}\sin 2\omega t = \frac{A}{2} - x$

or $\quad \sin 2\omega t = 1 - \frac{2}{A}x$

$$\therefore \quad \cos 2\omega t = \sqrt{1 - \left(1 - \frac{2}{A}x\right)^2}$$

So, $\quad v = -A\omega\sqrt{1 - \left(1 - \frac{2}{A}x\right)^2}$

$$v^2 = A^2\omega^2\left[1 - \left(1 - \frac{2}{A}x\right)^2\right]$$

$$= A^2\omega^2\left[1 - \left\{1 + \frac{4}{A^2}x^2 - \frac{4}{A}x\right\}\right]$$

$$= A^2\omega^2\left(-\frac{4}{A^2}x^2 + \frac{4}{A}x\right)$$

$$= \omega^2(-4x^2 + 4Ax) = 4\omega^2 x(A - x)$$

$$\therefore \quad v^2 = 4\omega^2 x(A - x)$$

18. (a,d) ***Case I*** $\quad v_{av} = \frac{2A}{T/2} = 4\frac{A}{T}$

$$v_{max} = A\omega = A\cdot\frac{2\pi}{T}$$

$$\therefore \quad r_1 = \frac{v_{av}}{v_{max}} = -\frac{4A/T}{2\pi A/T} = \frac{2}{\pi}$$

Case II $\quad a_{av} = \frac{v_{max} - 0}{T/4} = \frac{4A\omega}{T} = \frac{4A}{T}\left(\frac{2\pi}{T}\right)$ or $a_{av} = \frac{8A\pi}{T^2}$

$$a_{max} = A\omega^2 = A\left(\frac{2\pi}{T}\right)^2 = \frac{4\pi^2 A}{T^2}$$

$$\therefore \quad r_2 = \frac{a_{av}}{a_{max}} = \frac{8A\pi/T^2}{4\pi^2 A/T^2} = \frac{2}{\pi}$$

19. (a,b,c,d) At $t = 0$, it has zero velocity and is at $x = 2$ cm.

So, amplitude is $A = 2$ cm

$$f = 0.25\ s^{-1}$$

$$\Rightarrow \quad \omega = 2\pi f = 2\pi(0.25) = \frac{\pi}{2}\ \text{rad}s^{-1}$$

$$\therefore \quad v_{max} = A\omega = 2 \times \frac{\pi}{2} = \pi\text{cm s}^{-1} = 3.14\ \text{cm s}^{-1}$$

$$x = A\sin(\omega t + \delta),$$

At $t = 0$, $\quad x = +2$ cm

$$\therefore \quad 2 = 2\sin[\omega(0) + \delta] \quad \Rightarrow \quad \delta = \frac{\pi}{2}$$

$\therefore \quad x = A\cos\omega t$ and $v = -A\omega\sin\omega t$

At $t = 3$s, $\quad x = 2\cos\left(\frac{\pi}{2}\cdot 3\right) = 0$

At $t = 3$ s, $\quad v = -A\omega\sin\left(\frac{\pi}{2}\cdot 3\right) = A\omega$

$$\Rightarrow \quad v = 3.14\ \text{cms}^{-1}$$

20. (a,b,c,d) From the figure, it is clear that $A = 0.08$ m and initial phase is $\delta = 0$.

$\therefore \quad x = 0.08\sin\omega t$

From figure, $T = 4$s

so, $\omega = \frac{2\pi}{4} = \frac{\pi}{2}$ rad s^{-1} $= \frac{3.14}{2} = 1.57$ rad s^{-1}

So, $x = 0.08\sin(1.57)t$

$\therefore \quad v = 0.08(1.57)\cos(1.57)t = 0.1256\cos(1.57)t$

$(v)_{\text{at } t = 1s} = 0.1256\cos(1.57)(1) = 0$

$$a = \frac{dv}{dt} = -(0.1256)(1.57)\sin(1.57)t$$

$$= -0.1972\sin(1.57)t$$

$(a)_{t=1s} = -0.1972\sin(1.57)(1) = -0.1972$ ms^{-2}

$|(a)_{t=1s}| = 0.1972$ ms^{-2}

Spring constant is

$$k = m\omega^2 = (0.8)(1.57)^2 = 1.972 \text{ Nm}^{-1}$$

21. (c,d) $x = 5\sin\left(20t + \frac{\pi}{3}\right)$

$\therefore \quad v = 5(20)\cos\left(20t + \frac{\pi}{3}\right) = 100\cos\left(20t + \frac{\pi}{3}\right)$

and $a = -100(20)\sin\left(20t + \frac{\pi}{3}\right) = -2000\sin\left(20t + \frac{\pi}{3}\right)$

When $v = 0$, $100\cos\left(20t + \frac{\pi}{3}\right) = 0$

$\Rightarrow \quad 20t + \frac{\pi}{3} = \frac{\pi}{2} \Rightarrow 20t = \frac{\pi}{6}$

$\Rightarrow \quad t = \frac{\pi}{120}$ s

So, option (a) is incorrect.

When $a = 0$,

$-2000\sin\left(20t + \frac{\pi}{3}\right) = 0 \Rightarrow 20t + \frac{\pi}{3} = 0$ gives t negative

so, $20t + \frac{\pi}{3} = \pi \Rightarrow t = \frac{2\pi}{60} = \frac{\pi}{30}$ s

So, option (b) is incorrect.

When acceleration is zero, v is maximum.

So, option (c) is correct.

From the given equation,

$\Rightarrow \quad \omega = 20 \Rightarrow 2\pi f = 20$

$\Rightarrow \quad f = \frac{10}{\pi}$ s^{-1}

So, option (d) is correct.

22. (a,b,d) First two expression satisfy, $\frac{d^2y}{dt^2} \propto (-y)$

$$y = A\sin\omega t\cos\omega t \Rightarrow y = \frac{A}{2}\sin 2\omega t$$

This also satisfies, $\frac{d^2y}{dt^2} \propto (-y)$

So, options (a), (b), (d) represent SHM.

In case of option (c), $\frac{d^2y}{dt^2}$ is not proportional to $(-y)$, so it does not represent SHM.

23. (d) In the oscillatory system shown time period is not constant i.e. damped oscillatory are not periodic. In SHM, time period is constant.

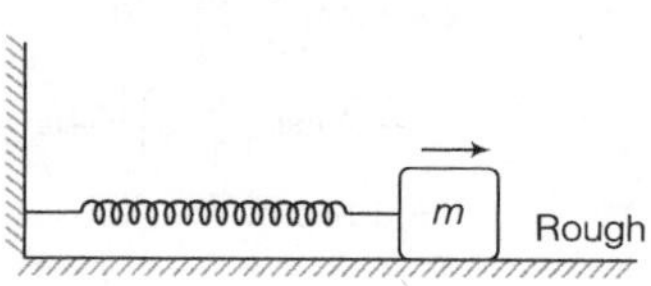

24. (d) $T = 2\pi\sqrt{\frac{l}{g}}$, when the boy stands centre of mass moves up, so effective length decreases and thus time period decreases.

Explanations for Q. Nos. (25-27)

When collision is perfectly inelastic

From momentum conservation, $(1 + 2)v = 1 \times 6$

$\Rightarrow \quad v = 2$ ms^{-1}

From conservation of energy,

$$\frac{1}{2}mv^2 = \frac{1}{2}kA^2$$

$\therefore \quad A = \sqrt{\frac{m}{k}}v = \sqrt{\frac{3}{600}} \times 2 = 0.141$ m $= 14.1$ cm

$$T = 2\pi\sqrt{\frac{m}{k}} = 2\pi\sqrt{\frac{3}{600}} = 0.44\text{s}$$

When the collision is perfectly elastic

$$v_2 = \left(\frac{m_2 - m_1}{m_1 + m_2}\right)u_2 + \left(\frac{2m_1}{m_1 + m_2}\right)u_1$$

$$= 0 + \left(\frac{2 \times 1}{1 + 2}\right)6 = 4\text{ms}^{-1}$$

$$A = \sqrt{\frac{m}{k}} \cdot v_2$$

$$= 4\sqrt{\frac{2}{600}} = 0.23\text{m} = 23 \text{ cm}$$

$$T = 2\pi\sqrt{\frac{m}{k}}$$

$$= 2\pi\sqrt{\frac{2}{600}} = 0.36\text{s}$$

Case I $\quad \omega = \frac{2\pi}{T} = 14.28$ rad s^{-1}

$\therefore \quad x = (14.1 \text{ cm})\sin(14.28t)$

Case II $\quad \omega = \frac{2\pi}{T} = \frac{2\pi}{0.36} = 17.45$ rads^{-1}

$\therefore \quad x = (23 \text{ cm})\sin(17.45t)$

25. (a) **26.** (b) **27.** (a)

28. (d) $\omega = \sqrt{\frac{k}{m}} = \sqrt{\frac{500}{2}} = 15.81$ rads^{-1}

29. (c) When elevator is accelerating, the FBD of block is

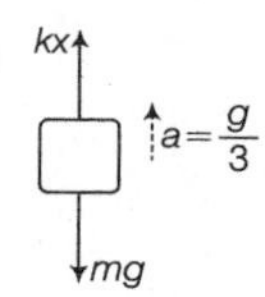

$$kx - mg = ma = \frac{mg}{3}$$

$$x = \frac{4mg}{3k} = \frac{4(2)(10)}{3(500)}$$

$$= 0.053 \text{ m} = 5.3 \text{ cm}$$

30. (b) When the elevator has no acceleration, $kx_0 = mg$

$$x_0 = \frac{mg}{k} = \frac{(2)(10)}{500} = 0.04 \text{ m} = 4\text{cm}$$

$\therefore$ Amplitude, $A = 5.3 - 4 = 1.3$cm

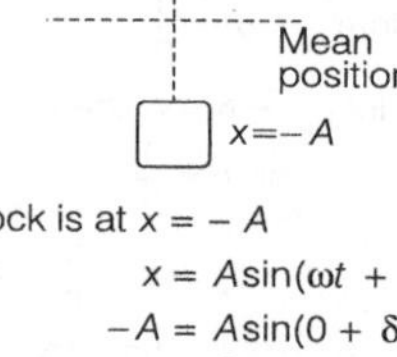

At time $t = 0$, block is at $x = -A$

$$x = A\sin(\omega t + \delta)$$

$$-A = A\sin(0 + \delta) \Rightarrow \sin\delta = -1$$

$\therefore \quad \delta = \frac{3\pi}{2}$

DPP-2 Energy in SHM

1. (b) $U = \beta x^4$ (given)

$\therefore \quad U_{max} = \beta \cdot A^4$

$$K(x) = U_{max} - U(x) = \beta A^4 - \beta x^4 = \beta(A^4 - x^4)$$

$$U(x) = \frac{1}{3}K(x) \qquad \text{(given)}$$

$\therefore \quad \beta x^4 = \frac{1}{3}\beta(A^4 - x^4)$

$\Rightarrow \quad 3\beta x^4 = \beta A^4 - \beta x^4$

$$4\beta x^4 = \beta A^4 \Rightarrow x^4 = \frac{A^4}{4}$$

$\Rightarrow \quad x = \pm \frac{A}{\sqrt{2}}$

2. (a) $U = \frac{1}{2}m\omega^2 x^2 = \frac{1}{2}m\omega^2 A^2 \cos^2 \omega t$

At $t = 0$, $\quad U \neq 0$

At $x = 0$, $\quad U \neq 0$

So, the correct graphs are (1) and (3).

3. (c) At displacement x, $U_1 = \frac{1}{2}m\omega^2 x^2$

$\Rightarrow \quad x = \sqrt{\frac{2U_1}{m\omega^2}}$

Similarly, $\quad y = \sqrt{\frac{2U_2}{m\omega^2}}$

When displacement is $(x + y)$, potential energy is

$$U = \frac{1}{2}m\omega^2 (x + y)^2 = \frac{1}{2}m\omega^2 (x^2 + y^2 + 2\sqrt{xy})$$

$$= \frac{1}{2}m\omega^2 x^2 + \frac{1}{2}m\omega^2 y^2 + m\omega^2\sqrt{xy}$$

$$= U_1 + U_2 + m\omega^2 \sqrt{\frac{2U_1}{m\omega^2} \cdot \frac{2U_2}{m\omega^2}}$$

$$= U_1 + U_2 + 2\sqrt{U_1 U_2}$$

4. (a) $y = \frac{Mg}{k}$, $\quad U = \frac{1}{2}ky^2 = \frac{1}{2}k\frac{M^2g^2}{k^2} = \frac{1}{2}\cdot\frac{M^2g^2}{k}$

$\therefore$ Total potential energy is $U = U_1 + U_2 + U_3$

or $\quad U = \frac{1}{2}M^2g^2\left(\frac{1}{k_1} + \frac{1}{k_2} + \frac{1}{k_3}\right)$

5. (c) $K = \frac{1}{2}mv^2 = \frac{1}{2}mA^2\omega^2\cos^2 \omega t$

Average value of $\cos^2 \omega t$ is $\frac{1}{2}$ over one cycle.

$\therefore \quad K_{av} = \frac{1}{4}m\omega^2 A^2 \qquad \ldots$(i)

$$U = \frac{1}{2}m\omega^2 x^2 = \frac{1}{2}m\omega^2 A^2 \sin^2 \omega t$$

Average value of $\sin^2 \omega t$ is $\frac{1}{2}$ over one cycle.

$\therefore \quad U_{av} = \frac{1}{4}m\omega^2 A^2 \qquad \ldots$(ii)

From Eqs. (i) and (ii), we get

$$K_{av} = U_{av}$$

6. (b) Total mechanical energy is $E_T = 160\text{ J}$

$\therefore \quad U_{max} = 160\text{ J}$

At extreme position KE is zero. Work done by spring force from extreme position to mean position is $W = \frac{1}{2}kA^2$

$\therefore \quad K_{max} = W = \frac{1}{2}(2 \times 10^6)(0.01)^2 = 100\text{ J}$

$\therefore \quad U_{min} = 160 - 100 = 60\text{ J}$

7. (d) $U = k - ke^{-x^2}$

$$\frac{dU}{dx} = (-k)(e^{-x^2})(-2x) = 2kxe^{-x^2}$$

$$F = -\frac{dU}{dx} = -2kxe^{-x^2}$$

$F = 0$, only when $x = 0$. So, there cannot be an equilibrium position other than zero (except at ∞). So option (a) is incorrect.

If x is positive, F is towards the origin. So, (b) is incorrect.

$$\frac{d^2U}{dx^2} = +2k[xe^{-x^2}(-2x) + e^{-x^2}(1)] = +2ke^{-x^2}(1 - 2x^2)$$

At $x = 0$, $\frac{d^2U}{dx^2}$ is positive.

So, at the origin U is minimum. So, KE cannot be minimum. So, option (c) is incorrect.

$F = -\left(\frac{2k}{e^{-x^2}}\right)x$, for small displacements from $x = 0$.

F is approximately proportional to $-x$. So, option (d) is correct.

8. (a) When a particle is executing SHM, at mean position $F = -\frac{dU}{dr} = 0$. Also, its potential energy is minimum at equilibrium position. So, $\frac{d^2U}{dr^2} > 0$.

$$\frac{dU}{dr} = -F \Rightarrow \frac{d^2U}{dr^2} = -\frac{dF}{dr}$$

At equilibrium, $\quad dF = -\left|\frac{d^2U}{dr^2}\right|_{r=0} dr$

Comparing this standard SHM equation, i.e.

$$F_{net} = -k_{eff}\, x \text{ we have } k_{eff} = \left|\frac{d^2U}{dr^2}\right|_{r=0}$$

In general equilibrium happens at $r = r_0$ (say)

$\therefore \quad k_{eff} = \left|\frac{d^2U}{dr^2}\right|_{r=r_0}$

Also, at equilibrium $F = -\frac{dU}{dr} = 0$

Now, coming to the situation given, when the bob is given a small horizontal displacement as shown.

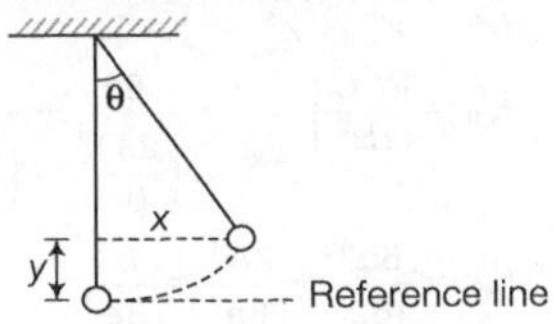

$$U = mgy$$

$$y = l(1 - \cos\theta) = 2l\sin^2\frac{\theta}{2}$$

$\therefore \quad U = 2mgl\sin^2\frac{\theta}{2}$

As θ is small $\quad \sin\frac{\theta}{2} \simeq \frac{\theta}{2}$.

So, $\quad U = \frac{1}{2}mgl\left(\frac{\theta}{2}\cdot\frac{\theta}{2}\right) \quad$ or $\quad U = \frac{1}{2}mgl\,\theta^2$

$$\theta \simeq \frac{x}{l} \qquad \Rightarrow U = \frac{mg}{2l}x^2$$

$$\frac{dU}{dx} = \frac{mg}{2l}\cdot 2x = \frac{mg}{l}\cdot x$$

$\Rightarrow \quad \frac{d^2U}{dx^2} = \frac{mg}{l}$

$\therefore \quad k_{eff} = \left|\frac{d^2U}{dx^2}\right|_{x=0} = \frac{mg}{l}$

9. (c) $U = U_0 \sin ax \Rightarrow \frac{dU}{dx} = U_0 a \cos ax$

At equilibrium, $-\frac{dU}{dx} = 0$

$\Rightarrow \quad U_0 a \cos ax = 0$

$\Rightarrow \quad ax = \frac{\pi}{2}$ or $x = \frac{\pi}{2a}$

$$\frac{d^2U}{dx^2} = -U_0 a^2 \sin ax$$

$$k_{eff} = \left|\frac{d^2U}{dx^2}\right|_{x=\frac{\pi}{2a}} = U_0 a^2 \sin\left(a \cdot \frac{\pi}{2a}\right) = U_0 a^2$$

10. (b) $U = \frac{a}{x^{12}} - \frac{b}{x^6}$

$\Rightarrow \quad \frac{dU}{dx} = \frac{-12a}{dx} = \frac{-12a}{x^{13}} + \frac{6b}{x^7}$

At equilibrium, $F = -\frac{dU}{dx} = 0.$

Let at equilibrium $x = x_0$

$\therefore \quad \frac{-12a}{x_0^{13}} + \frac{6b}{x_0^7} = 0 \quad \Rightarrow \quad \frac{2a}{x_0^{13}} = \frac{b}{x_0^7}$

$x_0^6 = \frac{2a}{b} \quad \Rightarrow \quad x_0 = \left(\frac{2a}{b}\right)^{1/6}$

Now, $\frac{d^2U}{dx^2} = \frac{156a}{x^{14}} - \frac{42b}{x^8}$

$$k_{eff} = \left|\frac{d^2U}{dx^2}\right|_{x=\left(\frac{2a}{b}\right)^{1/6}} = \frac{156a}{\left[\left(\frac{2a}{b}\right)^{1/6}\right]^{14}} - \frac{42b}{\left[\left(\frac{2a}{b}\right)^{1/6}\right]^{8}}$$

This gives, $k_{eff} = \frac{9b^2}{a}\left(\frac{b}{2a}\right)^{1/3}$

11. (d) $U = \frac{a}{x^2} - \frac{b}{x} \Rightarrow \frac{dU}{dx} = \frac{-2a}{x^3} + \frac{b}{x^2}$

At equilibrium,

$F = -\frac{dU}{dx} = 0$ and $x = x_0$

$\therefore \quad \frac{-2a}{x_0^3} + \frac{b}{x_0^2} = 0 \Rightarrow x_0 = \frac{2a}{b}$

$$\frac{d^2U}{dx^2} = \frac{6a}{x^4} - \frac{2b}{x^3}$$

$$k_{eff} = \left|\frac{d^2U}{dx^2}\right|_{x=x_0} = \frac{6a}{\left(\frac{2a}{b}\right)^4} - \frac{2b}{\left(\frac{2a}{b}\right)^3}$$

$$k_{eff} = \frac{6b^4}{16a^3} - \frac{2b^4}{8a^3} = \frac{b^4}{8a^3}$$

12. (c) The left half of the graph is usual SHM.

$\therefore \quad t_1 = \pi\sqrt{\frac{m}{k}}$

For the right part, $U = mgx$

At the origin, $U = 0$

$\therefore \quad K = E_0$

As we move right the potential energy increases linearly with x.

This is similar to a body that is thrown vertically upwards with initial velocity.

$\therefore \quad \frac{1}{2}mu^2 = E_0 \Rightarrow u = \sqrt{\frac{2E_0}{m}}$

Time of flight till it reaches point of projection (i.e. origin) is

$$t_2 = \frac{2u}{g} = \frac{2}{g}\sqrt{\frac{2E_0}{m}} = 2\sqrt{\frac{2E_0}{mg^2}}$$

Total time of oscillation is $T_1 = t_1 + t_2 = \pi\sqrt{\frac{m}{k}} + 2\sqrt{\frac{2E_0}{mg^2}}$

13. (b) Initial energy $= \frac{1}{2}kL^2$ (at the compression L)

At the mean position, m_1 and m_2 will have maximum velocity towards right. Later velocity of m_1 decreases due to spring force, while m_2 continues to move towards right with v_{max}.

$$\frac{1}{2}kL^2 = \frac{1}{2}(m_1 + m_2)v_{max}^2$$

$\Rightarrow \quad v_{max} = \sqrt{\frac{k}{m_1 + m_2}}L$

Now, let the amplitude of oscillation of m_1 is A.

$\therefore \quad \frac{1}{2}kA^2 = \frac{1}{2}m_1\left(\sqrt{\frac{k}{m_1 + m_2}}L\right)^2$

$$A^2 = \frac{m_1}{m_1 + m_2}L^2 \quad \Rightarrow \quad A = \sqrt{\frac{m_1}{m_1 + m_2}}L$$

14. (d) At mean position, KE = 0.008 J [given]

$\therefore \quad \frac{1}{2}mv_{max}^2 = 0.008$

$$v_{max}^2 = \frac{0.008 \times 2}{m} = \frac{0.008 \times 2}{0.1} = 0.16$$

$$v_{max} = 0.4 \text{ ms}^{-1}$$

Given, $A = 0.1$ m, $v_{max} = A\omega$

$\Rightarrow \quad \omega = \frac{v_{max}}{A} = \frac{0.4}{0.1} = 4\text{s}^{-1}$

$\delta = 45° = \frac{\pi}{4}$ (given)

The equation of motion is $x = 0.1\sin\left[(4\text{s}^{-1})t + \frac{\pi}{4}\right]$

$[\because x = A\sin(\omega t + \delta)]$

15. (a) Comparing the given equation with $x = A\sin(\omega t + \delta)$

$$A = 1.5\text{m}, \omega = \frac{\pi}{4}\text{s}^{-1}, m = 2\text{kg (given)}$$

$\therefore$ Total energy is $E_T = \frac{1}{2}m\omega^2 A^2 = \frac{1}{2}(2)\left(\frac{\pi}{4}\right)^2(1.5)^2 = 1.39$ J

16. (b,c) $E = Ax^2 + Bv^2$

When $v = 0$, x is maximum are equal to amplitude.

$\therefore \quad$ Amplitude $= \sqrt{\frac{E}{A}}$

When $x = 0$, v is maximum.

So, $v_{max} = \sqrt{\frac{E}{B}} \Rightarrow v_{max} = A\omega$

$$\Rightarrow \quad \omega = \frac{v_{max}}{A} = \frac{\sqrt{\frac{E}{B}}}{\sqrt{\frac{E}{A}}} = \sqrt{\frac{A}{B}}$$

$\therefore \quad T = \frac{2\pi}{\omega} = 2\pi\sqrt{\frac{B}{A}}$

From the given equation, it is clear that x is not directly proportional to v.

17. (a,b,c) At extreme position, $F \cdot 2A = \frac{1}{2}k(2A)^2 \Rightarrow A = \frac{F}{k}$

At mean position, work done by $F = \frac{1}{2}kA^2 + \frac{1}{2}mv^2$

$\therefore \quad F \cdot A = \frac{1}{2}kA^2 + (\text{KE})_{\text{mean position}}$

$$F \cdot \frac{F}{k} - \frac{1}{2}k\left(\frac{F}{k}\right)^2 = (\text{KE})_{\text{mean position}}$$

$$(\text{KE})_{\text{mean position}} = \frac{F^2}{2k}$$

$\therefore \quad (U)_{\text{mean position}} = \frac{F^2}{k} - \frac{F^2}{2k} = \frac{F^2}{2k}$

18. (b) $x = A\sin(\omega t + \delta)$

$$v = \frac{dx}{dt} = A\omega\cos(\omega t + \delta)$$

$$a = \frac{dv}{dt} = -A\omega^2(\omega t + \delta)$$

Statement II is not the correct explanation of Statement I.

19. (d) $x = A\sin\omega t \Rightarrow v = A\omega\cos\omega t$

Kinetic energy is $K = \frac{1}{2}mv^2 = \frac{1}{2}m\omega^2A^2\cos^2\omega t$

$$K = \frac{1}{2}m\omega^2A^2\left(\frac{1+\cos 2\omega t}{2}\right)$$

You can see that $\frac{d^2K}{dt^2}$ is not proportional to $-K$.

So, KE varies periodically with angular frequency 2ω.

Same is the case with PE.

20. (d) At mean position potential energy is minimum (may be zero, not necessarily zero.) U-x graph can be

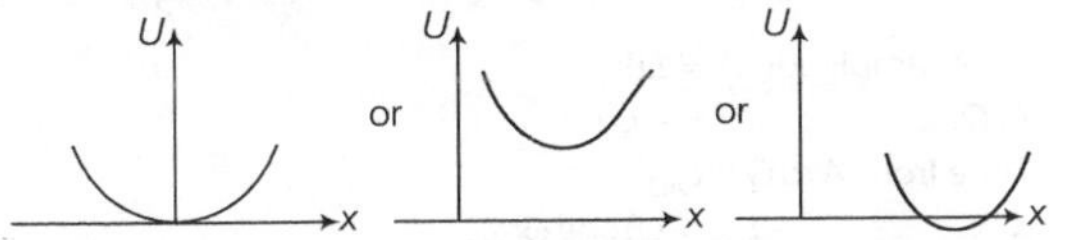

21. (a)

(i) For SHM, $v = A\sqrt{\omega^2 - x^2}$ given equation is similar to this.

(ii) $v = -kx$

If x is positive, v is negative and as x decreases, v decreases. When $x = 0$, $v = 0$. So, the object does not change its direction.

If x is negative, v is positive and as x decreases in magnitude, v also decreases, when $x = 0$, $v = 0$. So, the object does not change its direction. [Note that $a = v\frac{dv}{dx} = K^2x$]

(iii) When acceleration of elevator is zero, $kx_0 = mg$

When the elevator has an upward acceleration a_0, from elevator frame, $ma_0 + mg = kx$

$$ma_0 = kx - kx_0 = k(x - x_0)$$

As elevator accelerates from rest, the object performs SHM in elevator's frame.

(iv) The speed is $\sqrt{2}$ times the escape speed. Therefore, the object will leave the earth. It will therefore, not change the direction and its kinetic energy will keep on decreasing.

Hence, (i) → (p), (ii) → (q,r), (iii) → (b), (iv) → (q,r)

22. (c) $U_{max} = E_T = 20\,J$, $U_{min} = 4\,J$

$\therefore$ $K_{max} = (K)_{at\ x=0} = E_T - U_{min} = 20 - 4 = 16\,J$

$$U_{min} = \frac{1}{2}kA^2 \Rightarrow \frac{1}{2}k = \frac{20}{A^2}$$

$$U_{at\ x=\frac{A}{2}} = \frac{1}{2}k\left(\frac{A}{2}\right)^2 = \frac{20}{A^2}\cdot\frac{A^2}{4} = 5\,J$$

$$(K)_{at\ x=\frac{A}{2}} = 20 - 5 = 15\,J$$

$$(U)_{at\ x=\frac{A}{4}} = \frac{1}{2}k\left(\frac{A^2}{16}\right)$$

$$= \frac{20}{A^2}\cdot\frac{A^2}{16} = 1.25\,J$$

Hence, (i) → (s), (ii) → (s), (iii) → (q), (iv) → (s)

23. (d) $(U)_{at\ x=5m} = 30 + (5-3)^2 = 34\,J$

$(K)_{at\ x=5m} = 25\,J$ (given)

So, total mechanical energy,

$$K + U = 25 + 34 = 59\,J$$

24. (c) Particle will move between the points, where KE is zero.

$$K = E - U = 59 - 30 - (x-3)^2 = 29 - (x-3)^2$$

When $K = 0$, $29 - (x-3)^2 = 0$

$\Rightarrow$ $x^2 + 9 - 6x = 29$

$$x^2 - 6x - 20 = 0$$

$$x = \frac{6 \pm \sqrt{36 - 4(1)(-20)}}{2(1)}$$

$$x = \frac{6 \pm \sqrt{116}}{2} = 3 \pm \sqrt{29}$$

$$x = 8.39\,m \text{ and } x = -2.39\,m$$

So, the particle oscillates between −2.39 m and 8.39 m.

25. (b) $K = 29 - (x-3)^2$

KE is maximum when $x = 3m$ and $K_{max} = 29\,J$

26. (a) $U = 30 + (x-3)^2$

$$\frac{dU}{dx} = 2(x-3)$$

$$F = -\frac{dU}{dx} = -2x + 6$$

$$F = 0$$

when $-2x + 6 = 0$

or $x = 3m$

27. (2) $U = 2 - 20x + 5x^2$

$$\frac{dU}{dx} = -20 + 10x$$

$$F = -\frac{dU}{dx} = 20 - 10x = -10(x-2)$$

$\therefore$ The particle is executing SHM about mean position $x - 2 = 0$ or $x = 2$

$\therefore$ $k = 10$

$\Rightarrow$ $m\omega^2 = 10$

$$\omega^2 = \frac{10}{m} = \frac{10}{0.1} = 100$$

$\Rightarrow$ $\omega = 10\text{ rad s}^{-1}$

$$v_{max} = A\omega = 5(10) = 50\text{ ms}^{-1}$$

$\therefore$ $\beta = 2$

28. (5) In SHM, U-x graph is a parabola.

$\therefore$ $U = C_1x^2$, where C_1 is a constant.

From figure when, $x = 0.2\,m$, $U = 2\,J$

$\therefore$ $2 = C_1(0.2)^2$

$\Rightarrow$ $C_1 = \frac{2}{0.04} = 50$

$\therefore$ $U = 50x^2$

Also, in SHM $U = \frac{1}{2}m\omega^2x^2$

$\therefore$ $\frac{1}{2}m\omega^2 = 50$

$\Rightarrow$ $\omega^2 = \frac{100}{4} = 25$

$\Rightarrow$ $\omega = 5\text{s}^{-1}$

$\therefore$ $T = \frac{2\pi}{\omega} = \frac{2\pi}{5}$

$$T = \frac{2\pi}{5}\text{s}$$

29. (8) When x is small, from the given equation,

$$F \approx -8x$$

$\Rightarrow$ $k = 8$

DPP-3 **Simple Pendulum and Spring-Block System**

1. (a) In the absence of wall W_2, the time period of block would have been $2\pi\sqrt{\frac{m}{k}}$. But due to the presence of W_2, it alters. But for the left part of oscillation (from the mean position shown), the period will be $T_1 = \pi\sqrt{\frac{m}{k}}$.

For the right side part, $d = A\sin\omega t$ [from $x = A\sin\omega t$]

$$\therefore \quad d = \frac{5d}{3}\sin\omega t$$

$$\Rightarrow \quad \sin\omega t = \frac{3}{5}$$

$$\omega t = \sin^{-1}\left(\frac{3}{5}\right)$$

$$\Rightarrow \quad t = \frac{T}{2\pi}\sin^{-1}\left(\frac{3}{5}\right)$$

This is the time taken by the block to reach W_2 from mean position. Collision with W_2 is perfectly elastic (given).

Time taken for right side part of oscillation will be

$$T_2 = 2t = 2\left[\frac{T}{2\pi}\sin^{-1}\left(\frac{3}{5}\right)\right]$$

$$= \frac{2\cdot 2\pi\sqrt{\frac{m}{k}}}{2\pi}\sin^{-1}\left(\frac{3}{5}\right)$$

$$= 2\sqrt{\frac{m}{k}}\sin^{-1}\left(\frac{3}{5}\right)$$

$\therefore$ Total time period is

$$T = T_1 + T_2$$

$$= \pi\sqrt{\frac{m}{k}} + 2\sqrt{\frac{m}{k}}\sin^{-1}\left(\frac{3}{5}\right)$$

$$= \sqrt{\frac{m}{k}}\left[\pi + 2\sin^{-1}\left(\frac{3}{5}\right)\right]$$

2. (b) $T = 2\pi\sqrt{\frac{L}{g}}$

$\therefore T \propto \sqrt{L}$, as time period decreases when the length of pendulum decreases, the time period of shorter pendulum (T_s) is smaller than that of longer pendulum (T_l). That means shorter pendulum performs more oscillations in a given time.

It is given that after n oscillations of shorter pendulum, both are again in phase. So, by this time longer pendulum must have made $(n-1)$ oscillations.

$$\therefore \quad nT_s = (n-1)T_l$$

$$\Rightarrow \quad n\sqrt{1} = (n-1)\sqrt{25}$$

$$n = 5n - 5$$

$$\Rightarrow \quad 4n = 5$$

$$\Rightarrow \quad n = \frac{5}{4}$$

So, the two pendulums shall be in the same phase for the first time when the shorter pendulum has completed $\frac{5}{4}$ oscillation

3. (c) Half of oscillation (from mean position), the block performs SHM with spring of stiffness k and the remaining half with spring of stiffness $4k$.

$$t_1 = \frac{T_1}{2} = \pi\sqrt{\frac{M}{k}}$$

and $$t_2 = \frac{T_2}{2} = \pi\sqrt{\frac{M}{4k}} = \frac{\pi}{2}\sqrt{\frac{M}{k}}$$

$$\therefore \quad T = t_1 + t_2 = \frac{3}{2}\pi\sqrt{\frac{M}{k}}$$

4. (d) If $\alpha < \beta$, tha ball collides with the wall and rebounds with same speed. The motion of ball from A to Q is one part of oscillation of simple pendulum.

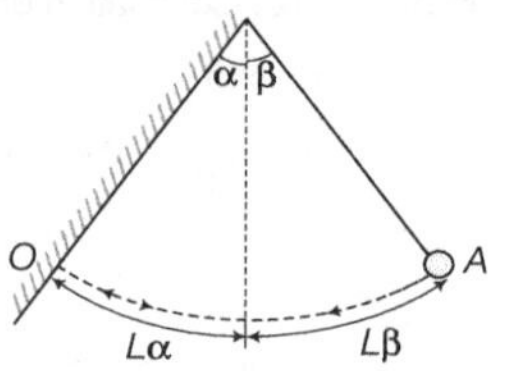

$$T' = 2(t_{AQ})$$

where, T' is time period for one complete oscillation, i.e. from A to Q and then from Q to A.

Consider A as starting point.

$$\therefore \quad x = A\sin(\omega t + \delta) = A\sin\left(\omega t + \frac{\pi}{2}\right)$$

Here, amplitude $A = L\beta$

At Q, $\quad x = -L\alpha$

Time from A to Q is t_{AQ}.

$$\therefore \quad -L\alpha = L\beta\cos\omega t_{AQ}$$

$$\therefore \quad \cos\omega t_{AQ} = -\frac{\alpha}{\beta}$$

$$\Rightarrow \quad t_{AQ} = \frac{1}{\omega}\cos^{-1}\left(-\frac{\alpha}{\beta}\right)$$

But $$\omega = \frac{2\pi}{T} = \frac{2\pi}{2\pi\sqrt{\frac{L}{g}}} = \sqrt{\frac{g}{L}}$$

$$\therefore \quad T' = 2t_{AQ} = 2\sqrt{\frac{L}{g}}\cos^{-1}\left(-\frac{\alpha}{\beta}\right)$$

$$= 2\sqrt{\frac{L}{g}}\left[\pi - \cos^{-1}\left(\frac{\alpha}{\beta}\right)\right]$$

$$= 2\pi\sqrt{\frac{L}{g}} - 2\sqrt{\frac{L}{g}}\cos^{-1}\left(\frac{\alpha}{\beta}\right)$$

5. (a) If $\alpha > \beta$, the ball does not collide with the wall.

Thus, $$T = 2\pi\sqrt{\frac{L}{g}}$$

6. (c) Let initial extension is x_0.

$$kx_0 = m_1 g$$

$$k(x_0 + x) = m_1 g + m_2 g$$

$$\Rightarrow \quad kx = m_2 g \quad \text{or} \quad k = \frac{m_2 g}{x}$$

$$T = 2\pi\sqrt{\frac{m_1 + m_2}{k}} = 2\pi\sqrt{\frac{(m_1 + m_2)x}{m_2 g}}$$

7. (d) When 500 g is removed, $m = (100 + 300)\text{g} = 0.4\text{kg}$

$$\therefore \quad T = 2\pi\sqrt{\frac{0.4}{k}} = 2\text{s}$$

$$\Rightarrow \quad \frac{2\pi}{\sqrt{k}} = \frac{2}{\sqrt{0.4}} \quad \text{... (i)}$$

When 300 g is also removed,

$$m' = 100\text{ g} = 0.1\text{ kg}$$

$$\therefore \quad T' = 2\pi\sqrt{\frac{0.1}{k}} = \frac{2}{\sqrt{0.4}}\sqrt{0.1} \quad \text{(using Eq. (i))}$$

$$T' = \frac{2}{2} = 1\text{ s}$$

8. (b) During oscillations, no external force is acting. So, centre of mass position remains fixed. An instance is shown in the figure when m is displaced by x towards right, M is displaced towards left by y, such that

$$mx = My \quad [\because \text{centre of mass does not change}]$$

$$y = \frac{m}{M}x$$

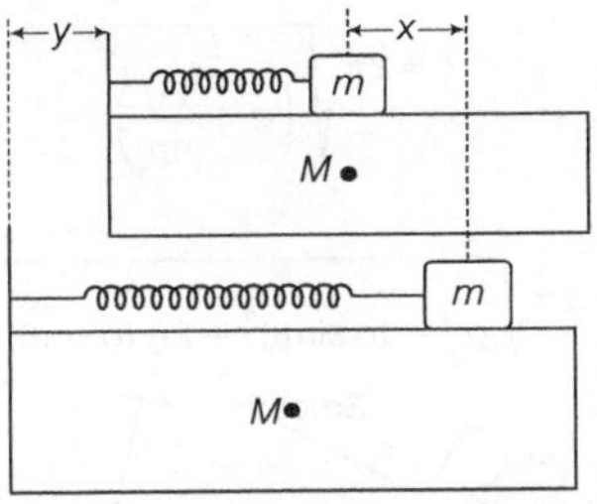

At the instant shown, elongation in the spring is $(x + y)$.

$$\therefore \quad m\frac{d^2x}{dt^2} = -k(x+y)$$

$$= -k\left[x + \frac{m}{M}x\right] = -k\left(\frac{M+m}{M}\right)x$$

or $$\frac{d^2x}{dt^2} = -k\left(\frac{M+m}{Mm}\right)x$$

Comparing with $a = -\omega^2 x$, we have

$$\omega = \sqrt{\frac{k(M+m)}{Mm}}$$

$$\therefore \quad T = \frac{2\pi}{\omega} = 2\pi\sqrt{\frac{Mm}{k(M+m)}}$$

Alternate Method

$$M\frac{d^2y}{dt^2} = -k(x+y)$$

$$= -k\left(\frac{M}{m}y + y\right) \quad \left[\because y = \frac{m}{M}x \Rightarrow x = \frac{M}{m}y\right]$$

$$= -k\left(\frac{M+m}{m}\right)y$$

or $$\frac{d^2y}{dt^2} = -k\left(\frac{M+m}{Mm}\right)y$$

Comparing with, $a = -\omega^2 y$, we have

$$\omega = \sqrt{\frac{k(M+m)}{Mm}}$$

$$\Rightarrow \quad T = \frac{2\pi}{\omega}$$

$$= 2\pi\sqrt{\frac{Mm}{(M+m)k}}$$

9. (a) $kx_0 = mg$

The spring is never compressed. Hence, spring shall exert least force on the block when the block is at top extreme.

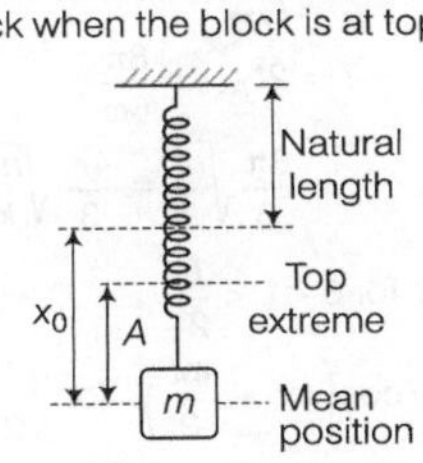

$$F_{min} = kx_0 - kA = mg - m\omega^2 A = mg - \frac{4\pi^2}{T^2}mA$$

10. (b) As the small block oscillates along the inclined plane with an amplitude A, the centre of mass of the system undergoes SHM along the horizontal direction. The position of centre of mass of at an instant is

$$X_{COM} = \frac{mx + M(0)}{m+M} = \frac{mA\sin\omega t\cos 60^\circ}{m+M}$$

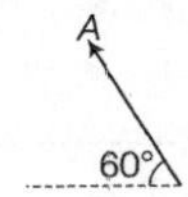

$$X_{COM} = \frac{1}{2}\cdot\frac{m}{m+M}A\sin\omega t$$

The acceleration of centre of mass in the horizantal direction is

$$a_{COM} = -\omega^2 X_{COM}$$

Along the horizontal net force (which is frictional force)

$$= (M+m)\,a_{COM} = (M+m)\omega^2\frac{1}{2}\cdot\frac{m}{m+M}A\sin\omega t$$

$$= \frac{1}{2}m\omega^2 A\sin\omega t$$

11. (c) In this case of spring-block system, a_0 and g do not effect time period.

$$\therefore \quad T = 2\pi\sqrt{\frac{m}{k}}$$

12. (a)

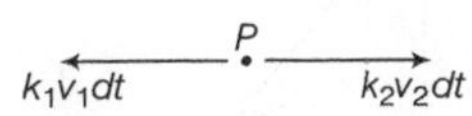

(k is spring constant)

$$k_1v_1dt = k_2v_2dt$$

and $$k_1 \propto \frac{1}{x}$$

and $$k_2 \propto \frac{1}{L_0 - x}$$

$$\therefore \quad \frac{v_1}{x} = \frac{v_2}{L_0 - x}$$

$$\Rightarrow \quad v_1L_0 - v_1x = v_2x$$

$$x(v_1 + v_2) = L_0v_1$$

$$\Rightarrow \quad x = \frac{L_0v_1}{v_1+v_2}$$

13. (c) $F = -\frac{dU}{dx} \Rightarrow \frac{dU}{dx} = 2x + x^3$

$$\int_0^U dU = \int_0^x (2x + x^3)\,dx$$

$$\Rightarrow \quad U = x^2 + \frac{x^4}{4}$$

No energy is lost.

$$\therefore \quad x^2 + \frac{x^4}{4} = \frac{1}{2}mu^2 = 3\text{J}$$

or $$x^4 + 4x^2 - 12 = 0$$

$$x^2 = \frac{-4 \pm \sqrt{16+48}}{2}$$

$$= \frac{-4 \pm 8}{2} = -6, +2$$

$$\therefore \quad x = \sqrt{2} = 1.41\text{ m}$$

14. (b) Initially left side block is moving towards left and right side block towards right, when they collide at the same place, one oscillation is complete. Again they are moving leftwards and rightwards.

$$\therefore \quad T = \pi\sqrt{\frac{m}{k}}$$

15. (c) Velocity of object $= v_{max}\frac{1}{\sqrt{2}}\hat{\mathbf{i}} + v_{max}\frac{1}{\sqrt{2}}\hat{\mathbf{j}}$

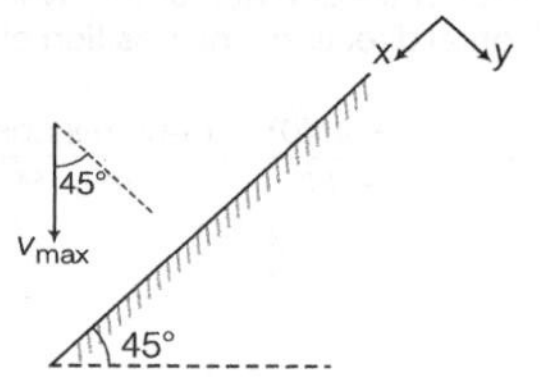

Velocity of image $= v_{max}\frac{1}{\sqrt{2}}\hat{\mathbf{i}} - v_{max}\frac{1}{\sqrt{2}}\hat{\mathbf{j}}$

$$\bar{v}_r = \bar{v}_{\text{object, image}} = \left(v_{max}\frac{1}{\sqrt{2}}\hat{\mathbf{i}} + v_{max}\frac{1}{\sqrt{2}}\hat{\mathbf{j}}\right) - \left(v_{max}\frac{1}{\sqrt{2}}\hat{\mathbf{i}} - v_{max}\frac{1}{\sqrt{2}}\hat{\mathbf{j}}\right)$$

$$= v_{max}\sqrt{2}\,\hat{\mathbf{j}}$$

$$|\bar{v}_r| = \sqrt{2}\,v_{max} = \sqrt{2}\,A\omega \qquad (\because v_{max} = A\omega)$$

$$= \sqrt{2}\,A\sqrt{\frac{k}{m}} \qquad \left(\because \omega = \sqrt{\frac{k}{m}}\right)$$

16. (a,b,c) In the figure, $F = -mg\sin(\theta_1 + \theta_2)$

For small amplitude, $\sin(\theta_1 + \theta_2) \simeq \theta_1 + \theta_2$

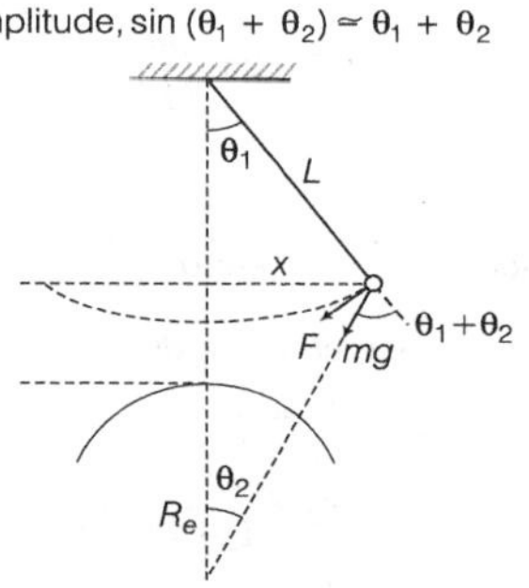

$\therefore \quad F = -mg(\theta_1 + \theta_2)$

$$F = -mg(\sin\theta_1 + \sin\theta_2) = -mg\left(\frac{x}{L} + \frac{x}{R_e}\right)$$

$$= -mg\left(\frac{1}{L} + \frac{1}{R_e}\right)x$$

$\therefore \quad a = -g\left(\frac{1}{L} + \frac{1}{R_e}\right)x$

Comparing with $a = -\omega^2 x$, $\omega = \sqrt{g\left(\frac{1}{L} + \frac{1}{R_e}\right)}$

$\therefore \quad T = 2\pi\sqrt{\dfrac{1}{g\left(\frac{1}{L} + \frac{1}{R_e}\right)}}$

When $L = R_e$ i.e. $L = 6400 \times 10^3$ m,

we get $T \simeq 59.8$ min

When $L \to \infty$, $\frac{R_e}{L} = 0$

$\therefore \quad T = 2\pi\sqrt{\frac{R_e}{g}} \simeq 84.6$ min

17. (d) Basically for a spring–block system, when there are no dissipative effects, time period is $T = 2\pi\sqrt{\frac{m}{k}}$.

Note *In case (iv), this is true only when the entire block is inside the liquid, during oscillations and no viscous effect is present.*

18. (c,d) When a constant force is superimposed on a system under going SHM, along the line of SHM, the time period does not change. The mean position, here shifts to the right, as at mean position net force has to be zero.

19. (d) Here, $\phi = 90°$

$\therefore \quad T = 2\pi\sqrt{\dfrac{L}{\sqrt{g^2 + a_0^2}}}$

20. (c) $F_{net} = mg + Eq$

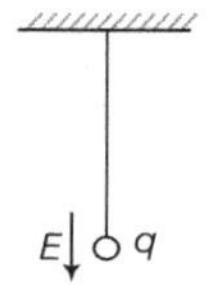

$$a_{net} = g + \frac{Eq}{m}$$

$\therefore \quad T = 2\pi\sqrt{\dfrac{L}{\left(g + \frac{Eq}{m}\right)}}$

21. (c)

$\therefore \quad T = 2\pi\sqrt{\dfrac{L}{[g^2 + (g\sin\theta)^2 + 2g(g\sin\theta)\cos(90°+\theta)]^{1/2}}}$

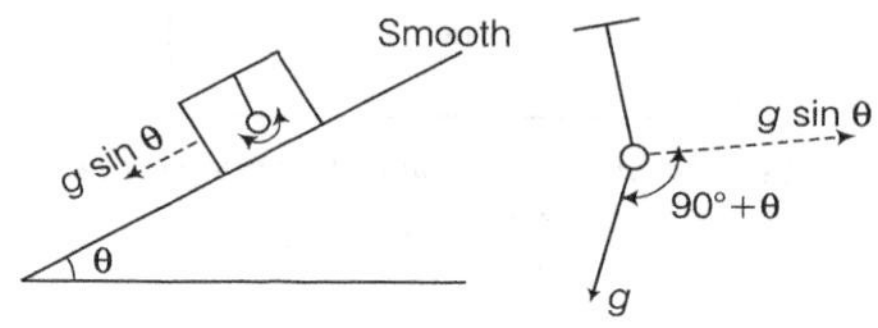

$$= 2\pi\sqrt{\frac{L}{g\cos\theta}}$$

22. (a) $F_{net} = \sqrt{(mg)^2 + (Eq)^2}$

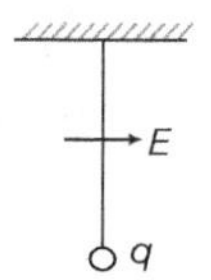

$$a_{net} = \frac{F_{net}}{m} = \sqrt{g^2 + \left(\frac{Eq}{m}\right)^2}$$

$\therefore \quad T = 2\pi\sqrt{\dfrac{L}{\sqrt{g^2 + \left(\frac{Eq}{m}\right)^2}}}$

23. (c) Had there been no wall W_2, the motion would have been SHM with amplitude $2d$.

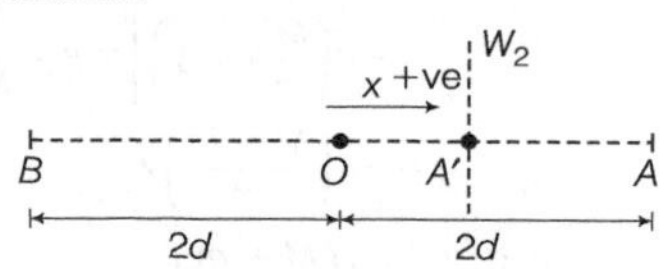

$$x = 2d\sin(\omega t + \delta)$$

Initially, the block is at point B.

$\therefore \quad \delta = \frac{3\pi}{2}$

$\therefore \quad x = -2d\cos\omega t$

For $t_{BA'}$, $\quad x = d$

$\therefore \quad d = -2d\cos\omega t_{BA'}$

$\Rightarrow \quad \omega t_{BA'} = \frac{\pi}{2} + \frac{\pi}{6} = \frac{4\pi}{6}$

$$t_{BA'} = \frac{4\pi}{6\omega}$$

$$t_{BA'} = t_{A'B}$$

$\therefore \quad T = 2t_{BA'} = \frac{8\pi}{6\omega}$

$$= \frac{8\pi}{6}\sqrt{\frac{m}{k}} = \frac{4\pi}{3}\sqrt{\frac{m}{k}}$$

24. (c) $x = -2d\cos\omega t$ for $0 < t < \frac{T}{2}$

$$x = -2d\cos\left(\omega t + \frac{4\pi}{3}\right) \quad \text{for} \quad \frac{T}{2} < t < T$$

Because when block is at A' and moving towards left the increase in phase (when compared with B) is $\frac{4\pi}{3}$.

25. (c)

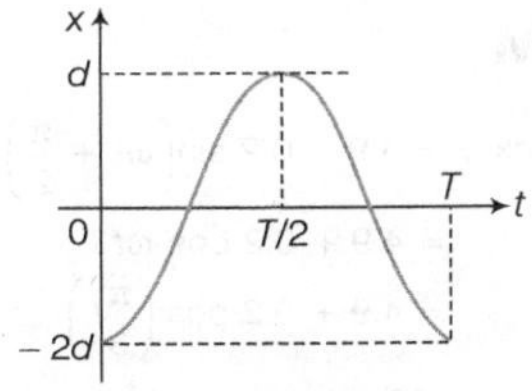

26. (a) $f = \frac{1}{2\pi}\sqrt{\frac{k}{m}}$, so f does not change.

So, ω does not change.

$$v_{max} = v = A\omega$$

and $$v' = A'\omega$$

$\Rightarrow$ $$A' = 2A \quad (\because v' = 2v)$$

$$a_{max} = A\omega^2$$

$\Rightarrow$ $$a'_{max} = 2a_{max}$$

KE_{max} increases by a factor 4. So, does PE.

Hence, (i) $\to$ (r), (ii) $\to$ (p), (iii) $\to$ (p), (iv) $\to$ (s)

27. (a)

(i) For equilibrium, $2mg = kA$

$\Rightarrow$ $$A = \frac{2mg}{k}$$

$$T = 2\pi\sqrt{\frac{m}{k}}$$

$$v_{max} = A\omega = \frac{2mg}{k}\sqrt{\frac{k}{m}} = 2g\sqrt{\frac{m}{k}}$$

$$a_{max} = A\omega^2 = \frac{2mg}{k}\cdot\frac{k}{m} = 2g$$

(ii) Same as above.

(iii) Initially, $mg = kx_0$

$$3mg = kx$$

$$x - x_0 = A = \frac{2mg}{k}$$

Spring will not acquire natural length.

(iv) $mg = kx_0$ (initially)

At equilibrium,

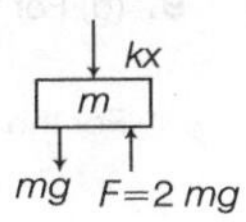

$$kx + mg = 2mg$$

$$x = \frac{mg}{k} \quad \text{(compression)}$$

$$A = x_0 + x = \frac{mg}{k} + \frac{mg}{k} = \frac{2mg}{k}$$

v'= velocity at natural length $= \omega\sqrt{A^2 - x^2}$

where, $$x = \frac{mg}{k}$$

$\therefore$ $$v' \neq 0$$

Hence, (i) $\to$ (p, q, r, s, t), (ii) $\to$ (p, q, r, s, t), (iii) $\to$ (p, q, r, s), (iv) $\to$ (p, q, r, s)

28. (1) If the sphere moves down along the plane by a distance x, extension in the spring = $x\cos 30°$

$\therefore$ Restoring force = $F\cos 30°$

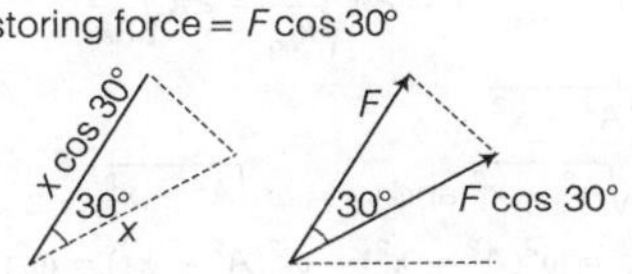

$$= (kx\cos 30°)\cos 30°$$

$$= kx\left(\frac{\sqrt{3}}{2}\right)^2 = \frac{3}{4}kx$$

As it is restoring in nature, $F_R = -\frac{3}{4}kx$

$\therefore$ $$a = -\frac{3}{4}\cdot\frac{k}{M}x$$

$\Rightarrow$ $$\omega = \sqrt{\frac{3}{4}\cdot\frac{k}{M}} = \sqrt{\frac{3}{4M}\cdot\frac{4M}{3}} = 1 \text{ rad s}^{-1}$$

29. (2) Velocity of the particle at the time of collision is $u = \sqrt{2gh}$

$$u = \sqrt{2g\cdot\frac{4.5\,mg}{k}} = 3g\sqrt{\frac{m}{k}}$$

Let velocity of combined mass after collision is v.

Then, $$mu = (2m + m)v$$

$\Rightarrow$ $$v = \frac{u}{3} = g\sqrt{\frac{m}{k}}$$

Now, the time period of SHM is $T = 2\pi\sqrt{\frac{3m}{k}}$.

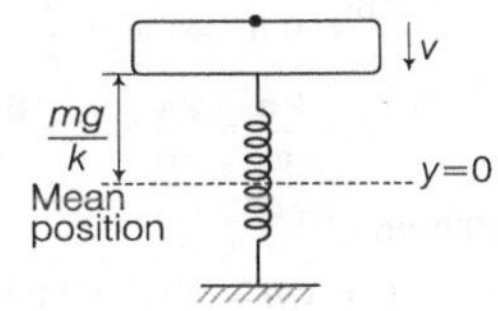

Mean position from the point of collision is at a distance $\frac{mg}{k}$ as shown in the figure.

$$y = A\sin(\omega t + \delta)$$

At $t = 0$, $$y = \frac{mg}{k}$$

$\therefore$ $$\frac{mg}{k} = A\sin\delta \quad \ldots\text{(i)}$$

Now, for amplitude, $v = \omega\sqrt{A^2 - y^2}$

$$g\sqrt{\frac{m}{k}} = \sqrt{\frac{k}{3m}}\sqrt{A^2 - \frac{m^2g^2}{k^2}}$$

$\Rightarrow$ $$\left(\sqrt{3}\,\frac{mg}{k}\right)^2 = A^2 - \frac{m^2g^2}{k}$$

$\therefore$ $$A = \frac{2mg}{k} \quad \ldots\text{(ii)}$$

From Eqs. (i) and (ii), $y = \frac{A}{2}$ to $y = 0$

$\Rightarrow$ $$t = \frac{T}{12}$$

$$y = 0 \text{ to } y = A$$

$\Rightarrow$ $$t' = \frac{T}{4}$$

$$\text{Total time} = \frac{T}{12} + \frac{T}{4} = \frac{T}{3} = \frac{1}{3}2\pi\sqrt{\frac{3m}{k}}$$

$$= \frac{2\pi}{3}\sqrt{\frac{3 \times 30 \times 10^{-3}}{100}}$$

$$= 2\pi \times 10^{-2} = 20\pi \text{ ms}$$

$\therefore$ $$\beta = 2$$

30. (7) $kx = mg$

$\Rightarrow$ $$k = \frac{0.5 \times 9.8}{0.2} = \frac{49}{2}\text{ Nm}^{-1}$$

$$T = 2\pi\sqrt{\frac{m'}{k}} = 2\pi\sqrt{\frac{0.25 \times 2}{49}}$$

$$= 2\pi\sqrt{\frac{1}{49 \times 2}} = \frac{\sqrt{2}\,\pi}{7}\text{ s}$$

$\Rightarrow$ $$\psi = 7$$

DPP-4 Time Period in SHM

1. (c) $A = 0.1\,m, v = A\omega \cos \omega t$

$\Rightarrow \quad 0.36 = 0.1\,\omega \cos\omega t$

$\cos \omega t = \frac{3.6}{\omega}$

$x = A \sin \omega t$

$\Rightarrow \quad 0.06 = 0.1 \sin \omega t$

$\Rightarrow \quad \sin \omega t = 0.6$

$\sin^2 \omega t + \cos^2 \omega t = 1$

$\Rightarrow \quad (0.6)^2 + \left(\frac{3.6}{\omega}\right)^2 = 1$

$\left(\frac{3.6}{\omega}\right)^2 = 0.64$

$\Rightarrow \quad \frac{3.6}{\omega} = 0.8 \Rightarrow \omega = \frac{36}{8}$

$T = \frac{2\pi}{\omega} = \frac{2\pi}{36} \times 8 = \frac{8\pi}{18} = \frac{4\pi}{9} = 1.4\text{ s}$

2. (a) $V = (8x^2 + 200)\text{ erg g}^{-1}$

$\therefore \quad U = Vm = (40x^2 + 800)\text{ erg}$

$\frac{dU}{dX} = 80\,x$

$F = -\frac{dU}{dx} = -80x$

$\Rightarrow \quad a = -\frac{80}{m}x = -16x$

$\therefore \quad \omega = \sqrt{16} = 4\text{ rad s}^{-1}$

$T = \frac{2\pi}{\omega} = \frac{2\pi}{4} = \frac{\pi}{2} = 1.57\text{ s}$

3. (b) $T \propto \sqrt{L}$

$\Rightarrow \quad \frac{T_1}{T_2} = \sqrt{\frac{100}{121}} = \frac{10}{11}$

Time period of shorter pendulum < Time period of longer pendulum

$\therefore$ 10 oscillations of longer pendulum

= 11 oscillations of shorter pendulum.

4. (b) $A = 1\text{ cm}, \quad a_{max} = \omega^2 A = (1.57)^2\text{cms}^{-2}$

$\therefore \quad \omega^2 = (1.57)^2$

$\Rightarrow \quad \omega = 1.57$

$\Rightarrow \quad T = \frac{2\pi}{\omega} = \frac{2\pi}{1.57} = 4\text{ s}$

5. (a) $k_{eq} = k + k = 2k$ [$\because$ two springs are parallel]

$T = 2\pi\sqrt{\frac{m}{k_{eq}}} = 2\pi\sqrt{\frac{m}{2k}}$

Time period is indepndent of θ.

6. (d) $T = 2\pi\sqrt{\frac{L}{g}} = 2\text{ s}$ (on earth)

$g \propto \frac{M}{R^2}$

$\Rightarrow \quad \frac{g'}{g} = \frac{M'}{M} \cdot \frac{R^2}{R'^2} = (2)\left(\frac{1}{4}\right) = \frac{1}{2}$

$g' = \frac{g}{2}$

$T' = 2\pi\sqrt{\frac{L}{g'}} = 2\pi\sqrt{\frac{L}{g/2}}$

$= \sqrt{2}T = 2\sqrt{2}\text{ s}$

7. (a) For the block, $x = 4.9 + 0.2 \sin\left(\omega t + \frac{\pi}{2}\right)$

$= 4.9 + 0.2 \cos \omega t$

$= 4.9 + 0.2 \cos\left(\frac{\pi}{3}t\right)$

For the particle, $x = 10 - (v \cos 45°)\,t$

$= 10 - \frac{v}{\sqrt{2}}t$

As, $t = 1$ s, they meet each other.

$\therefore \quad 4.9 + 0.2 \cos\frac{\pi}{3} = 10 - \frac{v}{\sqrt{2}}$

$10 - \frac{v}{\sqrt{2}} = 5$

$\Rightarrow \quad \frac{v}{\sqrt{2}} = 5$

$\Rightarrow \quad v = \sqrt{50}\text{ ms}^{-1}$

8. (d) For the wire, $Y = \frac{FL_0}{A\Delta L_0}$

$\Rightarrow \quad \frac{F}{\Delta L_0} = \frac{YA}{L_0}$

$\therefore$ For the wire, $k' = \frac{YA}{L_0}$

Spring and wire are in series

$k_{eq} = \frac{kk'}{k + k'} = \frac{kYA}{L_0\left(k + \frac{YA}{L_0}\right)}$

or $\quad k_{eq} = \frac{kYA}{YA + kL_0}$

$T = 2\pi\sqrt{\frac{m}{k_{eq}}} = 2\pi\sqrt{\frac{m\,(YA + kL_0)}{kYA}}$

9. (d) For the right (half) oscillation, $T_1 = \pi\sqrt{\frac{L_0}{g}}$

For the left (half) oscillation, $T_2 = \pi\sqrt{\frac{L_0/4}{g}} = \frac{1}{2}\pi\sqrt{L_0/g}$

$\therefore$ Time taken $= \pi\sqrt{\frac{L_0}{g}}\left(1 + \frac{1}{2}\right) = \frac{T}{2} \cdot \frac{3}{2} = \frac{3}{4}T \quad \left[\because T = 2\pi\sqrt{\frac{L_0}{g}}\right]$

10. (c) $T_1 = T_2 = 2\pi\sqrt{\frac{m}{k}}$

In cases (iii) and (iv),

$k_{eq} = 2k + 2k = 4k$

$\therefore \quad T_3 = T_4 = 2\pi\sqrt{\frac{m}{4k}} = \pi\sqrt{\frac{m}{k}}$

11. (a) k_{eq} = ($2k$ and $2k$ in series) parallel (k and k in parallel)

$= k||2k = 3k$

$\therefore \quad T = 2\pi\sqrt{\frac{m}{k_{eq}}} = 2\pi\sqrt{\frac{m}{3k}}$

12. (b) $v = \omega\sqrt{A^2 - x^2}$

$\Rightarrow \quad v_1 = \omega\sqrt{A^2 - x_1^2}$ and $v_2 = \omega\sqrt{A^2 - x_2^2}$

$\therefore \quad v_1^2 - v_2^2 = \omega^2(A^2 - x_1^2) - \omega^2(A^2 - x_2^2) = \omega^2(x_2^2 - x_1^2)$

$\omega = \sqrt{\frac{v_1^2 - v_2^2}{x_2^2 - x_1^2}}, \; T = \frac{2\pi}{\omega}$

$\therefore \quad T = 2\pi\sqrt{\frac{x_2^2 - x_1^2}{v_1^2 - v_2^2}}$

13. (b,d) For 1st particle, $x = A\sin\left(\omega t + \frac{3\pi}{2}\right) = -A\cos\omega t$.

For 2nd particle, $x = A\sin\omega t$

When they cross each other

$$-\cos\omega t = \sin\omega t \Rightarrow \tan\omega t = -1$$

$$\omega t = \frac{3\pi}{4}$$

$$\Rightarrow \quad t = \frac{3\pi}{4}\cdot\frac{T}{2\pi} = \frac{3T}{8}$$

When $t = \frac{3T}{8}$, $\quad x = A\sin\left(\frac{2\pi}{T}\cdot\frac{3T}{8}\right) = A\sin\left(\frac{3\pi}{4}\right) = \frac{A}{\sqrt{2}}$

14. (a,b,c) The situation is similar to the vertically loaded spring suddenly released from relaxed position.

$$\therefore \quad T = 2\pi\sqrt{\frac{m}{k}}$$

$$ma_0 = kA$$

$$\Rightarrow \quad A = \frac{ma_0}{k}$$

$$E = \frac{1}{2}m\omega^2A^2 = \frac{1}{2}m\cdot\frac{k}{m}\cdot\frac{m^2a_0^2}{k^2} = \frac{1}{2}\cdot\frac{m^2a_0^2}{k}$$

15. (a,b,d)

(i) $F = -[k(x + x_0) - mg]$

or $\quad F = -kx \quad$ ($\because$ at equilibrium $mg = kx_0$)

$$a = -\frac{k}{m}x$$

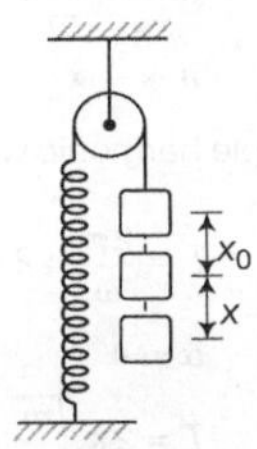

$$\omega = \sqrt{\frac{k}{m}} \Rightarrow T_1 = 2\pi\sqrt{\frac{m}{k}}$$

(ii) In this case, if the mass m moves down a distance x from its equilibrium position, then pulley will move down by $\frac{x}{2}$.

So, the extra force in spring will be $k\frac{x}{2}$.

Restoring force on the mass is

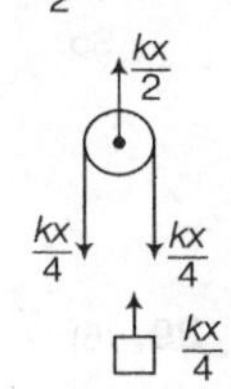

$$F = -\frac{k}{4}x$$

$$a = -\frac{k}{4m}x$$

$$\Rightarrow \quad \omega = \sqrt{\frac{k}{4m}}$$

$$\therefore \quad T_2 = 2\pi\sqrt{\frac{4m}{k}}$$

(iii) In this case, if the mass m moves down a distance x from its equilibrium position, then pulley will also move by x and so the spring will stretch by $2x$.

Restoring force on the block is

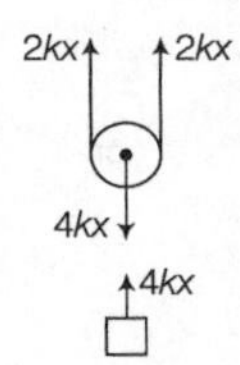

$$F = -4kx$$

$$\therefore \quad a = -\frac{4k}{m}x \Rightarrow \omega = \sqrt{\frac{4k}{m}}$$

$$T = 2\pi\sqrt{\frac{m}{4k}}$$

16. (a)

$$\frac{F}{m} = -\frac{GM}{R^3}x$$

$$a = -\frac{GM}{R^3}x$$

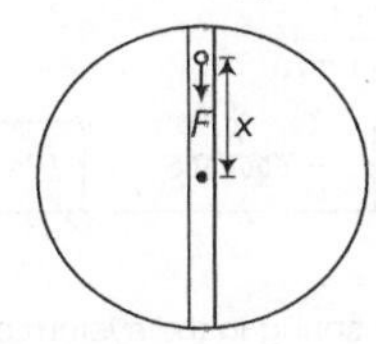

$\therefore$ The ball executes SHM,

$$\omega = \sqrt{\frac{GM}{R^3}}$$

$$T = \frac{2\pi}{\omega} = 2\pi\sqrt{\frac{R^3}{GM}}$$

$$\simeq 86.4 \text{ min}$$

$$t = \frac{T}{2} = 43.2 \text{ min}$$

17. (d) $T \propto \sqrt{L}$

$$\Rightarrow \quad \frac{\Delta T}{T} = \frac{1}{2}\times\frac{\Delta L}{L}$$

$$= \frac{1}{2}\times 4\% = 2\%$$

As length increases' time period will increase by 2%.

18. (b) $T = 2\pi\sqrt{\frac{L}{g}}$

$\Rightarrow$ Statement II is correct

When the water leaks out the centre of mass of the bob gets lowered and so effective length increases due to which time period increases. When the water level becomes less than half, the centre of mass of the bob rises up decreasing effective length which inturn decreases the time period. Finally, when the bob becomes empty, the centre of mass will have reached centre of the bob giving it original time period.

19. (a) $T = 2\pi\sqrt{\frac{m}{k}}$

For a hard spring, k is greater and time period is less.

20. (b) The centre of mass of the system remains at rest. The block moves a distance $5L$ relative to the plank when system comes to rest for the first time.

$$\therefore \quad m(5L - \Delta x) = M\Delta x$$

$$\Rightarrow \quad \Delta x(M + m) = 5mL$$

$$\Delta x = \frac{5mL}{M + m}$$

21. (a)

Note *If net external force on a system of two bodies (masses m_1 and m_2) is zero and only internal conservative forces are acting on the bodies, mechanical energy of the system will remain conserved. The relative speed between them can be found by putting reduced mass(μ) in energy equation, where*

$$\mu = \frac{m_1m_2}{m_1 + m_2}$$

So, here

$$\frac{1}{2}\mu v_{m/M}^2 = \frac{1}{2}kL^2$$

$$\frac{Mm}{M + m}v_{m/M}^2 = kL^2$$

$$\Rightarrow \quad v_{m/M} = L\sqrt{\frac{M + m}{Mm}k}$$

22. (c)

Note *If two blocks, connected by an ideal spring, as shown, are compressed and released, the time period of oscillation is given by* $T = 2\pi\sqrt{\frac{\mu}{k}}$, *where*

μ = *reduced mass* $= \frac{m_1 m_2}{m_1 + m_2}$

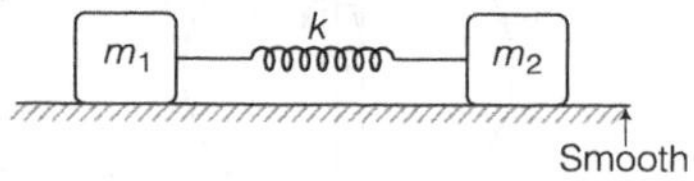

When the right side spring in the system given is relaxed, the time taken for relaxation is $\frac{1}{4}$th of the above said time period. Similarly compression of left spring, relaxation of left spring and compression of right spring, each will take $\frac{1}{4}$th of the same above said time period. So time taken for these four events to happen is

$$T_1 = 4\left[\frac{1}{4}2\pi\sqrt{\frac{Mm}{(M+m)k}}\right]$$

or
$$T_1 = 2\pi\sqrt{\frac{Mm}{(M+m)k}}$$

In addition to this, the relative distance traveled by the block in leftward journey is 3L and during rightward journey is 3L (when the springs are relaxed). Time taken for these is

$$T_2 = 2\left[\frac{3L}{V_{m/M}}\right] = 2\left[\frac{3L}{L\sqrt{\left(\frac{M+m}{Mm}\right)k}}\right]$$

or
$$T_2 = 6\sqrt{\frac{Mm}{(M+m)k}}$$

So total time period is

$$T = T_1 + T_2 = (2\pi + 6)\sqrt{\frac{Mm}{(M+m)}k}$$

23. (a) $f = \frac{1}{2\pi}\sqrt{\frac{k_{eff}}{\text{Total mass}}} = \frac{1}{2\pi}\sqrt{\frac{k_1 + k_2}{m_1 + m_2}}$

24. (b) Suppose the system is displaced towards left by a distance x. The restoring force on m_2 is

$$F = m_2\omega^2 x \quad \text{(towards right)}$$

$$F = m_2\left(\frac{k_1 + k_2}{m_1 + m_2}\right)x \quad \left[\because \omega = \sqrt{\frac{k_1 + k_2}{m_1 + m_2}}\right]$$

Friction f on m_2 will be towards left

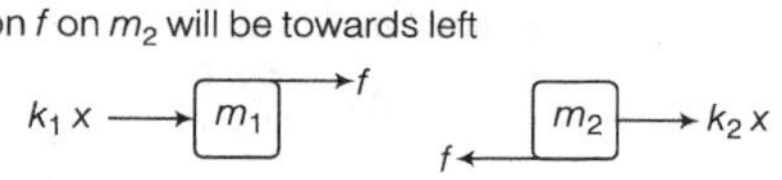

if $k_2 x > F$

$$k_2 x > m_2\left(\frac{k_1 + k_2}{m_1 + m_2}\right)x$$

$$\Rightarrow \quad \frac{k_1 + k_2}{k_2} < \frac{m_1 + m_2}{m_2}$$

$$\frac{k_1}{k_2} < \frac{m_1}{m_2}$$

25. (c)
$$k_1 A_m + \mu m_2 g = m_1\left(\frac{k_1 + k_2}{m_1 + m_2}\right)A_m$$

$$A_m\left[\frac{m_1 k_1 + m_1 k_2}{m_1 + m_2} - k_1\right] = \mu m_2 g$$

$$A_m\left[\frac{m_1 k_2 - m_2 k_1}{m_1 + m_2}\right] = \mu m_2 g$$

$$A_m = \frac{\mu(m_1 + m_2)m_2 g}{m_1 k_2 - m_2 k_1}$$

26. (9)
$$3 = 2\pi\sqrt{\frac{m}{k}} \text{ and } 4 = 2\pi\sqrt{\frac{m+1}{k}}$$

Dividing
$$\frac{3}{4} = \sqrt{\frac{m}{m+1}}$$

$$\Rightarrow \quad \frac{m}{m+1} = \frac{9}{16}$$

$$16m = 9m + 9$$

$$\Rightarrow \quad m = \frac{9}{7} \text{ kg}$$

$$\therefore \quad \beta = 9$$

27. (4) $\eta = \frac{F}{A\theta}$

Here, $A = L^2$ and $\theta = \frac{x}{L}$.

$\therefore$ Restoring force is $\quad F = -\eta A\theta = -\eta L x$

Acceleration is $\quad a = -\frac{\eta L}{m}x$

$\Rightarrow \quad a \propto -x$

$\therefore$ Oscillations are simple harmonic with $\omega = \sqrt{\frac{\eta L}{m}}$

So,
$$T = \frac{2\pi}{\omega} = 2\pi\sqrt{\frac{m}{\eta L}} = \pi\sqrt{\frac{4m}{\eta L}}$$

$$\Rightarrow \quad \alpha = 4$$

28. (2) Initially,
$$T = 2\pi\sqrt{\frac{m}{k}}$$

$$k \propto \frac{1}{L}$$

$\therefore \quad k_1 = 2k$

and $\quad k_2 = 2k$

So, $\quad k_{eq} = 4k$

$$\therefore \quad T' = 2\pi\sqrt{\frac{m}{4k}}$$

$$= \pi\sqrt{\frac{m}{k}} = \frac{T}{2}$$

29. (6)
$$g_{eff} = \frac{\text{Weight} - \text{Upthrust}}{\text{Mass}}$$

$$= \frac{v\rho g - v\frac{\rho}{3}g}{v\rho} = \frac{2}{3}g$$

$$T = 2\pi\sqrt{\frac{L}{g_{eff}}} = 2\pi\sqrt{\frac{3L}{2g}} = 2\pi\sqrt{\frac{3 \times 0.2}{2 \times 9.8}}$$

or
$$T \simeq 1.1 \text{ s} = \frac{5+6}{10} \text{ s}$$

$$\Rightarrow \quad x = 6$$

DPP-5 Time Period : Physical Systems-I

1. (c) Let the ball is pressed down by a small amount y, then the value of air decreases by Ay.

Then, excess pressure is

$$dp = -B\left(\frac{dV}{V}\right) = -B\left(\frac{yA}{V}\right)$$

A restoring force $F (= Adp)$ acts in the upward direction.

$$\therefore \qquad F = A\left[-B\left(\frac{yA}{V}\right)\right] = -\frac{BA^2}{V}\cdot y$$

$$\therefore \qquad a = -\frac{BA^2}{mV}y \text{ comparing this with } a = -\omega^2 y$$

$$\omega = \sqrt{\frac{BA^2}{mV}}$$

$$\Rightarrow \qquad T = \frac{2\pi}{\omega} = 2\pi\sqrt{\frac{mV}{BA^2}}$$

2. (b) An intermittent snap-shot of oscillating water column is shown below.

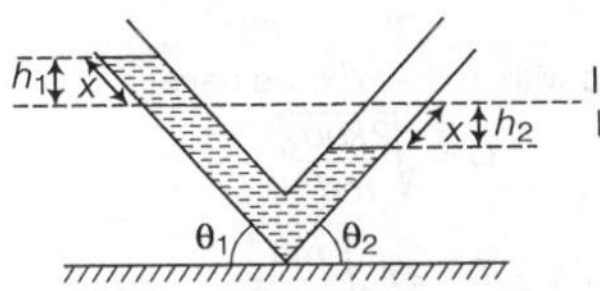

Pressure acting is

$$dp = \rho g\,(h_1 + h_2) = \rho g\,(\sin\theta_1 + \sin\theta_2)\,x$$

Restoring force is

$$F = -\rho g\,(\sin\theta_1 + \sin\theta_2)\,Ax$$

Acceleration is $\quad a = -\dfrac{\rho g\,(\sin\theta_1 + \sin\theta_2)}{m}Ax$

Comparing this with $a = -\omega^2 x$, we have

$$\omega = \sqrt{\frac{\rho g\,(\sin\theta_1 + \sin\theta_2)\,A}{m}}$$

$\therefore$ Time period of oscillation of water column is

$$T = 2\pi\sqrt{\frac{m}{\rho gA(\sin\theta_1 + \sin\theta_2)}}$$

3. (a) The equation of time period from the previous problem is

$$T = 2\pi\sqrt{\frac{m}{\rho gA\,(\sin\theta_1 + \sin\theta_2)}}$$

Here, mass of water oscillating is $m = \rho A\,(2L)$ and $\theta_1 = \theta_2 = 90°$

$$\therefore \qquad T = 2\pi\sqrt{\frac{\rho A(2L)}{\rho gA(1 + 1)}} = 2\pi\sqrt{\frac{L}{g}}$$

4. (a) When the pole is oscillating consider an intermediate snap-shot as shown. Let it is immersed by $L + y$.

Initially, when the rod is in equilibrium, weight of the pole = upthrust.

$$mg = AL\rho g \;\Rightarrow\; m = AL\rho$$

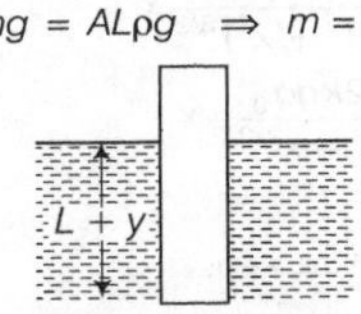

Additional upthrust in the situation shown is $Ay\rho g$. This provides restoring force.

$\therefore \qquad F = -A\rho g\, y$, so acceleration of the pole is

$$a = -\left(\frac{A\rho g}{m}\right)y = -\left(\frac{A\rho g}{AL\rho}\right)y = -\left(\frac{g}{L}\right)y$$

$$\therefore \qquad \omega = \sqrt{\frac{g}{L}} \Rightarrow T = \frac{2\pi}{\omega} = 2\pi\sqrt{\frac{L}{g}}$$

5. (d) Let us assume a force dF is applied at P in positive X-direction. This will stretch each spring by dl, inducing a spring force dF_s in each spring.

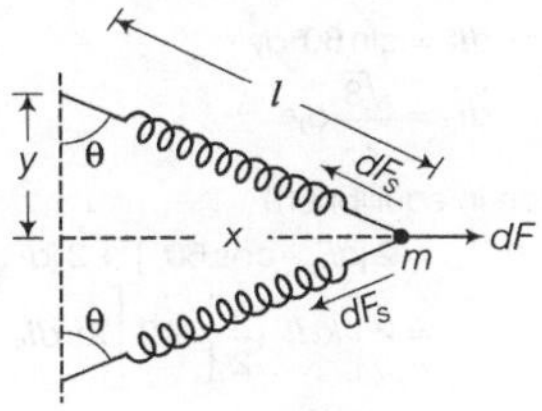

$$dF_s = kdl$$

Let the static deformation of the system is dx (along the x-direction). The particle is in equilibrium. So,

$$dF = 2dF_s\sin\theta = 2\,kdl\sin\theta$$

$$[\because dF_s = kdl]$$

$$\therefore \qquad dF = 2kdlk\sin\theta \qquad \ldots\text{(i)}$$

Using Pythagoras theorem,

$$x^2 + y^2 = l^2$$

$$\therefore \qquad 2\,xdx + 2\,ydy = 2\,ldl$$

Here y is constant.

So, $\qquad 2\,xdx = 2\,ldl$

$$\Rightarrow \qquad dl = \frac{x}{l}dx$$

Also, $\qquad \dfrac{x}{l} = \sin\theta$

$$\therefore \qquad dl = dx\sin\theta \qquad \ldots\text{(ii)}$$

From Eqs. (i) and (ii), we get

$$dF = 2k\,(dx\sin\theta)\sin\theta$$

or $\qquad \dfrac{dF}{dx} = 2k\sin^2\theta$

$$\therefore \qquad k_{\text{eff}} = 2k\sin^2\theta$$

Time period of oscillation is

$$T = 2\pi\sqrt{\frac{m}{k_{\text{eff}}}}$$

or $\qquad T = 2\pi\sqrt{\dfrac{m}{2k\sin^2\theta}}$

$$= \frac{\pi}{\sin\theta}\sqrt{\frac{2m}{k}}$$

6. (b) Let a force dF is applied on the particle along the Y-direction. Let the displacement of the particle is dy. Deformation of each upper spring is dl_1 and that of each lower spring is dl_2 (say). Let the static deformation of the system is dy.

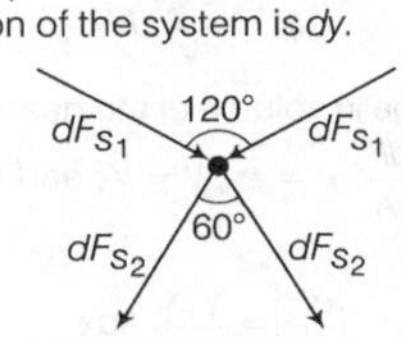

From the figure, $\qquad x^2 + y^2 = l_1^2$

or $\qquad 2\,xdx + 2\,ydy = 2\,l_1 dl_1$

x is constant. So, $\qquad dx = 0$

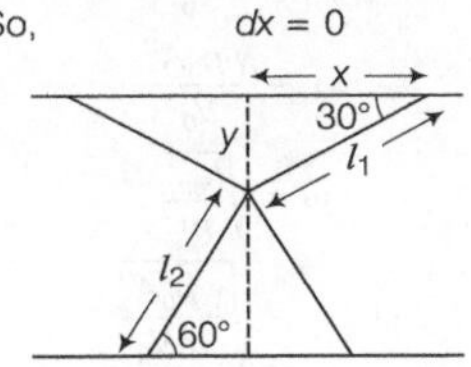

Thus, $y\,dy = l_1\,dl_1$

$$dl_1 = \frac{y}{l_1}dy = \sin 30^\circ dy$$

or $$dl_1 = \frac{dy}{2}$$

Similarly, $$dl_2 = \sin 60^\circ dy$$

$$\Rightarrow \quad dl_2 = \frac{\sqrt{3}}{2}dy$$

Net spring force in equilibrium

$$= 2[dF_{s_1}\cdot \cos 60^\circ] + 2[dF_{s_2}\cdot \cos 30^\circ]$$

$$= 2\left[k\,dl_1 \cdot \frac{1}{2}\right] + 2\left[2k\,dl_2 \cdot \frac{\sqrt{3}}{2}\right]$$

$$= k\,dl_1 + 4k\,dl_2\frac{\sqrt{3}}{2}$$

$$= k\frac{dy}{2} + 4k\left(\frac{\sqrt{3}}{2}dy\right)\frac{\sqrt{3}}{2}$$

$$= k\,dy\left[\frac{1}{2} + 3\right]$$

$$= \frac{7}{2}k\,dy$$

This force is in downward direction and is equal in magnitude to force applied dF.

$$\therefore \quad dF = \frac{7}{2}k\,dy$$

$$\Rightarrow \quad \frac{dF}{dy} = \frac{7}{2}k$$

So, $$k_{\text{eff}} = \frac{7}{2}k$$

$$\omega = \sqrt{\frac{k_{\text{eff}}}{m}} = \sqrt{\frac{7k}{2m}}$$

$$\Rightarrow \quad f = \frac{1}{2\pi}\sqrt{\frac{7k}{2m}}$$

7. (c) In adiabatic process,

$$pV^\gamma = C$$

$$\ln p + \gamma \ln V = \ln C$$

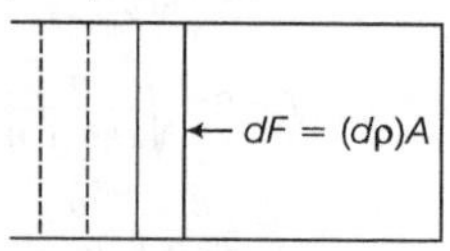

Taking differentials on both sides,

$$\frac{dp}{p} + \frac{\gamma dV}{V} = 0$$

or $$dp = \frac{-\gamma p}{V}dV$$

where, dV = change in volume of the gas = Adx.

Substituting $dp = \frac{dF}{A}$, $p = p_0$, $V = V_0$ and $dV = Adx$ in the above equation, we have

$$\left|\frac{dF}{A}\right| = \frac{\gamma p_0}{V_0}Adx$$

$$\Rightarrow \quad \left|\frac{dF}{dx}\right| = \frac{\gamma p_0 A^2}{V_0}$$

$$\therefore \quad k_{\text{eff}} = \frac{\gamma p_0 A^2}{V_0}$$

$$\omega = \sqrt{\frac{k_{\text{eff}}}{m}}$$

$$= \sqrt{\frac{\gamma p_0 A^2}{mV_0}}$$

8. (c) Restoring force is

$$F = -2F'\cos\theta$$

[negative sign because of restoring nature]

$$F = -2\left[\frac{kqq_0}{(x^2+d^2)}\right]\frac{x}{\sqrt{x^2+d^2}}$$

or $$F = -\frac{2kqq_0}{(d^2+x^2)^{3/2}}x$$

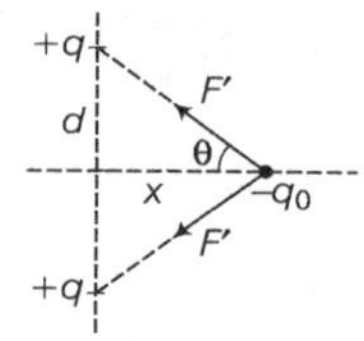

As $$x << d,\ F \simeq -\frac{2kqq_0}{d^3}x$$

Thus, acceleration of particle is

$$a = \frac{F}{m} = -\frac{2kqq_0}{md^3}x$$

Comparing this with $a = -\omega^2 x$, we have

$$\omega = \sqrt{\frac{2kqq_0}{md^3}}$$

$$\therefore \quad T = 2\pi\sqrt{\frac{md^3}{2kqq_0}}$$

Alternate Method

Note *For expression of a given mathematical function $y = f(x)$, Taylor's theorem is defined as*

$$f(x) = f(0) + xf'(0) + \frac{x^2}{2!}f''(0) + \cdots$$

If x is taken as the displacement of a particle from its mean position and restoring force is $F = f(x)$, then according to Taylor's theorem, restoring force is

$$F = f(x) = f(0) + xf'(0) + \frac{x^2}{2!}f''(0) + \cdots$$

At $x = 0$ (i.e. at mean position) restoring force is zero.

So, $f(0) = 0$ and for small displacements of particle, higher powers of x can be neglected. Thus, restoring force can be written as $F = -xf'(0)$ [negative sign shows restoring nature]

$\therefore$ Acceleration of particle is

$$a = \frac{F}{m} = -\left[\frac{f'(0)}{m}\right]x$$

$$\therefore \quad \omega = \sqrt{\frac{f'(0)}{m}}$$

$$\Rightarrow \quad T = 2\pi\sqrt{\frac{m}{f'(0)}}$$

Now, in the given problem, restoring force is

$$F = -\frac{2kqq_0}{(d^2+x^2)^{3/2}}x$$

$$\Rightarrow \quad |F| = \frac{2kqq_0}{(d^2+x^2)^{3/2}}x$$

$$f'(0) = \left[\frac{dF}{dx}\right]_{\text{at } x=0}$$

$$= 2kqq_0\left[\frac{(d^2+x^2)^{3/2}\cdot 1 - \frac{3}{2}(d^2+x^2)^{\frac{1}{2}}\cdot 2x\cdot x}{(d^2+x^2)^3}\right]_{x=0}$$

$$= 2kqq_0\left[\frac{d^3-0}{d^6}\right] = \frac{2kqq_0}{d^3}$$

According to Taylor's theorem,

$$T = 2\pi\sqrt{\frac{m}{f'(0)}}$$

$$\therefore \qquad T = 2\pi\sqrt{\frac{md^3}{2kqq_0}}$$

9. (b) $U = U_0(1-\cos\beta x)$ (given)

$$F = \left|\frac{dU}{dx}\right| = \beta U_0 \sin\beta x$$

$$f'(0) = \left|\frac{dF}{dx}\right|_{x=0} = U_0\beta^2\cos 0^\circ = U_0\beta^2$$

$$T = 2\pi\sqrt{\frac{m}{f'(0)}} = 2\pi\sqrt{\frac{m}{U_0\beta^2}}$$

10. (c) $U = \frac{k_1}{x^2} - \frac{k_2}{x}$

$$F = \left|\frac{dU}{dx}\right| = \left|-\frac{2k_1}{x^3} + \frac{k_2}{x^2}\right|$$

At equilibrium position,

$$x = 0$$

So, $$-\frac{2k_1}{x^3} + \frac{k_2}{x^2} = 0$$

$$\Rightarrow \qquad x = \frac{2k_1}{k_2}$$

Restoring force is

$$F = -\left[-\frac{2k_1}{x^3} + \frac{k_2}{x^2}\right]$$

(negative sign due to restoring nature)

According to Taylor's theorem,

$$T = 2\pi\sqrt{\frac{m}{f'\left(\frac{2k_2}{k_2}\right)}}$$

[$\because$ here, mean position is at $x = \frac{2k_1}{k_2}$]

$$f'\left(\frac{2k_1}{k_2}\right) = \left|\frac{dF}{dx}\right|_{x=\frac{2k_1}{k_2}}$$

$$= \left[\frac{6k_1}{x^4} - \frac{2k_2}{x^3}\right]_{x=\frac{2k_1}{k_2}}$$

$$= 6k_1\frac{k_2^4}{16k_1^4} - 2k_2\frac{k_2^3}{8k_1^3}$$

$$= \left(\frac{3}{8} - \frac{2}{8}\right)\frac{k_2^4}{k_1^3}$$

$$= \frac{k_2^4}{8k_1^3}$$

$$\therefore \qquad T = 2\pi\sqrt{\frac{8mk_1^3}{k_2^4}} = 4\pi\sqrt{\frac{2mk_1^3}{k_2^4}}$$

11. (d) Let initial length of spring is L. After charging its length becomes $2L$. For equilibrium of sphere -2,

$$\frac{1}{4\pi\varepsilon_0}\frac{Q^2}{(2L)^2} = k.L$$

$$\Rightarrow \qquad k = \frac{1}{4\pi\varepsilon_0}\frac{Q^2}{4L^3} = \frac{Q^2}{16\pi\varepsilon_0 L^3} \qquad \ldots\text{(i)}$$

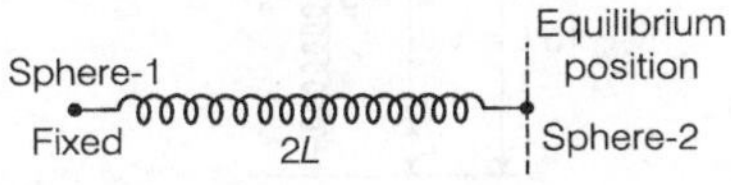

Now, if the sphere -2 is slightly displaced away from sphere -1, from the equilibrium position such that the separation between two spheres becomes x, then we have restoring force on sphere-2 as

$$F = k(x-L) - \frac{1}{4\pi\varepsilon_0}\frac{Q^2}{x^2} \qquad [\because \text{net extension} = x - L]$$

Now, $$T = 2\pi\sqrt{\frac{m}{f'(2L)}}$$

[$\because$ here, equilibrium position is at $x = 2L$]

$$f'(2L) = \left.\frac{dF}{dx}\right|_{x=2L} = \left[k + \frac{1}{4\pi\varepsilon_0}\frac{Q^2}{x^3}\cdot 2\right]_{x=2L}$$

$$= k + \frac{1}{2\pi\varepsilon_0}\frac{Q^2}{8L^3}$$

$$= k + \frac{Q^2}{16\pi\varepsilon_0 L^3}$$

Using Eq. (i) in this, we have

$$f'(2L) = k + k = 2k$$

$$\therefore \qquad T = 2\pi\sqrt{\frac{m}{2k}}$$

12. (a) Consider the situation when the block is at mean position and moving with maximum speed $v_{max} = A\omega$.

Velocity of the element considered is $\frac{v_{max}}{L}\cdot x$. Its mass is $\frac{m_s}{L}dx$

Its kinetic energy is dE_k

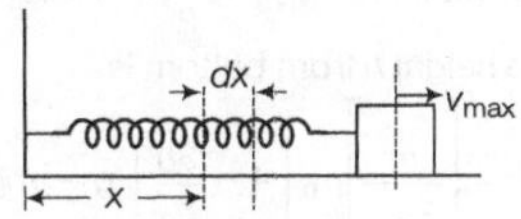

$$= \frac{1}{2}\left(\frac{v_{max}}{L}x\right)^2\frac{m_s}{L}dx$$

$$= \frac{1}{2}\frac{m_s v_{max}^2}{L^3}x^2 dx$$

Total kinetic energy of the spring is

$$E_k = \frac{1}{2}\frac{m_s v_{max}^2}{L^2}\int_0^L x^2 dx$$

or $$E_k = \frac{1}{2}\frac{m_s}{L^3}\cdot\frac{L^3}{3}\cdot v_{max}^2 = \frac{1}{2}\frac{m_s}{3}(A\omega)^2$$

$$= \frac{1}{2}\frac{m_s}{3}\omega^2 A^2$$

$$\therefore \qquad (\text{KE})_{max} = \frac{1}{2}mv_{max}^2 + E_k$$

$$= \frac{1}{2}\left(m + \frac{m_s}{3}\right)\omega^2 A^2$$

$$(\text{PE})_{max} = \frac{1}{2}kA^2$$

$$(\text{KE})_{max} = (\text{PE})_{max}$$

$$\Rightarrow \qquad \frac{1}{2}\left(m + \frac{m_s}{3}\right)\omega^2 A^2 = \frac{1}{2}kA^2$$

$$\omega = \sqrt{\frac{k}{m + \frac{m_s}{3}}}$$

$$\therefore \qquad T = \frac{2\pi}{\omega} = 2\pi\sqrt{\frac{m + \frac{m_s}{3}}{k}}$$

If the mass of the spring is to be considered, the same equation holds good for a vertically loaded spring also.

13. (a,b) When springs 1 and 2 share the weight equally :
Here, due to different elongations, the rod is no more horizontal. When the rod is in equilibrium, the stretch in each spring is

$$\Delta x_1 = \frac{mg}{2k_1}, \Delta x_2 = \frac{mg}{2k_2}$$

and $$\Delta x_3 = \frac{mg}{k_3}$$

The static deflection of mass m, in the equilibrium position is

$$\Delta x = \frac{\Delta x_1 + \Delta x_2}{2} + \Delta x_3 = \frac{1}{2}\left[\frac{mg}{2k_1} + \frac{mg}{2k_2}\right] + \frac{mg}{k_3}$$

or $$\Delta x = mg\left[\frac{4k_1k_2 + k_1k_3 + k_2k_3}{4k_1k_2k_3}\right]$$

Now, $f_1 = \frac{1}{2\pi}\sqrt{\frac{g}{\Delta x}}$, where Δx is static deflection of mass m.

or $$f_1 = \frac{1}{2\pi}\sqrt{\frac{4k_1k_2k_3}{m(4k_1k_2 + k_1k_3 + k_2k_3)}}$$

When the rod stays horizontally, [1 parallel 2] series with 3

$\therefore$ $$\frac{1}{k_{eq}} = \frac{1}{k_1 + k_2} + \frac{1}{k_3}$$

$\Rightarrow$ $$k_{eq} = \frac{k_1k_3 + k_2k_3}{k_1 + k_2 + k_3}$$

$$f_2 = \frac{1}{2\pi}\sqrt{\frac{k_{eq}}{m}} = \frac{1}{2\pi}\sqrt{\frac{k_1k_3 + k_2k_3}{m(k_1 + k_2 + k_3)}}$$

14. (a,c) Density of block is $\rho_{block} = \frac{5}{2}\rho_0$. Resultant force acting on the block at the height h from bottom is

$$F = \left[\frac{m}{\frac{5}{2}\rho_0}\right]\rho_0\left[4 - \frac{3h}{h_0}\right]g - mg$$

When the block is in equilibrium position, $F = 0$

$\therefore$ $$\frac{2}{5}\left(4 - \frac{3h}{h_0}\right) = 1$$

$\Rightarrow$ $$4 - \frac{3h}{h_0} = \frac{5}{2}$$

$$\frac{3h}{h_0} = 4 - \frac{5}{2} = \frac{3}{2}$$

$\Rightarrow$ $$h = \frac{h_0}{2}$$

Let the block has moved up by a distance x from mean position. Now, the resultant force is

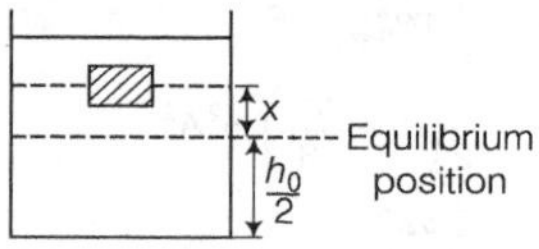

$$F_R = \left[\frac{m}{\frac{5}{2}\rho_0}\right]\rho_0\left\{4 - \frac{3\left(\frac{h_0}{2} + x\right)}{h_0}\right\}g - mg$$

$$= -\left[\frac{6mg}{5h_0}\right]x$$

Acceleration of the block is $a = \frac{F_R}{m} = -\left(\frac{6g}{5h_0}\right)x$. So, the block performs SHM and $\omega = \sqrt{\frac{6g}{5h_0}}$

$$f = \frac{1}{2\pi}\sqrt{\frac{6g}{5h_0}}$$

15. (a,d) In equilibrium of the cone,
Buoyant force = Weight

$$\frac{1}{3}\left[A\left(\frac{h}{H}\right)^2\right]h\rho_w g = \frac{1}{3}(AH)\rho_w sg$$

or $$h = H(s)^{1/3} = 4\left(\frac{27}{64}\right)^{1/3} = 3\text{m}$$

When the cone is pushed slightly downwards, it performs oscillations. Consider the situation, when the cone is displaced downwards by x. Then,
Buoyant force – Weight = – Mass × Acceleration

$$\frac{1}{3}A\left(\frac{h + x}{H}\right)^2(h + x)\rho_w g - \frac{1}{3}AH\rho_w sg = -\frac{1}{3}AH\rho_w sa$$

$$\frac{(h + x)^3}{H^2}g - Hsg = -Hsa$$

$$\frac{h^3}{H^2}\left(1 + \frac{x}{h}\right)^3 g - Hsg = -Hsa$$

$$\frac{h^3}{H^2}\left(1 + \frac{3x}{h}\right)g - Hsg = -Hsa$$

[$\because x << h$ and approximating using Binomial theorem]

$$\frac{3^3}{4^2}\left(1 + \frac{3x}{3}\right)(10) - (4)\left(\frac{27}{64}\right)(10) = -4\left(\frac{27}{64}\right)a$$

$$\frac{270}{16}x = -\frac{27}{16}a$$

$\Rightarrow$ $$a = -10x$$

$$\omega = \sqrt{10}$$

$\Rightarrow$ $$T = \frac{2\pi}{\omega} = \frac{2\pi}{\sqrt{10}} = 1.98 \text{ s}$$

16. (a) For a spring pendulum,

$$T = 2\pi\sqrt{\frac{m}{k}}$$

17. (c) $F = -m\omega^2 x$

$\Rightarrow$ Slope $= -m\omega^2$

$\therefore$ $-1 = -m\omega^2$

$\Rightarrow$ $\omega^2 = 1$

$\Rightarrow$ $\omega = 1$

$\therefore$ $T = 2\pi = 6.28$ s

If m changes and slope remains same, then T changes.

18. (c) $v = \omega\sqrt{A^2 - x^2}$

$\Rightarrow$ $$\frac{v^2}{\omega^2 A^2} + \frac{x^2}{A^2} = 1$$

So, v-x graph is an ellipse.
In SHM, $x = A\sin(\omega t + \delta)$ and $v = A\omega\cos(\omega t + \delta)$

19. (a) When the speed of the block is maximum, $F_{net} = 0$
$\therefore$ Spring force = mg (↑)

20. (d) $F_s = mg = kx$ (when the speed is maximum)

$\Rightarrow$ $$x = \frac{mg}{k}$$

$\therefore$ Compression in spring $= \frac{mg}{k}$

$$= \frac{mg}{4mg} L = \frac{L}{4}$$

$$h = 11L - \left(10\,L - \frac{L}{4}\right)$$

$$= L + \frac{L}{4} = \frac{5L}{4}$$

Applying conservation of energy,

$$mg\frac{5L}{4} = \frac{1}{2}k\left(\frac{L}{4}\right)^2 + \frac{1}{2}mv_{max}^2$$

$$mg\frac{5L}{4} = \frac{1}{2}\frac{4mg}{L}\cdot\frac{L^2}{16} + \frac{1}{2}mv_{max}^2$$

$$\frac{5g\,L}{2} = \frac{gL}{4} + v_{max}^2$$

$$\Rightarrow \quad v_{max}^2 = gL\left(\frac{5}{2} - \frac{1}{4}\right) = gL\left(\frac{10-1}{4}\right)$$

$$v_{max} = \frac{3}{2}\sqrt{gL}$$

21. (a) $mg\,(L + x') = \frac{1}{2}kx'^2$

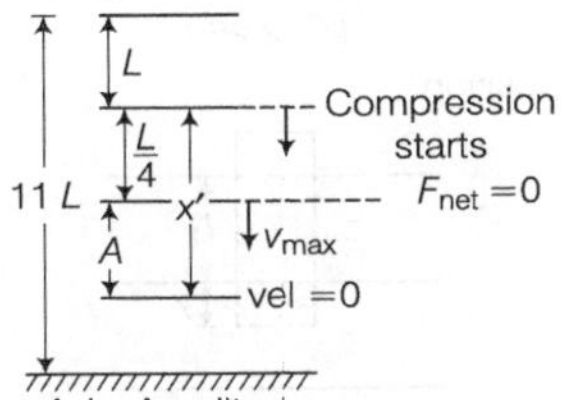

A is Amplitude

$$= \frac{1}{2}\frac{4mg}{L}x'^2$$

$$L + x' = \frac{2}{L}x'^2$$

$$\frac{2}{L}x'^2 - x' - L = 0$$

$$x' = \frac{1 + \sqrt{1 + 4\cdot\frac{2}{L}L}}{2\cdot\frac{2}{L}}$$

$$x' = \frac{1+3}{4}\cdot L = L$$

$\therefore$ Maximum compression $x' = L$

$$A = x' - \frac{L}{4} = L - \frac{L}{4} = \frac{3L}{4}$$

From equilibrium position to maximum compression,

$$t_1 = \frac{T}{4} = \frac{2\pi}{4}\sqrt{\frac{L/4}{g}} = \frac{\pi}{4}\sqrt{\frac{L}{g}}$$

$$x = A\sin\omega t_2$$

$$\Rightarrow \quad \frac{L}{4} = \frac{3L}{4}\sin\omega t_2$$

$$\omega t_2 = \sin^{-1}\left(\frac{1}{3}\right)$$

$$\frac{2\pi}{2\pi\sqrt{\frac{L/4}{g}}}\cdot t_2 = \sin^{-1}\left(\frac{1}{3}\right)$$

$$\Rightarrow \quad t_2 = \sqrt{\frac{L}{4g}}\sin^{-1}\left(\frac{1}{3}\right)$$

$\therefore$ Required time $= t_1 + t_2$

$$= \frac{\pi}{4}\sqrt{\frac{L}{g}} + \sqrt{\frac{L}{4g}}\sin^{-1}\left(\frac{1}{3}\right)$$

Note *(i) While calculating t_2, we calculated time taken from mean position to $x = +\frac{L}{4}$, which is same as required. (i.e. from $x = +\frac{L}{4}$ to $x = 0$).*

(ii) For a vertical spring-block system $T = 2\pi\sqrt{\frac{l}{g}}$, where l is deformation (extension or compression) in equilibrium position.

22. (6) In the figure, $\theta = 30°$.

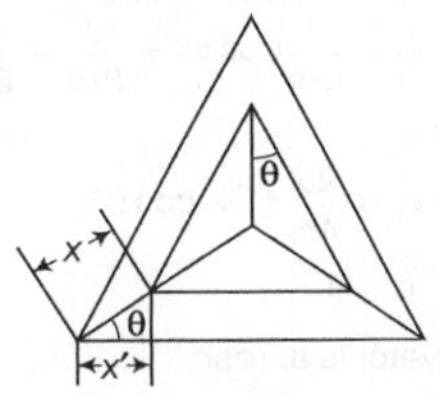

When displacement is each ball x, the extension in each spring is

$$x' = x\cos\theta$$

$$= x\frac{\sqrt{3}}{2}$$

Each spring is connected to two balls.

So, net extension of each spring $= 2x' = \sqrt{3}x$

Force along the direction of displacement on each ball is

$$F = k\sqrt{3}\,x\cos 30° + k\sqrt{3}\,x\cos 30° = 3kx$$

$F = -3kx$ [negative because restoring in nature]

$$\therefore \quad a = \frac{F}{m} = -\frac{3k}{m}x$$

$$\Rightarrow \quad \omega = \sqrt{\frac{3k}{m}}$$

$$\therefore \quad f = \frac{1}{2\pi}\sqrt{\frac{3k}{m}} = \frac{1}{2\pi}\sqrt{\frac{6k}{2m}}$$

$$\Rightarrow \quad \beta = 6$$

23. (4) $u_1 = \frac{J}{m}, u_2 = 0$

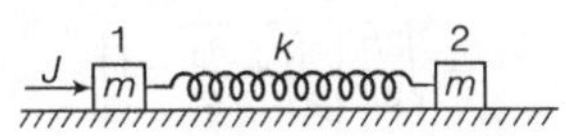

$$v_{COM} = \frac{m\cdot\frac{J}{m} + m\cdot(0)}{m + m}$$

$$v_{COM} = \frac{J}{2m}$$

In centre of mass frame of reference the blocks execute SHM only. (and do not slide rightwards in centre of mass frame)

Initially

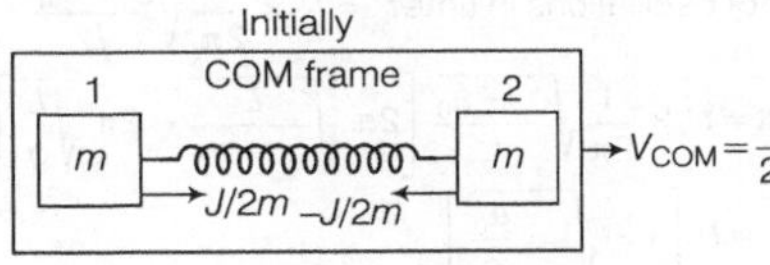

$$u_{1,\,COM} = u_1 - u_{COM}$$

$$= \frac{J}{m} - \frac{J}{2m}$$

$$= \frac{J}{2m}$$

$$u_{2\,COM} = u_2 - u_{COM}$$

$$= 0 - \frac{J}{2m} = -\frac{J}{2m}$$

In centre of mass frame, the blocks are executing SHM with

$$\omega = \sqrt{\frac{k}{\mu}} = \sqrt{\frac{k}{m \cdot m/(m+m)}} = \sqrt{\frac{2k}{m}}$$

At initial moment spring is relaxed. Therefore,

$$x_{1,\,COM} = A\sin\omega t$$

and $$x_{2,\,COM} = A\sin(\omega t + \pi)$$

$$v_{1,\,COM} = \frac{dx_{1,\,COM}}{dt} = \frac{J}{2m}\cos\omega t$$

$$v_{1,\,COM} = v_1 - v_{COM}$$

$\Rightarrow$ $$v_1 = v_{1,\,COM} + v_{COM}$$

or $$v_1 = \frac{J}{2m}\cos\omega t + \frac{J}{2m} = \frac{J}{2m}(1 + \cos\omega t)$$

or in ground frame

$$v_1 = \frac{2J}{4m}(1 + \cos\omega)$$

$\therefore$ $$\alpha = 4$$

24. (9) When the elevator is at rest

$$f = \frac{1}{2\pi}\sqrt{\frac{g}{L}}$$

When the elevator accelerates up

$$f_1 = \frac{1}{2\pi}\sqrt{\frac{g + a_0}{L}}$$

Time gain in one oscillation $= T - T_1$

$$= 2\pi\sqrt{\frac{L}{g}} - 2\pi\sqrt{\frac{L}{g + a_0}}$$

Time taken by the elevator to travel a distance h is $t = \sqrt{\frac{2h}{a_0}}$

Number of oscillations in time $t = t \times f_1$,

$$= \sqrt{\frac{2h}{a_0}} \cdot \frac{1}{2\pi}\sqrt{\frac{g + a_0}{L}}$$

$$\text{Total time gained} = \sqrt{\frac{2h}{a_0}} \cdot \frac{1}{2\pi}\sqrt{\frac{g + a_0}{L}}\left[2\pi\sqrt{\frac{L}{g}} - 2\pi\sqrt{\frac{L}{g + a_0}}\right]$$

$$= \sqrt{\frac{2h}{a_0}}\left[\frac{\sqrt{g + a_0}}{\sqrt{g}} - 1\right]$$

$$= \sqrt{\frac{2h}{a_0}}\left[\sqrt{1 + \frac{a_0}{g}} - 1\right]$$

Let t' be the time in which the pendulum will lose the time it gained during a_0 is upwards.

$$\text{Time lost in one oscillation} = 2\pi\sqrt{\frac{L}{g - a_0}} - 2\pi\sqrt{\frac{L}{g}}$$

$$f_2 = \frac{1}{2\pi}\sqrt{\frac{g - a_0}{L}}$$

Number of oscillations in time $t' = t' \times \frac{1}{2\pi}\sqrt{\frac{g - a_0}{L}}$

$$\text{Time lost} = t' \times \frac{1}{2\pi}\sqrt{\frac{g - a_0}{L}}\left[2\pi\sqrt{\frac{L}{g - a_0}} - 2\pi\sqrt{\frac{L}{g}}\right]$$

$$= t'\left[1 - \sqrt{1 - \frac{a_0}{g}}\right]$$

When the clock shows right time, time lost = time gain

$\therefore$ $$t'\left[1 - \sqrt{1 - \frac{a_0}{g}}\right] = \sqrt{\frac{2h}{a_0}}\left[\sqrt{1 + \frac{a_0}{g}} - 1\right]$$

$$t' = \sqrt{\frac{2h}{a_0}}\left[\frac{\sqrt{1 + \frac{a_0}{g}} - 1}{1 - \sqrt{1 - \frac{a_0}{g}}}\right]$$

Total time

$$= \sqrt{\frac{2h}{a_0}} + t' = \sqrt{\frac{2h}{a_0}}\left[1 + \frac{\sqrt{1 + \frac{a_0}{g}} - 1}{1 - \sqrt{1 - \frac{a_0}{g}}}\right]$$

$$= \sqrt{\frac{2h}{a_0}}\left[\frac{\sqrt{1 + \frac{a_0}{g}} - \sqrt{1 - \frac{a_0}{g}}}{1 - \sqrt{1 - \frac{a_0}{g}}}\right] = \sqrt{\frac{2h}{a_0}}\left[\frac{\sqrt{1 + \frac{2}{5}} - \sqrt{1 - \frac{2}{5}}}{1 - \sqrt{1 - \frac{2}{5}}}\right]$$

$$= \sqrt{\frac{2h}{a_0}}\left[\frac{\sqrt{7} - \sqrt{3}}{\sqrt{5} - \sqrt{3}}\right] = \frac{9}{5}\sqrt{\frac{2h}{a_0}}$$

$\therefore$ $$\psi = 9$$

25. (7) In equilibrium,

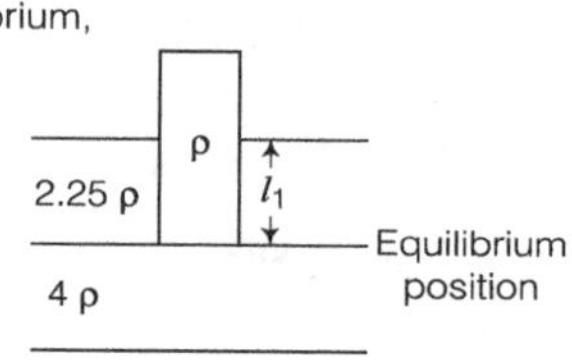

$$\rho ALg = 2.25\,\rho A l_1 g$$

$$L = 2.25\,l_1 \quad \text{...(i)}$$

In displaced position,

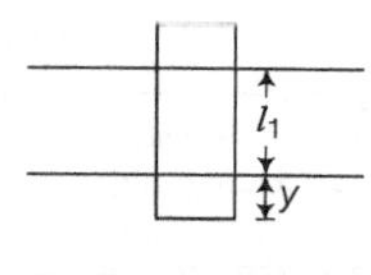

Displaced position

Net force $F_1 = \rho ALg - 2.25\,\rho A l_1 g - 4\rho Ayg$

Using Eq. (i), we get

$$F_1 = -4\rho Ag\,y$$

$$\frac{F_1}{m} = \frac{-4\rho Ag}{\rho AL}y = -\frac{4g}{L}y$$

$\therefore$ $$\omega_1 = \sqrt{\frac{4g}{L}}$$

$\Rightarrow$ $$\frac{T_1}{2} = \frac{1}{2}\frac{2\pi}{\omega_1} = \pi\sqrt{\frac{L}{4g}}$$

The pole remains in the bottom liquid for a time $\frac{T_1}{2}$.

When it moves to upper liquid.

$$F_2 = -\rho ALg + (l_1 - y')A(2.25\rho)g$$

or $$F_2 = -2.25\,\rho g Ay'$$

$\Rightarrow$ $$\frac{F_2}{m} = -\frac{2.25\rho g Ay'}{\rho AL} = -\frac{2.25g}{L}y'$$

$\therefore$ $$\omega_2 = \sqrt{\frac{2.25g}{L}}$$

$\Rightarrow$ $$T_2 = 2\pi\sqrt{\frac{L}{2.25g}}$$

$$\frac{T_2}{2} = \pi\sqrt{\frac{L}{2.25g}}$$

Required time period is

$$T = \frac{T_1}{2} + \frac{T_2}{2} = \pi\sqrt{\frac{L}{g}}\left[\frac{1}{\sqrt{4}} + \frac{1}{\sqrt{2.25}}\right]$$

$\Rightarrow$ $$T = \pi\sqrt{\frac{L}{g}}\left[\frac{1}{2} + \frac{2}{3}\right] = \frac{7}{6}\pi\sqrt{\frac{L}{g}}$$

$\therefore$ $$c = 7$$

DPP-6 Time Period : Physical Systems-II

1. (b) If block is in equilibrium $mg = kd$, where d is extension in the spring. Now, let the block is pulled down and released. It performs oscillations. The figure shows the snapshot of the system in which x is the displacement (downwards) from the mean position and v is the velocity (downwards), at that instant. Angular speed of the pulley at this instant $\omega = \frac{v}{R}$

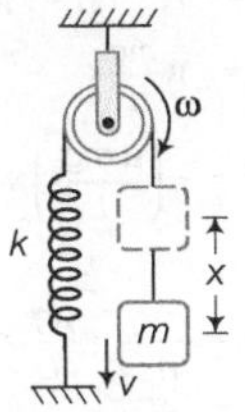

At this position total energy of oscillating system is

$$E_T = -mgx + \frac{1}{2}mv^2 + \frac{1}{2}I\omega^2 + \frac{1}{2}k(x+d)^2$$

$$\frac{dE_T}{dt} = -mg\left(\frac{dx}{dt}\right) + \frac{1}{2}m(2v)\frac{dv}{dt} + \frac{1}{2}\frac{I}{R^2}(2v)\frac{dv}{dt} + \frac{1}{2}k2(x+d)\frac{dx}{dt}$$

As total energy of the system remains constant, $\frac{dE_T}{dt} = 0$

$$\therefore \quad -mgv + mva + \frac{I}{R^2}va + k(x+d)v = 0$$

$$-mg + ma + \frac{I}{R^2}a + kx + kd = 0$$

$$a\left[m + \frac{I}{R^2}\right] = -kx \qquad [\because mg = kd]$$

$$a = -\left[\frac{k}{m + \frac{I}{R^2}}\right]x$$

Comparing with standard equation of SHM i.e. $a = -\omega^2 x$, we have

$$\omega = \sqrt{\frac{k}{m + \frac{I}{R^2}}}$$

$$\Rightarrow \quad T = 2\pi\sqrt{\frac{m + (I/R^2)}{k}}$$

2. (d) In the equilibrium position, let the tension in the string is T and extension in spring is x_0. Then, $T = kx_0$

$$\therefore \quad 2kx_0 = Mg$$

Now, let the pulley is slightly displaced from its equilibrium position (O) and released. It starts oscillating about the equilibrium position (O). The following figure shows the snapshot of the system during its oscillations.

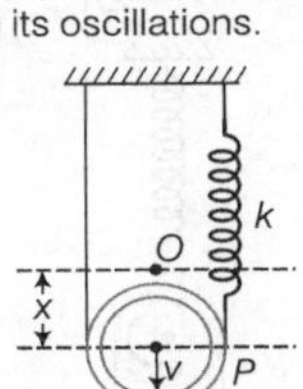

String does not slip over pulley's surface

$$\Rightarrow \quad \omega = \frac{v}{R}$$

[Particle of the string has velocity $2v(\downarrow)$ for no slipping $2v = v + R\omega \Rightarrow \omega = v/R$]

Total energy of the system at this instant is

$$E_T = -Mgx + \frac{1}{2}I\omega^2 + \frac{1}{2}Mv^2 + \frac{1}{2}k(x_0 + 2x)^2$$

$$= -Mgx + \frac{1}{2}I\frac{v^2}{R^2} + \frac{1}{2}Mv^2 + \frac{1}{2}k(x_0 + 2x)^2$$

$$\frac{dE_T}{dt} = -Mg\frac{dx}{dt} + \frac{1}{2}\frac{I}{R^2}2v\frac{dv}{dt} + \frac{1}{2}M2v\frac{dv}{dt} + \frac{1}{2}k\,2(x_0 + 2x)\cdot 2\frac{dx}{dt}$$

$$\frac{dE_T}{dt} = 0$$

$$\therefore -Mgv + \frac{I}{R^2}va + Mva + 2k(x_0 + 2x)v = 0$$

$$-Mg + a\left(M + \frac{I}{R^2}\right) + 2kx_0 + 4kx = 0$$

$$a\left(M + \frac{I}{R^2}\right) = -4kx \qquad [\because Mg = 2kx_0]$$

$$a = -\left(\frac{4k}{M + \frac{I}{R^2}}\right)x, \text{ comparing with } a = -\omega^2 x, \text{ we have}$$

$$\omega = \sqrt{\frac{4k}{M + \frac{I}{R^2}}}$$

$$\Rightarrow \quad T = 2\pi\sqrt{\frac{M + \frac{I}{R^2}}{4k}}$$

3. (a) In equilibrium, net torque on the pulley must be zero.

$$\therefore \quad MgR = mgR\sin\phi$$

$$\Rightarrow \quad M = m\sin\phi$$

Snapshot during oscillations is shown in the figure.

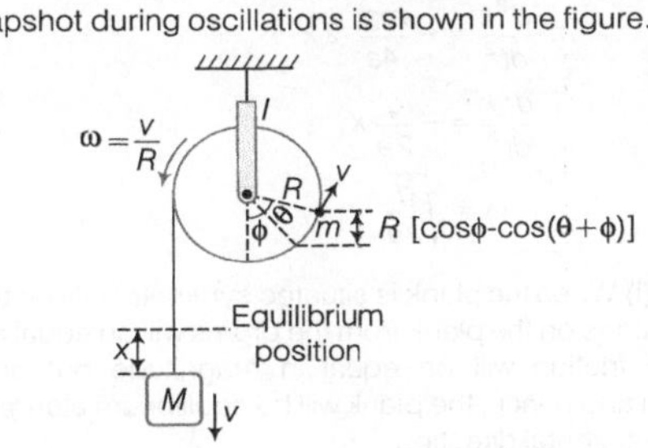

In the situation shown, total energy is

$$E_T = -Mgx + \frac{1}{2}Mv^2 + \frac{1}{2}I\left(\frac{v}{R}\right)^2 + \frac{1}{2}mv^2 + mgR[\cos\phi - \cos(\theta + \phi)]$$

$$\frac{dE_T}{dt} = -Mgv + \frac{1}{2}M(2v)a + \frac{1}{2}\frac{I}{R^2}(2v)(a) + \frac{1}{2}m\,2v(a) + mgR\sin(\theta + \phi)\frac{d\theta}{dt} = 0$$

$$\frac{d\theta}{dt} = \omega = \frac{v}{R}$$

$$-Mg + Ma + \frac{I}{R^2}a + ma + mg\sin(\theta + \phi) = 0$$

$$-Mg + a\left[M + m + \frac{I}{R^2}\right] + mg[\sin\theta\cos\phi + \cos\theta\sin\phi] = 0$$

θ is very small.

$$\therefore \sin\theta \simeq \theta = \frac{x}{R} \text{ and } \cos\theta = 1$$

$$-Mg + a\left[M + m + \frac{I}{R^2}\right] + mg\frac{x}{R}\cos\phi + mg\cdot\sin\phi = 0$$

$$a\left[M + m + \frac{I}{R^2}\right] = -\frac{mg\cos\phi}{R}x \qquad [\because M = m\sin\phi]$$

$$a = -\frac{mg\cos\phi}{R\left(M + m + \dfrac{I}{R^2}\right)}x$$

$$\Rightarrow \qquad \omega = \sqrt{\frac{mg\cos\phi}{R\left(M + m + \dfrac{I}{R^2}\right)}}$$

$$\therefore \qquad T = \frac{2\pi}{\omega} = 2\pi\sqrt{\frac{R\left(M + m + \dfrac{I}{R^2}\right)}{mg\cos\phi}}$$

4. (a) During its oscillations, at a particular instant let the coordinates of the particle are (x, y).

The total energy of the particle is

$$E_T = \frac{1}{2}m\left(\frac{dx}{dt}\right)^2 + \frac{1}{2}m\left(\frac{dy}{dt}\right)^2 + mgy$$

Equation of the curve is $x^2 = 4ay$

$$\Rightarrow \qquad 2x\frac{dx}{dt} = 4a\frac{dy}{dt}$$

$$\Rightarrow \qquad \frac{dy}{dt} = \frac{x}{2a}\frac{dx}{dt}$$

$$\therefore \qquad E_T = \frac{1}{2}m\left(\frac{dx}{dt}\right)^2 + \frac{1}{2}m\left(\frac{x}{2a}\right)^2\left(\frac{dx}{dt}\right)^2 + mg\frac{x^2}{4a}$$

$$\left[\because y = \frac{x^2}{4a}\right]$$

As oscillations are very small, we can ignore the middle term.

$$E_T \simeq \frac{1}{2}m\left(\frac{dx}{dt}\right)^2 + \frac{mg}{4a}x^2$$

$$\frac{dE_T}{dt} = \frac{1}{2}m\cdot 2\left(\frac{dx}{dt}\right)\left(\frac{d^2x}{dt^2}\right) + \frac{mg}{4a}\cdot 2x\frac{dx}{dt} = 0$$

$$m\cdot\frac{d^2x}{dt^2} = -\frac{mg}{4a}\cdot 2x$$

$$\frac{d^2x}{dt^2} = -\frac{g}{2a}x$$

$$\Rightarrow \qquad \omega = \sqrt{\frac{g}{2a}}$$

5. (a) **Fig. (i)** When the plank is situated symmetrically on the drums, the reactions on the plank from the drums will be equal and so the force of friction will be equal in magnitude but opposite in direction and hence, the plank will be equilibrium along vertical as well as horizontal direction.

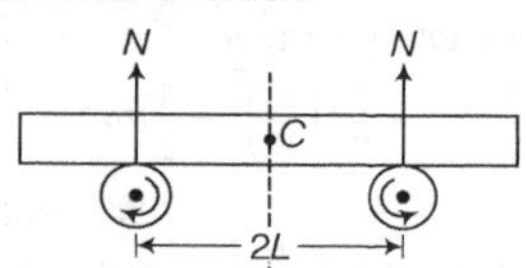

Fig (i)

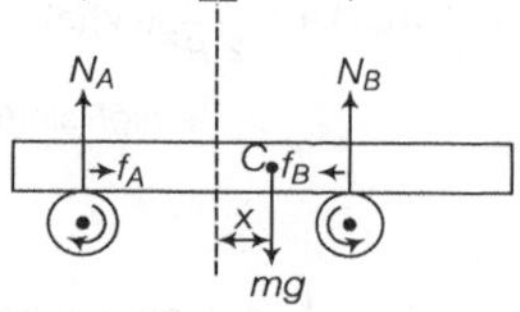

Fig (ii)

Fig. (ii) Now, if the plank is slightly displaced horizontally, the reactions will not be equal. Consider a moment at which the displacement of COM is x rightwards.

For vertical equilibrium of the plank, $N_A + N_B = mg$ $\qquad$ ...(i)

And taking moments about COM (i.e. C), we have

$$N_A(L + x) = N_B(L - x) \qquad \text{...(ii)}$$

From Eqs. (i) and (ii), we get

$$N_A = mg\left(\frac{L - x}{2L}\right)$$

$$N_B = mg\left(\frac{L + x}{2L}\right)$$

Note $N_B > N_A$

$f = \mu N$. So, friction at B will be more and tries to bring the plank back, i.e. restoring force, here is

$$F = -(f_B - f_A) = -\mu(N_B - N_A)$$

$$= -\mu\frac{mg}{L}x \qquad \left[\because N_B - N_A = \frac{mgx}{L}\right]$$

$$F = -\left(\frac{\mu mg}{L}\right)x$$

$\Rightarrow$ Acceleration is

$$a = \frac{F}{m} = -\frac{\mu g}{L}x$$

$$\therefore \qquad \omega = \sqrt{\frac{\mu g}{L}}$$

$$\Rightarrow \qquad T = \frac{2\pi}{\omega} = 2\pi\sqrt{\frac{L}{\mu g}}$$

6. (c) Initially, at equilibrium, let the extension in spring is x_0.

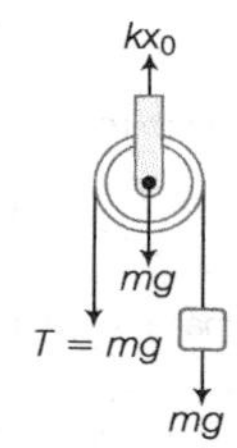

$$\therefore \qquad kx_0 = mg + mg + mg = 3mg \qquad \text{...(i)}$$

If the block is pulled down slightly and released, it performs oscillations.

During oscillations, when the displacement of block is x, the extension in spring will be $\frac{x}{2}$ (in addition to x_0). At this moment, total energy of the system is

$$E_T = \underbrace{\left[-mgx + \frac{1}{2}mv^2\right]}_{\text{Block}} + \underbrace{\left[-mg\frac{x}{2} + \frac{1}{2}m\left(\frac{v}{2}\right)^2 + \frac{1}{2}I\omega^2\right]}_{\text{Pulley}} + \underbrace{\frac{1}{2}k\left(x_0 + \frac{x}{2}\right)^2}_{\text{Spring}}$$

For no slipping,

$$\omega R + \frac{v}{2} = v$$

$$\Rightarrow \qquad \omega = \frac{v}{2R}$$

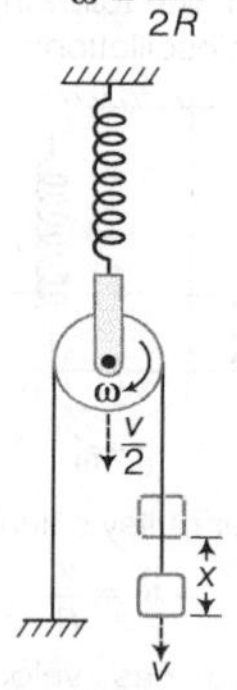

$$\therefore \quad E_T = -\frac{3}{2}mgx + \frac{5}{4}\left(\frac{1}{2}mv^2\right) + \frac{1}{2}I\frac{v^2}{4R^2} + \frac{1}{2}k\left(x_0 + \frac{x}{2}\right)^2$$

$$\frac{dE_T}{dt} = -\frac{3}{2}mgv + \frac{5}{8}m(2v)\frac{dv}{dt} + \frac{I}{8R^2}2v\left(\frac{dv}{dt}\right) + \frac{1}{2}k2\left(x_0 + \frac{x}{2}\right)\frac{v}{2}$$

$\frac{dE_T}{dt} = 0$, as E_T is constant.

$$\therefore \quad 0 = -\frac{3}{2}mg + \frac{5}{4}ma + \frac{I}{4R^2}a + \frac{kx_0}{2} + \frac{k}{4}x$$

$$\left(\frac{5}{4}m + \frac{I}{4R^2}\right)a = -\frac{k}{4}x \qquad \text{[after using Eq. (i)]}$$

$$\therefore \quad a = -\frac{k}{5m + \frac{I}{R^2}}x$$

$$\Rightarrow \quad \omega = \sqrt{\frac{k}{5m + \frac{I}{R^2}}}$$

$$\therefore \quad f = \frac{1}{2\pi}\sqrt{\frac{k}{5m + \frac{I}{R^2}}}$$

7. (b) When the particle is oscillating, a snapshot at an instant is shown.

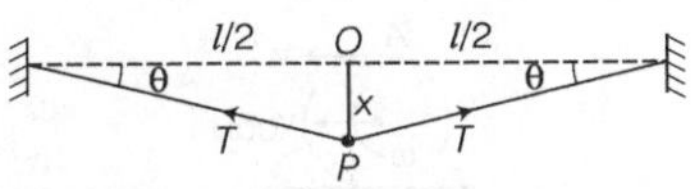

Restoring force is $F = -2T\sin\theta$

As oscillations are small,

$$\sin\theta \approx \theta = \frac{x}{l/2} = \frac{2x}{l}$$

$$\therefore \quad F = -2T\frac{2x}{l} = -\frac{4T}{l}x$$

$$a = -\frac{4T}{ml}x$$

$$\Rightarrow \quad \omega = \sqrt{\frac{4T}{ml}}$$

Time period is $\quad T' = \frac{2\pi}{\omega} = 2\pi\sqrt{\frac{ml}{4T}}$

$$= 3.14\sqrt{\frac{0.04 \times 1}{10}} = 0.2\,\text{s}$$

8. (c) A snapshot during oscillations of the cylinder is shown in the figure. Its displacement is x and velocity v. Its angular velocity is $\omega = \frac{v}{R}$ (due to pure rolling).

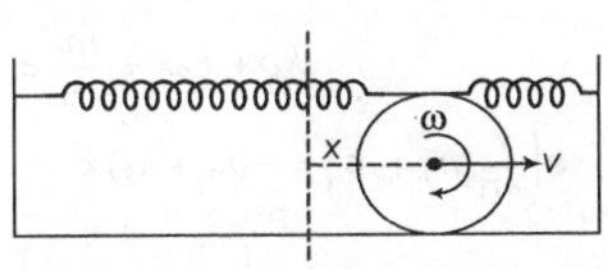

As the centre of the cylinder is at a distance x from the initial position, the springs which are connected at a point on its rim must be compressed and stretched by a distance $2x$. Thus, at this intermediate position total energy of the oscillating system can be given as

$$E_T = \frac{1}{2}mv^2 + \frac{1}{2}\left(\frac{mR^2}{2}\right)\omega^2 + \left[\frac{1}{2}k(2x)^2\right]2$$

$$= \frac{1}{2}mv^2 + \frac{1}{4}mv^2 + 4kx^2$$

$$= \frac{3}{4}mv^2 + 4kx^2$$

$$\frac{dE_T}{dt} = \frac{3}{4}m\cdot 2v\cdot a + 4k\cdot 2x\cdot v = 0 \quad [\because E_T \text{ is constant}]$$

$$\frac{3}{2}ma = -8kx$$

$$\Rightarrow \quad a = -\left(\frac{16x}{3m}\right)x$$

$$\therefore \quad \omega = \sqrt{\frac{16k}{3m}}$$

$$\Rightarrow \quad T = \frac{2\pi}{\omega} = 2\pi\sqrt{\frac{3m}{16k}} = \frac{\pi}{2}\sqrt{\frac{3m}{k}}$$

9. (a) $T_1 = 2\pi\sqrt{\frac{l}{g}}$ (elevator is at rest) ...(i)

$T_2 = 2\pi\sqrt{\frac{l}{g + a_0}}$ (elevator accelerates up) ... (ii)

$T_3 = 2\pi\sqrt{\frac{l}{g - a_0}}$ (elevator accelerates down) ...(iii)

From Eq. (ii), $\quad \frac{4\pi^2 l}{T_2^2} = g + a_0$

From Eq. (iii), $\quad \frac{4\pi^2 l}{T_3^2} = g - a_0$

Adding the above two equations gives,

$$\frac{4\pi^2 l}{T_2^2} + \frac{4\pi^2 l}{T_3^2} = g + a_0 + g - a_0$$

$$= 2g = 2\frac{4\pi^2 l}{T_1^2} \qquad \text{[from Eq. (i)]}$$

$$\therefore \quad \frac{1}{T_2^2} + \frac{1}{T_3^2} = \frac{2}{T_1^2}$$

$$\Rightarrow \quad T_1 = \frac{\sqrt{2}\,T_2 T_3}{\sqrt{T_2^2 + T_3^2}}$$

10. (d) A snapshot during oscillations is shown in the figure. The right side has moved down by x while the left side has moved up by x.

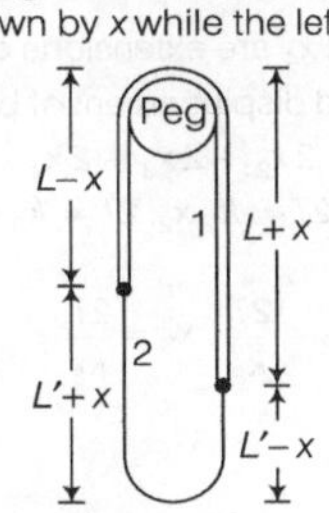

Let at the top of peg tension is T.

Mass of right side part = $m_R = \lambda(L + x) + \lambda'(L' - x)$

Mass of left side part = $m_L = \lambda(L - x) + \lambda'(L' + x)$

For right side part, $\quad m_R g - T = m_R a$

For left side part, $\quad T - m_L g = m_L a$

$$(m_R - m_L)g = (m_R + m_L)a$$

$$\therefore \quad a = \frac{m_R - m_L}{m_R + m_L}g$$

$$\therefore \quad a = \frac{\lambda L + \lambda x + \lambda' L' - \lambda' x - \lambda L + \lambda x - \lambda' L' - \lambda' x}{\lambda L + \lambda x + \lambda' L' - \lambda' x + \lambda L - \lambda x + \lambda' L' + \lambda' x}g$$

$$= \frac{2(\lambda - \lambda')}{2(\lambda L + \lambda' L')}g = -\frac{(\lambda' - \lambda)g}{(\lambda L + \lambda' L')}x$$

$$\frac{d^2x}{dt^2} = -\left[\frac{(\lambda' - \lambda)g}{(\lambda L + \lambda' L')}\right]x$$

$$\Rightarrow \quad \omega = \sqrt{\frac{(\lambda' - \lambda)g}{\lambda L + \lambda' L'}}$$

$$\therefore \quad T = \frac{2\pi}{\omega} = 2\pi\sqrt{\frac{\lambda L + \lambda' L'}{(\lambda' - \lambda)g}}$$

11. (a) Imagine pulley 2 is fixed and cannot change its position. Now if pulley 1 moves down by x_1, block will move down by $2x_1$. Similarly, if pulley 1 is assumed to be fixed and pulley 2 moves up by x_2, then block will move down by $2x_2$. But in the problem both pulleys are not fixed. So, if block m moves down by x_0 then

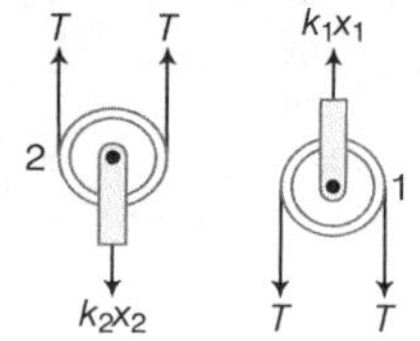

$$x_0 = 2x_1 + 2x_2$$

$$2T = k_2 x_2$$

and $$2T = k_1 x_1$$

$\therefore$ $$x_1 = \frac{2T}{k_1}$$

and $$x_2 = \frac{2T}{k_2}$$

If k_{eq} is equivalent spring constant of the system, then

$$x_0 = \frac{T}{k_{eq}}$$

$\Rightarrow$ $$2x_1 + 2x_2 = \frac{T}{k_{eq}}$$

$$2\cdot\frac{2T}{k_1} + 2\cdot\frac{2T}{k_2} = \frac{T}{k_{eq}}$$

$\Rightarrow$ $$\frac{1}{k_{eq}} = 4\left[\frac{1}{k_1} + \frac{1}{k_2}\right]$$

$\therefore$ $$k_{eq} = \frac{k_1 k_2}{4(k_1 + k_2)}$$

$$\omega = \sqrt{\frac{k_{eq}}{m}} = \sqrt{\frac{k_1 k_2}{4m(k_1 + k_2)}}$$

12. (c) Let x_1, x_2, x_3 and x_4 are extensions of respective springs. If x_0 is the downward displacement of block, then

$$x_0 = 2x_1 + 2x_2 + 2x_3 + 2x_4$$

Also, $2T = k_1 x_1, 2T = k_2 x_2, 2T = k_3 x_3$

and $2T = k_4 x_4$

$\therefore$ $$x_1 = \frac{2T}{k_1}, x_2 = \frac{2T}{k_2}, x_3 = \frac{2T}{k_3}$$

and $$x_4 = \frac{2T}{k_4}$$

$\therefore$ $$x_0 = 2\cdot\frac{2T}{k_1} + 2\cdot\frac{2T}{k_2} + 2\cdot\frac{2T}{k_3} + 2\cdot\frac{2T}{k_4}$$

$$= 4T\left[\frac{1}{k_1} + \frac{1}{k_2} + \frac{1}{k_3} + \frac{1}{k_4}\right]$$

Also, $$x_0 = \frac{T}{k_{eq}}$$

$\therefore$ $$\frac{T}{k_{eq}} = 4T\left[\frac{1}{k_1} + \frac{1}{k_2} + \frac{1}{k_3} + \frac{1}{k_4}\right]$$

$\therefore$ $$k_{eq} = \frac{1}{4\left[\frac{1}{k_1} + \frac{1}{k_2} + \frac{1}{k_3} + \frac{1}{k_4}\right]}$$

$$\omega = \sqrt{\frac{k_{eq}}{m}} = \sqrt{\frac{1}{4m\left(\frac{1}{k_1} + \frac{1}{k_2} + \frac{1}{k_3} + \frac{1}{k_4}\right)}}$$

$$T = \frac{2\pi}{\omega} = 2\pi\sqrt{4m\left(\frac{1}{k_1} + \frac{1}{k_2} + \frac{1}{k_3} + \frac{1}{k_4}\right)}$$

13. (b) From the figure it is clear that, if the displacement of block is x_0, extension in spring will be $4x_0$.

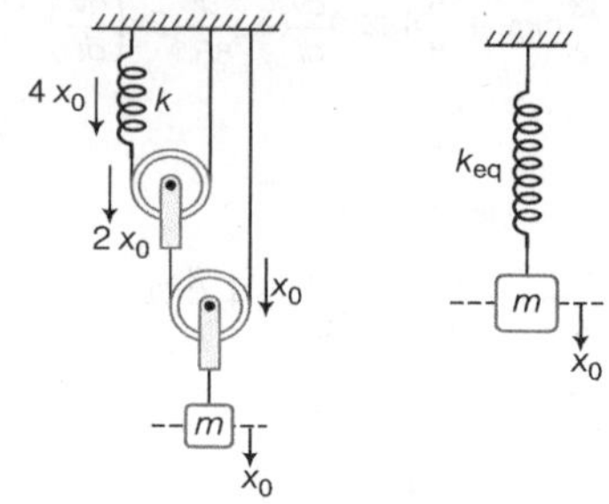

Force on the block will be $(8kx_0)\,2 = 16kx_0$

$\therefore$ $$16kx_0 = k_{eq}x_0$$

$\Rightarrow$ $$k_{eq} = 16k$$

$$\omega = \sqrt{\frac{k_{eq}}{m}} = \sqrt{\frac{16k}{m}} = 4\sqrt{\frac{k}{m}}$$

$\therefore$ $$f = \frac{\omega}{2\pi} = \frac{2}{\pi}\sqrt{\frac{k}{m}}$$

14. (c) Let the horizontal displacement of the plank, at an instant during its oscillations is x and its velocity is v.

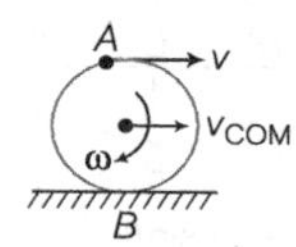

For pure rolling at A,

$$v = v_{COM} + \omega R$$

For pure rolling at B,

$$v_{COM} = \omega R$$

$\therefore$ $$v_{COM} = \frac{v}{2}, \omega = \frac{v}{2R}$$

$$E_T = \frac{1}{2}k_1x^2 + \frac{1}{2}k_2x^2 + \left(\frac{1}{2}I\omega^2\right)^2 + 2\left(\frac{1}{2}m_0v_{COM}^2\right) + \frac{1}{2}mv^2$$

I is MI of sphere along horizontal diameter.

$$E_T = \frac{1}{2}k_1x^2 + \frac{1}{2}k_2x^2 + \frac{1}{2}\cdot\frac{2}{5}m_0R^2\cdot\frac{v^2}{4R^2}\cdot 2$$

$$+ 2\cdot\frac{1}{2}m_0\cdot\frac{v^2}{4} + \frac{1}{2}mv^2$$

$$\frac{dE_T}{dt} = \frac{1}{2}k_1 2x\cdot v + \frac{1}{2}k_2\cdot 2x\cdot v + \frac{1}{10}m_0\,2v\cdot a + \frac{m_0}{4}2v\cdot a$$

$$+ \frac{1}{2}m\cdot 2v\cdot a = 0$$

$$k_1x + k_2x + \frac{m_0}{5}a + \frac{m_0}{2}\cdot a + ma = 0$$

$$a\left(\frac{7}{10}m_0 + m\right) = -(k_1 + k_2)x$$

$\Rightarrow$ $$a = -\left(\frac{k_1 + k_2}{m + \frac{7}{10}m_0}\right)x$$

$\therefore$ $$\omega = \sqrt{\frac{k_1 + k_2}{m + \frac{7}{10}m_0}}$$

$\Rightarrow$ $$f = \frac{1}{2\pi}\sqrt{\frac{k_1 + k_2}{m + \frac{7}{10}m_0}}$$

15. (d) $x' = x\cos(90° - \theta) = x\sin\theta$

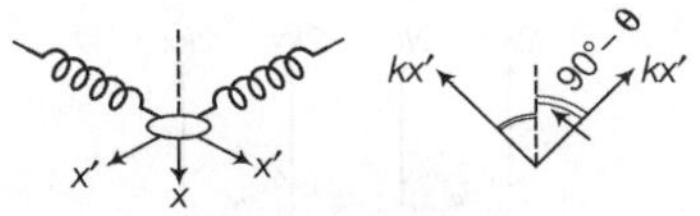

Restoring force is $F = -2kx'\sin\theta$

$$= -2kx\sin^2\theta$$

$$a = -\frac{2k\sin^2\theta}{m}x$$

$$\omega = \sqrt{\frac{2k\sin^2\theta}{m}}$$

$$T = \frac{2\pi}{\omega} = 2\pi\sqrt{\frac{m}{2k\sin^2\theta}}$$

$$T = \pi\sqrt{\frac{2m}{k\sin^2\theta}}$$

16. (b,c) $y^2 = (5l)^2 - x^2$

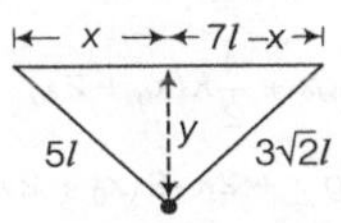

$$y^2 = (3\sqrt{2}\,l)^2 - (7l - x)^2$$

$\therefore$ $x = 4l$

Thus, $y = 3l$

$$T = 2\pi\sqrt{\frac{3l}{g}}$$

17. (b,d) For SHM $F \propto -x$. So, particle does not perform SHM. But it performs periodic motion.

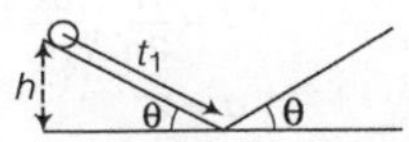

$$a = \frac{mg\sin\theta}{m} = g\sin\theta$$

$$\frac{h}{\sin\theta} = (0)t_1 + \frac{1}{2}g\sin\theta\,(t_1)^2 \quad \left[\because s = ut + \frac{1}{2}at^2\right]$$

$$t_1 = \sqrt{\frac{2h}{g}}\sqrt{\frac{1}{\sin^2\theta}}$$

$$T = 4t_1 = \frac{4}{\sin\theta}\sqrt{\frac{2h}{g}}$$

18. (a,b,c)

$$v = \omega\sqrt{A^2 - x^2}$$

$\Rightarrow$ $v^2 = \omega^2A^2 - \omega^2x^2$

Comparing this equation with given equation. i.e. $v^2 = 144 - 9x^2$

we have, $\omega = 3$

$\Rightarrow$ $\frac{2\pi}{T} = 3$

or $T = \frac{2\pi}{3}$ unit

Similarly, $A = \frac{v_{max}}{\omega} = \frac{A\omega}{\omega} = \sqrt{\frac{A^2\omega^2}{\omega^2}}$

$$= \sqrt{\frac{144}{9}} = \frac{12}{3}$$

or $A = 4$ unit

$$a = -\omega^2 x$$

$\Rightarrow$ $|a|_{x=3} = (3)^2(3)$

$$= 27 \text{ unit}$$

19. (c) For the motion to be periodic, the particle must reverse its path i.e. KE should become zero for a finite value of x.

$$K + U = E_T$$

$\Rightarrow$ $K = E_T - U$

From the graph, $U_{max} = V_0$

$\therefore$ $K_{min} = E_T - V_0$

If $K_{min} > 0$, the particle will escape.

So, $E_T - V_0 < 0$

$\Rightarrow$ $E_T < V_0$

Also, $E_T = K + U > 0$

20. (b) $V = Cx^4$

$\Rightarrow$ $[C] = [ML^{-2}T^{-2}]$

By using dimensional analysis in all options, we can see $\left[\frac{1}{A}\sqrt{\frac{m}{C}}\right] = [M^0L^0T]$

$\therefore$ Correct option is (b).

21. (d) For $x > x_0$, potential energy is constant.

$\therefore$ $F = 0$ $\left[\because F = -\frac{dU}{dx}\right]$

So, $a = 0$

22. (c) The forces on the disc when the displacement is x, are shown in the FBD of the disc.

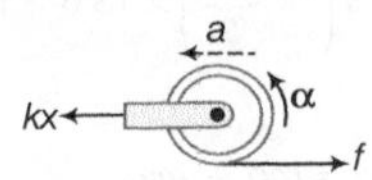

For pure rolling, $a = R\alpha$

$$\frac{kx - f}{m} = R\left[\frac{fR}{(mR^2/2)}\right]$$

$\therefore$ $f = \left(\frac{k}{3}\right)x$

The cylinder starts at $t = 0$ from $x = +A$. So,

$$x = A\sin\left(\omega t + \frac{\pi}{2}\right) = A\cos\omega t$$

Thus, $f = \frac{k}{3}A\cos\omega t$

So, the best answer is (c).

23. (a) $E_T = \frac{1}{2}kA^2 = \frac{1}{2}(10)(2)^2 = 20\,\text{J}$

At mean position, $E_T = K_{max} = K_T + K_R$

$$20 = \frac{1}{2}mv^2 + \frac{1}{2}\frac{mR^2}{2}\cdot\frac{v^2}{R^2} = \frac{3}{4}mv^2$$

$\therefore$ $v^2 = \frac{80}{3m} = \frac{80}{3\times 2} = \frac{40}{3}$

$\therefore$ $v = \sqrt{\frac{40}{3}}\ \text{ms}^{-1}$

24. (2) $\omega = \frac{v}{R}$ (for no slipping)

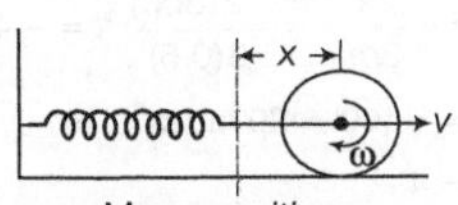

Mean position

$$E_T = \frac{1}{2}mv^2 + \frac{1}{2}I\omega^2 + \frac{1}{2}kx^2$$

$$= \frac{1}{2}mv^2 + \frac{1}{2}\left(\frac{mR^2}{2}\right)\left(\frac{v}{R}\right)^2 + \frac{1}{2}kx^2$$

$$E_T = \frac{3}{4}mv^2 + \frac{1}{2}kx^2$$

$\Rightarrow \quad \frac{dE_T}{dt} = \frac{3}{4}mv \cdot 2a + \frac{1}{2}k\,2x \cdot v = 0$

$\frac{3}{2}mva = -kxv$

$\Rightarrow \quad a = \frac{-2k}{3m}x$

$\therefore \quad \omega = \sqrt{\frac{2k}{3m}}$

$\Rightarrow \quad T = 2\pi\sqrt{\frac{3m}{2k}} = 2\pi\sqrt{\frac{3(4)}{2(3)}} = 2\pi\sqrt{2}\,\text{s}$

$\therefore \quad \beta = 2$

25. (8)

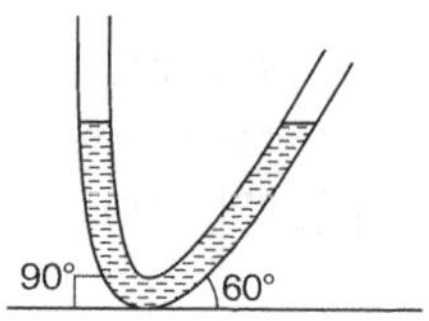

$$T = \sqrt{\frac{m}{A\rho g\,(\sin\theta_1 + \sin\theta_2)}}$$

$$= \sqrt{\frac{0.2}{0.5 \times 10^{-4} \times 9.8(\sin 90° + \sin 60°) \times 13.6 \times 10^3}}$$

$$T = \sqrt{\frac{0.2}{0.5 \times 9.8\left(1 + \frac{\sqrt{3}}{2}\right) \times 13.6 \times 10^{-1}}} = 0.8\,\text{s}$$

$\therefore \; 10T = 8$

26. (5) In equilibrium, $kx = mg + ma_0$

$$x = \frac{mg + ma_0}{k} = \frac{5(9.8 + 2.2)}{1200}$$

$$x = \frac{60}{1200}\,\text{m} = 5\,\text{cm}$$

27. (4) In equilibrium, $kx_0 = mg\sin 30° = \frac{mg}{2}$

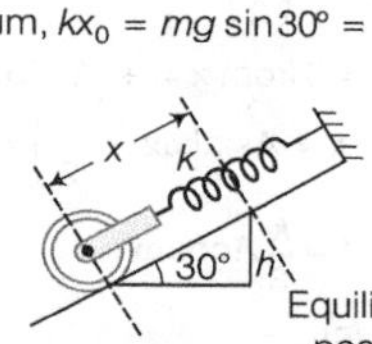

$$E_T = \frac{1}{2}k(x_0 + x)^2 - mgh + \frac{1}{2}mv^2 + \frac{1}{2}\left(\frac{mR^2}{2}\right)\omega^2$$

$$= \frac{1}{2}k(x_0 + x)^2 - mg\,x\sin 30° + \frac{1}{2}mv^2 + \frac{1}{2}\frac{mR^2}{2} \cdot \frac{v^2}{R^2}$$

$$\frac{dE_T}{dt} = \frac{1}{2}k \cdot 2(x_0 + x)v - mgv \cdot \frac{1}{2} + \frac{1}{2}m2v \cdot a + \frac{1}{4}m2va = 0$$

$$kx_0 + kx - \frac{mg}{2} + ma + \frac{m}{2}a = 0$$

$$\frac{3}{2}a \cdot m = -kx \qquad \left[\because k\,x_0 = \frac{mg}{2}\right]$$

$$a = -\frac{2k}{3m}x = \frac{-2(300)}{3(0.5)}x = -400x$$

$$\omega = \sqrt{400} = 20\,\text{rad s}^{-1}$$

$\Rightarrow \quad p = 4$

28. (2) If the mass m is displaced by x from its mean position, the spring further stretches by $2x$.

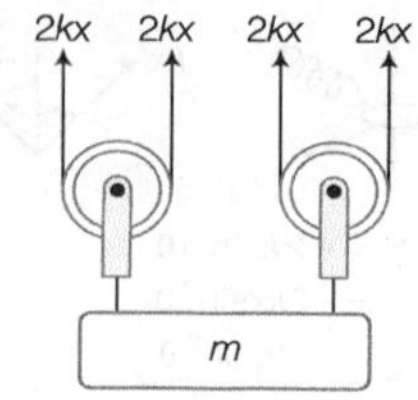

Net restoring force is $F = -8kx$

$$ma = -8k\,x \Rightarrow a = -\frac{8k}{m}x$$

$$\omega = \sqrt{\frac{8k}{m}}$$

$\Rightarrow \quad f = \frac{1}{2\pi}\sqrt{\frac{8k}{m}} = \frac{1}{\pi}\sqrt{\frac{2k}{m}}$

$\Rightarrow \quad q = 2$

Alternate Method

$$E_T = -mgx + \frac{1}{2}k(x_0 + 2x)^2\,4 + \frac{1}{2}mv^2$$

$$\frac{dE_T}{dt} = -mg\,v + 2k \cdot 2(x_0 + 2x) \cdot 2v + \frac{1}{2}m\,2va = 0$$

$$-mg + 4kx_0 + 8kx + ma = 0$$

Initially, $\quad 4kx_0 = mg$

$\therefore \quad 8k\,x + ma = 0$

$\Rightarrow \quad a = -\frac{8k}{m}x$

$\therefore \quad \omega = \sqrt{\frac{8k}{m}}$

$\Rightarrow \quad f = \frac{1}{2\pi}\sqrt{\frac{8k}{m}} = \frac{1}{\pi}\sqrt{\frac{2k}{m}}$

$\Rightarrow \quad q = 2$

29. (5) In the equilibrium of two blocks,

$$T = kx_0 + mg\sin\theta \qquad \ldots\text{(i)}$$

$$2T = mg \qquad \ldots\text{(ii)}$$

During oscillations if displacement of block 1 is x and its velocity is v, then the respective values for block 2 will be $\frac{x}{2}$ and $\frac{v}{2}$.

$$E_T = \frac{1}{2}k(x_0 + x)^2 + \frac{1}{2}m_1v^2 + \frac{1}{2}m_2\left(\frac{v}{2}\right)^2 + m_1 g h_1 - m_2 g h_2$$

$$E_T = \frac{1}{2}k(x_0 + x)^2 + \frac{5}{8}mv^2 + mgx\sin\theta - mg\frac{x}{2}$$

$$\frac{dE_T}{dt} = \frac{1}{2}k2(x_0 + x) \cdot v + \frac{5}{8}m2\,va + mg\,v\sin\theta - \frac{mg}{2}v = 0$$

$$kx_0 + kx + \frac{5}{4}ma + mg\sin\theta - \frac{mg}{2} = 0$$

In the light of Eqs. (i) and (ii), the above equation reduces to $\quad \frac{5}{4}ma = -kx$

$\Rightarrow \quad a = -\frac{4k}{5m}x$

$\Rightarrow \quad \omega = \sqrt{\frac{4k}{5m}}$

$\therefore \quad \beta = 5$

DPP-7 Angular SHM

1. (b) When the angular displacement is θ, the situation is shown in the figure.

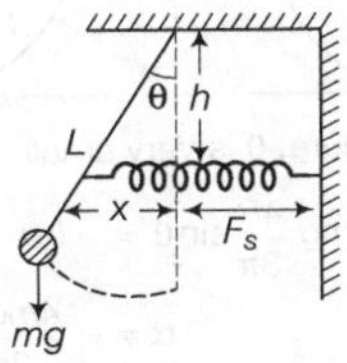

The extension in the spring is $x = h\tan\theta$ and spring force is $F_s = kx = kh\tan\theta$

$\therefore$ Restoring torque is

$$\tau_R = -[mg\,L\sin\theta + (kh\tan\theta)h]$$

As θ is small, $\sin\theta \simeq \tan\theta \simeq \theta$

$\therefore$ $\tau_R = -[mgL + kh^2]\theta$

Also, $\tau_R = I\alpha = mL^2 \cdot \alpha$

$\therefore$ $mL^2\alpha = -[mgL + kh^2]\,\theta$

$\Rightarrow$ $\alpha = -\left[\dfrac{mgL + kh^2}{mL^2}\right]\theta$

Comparing this with standard equation of angular SHM i.e. $\alpha = -\omega^2\theta$, we have

$$\omega = \sqrt{\frac{mgL + kh^2}{mL^2}}$$

$\therefore$ $f = \dfrac{1}{2\pi}\sqrt{\dfrac{mgL + kh^2}{mL^2}}$

2. (d) $\omega' = \dfrac{v}{R - r}$

For pure rolling, $v = r\omega$

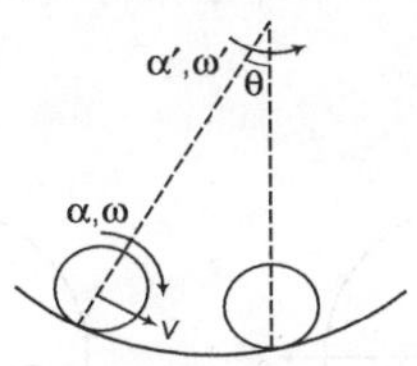

$\therefore$ $\omega' = \dfrac{r}{R - r}\omega$

$\dfrac{d\omega'}{dt} = \dfrac{r}{R - r}\dfrac{d\omega}{dt}$

$\Rightarrow$ $\alpha' = \dfrac{r}{R - r}\alpha$

For pure rolling, $\alpha = \dfrac{a}{r}$

where, $a = \dfrac{g\sin\theta}{1 + \dfrac{I}{mr^2}} = \dfrac{g\sin\theta}{1 + \dfrac{1}{2}} = \dfrac{2}{3}g\sin\theta$

$\therefore$ $\alpha = \dfrac{2g}{3r}\sin\theta$

$\alpha' = -\dfrac{r}{R - r}\cdot\dfrac{2g}{3r}\sin\theta$ [negative, since restoring torque]

$\alpha' = -\dfrac{2g}{3(R - r)}\theta$ [$\because$ θ is small]

$\therefore$ Angular frequency $= \sqrt{\dfrac{2g}{3(R - r)}}$

$\therefore$ $T = \dfrac{2\pi}{\text{Angular frequency}} = 2\pi\sqrt{\dfrac{3(R - r)}{2g}} = \pi\sqrt{\dfrac{6(R - r)}{g}}$

3. (b) Restoring torque is $\tau = -W\left(\dfrac{4r}{3\pi}\sin\theta\right)$

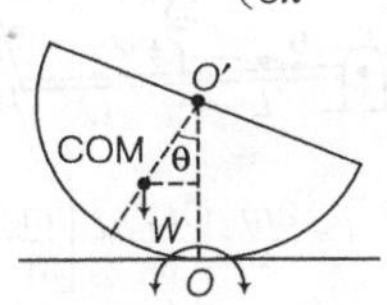

or $\tau = -\left(W\dfrac{4r}{3\pi}\right)\theta$ ($\because$ θ is small)

$I\alpha = -W\dfrac{4r}{3\pi}\theta$ (here, $I = I_0$)

$I_{O'} = \dfrac{1}{2}mr^2,\; I_{O'} = I_{COM} + m\left(\dfrac{4r}{3\pi}\right)^2$

$\therefore$ $I_{COM} = \dfrac{1}{2}mr^2 - m\left(\dfrac{4r}{3\pi}\right)^2$

$\therefore$ $I_0 = \dfrac{1}{2}mr^2 - m\left(\dfrac{4r}{3\pi}\right)^2 + m\left(r - \dfrac{4r}{3\pi}\right)^2$

$= \dfrac{1}{2}mr^2 - m\left(\dfrac{4r}{3\pi}\right)^2 + mr^2 + m\left(\dfrac{4r}{3\pi}\right)^2 - 2mr\cdot\dfrac{4r}{3\pi}$

$= \dfrac{3}{2}mr^2 - \dfrac{8}{3\pi}mr^2 = \dfrac{9\pi - 16}{6\pi}mr^2$

$= 0.651mr^2$

$\therefore$ $\alpha = -W\dfrac{4r}{3\pi}\cdot\dfrac{1}{0.65mr^2}\cdot\theta$

$= -50 \times \dfrac{4 \times 0.15}{3 \times 3.14} \times \dfrac{1}{0.65 \times 5 \times (0.15)^2}\theta$

$= -(43.55)\theta$

$\therefore$ $\omega = \sqrt{43.55} = 6.6\ \text{s}^{-1}$

$\Rightarrow$ $T = \dfrac{2 \times 3.14}{6.6} \simeq 0.96$ s

4. (a) Restoring torque is $\tau = -\left[-mg\dfrac{L}{2}\sin\theta + 2kxL\right]$

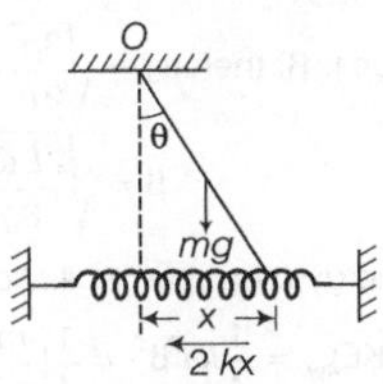

$\tau = -\left[\dfrac{mgL}{2}\theta + 2kL\cdot L\theta\right]$

($\because$ θ is small and $x = L\theta$)

$= -\left[\dfrac{mgL}{2} + 2kL^2\right]\theta$

$I\alpha = -\left[\dfrac{mgL}{2} + 2kL^2\right]\theta$

$\alpha = -\dfrac{\dfrac{mgL}{2} + 2kL^2}{mL^2/3}\theta = -\dfrac{3}{mL^2}\left(\dfrac{mgL}{mL^2} + 2kL^2\right)\theta$

or $\alpha = -\left(\dfrac{3g}{2L} + \dfrac{3}{m}\cdot 2k\right)\theta = -\left(\dfrac{3mg + 12kL}{2mL}\right)\theta$

Comparing with $\alpha = -\omega^2\theta$ given

$$\omega = \sqrt{\frac{3mg + 12kL}{2mL}}$$

$\therefore$ $f = \dfrac{1}{2\pi}\sqrt{\dfrac{3mg + 12kL}{2mL}}$

5. (c) Moment of inertia of T-bar about O is

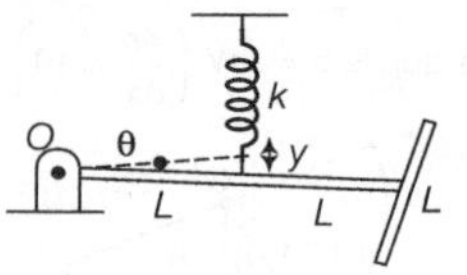

$$I = \frac{2m}{3} \cdot \frac{(2L)^2}{3} + \left[\frac{m}{3} \cdot \frac{L^2}{12} + \frac{m}{3}(2L)^2\right]$$

or $$I = \frac{81}{36} mL^2$$

In equilibrium, $ky_0 \cdot L = mg \cdot \frac{4}{3} L$ ...(i)

$$E_T = \frac{1}{2} I\omega^2 + \frac{1}{2} k(y_0 + y)^2 - mg \cdot \frac{4}{3} y$$

$$= \frac{1}{2} \cdot \frac{81}{36} mL^2\omega^2 + \frac{1}{2} k(y_0 + L\theta)^2 - mg \cdot \frac{4}{3} L\theta$$

$\because$ $$\frac{dE_T}{dt} = 0$$

$$\therefore \frac{1}{2} \cdot \frac{81}{36} mL^2 2\omega \cdot \alpha + \frac{1}{2} k \cdot 2(y_0 + L\theta) \cdot L\omega - mg \cdot \frac{4}{3} L\omega = 0$$

Using Eq. (i), we get

$$\frac{1}{2} \frac{81}{36} mL^2 2\, \alpha\, \omega + kL^2\theta\omega = 0$$

$$\frac{81}{36} m\alpha = -k\theta \Rightarrow \alpha = -\frac{36k}{81m}\theta$$

$\therefore$ $$\omega = \sqrt{\frac{36k}{81m}} = \frac{6}{9}\sqrt{\frac{k}{m}} = \frac{2}{3}\sqrt{\frac{k}{m}}$$

$\therefore$ $$T = 2\pi \frac{3}{2}\sqrt{\frac{m}{k}} = 3\pi\sqrt{\frac{m}{k}}$$

6. (c) Here the rod is a physical pendulum. So, angular frequency is

$$\omega = \sqrt{\frac{mgl}{I}}, \quad \text{where } I = \frac{mL^2}{3} \text{ and } l = \frac{L}{2}$$

$$\omega = \sqrt{\frac{mg\left(\frac{L}{2}\right)}{mL^2/3}} = \sqrt{\frac{3g}{2L}}$$

If angular amplitude is β, then $\omega_0 = \sqrt{\frac{3g}{2L}}\sqrt{\beta^2 - \theta_0^2}$

$\therefore$ $$\beta = \sqrt{\frac{2L\omega_0^2}{3g} + \theta_0^2}$$

Average kinetic energy over one oscillation is

$$(KE)_{av} = \frac{1}{4} I\omega^2\beta^2 = \frac{1}{4}\left(\frac{mL^2}{3}\right)\left(\frac{3g}{2L}\right)\left(\frac{2L\omega_0^2}{3g} + \theta_0^2\right)$$

$$= \frac{1}{8} mgL \left(\frac{2L\omega_0^2}{3g} + \theta_0^2\right)$$

7. (d) For a torsional pendulum, angular frequency of small oscillations is

$$\omega = \sqrt{\frac{C}{I}}, \quad \text{where } I = \frac{mR^2}{2}$$

$\therefore$ $$\omega = \sqrt{\frac{2C}{mR^2}}$$

It is given that at the mean position, the disc is imparted an angular speed ω_0. So, if the angular amplitude of oscillation is β, then

$$\omega_0 = \beta\omega = \beta\sqrt{\frac{2C}{mR^2}}$$

$\therefore$ $$\beta = \sqrt{\frac{mR^2}{2C}}\, \omega_0$$

$$E_T = \frac{1}{2} I\omega^2\beta^2 = \frac{1}{2}\left(\frac{mR^2}{2}\right)\left(\frac{2C}{mR^2}\right)\left(\frac{mR^2}{2C}\omega_0^2\right) = \frac{1}{4} mR^2\omega_0^2$$

8. (b) Restoring torque is $\tau = mg\,(O'P) = -I_O\alpha$

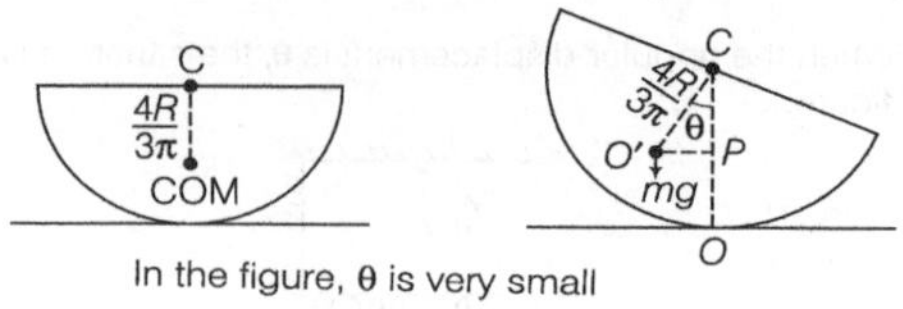

In the figure, θ is very small

or $$mg\frac{4R}{3\pi}\sin\theta = -I_O\alpha$$

$\Rightarrow$ $$\alpha = -\frac{4mgR}{3\pi} \cdot \frac{1}{I_O}\theta \quad \text{...(i)}$$

$$I_C = \frac{1}{2}mR^2, \; I_C = I_{O'} + m\left(\frac{4R}{3\pi}\right)^2$$

$$I_{O'} = \frac{1}{2}mR^2 - m\left(\frac{4R}{3\pi}\right)^2$$

(here, O' is centre of mass)

Now,

$$I_O = I_{O'} + m\left(R - \frac{4R}{3\pi}\right)^2 = \frac{1}{2}mR^2 - m\left(\frac{4R}{3\pi}\right)^2 + m\left(R - \frac{4R}{3\pi}\right)^2$$

or $$I_O = \left(\frac{9\pi - 16}{6\pi}\right) mR^2$$

Using this in Eq. (i) gives,

$$\alpha = -\frac{4mgR}{3\pi} \cdot \frac{6\pi}{(9\pi - 16)\, mR^2} \cdot \theta = -\frac{8g}{R(9\pi - 16)} \cdot \theta$$

$\therefore$ $$\omega = \sqrt{\frac{8g}{R(9\pi - 16)}}$$

$\Rightarrow$ $$T = \frac{2\pi}{\omega} = \pi\sqrt{\frac{(9\pi - 16)R}{2g}}$$

9. (a) Let us locate the COM of the body. It is given that mass of the body is m. So, mass of the cylinder (if present) that fills the cavity $= \frac{m}{15}$

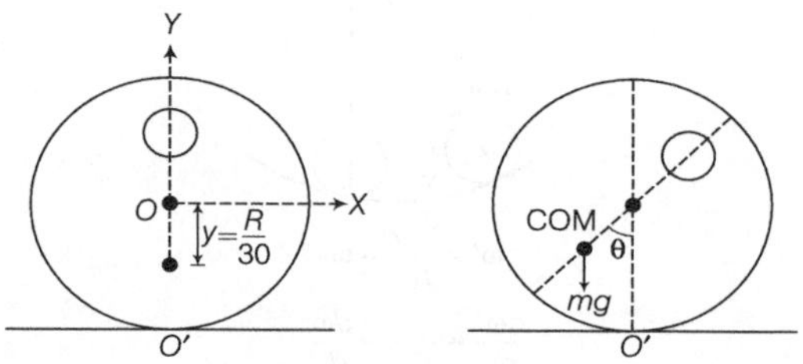

Mass of the cylinder without cavity $= m + \frac{m}{15} = \frac{16}{15}m$

$$Y_{COM} = \frac{\frac{16}{15}m(0) - \frac{m}{15}\left(\frac{R}{2}\right)}{\frac{16m}{15} - \frac{m}{15}} = -\frac{R}{30}$$

MI of cylinder (without cavity) about contact point is

$$I_1 = \frac{16}{15}m\frac{R^2}{2} + \frac{16}{15}m\,R^2 = \frac{16}{15}m\frac{3}{2}R^2 = \frac{8}{5}mR^2$$

MI of cylinder (if present) that fills the cavity about O' is

$$I_2 = \frac{m}{15}\frac{(R/4)^2}{2} + \frac{m}{15}\left(\frac{3R}{2}\right)^2 = mR^2\left[\frac{1}{15} \cdot \frac{1}{32} + \frac{1}{15} \cdot \frac{9}{4}\right]$$

$$I_2 = mR^2\left[\frac{1 + 72}{15 \cdot 32}\right] = mR^2\left(\frac{73}{480}\right)$$

MI of remaining body about O' is $I = I_1 - I_2$

$$I = \left(\frac{8}{5} - \frac{73}{480}\right)mR^2 = \left(\frac{768 - 73}{480}\right)mR^2$$

$$I = \frac{695}{480}mR^2 = \frac{139}{96}mR^2$$

In the second figure, restoring torque is $\tau = mgy\sin\theta$

or $$\tau = -mg\frac{R}{30}\theta$$

Also, $$\tau = I\alpha = \frac{139}{96}mR^2\alpha$$

$\therefore$ $$\frac{139}{96}mR^2\alpha = -\frac{mgR}{30}\theta$$

$$\alpha = -\left(\frac{g}{R}0.023\right)\theta$$

$\Rightarrow$ $$\omega = \sqrt{(0.023)\frac{g}{R}}$$

$$T = \frac{2\pi}{\omega} = 2\pi\sqrt{\frac{R}{(0.023)g}} = 41.4\sqrt{R/g}$$

10. (c) Case (i) is similar to

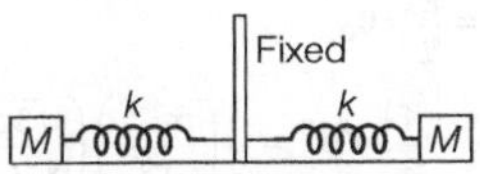

$\therefore$ $$\omega_1 = \sqrt{\frac{k}{M}}$$

In case(ii), let us assume the system is at rest initially.

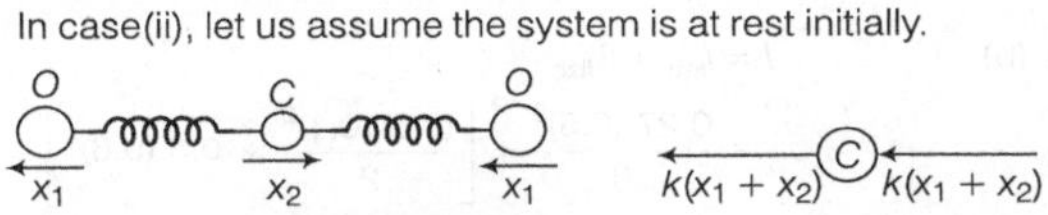

$$\Delta x_1 = -x_1$$
$$\Delta x_2 = +x_2$$
$$\Delta x_3 = -x_1$$
$$M\Delta x_1 + m\Delta x_2 + M\Delta x_3 = 0$$

$\Rightarrow$ $$M(-x_1) + mx_2 + M(-x_1) = 0$$
$$x_1 = \frac{m}{2M}x_2$$

The stretch in the left spring $= x_1 + x_2$

The compression in the right spring $= x_2 + x_1$

$\therefore$ Net force on carbon atom is $F = -k\,(2x_1 + 2x_2)$

$$a = -\frac{k}{m}\cdot 2(x_1 + x_2) = -\frac{2k}{m}\left(\frac{m}{2M}x_2 + x_2\right)$$

$$\frac{d^2x_2}{dt^2} = -\frac{2k}{m}\left(\frac{m+2M}{2M}\right)x_2 = -\frac{k}{m}\left(\frac{m+2M}{M}\right)x_2$$

$\therefore$ $$\omega_2 = \sqrt{\frac{k}{m}\left(\frac{m+2M}{M}\right)}$$

$\therefore$ $$\frac{\omega_1}{\omega_2} = \frac{\sqrt{\frac{k}{M}}}{\sqrt{\frac{k}{m}\left(\frac{m+2M}{M}\right)}} = \sqrt{\frac{m}{m+2M}}$$

11. (a) The time period of a physical pendulum is

$$T = 2\pi\sqrt{\frac{I}{Mgl}}$$

where, I = MI of the system about point of suspension

M = mass of the system and

l = distance between COM of the system and point suspension

Now, $$M = m + m = 2m$$

$$l = \frac{m(R) + m(3R)}{m + m} = 2R$$

$$I = \left(\frac{mR^2}{2} + mR^2\right) + [mR^2 + m(3R)^2]$$

$$= mR^2\left[\frac{1}{2} + 1 + 1 + 9\right] = \frac{23}{2}mR^2$$

Now, $$T = 2\pi\sqrt{\frac{\frac{23}{2}mR^2}{(2m)g(2R)}} = \pi\sqrt{\frac{23R}{2g}}$$

12. (a,c) When the rod is displaced by a small angle θ, the restoring torque is

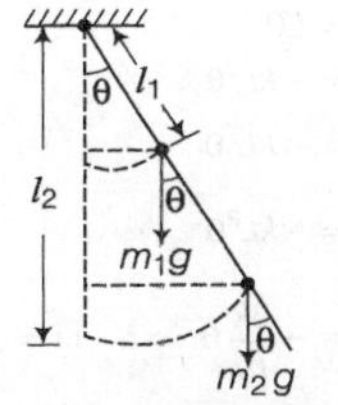

$$\tau_R = -(m_1gl_1\sin\theta + m_2gl_2\sin\theta)$$

$$\tau_R = -(m_1l_1 + m_2l_2)\,g\theta \quad \ldots\text{(i)}$$

($\because$ θ is small and so $\sin\theta \simeq \theta$)

$$\tau = I\alpha = (m_1l_1^2 + m_2l_2^2)\,\alpha \quad \ldots\text{(ii)}$$

From Eqs. (i) and (ii),

$$\alpha = -\left[\frac{m_1l_1 + m_2l_2}{m_1l_1^2 + m_2l_2^2}\cdot g\right]\theta$$

Comparing this with the standard equation of angular SHM i.e. $\alpha = -\omega^2\theta$, we have

$$\omega = \sqrt{\frac{m_1l_1 + m_2l_2}{m_1l_1^2 + m_2l_2^2}\cdot g}$$

$\therefore$ $$f = \frac{\omega}{2\pi} = \frac{1}{2\pi}\left[\frac{m_1l_1 + m_2l_2}{m_1l_1^2 + m_2l_2^2}\cdot g\right]^{1/2}$$

For a simple pendulum, $f = \frac{1}{2\pi}\sqrt{\frac{g}{L}}$

From the above two equations,

$$L = \frac{m_1l_1^2 + m_2l_2^2}{m_1l_1 + m_2l_2}$$

13. (c,d) (i) In this case, spring exerts no restoring torque and restoring torque is only due to the weight of the bob. So, time periold is as usual $T_1 = 2\pi\sqrt{\frac{L}{g}}$.

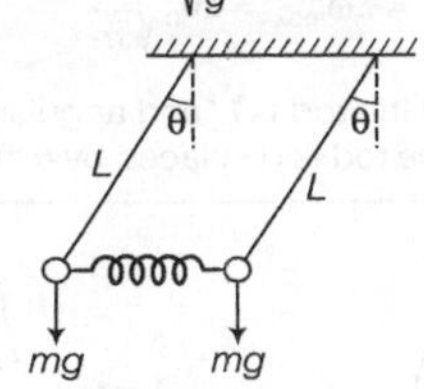

(ii) In this case, restoring torque on one pendulum is

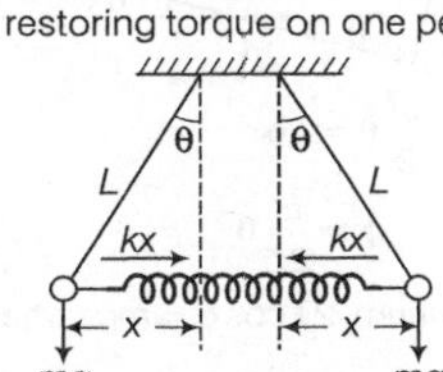

$$X = L\sin\theta$$

$$\tau = -[mgL\sin\theta + k(2L\sin\theta)L]$$
$$= -(mgL + 2kL^2)\,\theta$$

$\therefore$ $$I\alpha = -(mgL + 2kL^2)\theta$$

so, $$\alpha = -\frac{(mgL + 2kL^2)\theta}{mL^2} = -\left(\frac{g}{L} + \frac{2k}{m}\right)\theta$$

or $$\alpha = -\left(\frac{mg + 2kL}{mL}\right)\theta, \text{ comparing with } \alpha = -\omega^2\theta$$

$$\omega = \sqrt{\frac{mg + 2kL}{mL}}$$

$\Rightarrow$ $$T_2 = \frac{2\pi}{\omega} = 2\pi\sqrt{\frac{mL}{mg + 2kL}}$$

14. (b,c) Restoring torque is

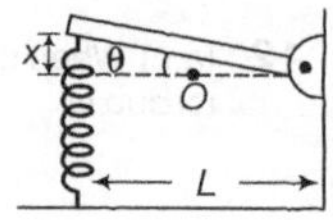

$$\tau = -F \cdot L = -kxL$$

$$x = L\theta$$

$$\therefore \quad \tau = -kL^2\theta$$

$$I\alpha = -kL^2\theta$$

$$\Rightarrow \quad \frac{mL^2}{3}\alpha = -kL^2\theta$$

$$\alpha = -\frac{3k}{m}\theta \Rightarrow \omega = \sqrt{\frac{3k}{m}}$$

$$\therefore \quad T = 2\pi\sqrt{\frac{m}{3k}}$$

Alternate Method

$$E_T = \frac{1}{2}I\omega'^2 + \frac{1}{2}kx^2$$

(ω' here is angular velocity of the rod $= \frac{d\theta}{dt}$)

$$\frac{dE_T}{dt} = \frac{1}{2}I2\omega'\alpha + \frac{1}{2}k2xv = 0$$

$$\frac{1}{2}\frac{mL^2}{3}2\omega'\alpha + \frac{1}{2}k2x\,\omega' L = 0$$

$$\left(\because I = \frac{mL^2}{3} \text{ and } v = L\omega'\right)$$

$$\frac{mL^2}{3}\alpha = -kL\,L\theta \qquad (\because x = L\theta)$$

$$\alpha = -\frac{3k}{m}\cdot\theta$$

$$\Rightarrow \quad \omega = \sqrt{\frac{3k}{m}} \Rightarrow T = 2\pi\sqrt{\frac{m}{3k}}$$

Now, $\left(\frac{d\theta}{dt}\right)_{max} = \theta_0\omega$ i.e. $\omega'_{max} = \theta_0\omega$

$$\omega'_{max} = \theta_0\sqrt{\frac{3k}{m}}$$

$$v_{max} = L\omega'_{max} = L\theta_0\sqrt{\frac{3k}{m}}$$

15. (b) If the length of the rod is L' and angular displacement of each string is ϕ when the rod is displaced by θ then

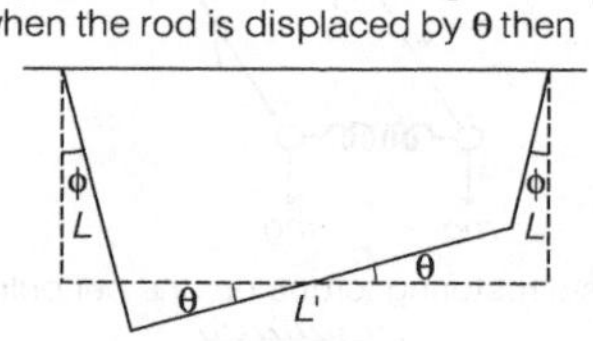

$$\frac{L'}{2}\theta = L\phi$$

$$\Rightarrow \quad \phi = \frac{L'}{2L}\theta \qquad \text{...(i)}$$

For vertical equilibrium, $2T\cos\phi = mg$, where T is tension in each string.

When ϕ is small, $\quad 2T = mg$

$$\Rightarrow \quad T = \frac{mg}{2}$$

Restoring torque on the rod is $\tau = -\left(T\sin\phi\frac{L'}{2}\right)2$

$$\text{or} \quad \tau = -\frac{1}{2}mgL'\phi \qquad (\because \phi \text{ is small})$$

$$\text{or} \quad \tau = -\frac{1}{2}mgL'\left[\frac{L'}{2L}\theta\right] \qquad \text{[from Eq. (i)]}$$

$$= -\frac{mgL'^2}{4L}\theta$$

$$I\alpha = -\frac{mgL'^2}{4L}\theta$$

$$\Rightarrow \quad \frac{mL'^2}{12}\alpha = -\frac{mgL'^2}{4L}\theta$$

$$\alpha = -\left(\frac{3g}{L}\right)\theta$$

$$\Rightarrow \quad \text{Angular frequency} = \sqrt{\frac{3g}{L}}$$

$$\therefore \quad T = 2\pi\sqrt{\frac{L}{3g}}$$

16. (a) If β is the angular amplitude of oscillations of the rod, then using Eq. (i) $\frac{L'}{2}\beta = L\theta_0$ (it is given that initially strings are deflected by angle θ_0)

$$\therefore \quad \beta = \frac{2L}{L'}\theta_0$$

$$E_T = \frac{1}{2}I\omega^2\beta^2 = \frac{1}{2}\left(\frac{mL'^2}{12}\right)\left(\sqrt{\frac{3g}{L}}\right)^2\left(\frac{2L}{L'}\theta_0\right)^2$$

$$E_T = \frac{1}{2}mgL\theta_0^2$$

17. (b)

$$I = I_{rod} + I_{disc}$$

$$= \frac{0.27\,(0.5)^2}{3} + \left[\frac{(0.5)\,(0.1)^2}{2} + 0.5\,(0.6)^2\right]$$

$$= 0.205 \text{ kg-m}^2$$

18. (c) $l = \frac{m_1y_1 + m_2y_2}{m_1 + m_2} = \frac{(0.27)\,(0.25) + (0.5)\,(0.6)}{0.27 + 0.5}$

$$= 0.477\,\text{m} = 47.7\,\text{cm}$$

19. (a) Time period of physical pendulum is

$$T = 2\pi\sqrt{\frac{I}{mgl}}$$

$$= 2 \times 3.14\sqrt{\frac{0.205}{0.77 \times 9.8 \times 0.477}}$$

$$\text{or} \quad T = 1.5\text{ s}$$

20. (a) For small θ, restoring torque is

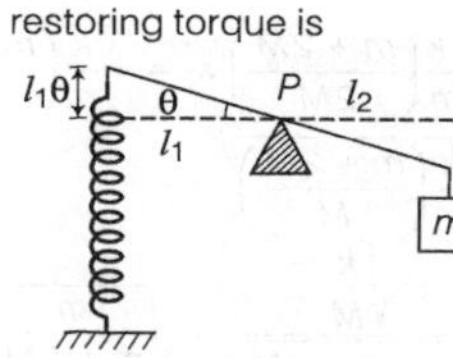

$$\tau = -(kl_1\theta)l_1 = -kl_1^2\theta$$

$$\alpha = \frac{\tau}{I} = \frac{-kl_1^2\theta}{ml_2^2}$$

On comparing with $\alpha = -\omega^2\theta$,

$$\omega = \sqrt{\frac{kl_1^2}{ml_2^2}}$$

$$T = \frac{2\pi}{\omega} = 2\pi\frac{l_2}{l_1}\sqrt{\frac{m}{k}}$$

Energy of oscillation is

$$E = \frac{1}{2}kA^2$$

$$= \frac{1}{2}k(l_1\theta)^2$$

Hence, (i) → (q), (ii) → (r), (iii) → (p), (iv) → (s)

21. (b) Suppose the plank is displaced by x and at that instant its speed is v then speed of COM of cylinder will be $\frac{v}{2}$. The total energy is

$$E_T = \frac{1}{2}mv^2 + \frac{1}{2}kx^2 + \frac{1}{2}(2m)\left(\frac{v}{2}\right)^2 + \frac{1}{2}\cdot\frac{2mR^2}{2}\left(\frac{v/2}{R}\right)^2$$

$$= \left(\frac{1}{2} + \frac{1}{4} + \frac{1}{8}\right)mv^2 + \frac{1}{2}kx^2$$

$$= \frac{4+2+1}{8}mv^2 + \frac{1}{2}kx^2$$

or $\quad E_T = \frac{7}{8}mv^2 + \frac{1}{2}kx^2$

Also, $\quad E_T = \frac{1}{2}kx_0^2$

$\frac{dE_T}{dt} = 0$, after calculating we get

$$a = -\frac{4k}{7m}x$$

Comparing with $a = -\omega^2 x$, we get

$$\omega = \sqrt{\frac{4k}{7m}}$$

Hence, (i) $\rightarrow$ (q), (ii) $\rightarrow$ (p,r), (iii) $\rightarrow$ (s), (iv) $\rightarrow$ (t)

22. (5) In equilibrium, $mg\left(\frac{L}{2}\right) = kyL \qquad \ldots(i)$

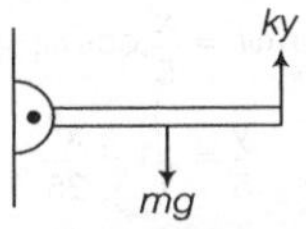

$\therefore \quad y = \frac{mg}{2k}$ (initial extension of spring)

During oscillations, consider a situation when the angular displacement of rod is θ as shown in the figure.

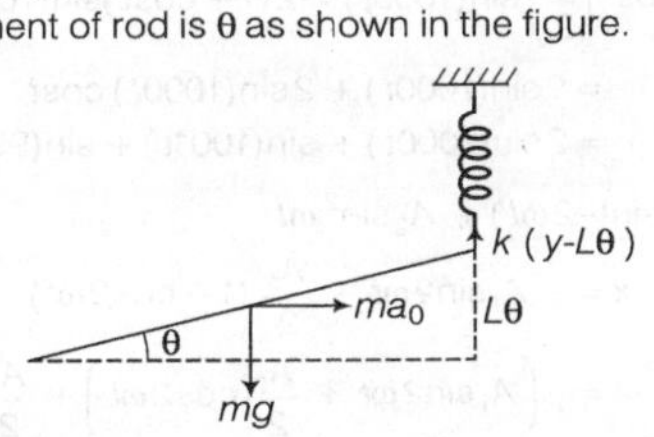

ma_0 = Pseudo force

In the reference frame of car, restoring torque is

$$\tau = -\left[ma_0\frac{L}{2}\sin\theta + my\left(\frac{L}{2}\cos\theta\right) - k(y - L\theta)L\cos\theta\right]$$

Using Eq. (i), $\quad \tau = -\left[\frac{ma_0 L}{2}\theta + kL^2\theta\right] \qquad (\because \theta \text{ is small})$

$$\tau = -\left[\frac{ma_0 L}{2} + kL^2\right]\theta$$

$$\therefore \quad I\alpha = -\left[\frac{ma_0}{2} + kL\right]\cdot L\theta$$

$$\alpha = -\frac{\left[\frac{ma_0}{2} + kL\right]}{mL^2/3}\theta$$

$$= -\left[\frac{3a_0}{2L} + \frac{3k}{m}\right]\theta$$

Using the given values, we get

$$\alpha = -75\theta$$

$$\therefore \quad \omega = \sqrt{75} = 5\sqrt{3}$$

$$\Rightarrow \quad T = \frac{2\pi}{\omega} = \frac{2\pi}{5\sqrt{3}}$$

$$\therefore \quad D = 5$$

23. (4) The time period of a physical pendulum (or compound pendulum) is minimum when the distance between point of suspension and COM (i.e. l) is equal to the radius of gyration about its COM i.e. $l = k$

The moment of inertia of a disc about an axis perpendicular to its plane and passing through its COM is

$$I = mk^2 = \frac{1}{2}mR^2$$

$$\Rightarrow \quad k = \frac{R}{\sqrt{2}}$$

$$T = 2\pi\sqrt{\frac{m(k^2 + l^2)}{mgl}}$$

$$\Rightarrow \quad T_{\min} = 2\pi\sqrt{\frac{\frac{R^2}{2} + \frac{R^2}{2}}{g\,R/\sqrt{2}}}$$

$$T_{\min} = 2\pi\sqrt{\frac{\sqrt{2}R}{g}} = \pi\sqrt{\frac{4\sqrt{2}R}{g}}$$

$$\therefore \quad B = 4$$

24. (8) $T = 2\pi\sqrt{\frac{I_0}{mg\,(OG)}} = 2\pi\sqrt{\frac{mL^2/3}{mg\cdot L/2}}$

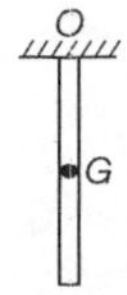

or $\quad T = 2\pi\sqrt{\frac{2L}{3g}} = \pi\sqrt{\frac{8L}{3g}}$

$$\therefore \quad \psi = 8$$

25. (7) $T = 2\pi\sqrt{\frac{I}{mgl}} = 2\pi\sqrt{\frac{l_{eq}}{g}}$

$$\therefore \quad \frac{I}{200 \times 20} = 35$$

$$\Rightarrow \quad I = 140 \times 10^3 \text{ g-cm}^2 = \frac{14}{1000}\text{ kg-m}^2$$

$$\therefore \quad \beta = 7$$

26. (6) In the equilibrium for block,

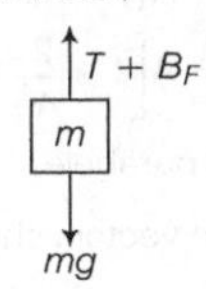

$$T + B_F = mg \qquad \ldots(i)$$

When the block is further depressed by y, upthrust B_F increases by $\rho_L Ayg$ and let ΔT be increase in tension.

$$\Delta T + \rho_L Ayg = ma \qquad \ldots(ii)$$

Now, the restoring torque on the cylinder is

$$\tau = \left[k\left(\frac{y}{2}\right)\left(\frac{R}{2}\right) - \Delta T R\right] = \left[\frac{kyR}{4} - (ma - \rho_L\,Ayg)R\right]$$

$$\frac{1}{2}mR^2\alpha = \left[\frac{kR^2\theta}{4} - (mR\alpha - \rho_L AgR\theta)\,R\right] \qquad (\because y = R\theta)$$

$$\frac{3}{2}mR^2\alpha = \left[\frac{kR^2}{4} + \rho_L AgR^2\right]\theta$$

$$\alpha = \frac{-\left[\frac{k}{4} + \rho_L Ag\right]}{\frac{3}{2}m}\theta = -\left[\frac{k}{6m} + \frac{2\rho_L Ag}{3m}\right]\theta$$

Here, negative sign is used for restoring nature.

$$\therefore \quad \omega = \sqrt{\frac{k}{6m} + \frac{2\rho_L Ag}{3m}} = \sqrt{\frac{k + 4\rho_L Ag}{6m}}$$

$$\therefore \quad f = \frac{1}{2\pi}\sqrt{\frac{k + 4\rho_L Ag}{6m}}$$

$$\therefore \quad Q = 6$$

DPP-8 Superposition of SHM

1. (c) $A_1 = 4\text{cm}$, $A_2 = 3\text{cm}$

(i) $\delta = 0°$,

$$A = \sqrt{A_1^2 + A_2^2 + 2A_1A_2\cos\delta} = \sqrt{4^2 + 3^2 + 2\cdot 4\cdot 3\cos 0°}$$

$$\Rightarrow \quad A = 7\text{cm}$$

(ii) $\delta = 60°$, $A = \sqrt{16 + 9 + 2(4)(3)\cos 60°} = \sqrt{37}$ cm

(iii) $\delta = 90°$, $A = \sqrt{16 + 9 + 0} = 5\text{cm}$

(iv) $\delta = 180$; $A = \sqrt{16 + 9 - 24} = 1\text{cm}$

2. (b) The resultant equation is

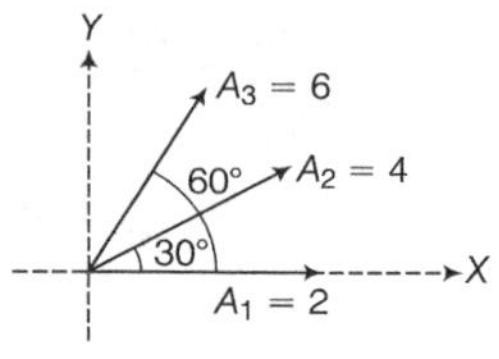

$$x = A\sin(\omega t + \varepsilon)$$

$$\Sigma A_x = 2 + 4\cos 30° + 6\cos 60° = 8.46$$

$$\Sigma A_y = 4\sin 30° + 6\sin 60° = 7.2$$

$$\therefore \quad A = \sqrt{(\Sigma A_x)^2 + (\Sigma A_y)^2} = \sqrt{(8.46)^2 + (7.2)^2} = 11.25$$

$$\varepsilon = \tan^{-1}\left(\frac{\Sigma A_y}{\Sigma A_x}\right) = \tan^{-1}\left(\frac{7.2}{8.46}\right) = \tan^{-1}(0.85)$$

So, the resultant equation assumes the form

$$x = 11.25\sin[\omega t + \tan^{-1}(0.85)]$$

3. (a) $x = A\sin\omega t$, $y = A\cos 2\omega t$ [$\because$ Phase difference $= \pi/2$]

Let us eliminate t from the above equations to find the path.

$$\therefore \quad \sin\omega t = \frac{x}{A}$$

and

$$y = A[1 - 2\sin^2\omega t]$$

$$\therefore \quad y = A\left[1 - \frac{2}{A^2}x^2\right]$$

This is the equation of a parabola.

4. (d) The resultant of three vectors shown in the figure is $2A$ (along the direction of 2).

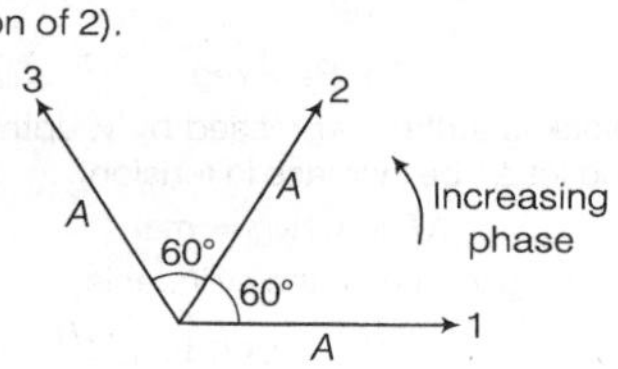

5. (d) $x_1 = 3\sin\omega t$, $x_2 = 5\sin(\omega t + 37°)$

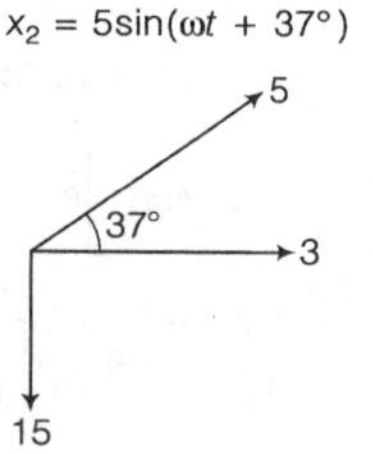

$$x_3 = 15\sin(\omega t + 270°)$$

$$\Sigma A_x = 3 + 5\cos 37°$$

$$= 3 + 5\times\frac{4}{5} = 7$$

$$\Sigma A_4 = 5\sin 37° - 15$$

$$= 3 - 15 = -12$$

$$A = \sqrt{144 + 49} = \sqrt{189}$$

6. (c) When $x = A_1\sin(\omega t + \phi_1)$

and $\quad y = A_2\sin(\omega t + \phi_2)$ then

$$\frac{x^2}{A_1^2} + \frac{y^2}{A_2^2} - \frac{2xy}{A_1A_2}\cos\delta = \sin^2\delta$$

Here phase difference i.e. $\delta = 0$

$$\therefore \quad \frac{x^2}{A_1^2} + \frac{y^2}{A_2^2} - \frac{2xy}{A_1A_2} = 0$$

$$\Rightarrow \quad \left(\frac{x}{A_1} - \frac{y}{A_2}\right)^2 = 0$$

$$\Rightarrow \quad y = \frac{A_2}{A_1}x$$

Resultant amplitude is $A = \sqrt{A_1^2 + A_2^2} = \sqrt{(0.1)^2 + (0.05)^2}$

$$= 0.112\text{m}$$

$$\omega = 2\pi f_1 = 2\pi f_2 = 10\pi$$

Thus the equation of resulting SHM is

$$r = 0.112\sin\left(10\,\pi t + \frac{\pi}{3}\right)\text{m}$$

7. (c) To find the equation of trajectory we have to eliminate time (t) term in the given equations

$$\sin\omega t = \frac{x}{5}, \cos\omega t = \frac{y}{5}$$

$$\therefore \quad \frac{y}{5} = \sqrt{1 - \frac{x^2}{25}} \Rightarrow \frac{y^2}{25} = 1 - \frac{x^2}{25}$$

$$x^2 + y^2 = 25$$

This is the equation of a circle.

8. (b) $x = 4\cos^2\left(\frac{t}{2}\right)\sin(1000t) = 2(1 + \cos t)\sin(1000t)$

$$= 2\sin(1000t) + 2\sin(1000t)\cos t$$

$$= 2\sin(1000t) + \sin(1001t) + \sin(999t)$$

9. (a) $x = A_1\sin(-2\omega t) + A_2\sin^2\omega t$

$$x = -A_1\sin 2\omega t + \frac{A_2}{2}(1 - \cos 2\omega t)$$

$$= -\left(A_1\sin 2\omega t + \frac{A_2}{2}\cos 2\omega t\right) + \frac{A_2}{2}$$

$$= A_1\sin(2\omega t + 180°) + \frac{A_2}{2}\sin(2\omega t + 270°) + \frac{A_2}{2}$$

Phase difference $= 90°$

$$\therefore \quad A_R = \sqrt{A_1^2 + \frac{A_2^2}{4}}$$

Mean position is $\frac{A_2}{2}$.

$\therefore$ The resultant motion is SHM with amplitude $\sqrt{A_1^2 + \frac{A_2^2}{4}}$

10. (d) When two SHM 's (which are perpendicular to each other) are given by $x = A_1\sin\omega t$ and $y = A_2\sin(\omega t + \delta)$ and when they are combined the trajectory of the particle is given by

$$\frac{x^2}{A_1^2} + \frac{y^2}{A_2^2} - \frac{2xy}{A_1A_2}\cos\delta = \sin^2\delta$$

(This is equation of ellipse)

In the given equations, phase difference is $\delta = \frac{\pi}{6}$ and $A_1 = A_2 = A$

$$\therefore \quad \frac{x^2}{A^2} + \frac{y^2}{A^2} - \frac{2xy}{A^2}\cdot\frac{\sqrt{3}}{2} = \frac{1}{4}$$

$x^2 + y^2 - \sqrt{3}\,xy = \frac{1}{4}A^2$ and this is an ellipse.

11. (a,b,c) $x = A_1 \sin^2 \omega t + A_2 \cos^2 \omega t + A_3 \sin \omega t \cos \omega t$

$$= \frac{A_1}{2}(2\sin^2 \omega t) + \frac{A_2}{2}(2\cos^2 \omega t) + \frac{A_3}{2}(2\sin \omega t \cos \omega t)$$

$$= \frac{A_1}{2}(1 - \cos 2\omega t) + \frac{A_2}{2}(1 + \cos 2\omega t) + \frac{A_3}{2}\sin(2\omega t)$$

(a) The above equation represents SHM.

If $A_1 = A_2 = 0$, then $x = \frac{A_3}{2}\sin(2\omega t)$

$\therefore$ Option (a) is correct.

(b) If $A_1 = A_2$ and $A_3 = 2A_2$

$$x = A_1 \sin^2 \omega t + A_1 \cos^2 \omega t + 2A_1 \sin \omega t \cos \omega t$$

$$= A_1 + A_1 \sin 2\omega t$$

This represents SHM. So, option (b) is correct.

(c) If $A_1 = -A_2$ and $A_3 = 2A_2$

$$x = A_1 \sin^2 \omega t - A_1 \cos^2 \omega t + 2A_1 \sin \omega t \cos \omega t$$

$$= -A_1 \cos 2\omega t + A_1 \sin 2\omega t$$

$$= A_1 \sin(270° + 2\omega t) + A_1 \sin \omega t$$

This again represents SHM. So, option (c) is correct.

(d) If $A_1 = A_2$ and $A_3 = 0$

$$x = A_1 \sin^2 \omega t + A_1 \cos^2 \omega t + 0 = A_1$$

This does not represent SHM.

12. (a,c,d) We can write the given equation as $x = A\cos \omega t$ and $y = 2A\cos \omega t$

$\therefore$ Phase difference is zero and $A_1 = A$, $A_2 = 2A$. Substituting in the standard equation,

$$\frac{x^2}{A^2} + \frac{y^2}{4A^2} - \frac{2xy}{2A^2} = 0$$

$$\Rightarrow \left(\frac{x}{A} - \frac{y}{2A}\right)^2 = 0$$

$$y = 2x$$

and $$r = \sqrt{A^2 + 4A^2}\cos \omega t = \sqrt{5}A\cos \omega t$$

13. (a,b,c,d) Phase difference is $\delta = \frac{\pi}{3} = 60°$

$\therefore$ Resultant amplitude is

$$A = \sqrt{2^2 + 4^2 + 2 \cdot 2 \cdot 4 \cdot \cos 60°}$$

$$= \sqrt{28}\text{ m}$$

Phase difference between resultant SHM and first SHM

$$\varepsilon = \tan^{-1}\left(\frac{A_2 \sin \delta}{A_1 + A_2 \cos \delta}\right)$$

$\therefore$ $$\varepsilon = \tan^{-1}\left(\frac{4\sin 60°}{2 + 4\cos 60°}\right) = \tan^{-1}\left(\frac{2\sqrt{3}}{4}\right)$$

$$= \tan^{-1}\left(\frac{\sqrt{3}}{2}\right)$$

$\therefore$ Resultant SHM is $x = \sqrt{28}\sin\left(10t + \tan^{-1}\left(\frac{\sqrt{3}}{2}\right)\right)$

At $$t = 0,\ x = \sqrt{28}\sin\left(\tan^{-1}\frac{\sqrt{3}}{2}\right)$$

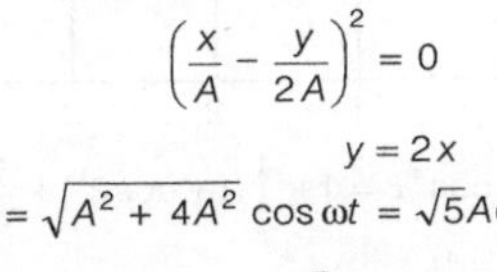

$\therefore$ $$x = \sqrt{28}\sqrt{\frac{3}{7}} = 2\sqrt{3}\text{ m}$$

$$v_{max} = A\omega = \sqrt{28} \times 10 = 20\sqrt{7}\text{ ms}^{-1}$$

$$a_{max} = A\omega^2 = \sqrt{28} \times 100 = 200\sqrt{7}\text{ ms}^{-2}$$

$$E_T = \frac{1}{2}m\omega^2 A^2 = \frac{1}{2} \times 0.02 \times 100 \times 28 = 28\text{ J}$$

14. (a,b) Let r_1 is along X-axis and r_2 is making 37° with positive X-axis.

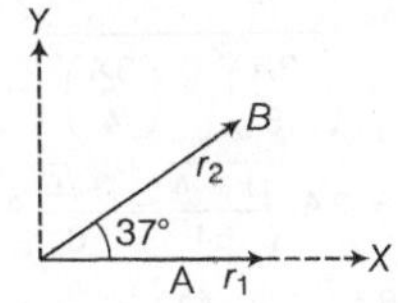

$$(r_2)_x = B\sin \omega t \cos 37° = \frac{4}{5}B\sin \omega t$$

$$(r_2)_y = B\sin \omega t \sin 37° = \frac{3}{5}B\sin \omega t$$

r_1 and $(r_2)_x$ along the same line. Their phase difference is zero. So, their resultant has amplitude $A' = \sqrt{A^2 + \frac{16}{25}B^2}$

and resultant SHM is $r' = \sqrt{A^2 + \frac{16}{25}B^2}\sin \omega t$

Now r' and $(r_2)_y$ are perpendicular to each other with phase difference = 0. Substituting in the standard equation,

$$\frac{x^2}{A_1^2} + \frac{y^2}{A_2^2} - \frac{2xy}{A_1 A_2}\cos \delta = \sin^2 \delta, \text{ we get}$$

$$\frac{x^2}{A^2 + \frac{16}{25}B^2} + \frac{y^2}{\frac{9}{25}B^2} - \frac{2xy}{\sqrt{A^2 + \frac{16}{25}B^2}\,\frac{3}{5}B} = 0$$

$$\left(\frac{x}{\sqrt{A^2 + \frac{16}{25}B^2}} - \frac{y}{\frac{3}{5}B}\right)^2 = 0$$

$$y = \frac{3}{5}\frac{B}{\sqrt{A^2 + \frac{16}{25}B^2}}x$$

This is a straight line and the motion is SHM with amplitude $\sqrt{A_1^2 + A_2^2}$ where A_1^2 and A_2^2 are $A^2 + \frac{16}{25}B^2$ and $\frac{9}{25}B^2$ respectively.

15. (a) At $t = 0.0125$ s

$$x_1 = 2\sin[100\pi \times 0.0125]$$

$$= 2\sin\left(\frac{5}{4}\pi\right) = 2\sin\left(\pi + \frac{\pi}{4}\right) = -2\sin\frac{\pi}{4} = -\sqrt{2}\text{ cm}$$

and $$x_2 = 2\sin\left[120\pi \times 0.0125 + \frac{\pi}{3}\right] = 2\sin\left[\frac{3}{2}\pi + \frac{\pi}{3}\right]$$

$$= 2\sin\left(2\pi - \frac{\pi}{6}\right) = 2\left(-\frac{1}{2}\right) = -1$$

$\therefore$ $$x = -(\sqrt{2} + 1)\text{ cm} = -2.41\text{ cm}$$

16. (d) At $t = 0.025$s

$$x_1 = 2\sin[100\pi \times 0.025] = 2\sin\left(2\pi + \frac{\pi}{2}\right) = 2\text{ cm}$$

$$x_2 = 2\sin\left[120\pi \times 0.025 + \frac{\pi}{3}\right] = 2\sin\left[3\pi + \frac{\pi}{3}\right]$$

$$= 2\left[-\frac{\sqrt{3}}{2}\right] = -1.73\text{ cm}$$

$\therefore$ $$x = 2 - 1.73 = 0.27\text{ cm}$$

17. (c)

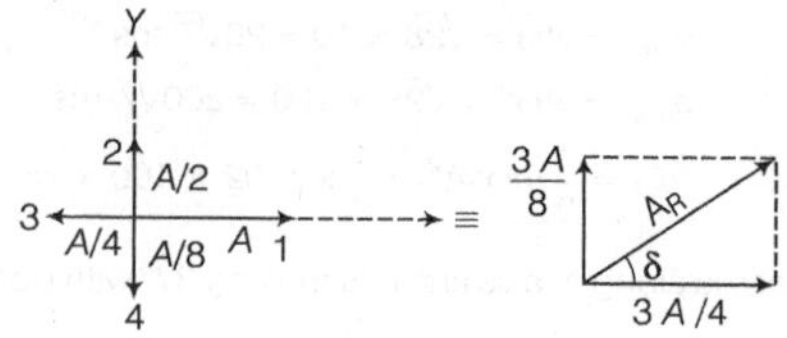

$$A_R = \sqrt{\left(\frac{3A}{8}\right)^2 + \left(\frac{3A}{4}\right)^2} = 3A\sqrt{\frac{1}{64} + \frac{1}{16}}$$

$$= 3A\sqrt{\frac{1+4}{64}} = \frac{3\sqrt{5}}{8}A$$

18. (d) $\delta = \tan^{-1}\left(\frac{3A/8}{3A/4}\right) = \tan^{-1}\left(\frac{1}{2}\right)$

19. (3) $x_1 = 10\sin\left(4\pi t + \frac{\pi}{2}\right)$

$\Rightarrow \quad A_1 = 10$ units

$$x_2 = 5\sin(2\pi t) + 5\sqrt{8}\cos(2\pi t)$$

$$= 5\sin(2\pi t) + 5\sqrt{8}\sin\left(2\pi t + \frac{\pi}{2}\right)$$

Phase difference between SHM's of second equation is $\frac{\pi}{2}$.

$\therefore$ Resultant amplitude is

$$A_2 = \sqrt{(5)^2 + (5\sqrt{8})^2 + 2\cdot 5\cdot 5\sqrt{8}\cdot\cos\frac{\pi}{2}}$$

$$= \sqrt{225} = 15 \text{ units}$$

$\therefore \quad \frac{A_1}{A_2} = \frac{10}{15} = \frac{2}{3} \Rightarrow \beta = 3$

20. $1 + \cos^2 t + \sin^4 t = 1 + \frac{1}{2}(1 + \cos 2t) + \frac{1}{4}(1 - \cos 2t)^2$

$$= \frac{1}{4}[4 + 2 + 2\cos 2t + 1 + \cos^2 2t - 2\cos 2t]$$

$$= \frac{1}{4}[7 + \cos^2 2t] = \frac{1}{4}\left[7 + \frac{1}{2}(1 + \cos 4t)\right]$$

$$= \frac{1}{8}[15 + \cos 4t]$$

$\therefore \quad x = \frac{1}{8}(15 + \cos 4t)\sin 500t$

$$= \frac{15}{8}\sin 500t + \frac{1}{8}\sin 500t\cdot\cos 4t$$

$$= \frac{15}{8}\sin 500t + \frac{1}{16}\sin 504t + \frac{1}{16}\sin 496t$$

The spectrum is shown below.

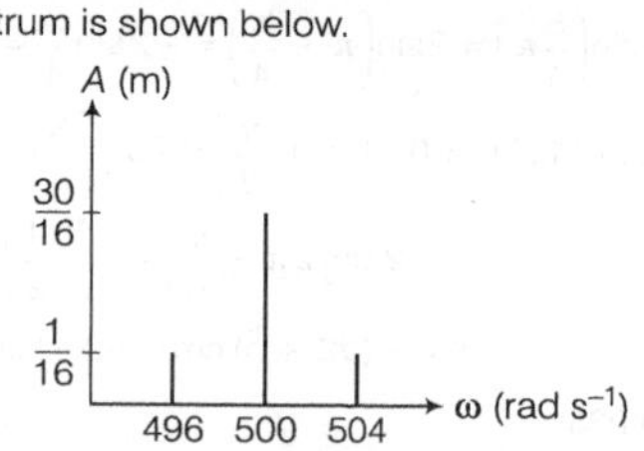

21. (i) From the given spectrum of oscillations, we can see that

Oscillation	Amplitude	Frequency	Equation
First	$A_1 = 0.03$m	$f_1 = 0.2\ s^{-1}$	$x = 0.03\sin(0.4\pi t)$m
Second	$A_2 = 0.02$m	$f_2 = 0.5\ s^{-1}$	$x = 0.02\sin(\pi t)$m
Third	$A_3 = 0.01$m	$f_3 = 1\ s^{-1}$	$x = 0.01\sin(2\pi t)$m

(ii) The diagram of oscillations is

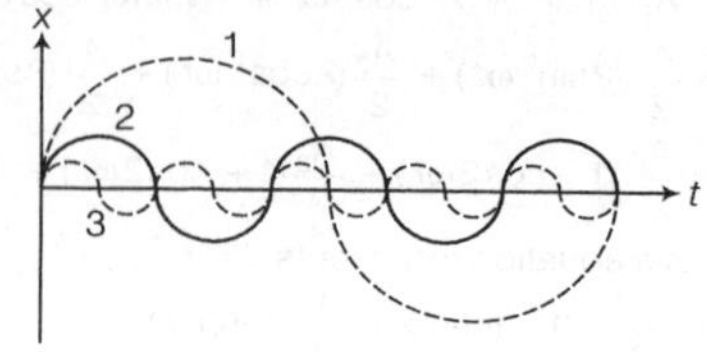

22. (i) $x = A\sin 2\pi f_1 t$

$$= A_0(1 + \cos 2\pi f_2 t)\sin 2\pi f_1 t$$

$$= A_0\sin 2\pi f_1 t + A_0\cos 2\pi f_2 t\sin 2\pi f_1 t$$

$$= A_0\sin 2\pi f_1 t + \frac{A_0}{2}\sin[2\pi(f_1 + f_2)t] + \frac{A_0}{2}\sin[2\pi(f_1 - f_2)t]$$

Thus, the oscillations being considered can be resolved into three harmonic oscillatory motions with frequencies f_1, $(f_1 - f_2)$, $(f_1 + f_2)$ and amplitudes of A_0, $\frac{A_0}{2}$, $\frac{A_0}{2}$ respectively.

(ii) $A_1 = A_0 = 4\text{cm}, f' = f_1 = 2\text{s}^{-1}$

$$A_2 = \frac{A_0}{2} = 2\text{cm}, f'' = f_1 + f_2 = 3\text{s}^{-1}$$

$$A_3 = \frac{A_0}{2} = 2\text{cm}, f''' = f_1 - f_2 = 1\text{s}^{-1}$$

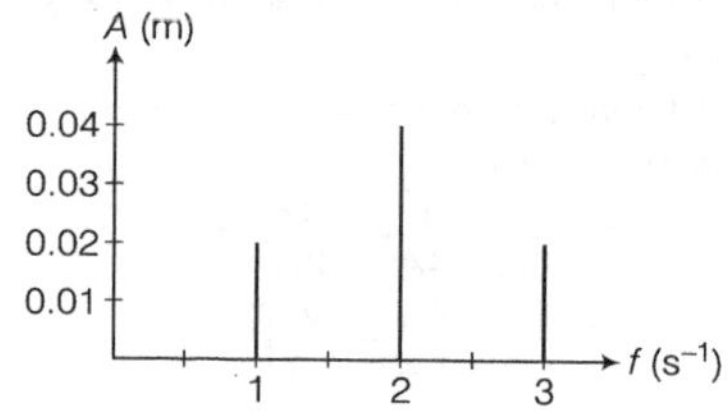

23. (d) $1 + \cos^2 t + \cos^4 t = 1 + \frac{1}{2}(1 + \cos 2t) + \frac{1}{4}(1 + \cos t)^2$

$$= \frac{1}{4}(4 + 2 + 2\cos 2t + 1 + \cos^2 2t + 2\cos 2t)$$

$$= \frac{1}{4}(7 + 4\cos 2t + \frac{1}{2}(1 + \cos 4t))$$

$$= \frac{1}{8}(15 + 8\cos 2t + \cos 4t)$$

$\therefore \quad x = \frac{1}{8}(15 + 8\cos 2t + \cos 4t)\sin 500t$

$$= \frac{15}{8}\sin 500t + \cos 2t\sin 500t + \frac{1}{8}\cos 4t\sin 500t$$

or $\quad x = \frac{15}{8}\sin 500t + \frac{1}{2}\sin 502t + \frac{1}{2}\sin 498t + \frac{1}{16}\sin 504t + \frac{1}{16}\sin 496t$

The spectrum of the vibrations is shown below.

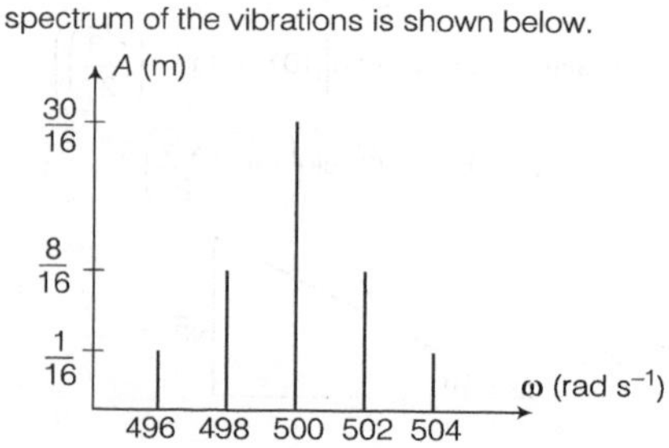

DPP-9 Damped Oscillations and Resonance

1. A body set into oscillation, continues to oscillate for ever with the same amplitude if damping forces are not acting, because mechanical energy is conserved. If damping forces are present, the amplitude of oscillation gradually decreases because energy is dissipated. Such oscillations are called "damped oscillations". In most cases, the damping force is directly proportional to the speed e.g. viscous drag due to air. Thus damping force is $-bv$, where b is called the damping constant. The net force on mass m with displacement x and velocity $v = \frac{dx}{dt}$ is

$$F_{net} = -kx - bv$$

Newton's second law gives $-kx - bv = ma$

or
$$m\frac{d^2x}{dt^2} + b\frac{dx}{dt} + kx = 0$$

(where, b is damping constant)

The solution for this equation is $x(t) = A_0 e^{-\frac{b}{2m}t}\cos(\omega t + \delta)$ ...(i)

where,
$$\omega = \sqrt{\omega_0^2 - \left(\frac{b}{2m}\right)^2} \quad \text{...(ii)}$$

and δ is phase constant.

$\left[\omega_0 = \sqrt{\frac{k}{m}}\right.$, the angular frequency for the case with no damping]

Note *Here amplitude is* $\left(A_0 e^{-\frac{b}{2m}t}\right)$ *and a function of time.*

2. From Eq. (ii) when $b = 0$, $\omega = \omega_0$ and from Eq. (i) $x = A_0\cos(\omega t + \delta)$. In such cases x-t graph is as usual and is shown below.

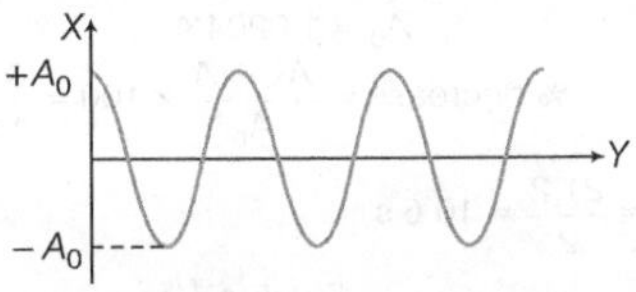

(i) Underdamped motion In this case b is small enough so that $\frac{b}{2m} < \sqrt{\frac{k}{m}}$. The motion is oscillatory, but the amplitude decreases exponentially. The SHM has a new angular frequency ω and an amplitude $A_0^{-\frac{b}{2m}t}$ that decreases exponentially. Exponential fall means that, if amplitude becomes $\frac{1}{n}$ of initial value in t time, then in next t time, it will become $\frac{1}{n^2}$ of its initial value.

The frequency is smaller than ω_0 because damping slows down the motion.

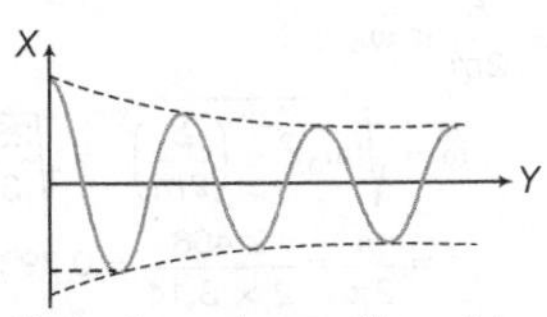

Under damped case with small b
fig. (a)

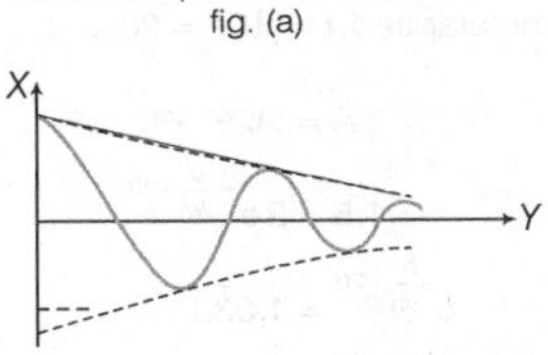

Under damped case with large b
fig. (b)

In Fig. (b) the damping constant b is larger than that in Fig. (a), yet small enough that $\frac{b}{2m} < \omega_0$.

The energy of an oscillator is proportional to the square of the amplitude. So, the energy of an underdamped oscillator (averaged over a cycle) also decreases exponentially with time.

$$E = \frac{1}{2}m\omega^2A^2$$
$$= \frac{1}{2}m\omega^2\left[A_0 e^{-\left(\frac{b}{2m}\cdot t\right)}\right]^2$$
$$= \frac{1}{2}m\omega^2A_0^2 e^{-\left(\frac{b}{m}\right)t}$$

or
$$E = E_0 e^{-\frac{b}{m}\cdot t} = E_0 e^{-t/\tau} \quad \text{...(iii)}$$

where,
$$E_0 = \frac{1}{2}m\omega^2A_0^2$$

is the initial energy of the oscillator. Decay time (τ) is the time for energy to decrease to e^{-1} times its original value.

A damped oscillator is described by its Q-factor i.e. quality factor,

$$Q = \omega_0\tau = \frac{\omega_0 m}{b}$$

We can relate Q to the fractional energy loss per cycle. Differentiating Eq. (iii) gives

$$dE = -\frac{1}{\tau}(E_0e^{-t/\tau})dt$$
$$= -\frac{1}{\tau}E\,dt$$

If the energy loss per period is small, we can replace dE by ΔE and dt by the period T. Then, $\frac{|\Delta E|}{E}$ in one period is given by

$$\frac{|\Delta E|}{E} = \frac{T}{\tau} = \frac{1}{\tau}\cdot\frac{2\pi}{\omega_0} = \frac{2\pi}{Q}$$

or
$$Q = \frac{2\pi}{(|\Delta E|/E)_{cycle}}$$

Q is thus inversely proportional to the fractional energy loss per cycle.

The power dissipated by the damping force equals the instantaneous rate of change of the total mechanical energy i.e.

$$P = \frac{dE}{dt} = \vec{F}_d \cdot \vec{v} = -bv^2$$

[where, $\vec{F}_d$ is damping force]

(ii) Overdamped motion In this case $\frac{b}{2m} > \omega_0$ i.e. in this case damping constant (b) is large and the motion is not oscillatory. The $x(t)$ function is given by

$$x(t) = (Ae^{-\omega' t} + Be^{-\omega' t})e^{-\left(\frac{b}{2m}\right)t}$$

where,
$$\omega' = \sqrt{\left(\frac{b}{2m}\right)^2 - \omega_0^2}$$

A and B are constants whose values depend on the initial position and velocity of the mass. A graph that displays one form of this solution is given in the figure shown below.

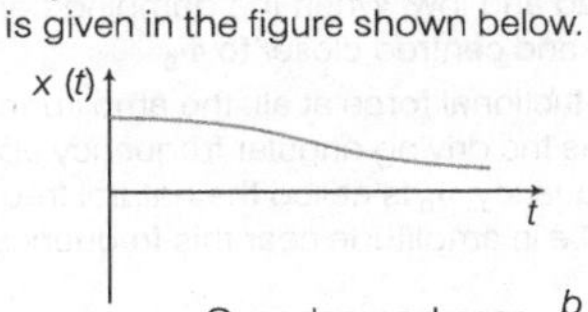

Over damped case $\frac{b}{2m} > \omega_0$

(iii) Critically damped motion In this case $\frac{b}{2m} = \omega_0$. The solution in this case is given by

$$x(t) = (C + Dt)e^{-\left(\frac{b}{2m}\right)t}$$

where, C and D are constants, whose values depend on the initial position and velocity of the mass. The following figure shows a plot of $x(t)$ for a critically damped case.

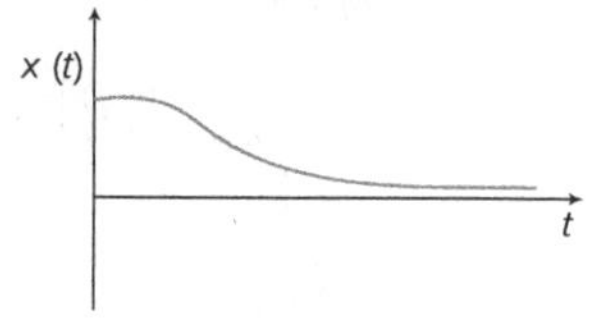

Note *The decay of the motion towards $x = 0$, is more rapid in the critically damped case than for the overdamped case.*

3. Suppose in addition to the spring and damping forces, the mass m is acted on by a periodic force

$$F(t) = F_0 \cos(\omega_d t)$$

where, F_0 is a constant and ω_d is the angular velocity of this driving force. The differential equation for this motion is

$$m\frac{d^2x}{dt^2} + b\frac{dx}{dt} + kx = F_0 \cos(\omega_d t)$$

After a sufficiently long time, this driving force is performing work on the mass at the same rate at which energy is dissipated by damping force. In this case steady-state condition is reached and $x(t)$ settles down to a simple harmonic oscillation described by

$$x(t) = A\cos(\omega_d t - \phi_0),$$

where $A = \dfrac{F_0/m}{\sqrt{(\omega_d^2 - \omega_0^2) + \left(\dfrac{b\omega_d}{m}\right)^2}}$

and $\phi_0 = \tan^{-1}\left[\dfrac{b\omega_d/m}{\omega_0^2 - \omega_d^2}\right]$

This angle specifies how far the displacement lags behind the driving force.

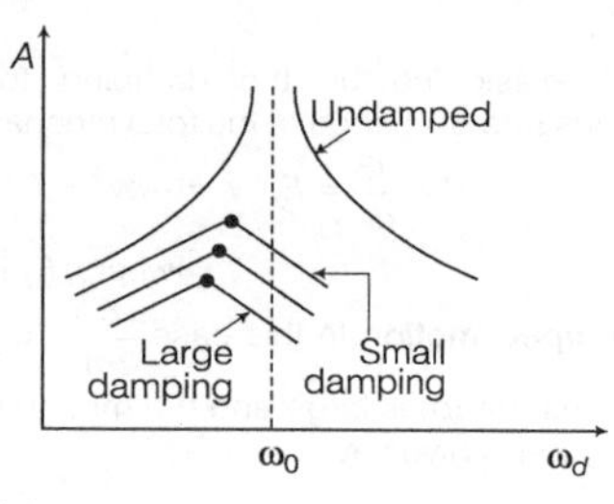

A graph of A *versus* driving frequency ω_d is shown in the figure. The steady -state oscillation amplitude depends on the driving frequency ω_d. As the driving frequency is changed, the amplitude peaks dramatically at a particular value of ω_d known as the resonance frequency.

For a large frictional force i.e. large damping force (thus large b), the peak is broad and low. When the damping constant is small, the peak is high and centred closer to ω_0.

If there were no frictional force at all, the amplitude would grow without bound as the driving angular frequency approaches ω_0. The angular frequency ω_0 is called the natural frequency and the dramatic increase in amplitude near this frequency is called resonance.

At resonance $\omega_d = \sqrt{\omega_0^2 - \left(\dfrac{b}{2m}\right)^2}$

4. (b) $A = A_0 e^{-\frac{b}{2m}t} \Rightarrow 5.5 = 15e^{-\frac{b}{2m}\cdot 1000}$

$$e^{\frac{b}{2m}\cdot 1000} = 2.73$$

$$\frac{b}{2m}\cdot 1000 = \ln 2.73 = 1$$

$$\frac{b}{2m} = 10^{-3}\,s^{-1} = 0.001\,s^{-1}$$

5. (d) $\omega_0 = \sqrt{\dfrac{k}{m}} = \sqrt{\dfrac{20500}{10.6}} \simeq 44\,s^{-1}$

$$\frac{b}{2m} = \frac{3}{2\times 10.6} = 0.1415\,s^{-1}$$

$$\frac{b}{2m} << \omega_0$$

∴ Frequency will be approximately same as that of undamped oscillation.

∴ $f = \dfrac{1}{2\pi}\sqrt{\dfrac{k}{m}} = \dfrac{44}{2\times 3.14} \simeq 7\,Hz$

6. (b) $T = \dfrac{1}{f} = \dfrac{1}{7}\,s$

$$A = A_0 e^{-\frac{b}{2m}\cdot t}$$

$\Rightarrow$ $\dfrac{A_0}{A} = e^{+\frac{b}{2m}\cdot t}$

$$\frac{A_0}{A} = e^{+\frac{0.1415}{7}}$$

$\Rightarrow$ $\dfrac{A_0}{A} = 1.0204$

or $A_0 = 1.0204\,A$

$$\%\text{ decrease} = \frac{A_0 - A}{A_0}\times 100 = \frac{0.0204}{1.0204}\times 100 \simeq 2\%$$

7. (c) $\tau = \dfrac{m}{b} = \dfrac{21.2}{2} = 10.6\,s$

$$E = E_0 e^{-t/\tau}$$

$\Rightarrow$ $0.1 = e^{-t/10.6}$

$$e^{t/10.6} = 10$$

$\Rightarrow$ $\dfrac{t}{10.6} = \ln 10$

$$t = (10.6)(2.302) = 24.4\,s$$

8. (b) $\omega_0 = \sqrt{\dfrac{k}{m}} = \sqrt{\dfrac{2.5}{0.3}} = \sqrt{\dfrac{25}{3}}\,s^{-1} = 2.887\,s^{-1}$

$$\frac{b}{2m} = \frac{0.9}{2\times 0.3} = 1.5\,s^{-1}$$

$$\frac{b}{2m} < \omega_0$$

∴ $\omega = \sqrt{(\omega_0)^2 - \left(\dfrac{b}{2m}\right)^2} = \sqrt{\dfrac{25}{3} - 2.25} = 2.466\,s^{-1}$

$$f = \frac{\omega}{2\pi} = \frac{2.466}{2\times 3.14} = 0.393\text{ Hz}$$

9. (b) $T = 2\,s$, Time elapsed, $t = 10T = 20\,s$

$$A = A_0 e^{-\frac{b}{2m}\cdot t}$$

$\Rightarrow$ $1.5 = 2e^{-\frac{b}{2m}\cdot 20}$

$$e^{\frac{b}{2m}\cdot 20} = 1.333$$

$\Rightarrow$ $\dfrac{b}{2m}\cdot 20 = \ln(1.333)$

$$\frac{b}{2m} = \frac{0.2877}{20} = 0.0144\,s^{-1}$$

10. (b) We know $E = E_0 e^{-t/\tau}$

$$E = \frac{E_0}{2} \quad \text{(given)}$$

$$\therefore \quad \frac{E_0}{2} = E_0 e^{-4/\tau}$$

$$\Rightarrow \quad e^{4/\tau} = 2$$

Taking natural logarithms on both the sides

$$\frac{4}{\tau} = \ln 2$$

$$\Rightarrow \quad \tau = \frac{4}{\ln 2} = 5.77\,\text{s}$$

Now, $Q = \omega_0 \tau = (2\pi f)\tau = 2 \times 3.14 \times 262 \times 5.77 = 9.5 \times 10^3$

11. (a) $\frac{|\Delta E|}{E} = \frac{T}{\tau} = \frac{1}{f\tau} = \frac{1}{(262)(5.77)} = 6.67 \times 10^{-4}$

12. (a) In case of critical damping, $\frac{b}{2m} = \omega_0$

$$\therefore \quad \frac{b}{2m} = \sqrt{\frac{k}{m}}$$

$$\Rightarrow \quad b = 2\sqrt{mk}$$

$$b = 2\sqrt{0.5 \times 200} = 20\,\text{kgs}^{-1}$$

13. (d) Amplitude will be maximum when $\omega = \omega_0$

$$\omega_0 = \sqrt{\frac{k}{m}} = \sqrt{\frac{4300}{12.5}} = 18.55\,\text{rads}^{-1}$$

$$f = \frac{1}{2\pi} \times 18.55 = 2.95\,\text{Hz}$$

14. (b) In case of driven oscillations $A = \frac{F_0/m}{\sqrt{(\omega_d^2 - \omega_0^2)^2 + \left(\frac{b\omega_d}{m}\right)^2}}$

Here, $b = 0$ ($\because$ no damping)

ω_0 = angular frequency of driving force = 2π rads^{-1}

$$\omega_0 = \sqrt{\frac{k}{m}} = \sqrt{\frac{20}{2}} = \sqrt{10}\,\text{rads}^{-1}$$

F_0 = Amplitude of driving force = 3N

$$\therefore \quad A = \frac{3/2}{\sqrt{(4\pi^2 - (\sqrt{10})^2 + 0}} = 0.0509 = 5.09\,\text{cm}$$

15. (b) The motion happens with the angular frequency of driving force.

$$\therefore \quad \omega = \omega_d = 2\pi$$

$$T = \frac{2\pi}{\omega} = 1\,\text{s}$$

16. (c) Mass of block is $m = \frac{40}{9.8} = 4.082\,\text{kg}$

Angualr frequency of driving force is $\omega_d = 2\pi \times 10 = 62.8\,\text{rads}^{-1}$

$$\omega_0 = \sqrt{\frac{k}{m}} = \sqrt{\frac{200}{4.082}} \simeq 7\,\text{rads}^{-1}$$

$$b = 0 \text{ and } A = 0.02\,\text{m} \quad \text{(given)}$$

Now, $$A = \frac{F_0/m}{\sqrt{(\omega_d^2 - \omega_0^2)^2 + \left(\frac{b\omega_d}{m}\right)^2}}$$

$$\Rightarrow \quad 0.02 = \frac{F}{4.082\sqrt{[(62.8)^2 - 7^2]^2 + 0}}$$

$$\Rightarrow \quad F_0 = 318\text{N}$$

17. (c) $v = 16 \times \frac{5}{8} = 4.444\,\text{ms}^{-1}$

We can take $\lambda = 4$ m.

Here, $$v = f\lambda$$

$$\Rightarrow \quad f = \frac{4.444}{4} = 1.111\,\text{Hz}$$

$$\omega_0 = \sqrt{\frac{k}{m}} = \sqrt{\frac{k}{(1000 + 4 \times 82)}} = \sqrt{\frac{k}{1328}}$$

Amplitude will be maximum when $\omega_0 = 2\pi f$

i.e. $$\sqrt{\frac{k}{1328}} = 2 \times 3.14 \times 1.111$$

$$k = 64647\,\text{Nm}^{-1}$$

Now four people (i.e. $82 \times 4 = 328$ kg) get out of the car.

$$\therefore \quad 328 \times 9.8 = kx$$

$$\Rightarrow \quad x = \frac{328 \times 9.8}{64647} \simeq 0.05\,\text{m} = 5\,\text{cm}$$

18. (a) $A = \frac{F_0/m}{\sqrt{(\omega_d^2 - \omega_0^2)^2 + \left(\frac{b\omega_d}{m}\right)^2}}$

When $\omega_d << \omega_0$,

$$A = \frac{F_0/m}{\sqrt{(\omega_0^2)^2}} = \frac{F_0}{m\omega_0^2}$$

When $\omega_d >> \omega_0$,

$$A = \frac{F_0/m}{\sqrt{\omega_d^4 + \left(\frac{b}{m}\right)^2 \omega_d^2}} = \frac{F}{m\omega_d\sqrt{\omega_d^2 + \left(\frac{b}{m}\right)^2}}$$

19. (b) $\frac{b}{2m} = \frac{70}{2 \times 250} = 0.14\,\text{s}^{-1}$

$$\omega_0 = \sqrt{\frac{k}{m}} = \sqrt{\frac{85}{0.25}} = 18.44\,\text{s}^{-1}$$

$\therefore \frac{b}{2m} << \omega_0$. The time period is approximately same as that for undamped motion.

$$\therefore \quad T = 2\pi\sqrt{\frac{m}{k}} = \frac{2\pi}{18.44} = 0.34\,\text{s}$$

20. (c) $A = A_0 e^{-\frac{b}{2m}t}$

$$\Rightarrow \quad \frac{A_0}{2} = A_0 e^{-\frac{b}{2m}\cdot t}$$

$$\frac{1}{2} = e^{-(0.14)t}$$

$$\Rightarrow \quad e^{(0.14)t} = 2$$

$$\text{or} \quad 0.14t = \ln 2$$

$$t = \frac{0.693}{0.14} \simeq 5\,\text{s}$$

21. (d) $E = E_0 e^{-t/\tau}$, where $\tau = \frac{m}{b} = \frac{1}{0.28}\text{s}$

$$\frac{E_0}{2} = E_0 e^{-(0.28)t}$$

$$\Rightarrow \quad e^{(0.28)t} = \frac{1}{2}$$

$$e^{(0.28)t} = 2$$

$$\Rightarrow \quad (0.28)t = \ln 2$$

$$\therefore \quad t = \frac{0.693}{0.28} \simeq 2.5\,\text{s}$$

Revisal Problems (JEE Main)

1. (a) $x = A\sin(\omega t + \delta)$,

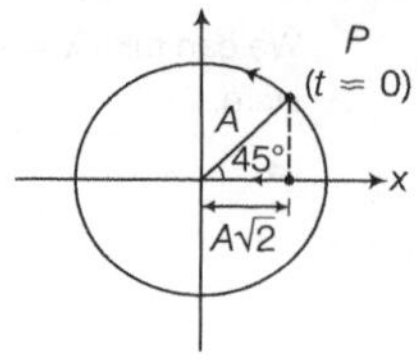

$$v = A\omega\cos(\omega t + \delta)$$

At $t = 0$, $\quad \frac{A}{\sqrt{2}} = A\sin\delta$

$$\sin\delta = \frac{1}{\sqrt{2}}$$

$\Rightarrow \quad \delta = 45°, 135°$

But at $t = 0$, v has to be negative (according to the figure)

$\therefore \quad \delta = 135°$

$\therefore \quad x = A\sin(\omega t + 135°) = A\sin\left(90° + \omega t + \frac{\pi}{4}\right)$

$$\omega = \frac{2\pi}{T} = \frac{2\pi}{4} = \frac{\pi}{2}$$

$\therefore \quad x = A\cos\left(\frac{\pi}{2}t + \frac{\pi}{4}\right)$

2. (c) Restoring force is $F = -\Delta p \cdot A$

$$= -\rho g(h_1 + h_2)A$$
$$= -\rho g A(\sin\theta_1 + \sin\theta_2)\, y$$
$$= -\rho g A(\sin 30° + \sin 30°)\, y = -\rho g A\, y$$

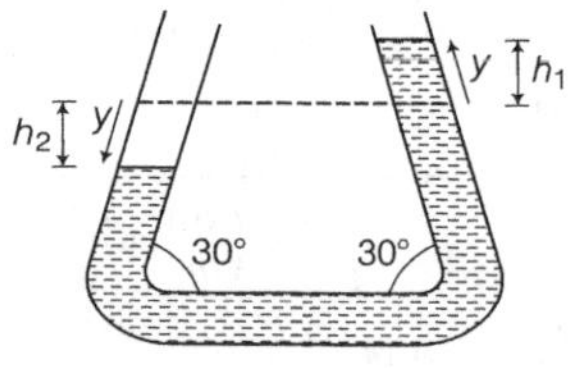

$\therefore \quad a = -\left(\frac{\rho g A}{m}\right) y \quad \Rightarrow \quad \omega = \sqrt{\frac{\rho g A}{m}}$

$$T = \frac{2\pi}{\omega} = 2\pi\sqrt{\frac{m}{\rho g A}}$$

3. (b) If Δx_1 and Δx_2 are their respective amplitudes of motion, then

$$m(\Delta x_1) + 2m(-\Delta x_2) = 0 \quad \text{...(i)}$$

and $\quad \Delta x_1 + \Delta x_2 = x_0 \quad$ ...(ii)

Solving the above equations gives $\Delta x_1 = \frac{2x_0}{3}$.

4. (d) $T_1 = 2\pi\sqrt{\frac{m}{k_1}}, T_2 = 2\pi\sqrt{\frac{m}{k_2}}$

$$\frac{1}{k_{eq}} = \frac{1}{k_1} + \frac{1}{k_2}$$

$$T = 2\pi\sqrt{\frac{m}{k_{eq}}} = 2\pi\sqrt{m\left(\frac{1}{k_1} + \frac{1}{k_2}\right)}$$

$$T^2 = 4\pi^2\left[m\left(\frac{1}{k_1} + \frac{1}{k_2}\right)\right]$$

$$T_1^2 = 4\pi^2\left(\frac{m}{k_1}\right) \quad \text{and} \quad T_2^2 = 4\pi^2\left(\frac{m}{k_2}\right)$$

$$T^2 = T_1^2 + T_2^2$$

5. (b) Here both the springs are in parallel

$\therefore \quad k_{eq} = k_1 + k_2 = 0.1 + 0.1 = 0.2\,\text{Nm}^{-1}$

$$\omega = \sqrt{\frac{k_{eq}}{\mu}}$$

where, $\mu = \frac{m_1 m_2}{m_1 + m_2} = \frac{(0.1)(0.1)}{0.1 + 0.1} = \frac{1}{20}\text{kg}$

$\therefore \quad 2\pi f = \sqrt{\frac{0.2}{1/20}} \quad \Rightarrow \quad f = \frac{1}{\pi}\text{Hz}$

6. (a) The given system is equivalent to as shown in the figure.

$$\tau_R = -\left(2mg \cdot \frac{L}{\sqrt{2}}\sin\theta \cdot \frac{1}{2}\right) \simeq -\frac{mgL}{\sqrt{2}}(\theta)$$

$$I\alpha = -\frac{mgL}{\sqrt{2}}(\theta)$$

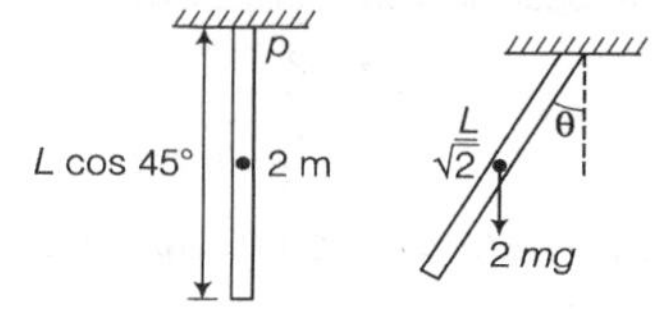

$$\frac{2m\left(\frac{L}{\sqrt{2}}\right)^2}{3}\alpha = -\frac{mgL}{\sqrt{2}}(\theta)$$

$$\alpha = -\frac{3}{mL^2} \cdot \frac{mgL}{\sqrt{2}}\theta \Rightarrow \alpha = -\frac{3g}{\sqrt{2}L}\theta$$

$\Rightarrow \quad \omega = \sqrt{\frac{3g}{\sqrt{2}L}}$

$$T = \frac{2\pi}{\omega} = 2\pi\sqrt{\frac{\sqrt{2}L}{3g}}$$

7. (d) $\tau_R = -\left[k.L\theta.L + k\frac{L}{2}\theta \cdot \frac{L}{2} - mg\frac{L}{2}\theta\right] = -\left[\frac{5}{4}kL^2 - \frac{mgL}{2}\right] \cdot \theta$

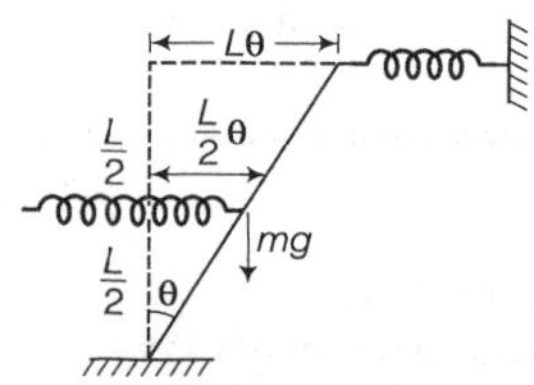

For the given requirement,

$$\frac{5}{4}kL^2 - \frac{mgL}{2} > 0$$

$$\frac{5}{4}kL^2 > \frac{mgL}{2} \quad \Rightarrow \quad k > \frac{2mg}{5L}$$

8. (b) In equilibrium $kx_0 = 2mg + Mg$ ...(i)

When the block descends by x, pulley moves down by a distance $\frac{x}{2}$. Now the total energy of the system is

$$-mgx - Mg \cdot \frac{x}{2} + \frac{1}{2}mv^2 + \frac{1}{2}M\left(\frac{v}{2}\right)^2 + \frac{1}{2}\frac{MR^2}{2}\omega^2 + \frac{1}{2}k\left(x_0 + \frac{x}{2}\right)^2 = E_T$$

For no slipping, $\quad v = \frac{v}{2} + \omega R$

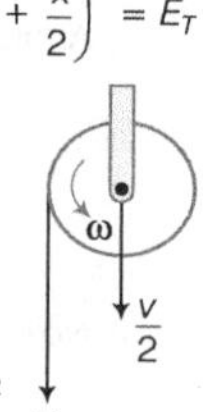

$\Rightarrow \quad \omega = \frac{v}{2R}$

$\therefore \quad E_T = -mgx - \frac{Mg}{2}x + \frac{1}{2}mv^2 + \frac{1}{8}Mv^2 + \frac{1}{4}MR^2\frac{v^2}{4R^2} + \frac{1}{2}k\left(x_0 + \frac{x}{2}\right)^2$

$$\frac{dE_T}{dt} = -mg \cdot v - \frac{Mg}{2} \cdot v + \frac{1}{2}m \cdot 2v \cdot a + \frac{1}{8}M \cdot 2v \cdot a + \frac{1}{16}M \cdot 2v \cdot a + \frac{1}{2}k \cdot 2\left(x_0 + \frac{x}{2}\right) \cdot \frac{1}{2} \cdot v = 0$$

$$-mg - \frac{Mg}{2} + ma + \frac{1}{4}Ma + \frac{1}{8}Ma + \frac{kx_0}{2} + \frac{kx}{4} = 0$$

Using Eq. (i), reduces the above equation to

$$ma + \frac{3M}{8}a = -\frac{k}{4}x$$

$$a\left[4m + \frac{3}{2}M\right] = -kx$$

$$a = -\left(\frac{k}{4m + \frac{3}{2}M}\right)x \Rightarrow \omega = \sqrt{\frac{k}{4m + \frac{3}{2}M}}$$

9. (b) At A, velocity is zero. So, it cannot be at mean position.
At B, slope of v-t graph is zero. So, acceleration is zero.
At C, slope of v-t graph is maximum and negative.

10. (c) $T = 2\pi\sqrt{\frac{L}{g_{\text{eff}}}}$

In the given situation, downward force on the bob and thus g_{eff} increases. So, T decreases.

11. (c) $\omega = \sqrt{\frac{k}{m}} = \sqrt{\frac{54}{1+2+3}} = 3 \text{ rads}^{-1}$

Maximum acceleration that 1 kg mass can have is

$$a_1 = \mu_1 g = 0.6 \times 10 = 6 \text{ ms}^{-2}$$

Similarly, $a_2 = 0.4 \times 10 = 4 \text{ms}^{-2}$

$\therefore \quad A\omega^2 = 4 \Rightarrow A = \frac{4}{9}\text{m}$

12. (a) The horizontal velocity of m after hitting the wedge is u (say).

Then, $R = u\sqrt{\frac{2h}{g}} \Rightarrow u = R\sqrt{\frac{g}{2h}}$

In the horizontal direction momentum is conserved.

i.e. $mu = Mv$

$$\Rightarrow \quad v = \frac{m}{M}\cdot R\sqrt{\frac{g}{2h}}$$

$$\frac{1}{2}Mv^2 = \frac{1}{2}kA^2$$

$$\Rightarrow \quad M\cdot\frac{m^2}{M^2}R^2\cdot\frac{g}{2h} = kA^2$$

$$A^2 = m^2R^2\frac{g}{2Mhk}$$

$$\Rightarrow \quad A = \frac{mR}{M}\sqrt{\frac{Mg}{2kh}}$$

13. (d) At $t = 0$, particle is at mean position and moving towards positive extreme.

$$T = 8\text{s}$$

$$\Rightarrow \quad \omega = \frac{2\pi}{8} = \frac{\pi}{4} \text{ rad s}^{-1}$$

$$\therefore \quad x = A\sin\omega t = 1\sin\frac{\pi}{4}t$$

$$\therefore \quad x = \sin\left(\frac{\pi}{4}t\right) \Rightarrow v = \frac{\pi}{4}\cos\left(\frac{\pi}{4}t\right)$$

$$\Rightarrow \quad a = -\left(\frac{\pi}{4}\right)^2\sin\left(\frac{\pi}{4}t\right)$$

At $t = \frac{4}{3}$s, $a = -\frac{\pi^2}{16}\sin\left(\frac{\pi}{4}\cdot\frac{4}{3}\right)$

$$= -\frac{\pi^2}{16}\cdot\frac{\sqrt{3}}{2} = -\frac{\sqrt{3}}{32}\pi^2 \text{ cm s}^{-2}$$

14. (d) Force in the spring 1 and spring 2 is same.

$\therefore \quad k_1A_1 = k_2A_2 \quad$ ...(i)

and $\quad A_1 + A_2 = A \quad$ ...(ii)

$\therefore \quad k_1A_1 = k_2A - k_2A_1$

$$\Rightarrow \quad A_1 = \frac{k_2}{k_1 + k_2}\cdot A$$

15. (c)

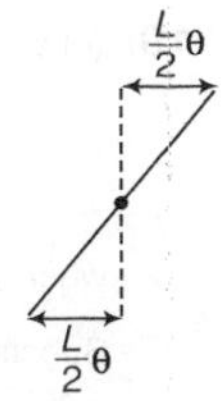

$$E_T = \frac{1}{2}k\left(\frac{L\theta}{2}\right)^2\cdot 2 + \frac{1}{2}\frac{mL^2}{12}\omega^2$$

$$= \frac{kL^2}{4}\theta^2 + \frac{mL^2}{12}\cdot\omega^2$$

$$\Rightarrow \quad \frac{dE_T}{dt} = \frac{kL^2}{4}\cdot 2\theta\cdot\omega + \frac{mL^2}{24}\cdot 2\omega\alpha = 0$$

$$\frac{k}{2}\theta + \frac{m}{12}\alpha = 0$$

$$\Rightarrow \quad \alpha = -\frac{6k}{m}\theta$$

$$\therefore \quad \omega = \sqrt{\frac{6k}{m}}$$

$$\Rightarrow \quad 2\pi f = \sqrt{\frac{6k}{m}}$$

$$\Rightarrow \quad f = \frac{1}{2\pi}\sqrt{\frac{6k}{m}}$$

16. (b) $T_{\min} = mg\cos\theta_0 \quad$ ...(i)

$T_{\max} - mg = \frac{mv^2}{L} \quad$ ...(ii)

$$mgL(1 - \cos\theta_0) = \frac{mv^2}{2}$$

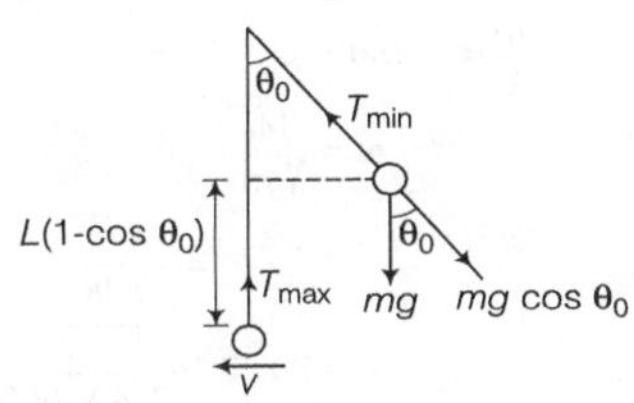

$$\therefore \quad \frac{mv^2}{L} = 2mg(1 - \cos\theta_0)$$

Using this in Eq. (ii),

$$T_{\max} - mg = 2mg(1 - \cos\theta_0)$$

$$T_{\max} = 3mg - 2mg\cos\theta_0$$

$T_{\max} = 2T_{\min} \quad$ (given)

$$\therefore \quad 3mg - 2mg\cos\theta_0 = 2mg\cos\theta_0$$

$$\Rightarrow \quad \cos\theta_0 = \frac{3}{4}$$

17. (d) $mg\sin\theta \simeq mg\theta = mg\frac{x}{R}$

$\therefore$ Restoring force is $F = -\frac{mg}{R}x$

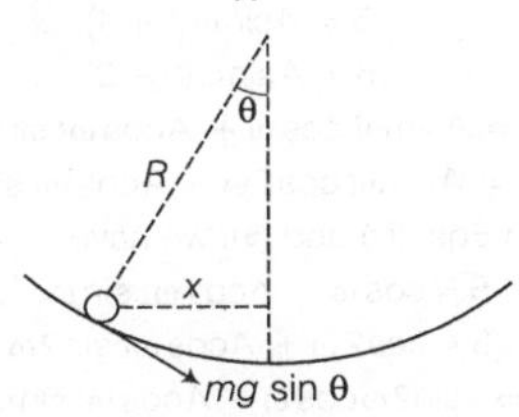

$$\therefore \quad a = \frac{F}{m} = -\frac{g}{R}x$$

$$\Rightarrow \quad \omega = \sqrt{\frac{g}{R}}$$

$$T = 2\pi\sqrt{\frac{R}{g}} = 2\pi\sqrt{\frac{5}{10}} = \pi\sqrt{2}\text{ s}$$

18. (a) $T = 2\pi\sqrt{\frac{L}{\sqrt{g^2 + a_0^2}}} = 2\pi\sqrt{\frac{0.5}{\sqrt{(9.8)^2 + 7^2}}}$

$$= 2\pi\sqrt{\frac{0.5}{12}} = 12.8\text{s}$$

19. (c) $\tau = mgr\sin\theta$

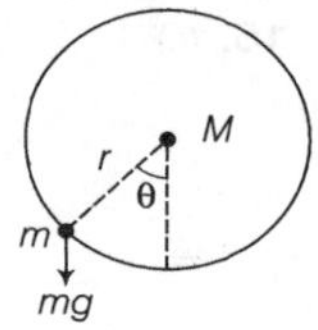

$$I\alpha \simeq -mgr\theta$$

$$T = 2\pi\sqrt{\frac{I}{mgr}}$$

where, $I = mr^2 + \frac{1}{2}Mr^2$

Substituting these values, we get

$$T = 2\pi\sqrt{\frac{\left(m + \frac{M}{2}\right)r}{mg}} \quad \text{or } T = 2\pi\sqrt{\frac{\left(1 + \frac{M}{2m}\right)r}{g}}$$

Comparing this with $T = 2\pi\sqrt{\frac{L_{eq}}{g}}$

$$L_{eq} = \left(1 + \frac{M}{2m}\right)r$$

20. (b) When the angular displacement of the disc is θ, the deformation (elongation and compression) in each spring is $x = R\theta$.

$$E_T = \frac{1}{2}I\omega^2 + 2\left(\frac{1}{2}kx^2\right) = \frac{1}{2}\frac{mR^2}{2}\cdot\omega^2 + kR^2\theta^2$$

$$\frac{dE_T}{dt} = \frac{1}{2}\cdot\frac{mR^2}{2}\cdot 2\omega\cdot\alpha + k\cdot R^2\cdot 2\theta\cdot\omega = 0$$

$$\frac{m}{2}\alpha + 2k\theta = 0 \quad \Rightarrow \quad \alpha = -\frac{4k}{m}\theta$$

$\therefore$ $$\omega = \sqrt{\frac{4k}{m}}$$

$\Rightarrow$ $$T = \frac{2\pi}{\omega} = 2\pi\sqrt{\frac{m}{4k}}$$

or $$T = 2 \times 3.14\sqrt{\frac{7}{4 \times 600}} = 0.34 \text{ s}$$

21. (a) $U = bx^2 - ax \Rightarrow \frac{dU}{dx} = 2bx - a$

$$F = -\frac{dU}{dx} = a - 2bx$$

When $$F = 0, x = \frac{a}{2b}$$

Now, $$k = \left|\frac{dF}{dx}\right|_{x = \frac{a}{2b}} = 2b$$

$\therefore$ $$\omega = \sqrt{\frac{k}{m}} \Rightarrow f = \frac{1}{2\pi}\sqrt{\frac{k}{m}} = \frac{1}{2\pi}\sqrt{\frac{2b}{m}}$$

22. (d) $x = A\sin\omega t$

$\therefore$

$$1 = A\sin\omega(t) \quad \ldots\text{(i)}$$

$$5 = A\sin\omega(t+1) \quad \ldots\text{(ii)}$$

$$5 = A\sin\omega(t+2) \quad \ldots\text{(iii)}$$

From Eq. (ii), $5 = A\sin\omega t\cos\omega + A\cos\omega t\sin\omega$ ...(iv)

From Eq. (iii), $5 = A\sin\omega t\cos 2\omega + A\cos\omega t\sin 2\omega$ ...(v)

Using Eq. (i), in Eqs. (iv) and (v), we have

$$5 = \cos\omega + A\cos\omega t\sin\omega \quad \ldots\text{(vi)}$$

$$5 = \cos 2\omega + A\cos\omega t\sin 2\omega \quad \ldots\text{(vii)}$$

(vi) $\times \sin 2\omega \Rightarrow \sin 2\omega\cos\omega + A\cos\omega t\sin\omega\sin 2\omega = 5\sin 2\omega$

(vii) $\times \sin\omega \Rightarrow \sin\omega\cos 2\omega + A\cos\omega t\sin\omega\sin 2\omega = 5\sin\omega$

$$\sin 2\omega\cos\omega - \sin\omega\cos 2\omega = 5\sin 2\omega - 5\sin\omega$$

$$\sin(2\omega - \omega) = 5(2\sin\omega\cos\omega) - 5\sin\omega$$

$$6\sin\omega = 10\sin\omega\cos\omega$$

$\Rightarrow$ $$\omega = \cos^{-1}\left(\frac{6}{10}\right)$$

$$\omega = \cos^{-1}\left(\frac{3}{5}\right)$$

$\Rightarrow$ $$T = \frac{2\pi}{\cos^{-1}\left(\frac{3}{5}\right)}$$

23. (a) $F_x = -kx$

$\Rightarrow$ $$a_x = -\frac{k}{m}x$$

$\Rightarrow$ $$\frac{d^2x}{dt^2} = -\omega^2 x \quad \ldots\text{(i)}$$

$$F_y = -ky$$

$\Rightarrow$ $$a_y = -\frac{k}{m}y$$

$\Rightarrow$ $$\frac{d^2y}{dt^2} = -\omega^2 y \quad \ldots\text{(ii)}$$

The solutions of above two equations are

$$x = A\sin\omega t + B\cos\omega t \quad \ldots\text{(iii)}$$

$$y = A'\sin\omega t + B'\cos\omega t \quad \ldots\text{(iv)}$$

where, A, A', B and B' are constants.

Differentiating Eqs. (iii) and (iv), we get

$$\frac{dx}{dt} = A\omega\cos\omega t - B\omega\sin\omega t \quad \ldots\text{(v)}$$

$$\frac{dy}{dt} = A'\omega\cos\omega t - B'\omega\sin\omega t \quad \ldots\text{(vi)}$$

At $t = 0$, $x = 2$, $y = 3$

and $$\frac{dx}{dt} = 0, \frac{dy}{dt} = 0 \quad \text{(given)}$$

From Eqs. (v) and (vi), $A = 0$ and $A' = 0$

Now, putting $t = 0$ and $x = 2$, $y = 3$ in Eqs. (iii) and (iv)

$$2 = B \text{ and } 3 = B'$$

$\therefore$ $$x = 2\cos\omega t$$

and $$y = 3\cos\omega t$$

$$\frac{x}{y} = \frac{2}{3}$$

$\Rightarrow$ $$2y - 3x = 0$$

24. (b) $T = 2\pi\sqrt{\frac{L}{g}}$

$\Rightarrow$ $$T = 2\pi\sqrt{L}\, g^{-\frac{1}{2}}$$

$\Rightarrow$ $$dT = 2\pi\sqrt{L}\left(-\frac{1}{2}\right)g^{-\frac{3}{2}}dg$$

$\therefore$ $$dg = -\frac{g\sqrt{g}\,dT}{\pi\sqrt{L}}$$

25. (c) $\tau = mg\cdot L\sin\theta + m_0 g\frac{L\sin\theta}{2}$

$$-I\alpha = gL\left(m + \frac{m_0}{2}\right)\theta$$

$$\left(mL^2 + \frac{m_0L^2}{3}\right)\alpha = -gL\left(m + \frac{m_0}{2}\right)\theta$$

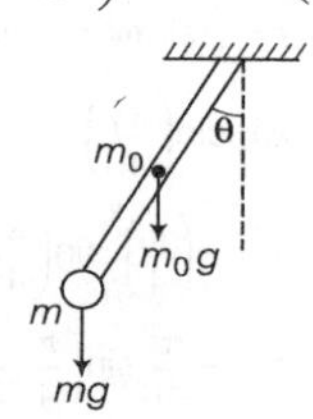

$$\alpha = -\frac{g}{L}\frac{\left(m + \frac{m_0}{2}\right)}{\left(m + \frac{m_0}{3}\right)}\theta$$

$\therefore$ $$\omega = \sqrt{\frac{g}{L}\frac{\left(m + \frac{m_0}{2}\right)}{\left(m + \frac{m_0}{3}\right)}} \Rightarrow T = 2\pi\sqrt{\frac{\left(m + \frac{m_0}{3}\right)L}{\left(m + \frac{m_0}{2}\right)g}}$$

26. (d) When x is small, $\cos\theta_1 \simeq \frac{x}{b}$ and $\cos\theta_2 \simeq \frac{x}{c}$

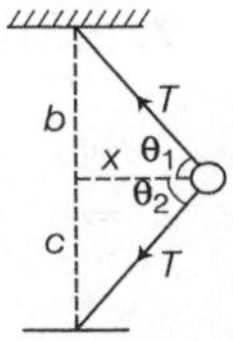

$\therefore \quad F = -(T\cos\theta_1 + T\cos\theta_2)$

$$= -T\left(\frac{1}{b} + \frac{1}{c}\right)x$$

$$= -T\left(\frac{b+c}{bc}\right)x$$

$$\therefore \quad a = -\frac{T}{m}\left(\frac{b+c}{bc}\right)x$$

$$\therefore \quad \omega = \sqrt{\frac{T}{m}\left(\frac{b+c}{bc}\right)}$$

$$\Rightarrow \quad f = \frac{1}{2\pi}\sqrt{\frac{T}{m}\left(\frac{b+c}{bc}\right)}$$

27. (b) In equilibrium, $B_F + kx_0 = w$

$$kx_0 = w - B_F \quad \ldots\text{(i)}$$

When the sphere is depressed by x (during oscillations), the total energy of the system is

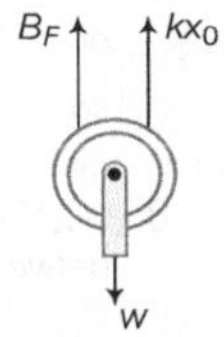

$$E_T = -(W - B_F)\,x + \frac{1}{2}k(x_0 + x)^2 + \frac{1}{2}mv^2 + \frac{1}{4}\rho V v^2$$

As the friction is absent, E_T, remains constant.

$$\therefore \quad \frac{dE_T}{dt} = -(w - B_F)v + \frac{1}{2}k2(x_0 + x)\cdot v + \frac{1}{2}m\cdot 2v\cdot a + \frac{1}{4}\rho V\cdot 2v\cdot a = 0$$

$$\Rightarrow \quad -(w - B_F) + k(x_0 + x) + ma + \frac{1}{2}\rho Va = 0$$

Using Eq. (i), in the above equation, we have

$$kx + a\left(m + \frac{1}{2}\rho V\right) = 0$$

$$\therefore \quad a = -\frac{k}{\left(m + \frac{1}{2}\rho V\right)}x$$

$$\Rightarrow \quad \omega = \sqrt{\frac{k}{m + \frac{1}{2}\rho V}}$$

$$\therefore \quad T = \frac{2\pi}{\omega} = 2\pi\sqrt{\frac{m + \frac{1}{2}\rho V}{k}}$$

$$= 2 \times 3.14\sqrt{\frac{0.5 + \frac{1}{2} \times 1000 \times \frac{4}{3}\pi\,(0.08)^3}{500}}$$

$$\Rightarrow \quad T = 0.352\,\text{s}$$

28. (b) Here, $B'_F = \dfrac{B_F\,(g + a_0)}{g}$

Now, $\quad B'_F + kx_0 - w = \dfrac{w}{g}a_0$

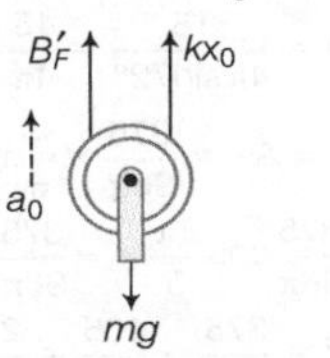

$$kx_0 = \frac{w}{g}\cdot a_0 + w - \frac{B_F\,(g - a_0)}{g}$$

$$kx_0 = (w - B_F) + \frac{a_0}{g}(w - B_F)$$

$$\Rightarrow \quad kx_0 = (w - B_F)\left(1 + \frac{a_0}{g}\right) \quad \ldots\text{(i)}$$

When the displacement is downwards, total energy is

$$E_T = -(w - B_F)\frac{(g + a_0)}{g}x + \frac{1}{2}k(x_0 + x^2) + \frac{1}{2}mv^2 + \frac{1}{2}\rho V v^2$$

$$\frac{dE_T}{dt} = -(w - B_F)\left(1 + \frac{a_0}{g}\right)v + \frac{1}{2}k\cdot 2(x_0 + x)\cdot v + \frac{1}{2}m\cdot 2v\cdot a + \frac{1}{2}\rho V\cdot 2va = 0$$

Using Eq. (i) in the above equation, we get the same result as that of the previous problem i.e. $T = 0.352$ s

29. (a) $y = \sin\omega t + \sqrt{3}\sin\left(\omega t + \frac{\pi}{2}\right)$

$\therefore \quad \delta = \frac{\pi}{2}$ (i.e. phase difference), so resultant

amplitude is $\quad A = \sqrt{1 + (\sqrt{3})^2} = 2$

and phase difference between first and resultant SHM is

$$\varepsilon = \tan^{-1}\left(\frac{\sqrt{3}\sin\frac{\pi}{2}}{1 + \sqrt{3}\cos\frac{\pi}{2}}\right) = \tan^{-1}(\sqrt{3}) = \frac{\pi}{3}$$

So resultant SHM is $\; y = 2\sin\left(\omega t + \frac{\pi}{3}\right)$

$$a = -2\omega^2\sin\left(\omega t + \frac{\pi}{3}\right)$$

The mass breaks of when the plank is at positive extreme and $a = g$

$$\therefore \quad |a_{\text{max}}| = |-2\omega^2| = g$$

$$\therefore \quad \omega = \sqrt{\frac{g}{2}}$$

This will happen for the first time when

$$\omega t + \frac{\pi}{3} = \frac{\pi}{2} \quad \Rightarrow \quad \omega t = \frac{\pi}{6}$$

$$t = \sqrt{\frac{2}{g}}\cdot\frac{\pi}{6}$$

30. (b) $x = 0.4\left[\cos^2\left(\frac{\pi t}{2}\right) - \sin^2\left(\frac{\pi t}{2}\right)\right] \Rightarrow x = 0.4\cos(\pi t)$

$$(\because \cos^2\theta - \sin^2\theta = \cos 2\theta)$$

$\therefore$ Amplitude is 0.4 m and motion is SHM.

31. (d) $y_2 = \frac{A}{2}\sin\omega t + \frac{A}{2}\sin\left(\omega t + \frac{\pi}{2}\right)$

$$\therefore \quad A_R = \sqrt{\frac{A^2}{4} + \frac{A^2}{4}} = \frac{A}{\sqrt{2}}$$

$$\varepsilon = \frac{\pi}{4}$$

$$\therefore \quad y_2 = \frac{A}{\sqrt{2}}\sin\left(\omega t + \frac{\pi}{4}\right)$$

$y_1 = A\sin\omega t$ (given)

$$\therefore \quad \frac{A_1}{A_2} = \frac{A}{A/\sqrt{2}} = \sqrt{2}$$

32. (d) In SHM, where acceleration is maximum (at the extremes) velocity is zero. $x = A\sin\omega t$ and $v = A\omega\cos\omega t$. So, Statement II is correct.

33. (b) $v = \omega\sqrt{A^2 - x^2}$. As x changes v changes. So, SHM is not a uniform motion. Statement II can also be called the definition of SHM.

34. (a) A spring wound watch runs on the basis of potential energy stored in the wound spring, which is not affected by g.

35. (c) $I = \frac{2}{5}mR^2$; As R decreases I decreases.

$$I_1\omega_1 = I_2\omega_2$$

$\Rightarrow \quad \omega_1 < \omega_2$ so, T decreases.

Revisal Problems (JEE Advanced)

1. (d) When the mass is released slowly, let the equilibrium position is at a depth x_0. This distance has to be accomodated between pulleys 3, 4 and 4, 5. So, pulley 4 descends by $\frac{x_0}{2}$. Then, upward force on pulley 4 is $k\frac{x_0}{2}$ and downward force on it is $kx_0 + kx_0 = 2kx_0$. So net force on it cannot be zero. Thus, the system cannot be in equilibrium.

2. (a) $k_{eq} = k + k = 2k$

$$T = 2\pi\sqrt{\frac{m}{k_{eq}}} = 2\pi\sqrt{\frac{m}{2k}} = \sqrt{2}\,\pi\sqrt{\frac{m}{k}}$$

3. (d) $T = 2\pi\sqrt{\frac{\mu}{k_{eq}}}$

Here, $\mu = \frac{m \cdot m}{m + m} = \frac{m}{2}$

and $k_{eq} = \frac{k \times 2k}{k + 2k} = \frac{2}{3}k$

$\therefore$ $T = 2\pi\sqrt{\frac{m/2}{\frac{2}{3}k}} = 2\pi\sqrt{\frac{3m}{4k}}$

4. (a) Decrease in GPE of B_1 = Increase in Gravitational PE of B_2 + Increase in elastic PE of spring.

$$mg(2x_0) = mgx_0 + \frac{1}{2}kx_0^2 \quad \ldots(i)$$

where, x_0 = displacement amplitude of B_1

Solving Eq. (i) gives $x_0 = \frac{2mg}{k}$

Again applying law of conservation of mechanical energy at the mean position we get

$$mgx_0 = mg\frac{x_0}{2} + \frac{1}{2}k\left(\frac{x_2}{2}\right)^2 + \frac{1}{2}mv_{max}^2 + \frac{1}{2}m\left(\frac{v_{max}}{2}\right)^2$$

where, v_{max} = velocity amplitude of B_1

$\Rightarrow$ $v_{max} = 2g\sqrt{\frac{m}{5k}}$

5. (c) Maximum velocity of B_2 = ω (maximum displacement of B_2)

$\therefore$ $\frac{v_{max}}{2} = \omega\left(\frac{x_0}{2}\right)$

$\Rightarrow$ $\omega = \frac{2g\sqrt{\frac{m}{5k}}}{\frac{2mg}{k}} = \sqrt{\frac{k}{5m}}$

$$f = \frac{\omega}{2\pi} = \frac{1}{2\pi}\sqrt{\frac{k}{5m}}$$

6. (c) $y = \sin\omega t + \sin 2\omega t + \sin 3\omega t = y_1 + y_2 + y_3$

$$T_1 = \frac{2\pi}{\omega}, T_2 = \frac{2\pi}{2\omega}$$

and $T_3 = \frac{2\pi}{3\omega}$

In time $T = \frac{2\pi}{\omega}$, first function completes one oscillation, second function two oscillations and the third three.

So, for the combination $T = \frac{2\pi}{\omega}$

7. (a,c) Net force on the ball will be zero when

$$\rho_L = \rho_0$$

i.e., $\beta h_0 = \rho_0$

$\Rightarrow$ $h_0 = \frac{\rho_0}{\beta}$

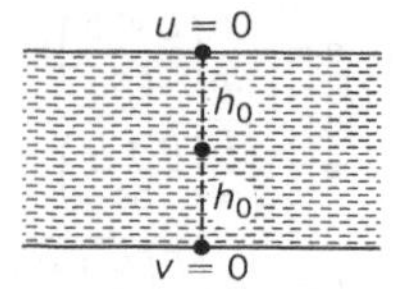

$\therefore$ Mean position is at a depth $\frac{\rho_0}{\beta}$.

Net force at a depth $(h_0 + x)$ is $F = (\rho_L - \rho_0)Vg(\uparrow)$

$F[\beta(h_0 + x) - \beta h_0]Vg(\uparrow)$

or $F = -\beta Vgx$ $\therefore$ SHM

so, $h_{max} = \frac{2\rho_0}{\beta}$ $(\because h_{max} = 2h_0)$

8. (a,c) The resultant amplitude 1 and 3 is $A\sqrt{2}$ and is mid-way between 1 and 3.

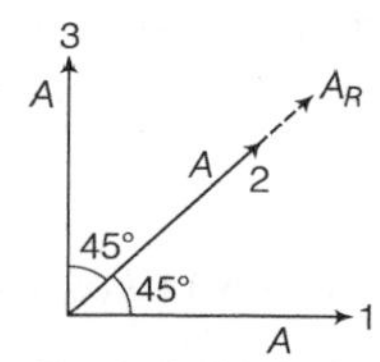

$\therefore$ $A_R = A + A\sqrt{2} = A(1 + \sqrt{2})$

From the figure it is clear that resultant SHM differs from the first in phase, by 45°.

$$E_1 = \frac{1}{2}kA^2$$

$$E_R = \frac{1}{2}k[A(1 + \sqrt{2})]^2 = \frac{1}{2}kA^2(1 + 2 + 2\sqrt{2})$$

$\therefore$ $E_R = (3 + 2\sqrt{2})E_1$

9. (b,d) The period is T,

$$T = (t_{PP_1} + t_{P_1P}) + (t_{PP_2} + t_{P_2P}) = (0.5 + 0.5) + (0.75 + 0.75) = 2.5\text{ s}$$

$\therefore$ $\omega = \frac{2\pi}{T} = \frac{2\pi}{2.5} = \frac{4}{5}\pi$

$x = -A\cos\omega t,\ v = +A\omega\sin\omega t$

$$3 = A\left(\frac{4}{5}\pi\right)\sin\omega\left(\frac{1}{2}\right) = \frac{4}{5}\pi A\sin\left(\frac{2}{5}\pi\right) \quad \ldots(i)$$

$$-3 = A\left(\frac{4}{5}\pi\right)\sin\omega(2) = \frac{4}{5}\pi A\sin\left(\frac{8}{5}\pi\right) \quad \ldots(ii)$$

From Eq. (i), $A = \frac{15}{4\pi\sin 72°} = \frac{15 \times 25}{4\pi \times 24} = \frac{375}{96\pi}$

$$v_{max} = A\omega = \frac{375}{96\pi} \times \frac{4}{5}\pi = \frac{75}{24}\text{ ms}^{-1}$$

At $t = \frac{1}{2}$ s, $x = -\frac{375}{96\pi}\cos\frac{2\pi}{5} = -\frac{375}{96\pi}\cdot\frac{7}{25} = -\frac{105}{96\pi}$

$\therefore$ $P_1P = \frac{375}{96\pi} - \frac{105}{96\pi} = \frac{270}{96\pi} = \frac{90}{32\pi}$

$$P_2P = \frac{375}{96\pi} + \frac{105}{96\pi} = \frac{480}{96\pi} = \frac{160}{32\pi}$$

$$\frac{P_1P}{P_2P} = \frac{90}{160} = \frac{3}{4}$$

10. (b,c) $T_1 = 2\pi\sqrt{\frac{m}{k}}$ is independent of g

$T_2 = 2\pi\sqrt{\frac{L}{g}}$, As height increases g decreases

So, $T_2' > T_2$

11. (b,c,d) $x = 2A - A\cos\omega t$ $\Rightarrow$ At $t = 0$, it is at left extreme and momentarily at rest and this point is $x = A$.
Its amplitude is A. So, the particle oscillates between $x = +A$ to $x = +3A$, $x = 2A$ being its mean position.

$$T = \frac{2\pi}{\omega}$$

$$\Rightarrow \quad t_{A\text{ to }3A} = \frac{T}{2} = \frac{\pi}{\omega}$$

$$\text{Also,} \quad t_{A\text{ to }2A} = \frac{T}{4} = \frac{\pi}{2\omega}$$

12. (d) $\frac{1}{2}mv^2 + mgx = \frac{1}{2}mu^2$

$$\frac{p^2}{2m} + mg\,x = \frac{p_i^2}{2m}$$

$$p^2 = p_i^2 - (2m^2g)x$$

13. (c) $E_1 = \frac{1}{2}k(2a)^2 = 4\cdot\left(\frac{1}{2}ka^2\right)$

$$E_2 = \frac{1}{2}ka^2$$

$$\therefore \quad E_1 = 4E_2$$

14. (b) Due to the viscous drag offered by the water, the momentum will not be same when it returns. In all the figures, the starting point is top point. Initially as the time runs the momentum (in negative direction) increases. Later in negative direction decreases. When it reaches the bottom most point momentum is zero. And position here is maximum negative. Later momentum in positive direction increases. Now the maximum momentum achieved is smaller in magnitude when compared to the previous maximum momentum. Final position is at a smaller height than the initial position.

15. (d) $\theta = \omega t$

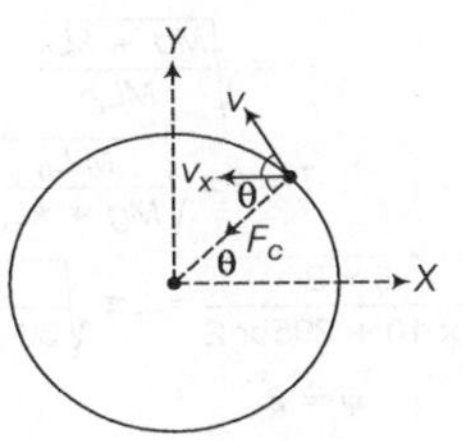

$$v_x = -v\cos(90° - \omega t)$$
$$= -v\sin\omega t$$

16. (b)

$$F_y = -F_c\sin\omega t$$
$$= -\frac{mv^2}{R}\sin\omega t$$

17. (a) $\theta = \omega t = \frac{v}{R}t$.

18. (c) $kx_0 + mg = p_0A_0$ ...(i)

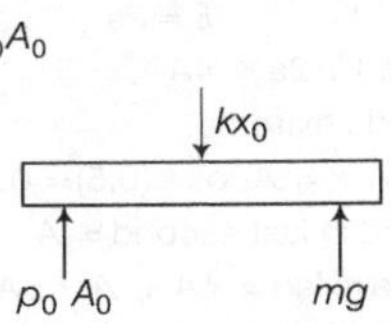

$$x_0 = \frac{p_0A_0 - mg}{k}$$

19. (d) $k(x + x_0) + mg - pA_0 = ma$ ...(ii)

For an adiabatic process

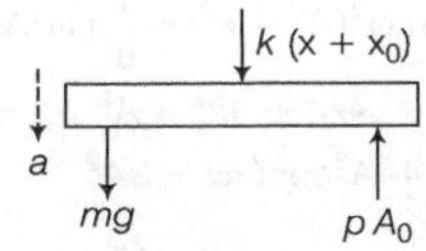

$$pV^\gamma = \text{constant}$$

$$\therefore \quad p_0(L_0A_0)^\gamma = p[(L_0 + x)A_0]^\gamma$$

$$\Rightarrow \quad p = p_0\left(1 - \frac{\gamma x}{L_0}\right)$$

using this in Eq. (ii), we get

$$kx + kx_0 + mg - p_0\left(1 - \frac{\gamma x_0}{L_0}\right)A_0 = ma$$

$$kx_0 + mg - p_0A_0 + \left(k + \frac{\gamma p_0A_0}{L_0}\right)x = ma$$

Using Eq. (i), we get $ma = \left(k + \frac{\gamma p_0A_0}{L_0}\right)k$

Here, a and x are opposite to each other.

$$\therefore \quad a = -\left(\frac{kL_0 + \gamma p_0A_0}{mL_0}\right)x$$

$$\therefore \quad \omega = \sqrt{\frac{kL_0 + \gamma p_0A_0}{mL_0}}$$

20. (a) Here, restoring force is only due to pressure.

So, $\omega = \sqrt{\frac{\gamma p_1A_0}{mL_1}}$

21. (a) At $t = 0$, $x = -A$

$$\therefore \quad x = A\sin\left(\omega t + \frac{3\pi}{2}\right) \quad \text{or} \quad x = -A\cos\omega t$$

$$\Rightarrow \quad x = -10\cos\omega t$$

(i) When $x = -5$cm,

$$-5 = -10\cos\frac{2\pi}{T}t$$

$$\frac{2\pi}{T}\cdot t = \frac{\pi}{3}$$

$$\Rightarrow \quad t = \frac{1}{6}\cdot T$$

$$T = 2\pi\sqrt{\frac{2}{3200}} = \frac{2\pi}{40} = \frac{\pi}{20}\text{s}$$

$$t = \frac{1}{6}\cdot\frac{\pi}{20} = \frac{\pi}{120}\text{s}$$

(ii) When $x = 0$, $0 = -10\cos\omega t'$

$$\omega t' = \frac{\pi}{2} \Rightarrow \frac{2\pi}{T}t' = \frac{\pi}{2}$$

$$t' = \frac{1}{4}\cdot T = \frac{\pi}{80}\text{s}$$

$$\text{Required time} = \frac{\pi}{80} - \frac{\pi}{120} = \frac{\pi}{240}\text{s}$$

(iii) PE = KE $\Rightarrow$ $\frac{1}{2}m\omega^2x^2 = \frac{1}{2}m\omega^2(A^2 - x^2)$

$$x^2 = A^2 - x^2$$

$$\Rightarrow \quad 2x^2 = A^2$$

$$2A^2\cos^2\omega t = A^2 \Rightarrow \cos^2\omega t = \frac{1}{2}$$

$$\cos\omega t = \frac{1}{\sqrt{2}} \Rightarrow \omega t = \frac{\pi}{4}$$

$$t = \frac{\pi}{4}\cdot\frac{T}{2\pi} = \frac{T}{8} = \frac{\pi}{160}\text{s}$$

(iv) $KE = \frac{1}{4}(KE)_{max}$

$$\Rightarrow \quad \frac{1}{2}m\omega^2(A^2 - x^2) = \frac{1}{8}m\omega^2A^2$$

$$4A^2 - 4x^2 = A^2 \quad \Rightarrow \quad 3A^2 = 4x^2$$

$$4\cdot A^2\cos^2\omega t = 3A^2$$

$$\Rightarrow \quad \cos\omega t = \frac{\sqrt{3}}{2}$$

$$\frac{2\pi}{T}\cdot t = \frac{\pi}{6} \quad \Rightarrow \quad t = \frac{T}{12} = \frac{\pi}{240}\,s$$

Hence, (i) → (r), (ii) → (q), (iii) → (p), (iv) → (q)

22. (a) $\delta = \frac{2\pi}{3}$

$$\therefore \quad A_R = \sqrt{A^2 + A^2 + 2A^2\cos\frac{2\pi}{3}} = \sqrt{2A^2 - 2A^2\cdot\frac{1}{2}} = A$$

$$\varepsilon = \tan^{-1}\left(\frac{A\sin\frac{2\pi}{3}}{A + A\cos\frac{2\pi}{3}}\right) = \tan^{-1}\left(\frac{\sqrt{3}/2}{1 + (-1/2)}\right)$$

$$= \tan^{-1}(\sqrt{3}) = \frac{\pi}{3}$$

∴ Resulting SHM is $y = A_R\sin(\omega t + \varepsilon) = A\sin\left(\omega t + \frac{\pi}{3}\right)$

Hence, (i) → (q), (ii) → (r), (iii) → (s), (iv) → (t)

23. (a) $U = \frac{a}{x^{12}} - \frac{b}{x^6}$

$$\Rightarrow \quad \frac{dU}{dx} = -12\frac{a}{x^{13}} + 6\frac{b}{x^7} \quad \ldots(i)$$

$$\frac{d^2U}{dx^2} = (12)(13)\frac{a}{x^{14}} - (6)(7)\frac{b}{x^8} \quad \ldots(ii)$$

When $F = 0$, $\frac{dU}{dx} = 0 \quad \Rightarrow \quad x^6 = \frac{2a}{b}$

Using this in Eq. (ii), we get

$$\frac{d^2U}{dx^2} > 0$$

So, when $x^6 = \frac{2a}{b}$, U is minimum.

If we use $x^6 = \frac{2a}{b}$ in $U = \frac{a}{x^{12}} - \frac{b}{x^6}$ we get

$$U = -\frac{b^2}{2a}$$

∴ When U is minimum, it is negative. So, U-x graph will be (p).

When $\frac{dU}{dx} = 0$, F has to be zero. $\left(\because F = -\frac{dU}{dx}\right)$ so, F-x graph will be (q).

As $U_{min} = -\frac{b^2}{2a}$, dissociation energy $= +\frac{b^2}{4a}$

Hence, (i) → (p), (ii) → (q), (iii) → (t)

24. (7) $T = 2\pi\sqrt{\frac{I}{mgl}}$, where $I = \frac{2}{5}mR^2 + mR^2$

$$= \frac{7}{5}mR^2$$

and $l = R$

$$\therefore \quad T = 2\pi\sqrt{\frac{\frac{7}{5}mR^2}{mgR}} = 2\pi\sqrt{\frac{7}{5}\cdot\frac{R}{g}}$$

For a simple pendulum, $T = 2\pi\sqrt{\frac{L}{g}}$

$$\therefore \quad L = \frac{7}{5}\times R = 7\,\text{cm}$$

25. (5) $\omega = \sqrt{\frac{mgl/2}{I}}$, where $I = \frac{ml^2}{3}$

$$\therefore \quad \omega = \sqrt{\frac{mgl/2}{ml^2}\cdot 3} = \sqrt{\frac{3\times g}{2l}} = \sqrt{\frac{3\times 9.8}{0.147\times 2}} = 10\,\text{rads}^{-1}$$

$$\therefore \quad \beta = 5$$

26. (2) At equilibrium, $kx_0 = mg$ ...(i)

When the displacement of the block is x downwards (from equilibrium, position)

$$E_T = -mgx + \frac{1}{2}k(x_0 + x)^2 + \frac{1}{2}(0.6\,MR^2)\,\omega^2 + \frac{1}{2}mv^2$$

$$= -mgx + \frac{1}{2}k(x_0 + x)^2 + 0.3\,MR^2\cdot\frac{v^2}{R^2} + \frac{1}{2}mv^2$$

$$\frac{dE_T}{dt} = -mgv + \frac{1}{2}k\cdot 2(x_0 + x)v + (0.3)M\cdot 2v\cdot a + \frac{1}{2}m\cdot 2v\cdot a = 0$$

$$\Rightarrow \quad -mg + kx_0 + kx + 0.6Ma + ma = 0$$

Using Eq. (i), reduces the above equation to

$$a(0.6\,M + m) = -kx$$

$$a = -\frac{k}{0.6M + m}\cdot x$$

$$\Rightarrow \quad \omega = \sqrt{\frac{k}{0.6M + m}}$$

$$\therefore \quad \omega = \sqrt{\frac{20}{(0.6)(5) + 2}} = \sqrt{4} = 2\,\text{rads}^{-1}$$

27. (8) $p\Delta V + V\Delta p = 0$

$$\Rightarrow \quad \Delta p = \frac{p}{V}\Delta V = \frac{p(Ax)}{AL_0}$$

$$(\Delta p)A = -\frac{pA}{L_0}x = -\frac{Mg}{L_0}x$$

∴ New restoring force is $F = -\left(\frac{Mg}{L_0} + k\right)x$

$$= -\left(\frac{Mg + kL_0}{L_0}\right)x$$

$$\therefore \quad a = -\frac{Mg + kL_0}{ML_0}\cdot x$$

$$\therefore \quad \omega = \sqrt{\frac{Mg + kL_0}{ML_0}}$$

$$T = 2\pi\sqrt{\frac{ML_0}{Mg + kL_0}}$$

$$\Rightarrow \quad T = 2\pi\sqrt{\frac{5\times 2}{5\times 10 + 295\times 2}} = 2\pi\sqrt{\frac{10}{50 + 590}} = \frac{2\pi}{8}$$

$$\therefore \quad \psi = 8$$

28. (3) $g_{eff} = \frac{V\rho g - V\frac{\rho}{4}g}{V\rho} = \frac{3}{4}g$

$$T = 2\pi\sqrt{\frac{L}{g_{eff}}} = 2\pi\sqrt{\frac{4L}{3g}} = 4\pi\sqrt{\frac{L}{3g}}$$

$$\therefore \quad \beta = 3$$

29. (5) $\omega = \pi \quad \Rightarrow \quad \frac{2\pi}{T} = \pi$

$$\Rightarrow \quad T = 2s$$

∴ Distance travelled in 2s = $4A$

Now it is at right side extreme.

$$x = A\cos\pi(0.5) = 0$$

∴ Distance travelled in last second = A

∴ Total distance travelled = $4A + A = 5A$

∴ * = 5

JEE Main & AIEEE Archive

1. (b) Given time period

$$T = 2\pi\sqrt{L/g}$$

Thus, changes can be expressed as

$$\pm\frac{2\Delta T}{T} = \pm\frac{\Delta L}{L} \pm \frac{\Delta g}{g}$$

According to the question, we can write

$$\frac{\Delta L}{L} = \frac{0.1\text{cm}}{20.0\text{cm}} = \frac{1}{200}$$

Again time period

$$T = \frac{90}{100}\text{ s}$$

and $$\Delta T = \frac{1}{100}\text{ s}$$

$$\Rightarrow \quad \frac{\Delta T}{T} = \frac{1}{90}$$

Now,

$$\because \quad T = 2\pi\sqrt{\frac{L}{g}}$$

$$\because \quad g = 4\pi^2\frac{L}{T^2}$$

$$\therefore \quad \frac{\Delta g}{g} = \frac{\Delta L}{L} + \frac{2\Delta T}{T}$$

or $$\frac{\Delta g}{g} \times 100\% = \left(\frac{\Delta L}{L}\right) \times 100\% + \left(\frac{2\Delta T}{T}\right) \times 100\%$$

$$= \left(\frac{1}{200} \times 100\right)\% + 2 \times \frac{1}{90} \times 100\%$$

$$\simeq 2.72\% \simeq 3\%$$

Thus, accuracy in the determination of g is approx 3%.

2. (a) We know that time period,

$$T = 2\pi\sqrt{\frac{L}{g}}$$

When additional mass M is added to its bob

$T_M = 2\pi\sqrt{\frac{L + \Delta L}{g}}$, where ΔL is increase in length.

We know that,

$$Y = \frac{Mg/A}{\Delta L/L} = \frac{MgL}{A\Delta L}$$

$$\Rightarrow \quad \Delta L = \frac{MgL}{AY}$$

$$\Rightarrow \quad T_M = 2\pi\sqrt{\frac{L + \frac{MgL}{AY}}{g}}$$

$$\Rightarrow \quad \left(\frac{T_M}{T}\right)^2 = 1 + \frac{Mg}{AY}$$

or $$\frac{Mg}{AY} = \left(\frac{T_M}{T}\right)^2 - 1$$

or $$\frac{1}{Y} = \frac{A}{Mg}\left[\left(\frac{T_M}{T}\right)^2 - 1\right]$$

3. (b) During oscillation, motion of a simple pendulum KE is maximum of mean position where PE is minimum. At extreme position, KE is minimum and PE is maximum. Thus, correct graph is depicted in option (b).

4. (b) Consider free body diagram of the wire. As the wires are in equilibrium, they must carry current in opposite direction .

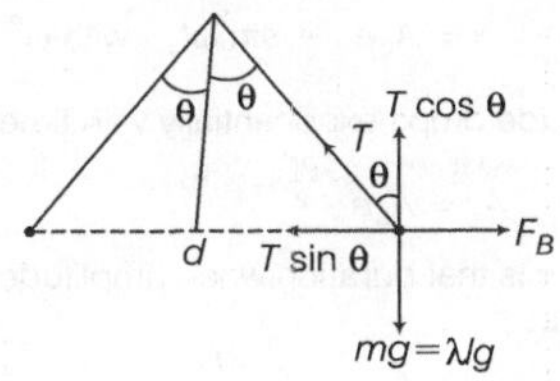

Here, $F_B = \frac{\mu_0 I^2 l}{2\pi d}$, where l is length of each wire are d is separation between wires.

From figure, $d = 2L\sin\theta$

$$T\cos\theta = mg = \lambda lg \quad \text{(in vertical direction)} \ldots\text{(i)}$$

$$T\sin\theta = F_B = \frac{\mu_0 I^2 l}{4\pi L\sin\theta} \quad \text{(In horizontal direction)} \ldots\text{(ii)}$$

From Eqs. (i) and (ii), we get

$$\frac{T\sin\theta}{T\cos\theta} = \frac{\mu_0 I^2 l}{4\pi L\sin\theta \times \lambda lg}$$

$$\therefore \quad I = \sqrt{\frac{4\pi\lambda Lg\sin^2\theta}{\mu_0\cos\theta}} = 2\sin\theta\sqrt{\frac{\pi\lambda Lg}{\mu_0\cos\theta}}$$

5. (d) In SHM, a particle starts from rest, we have

$$x = A\cos\omega t, \text{ at } t = 0, x = A$$

When $t = t$, then $x = A - a$

When $t = 2t$, then $x = A - 3a$

So, $$A - a = A\cos\omega t$$

$$A - 3a = A\cos 2\omega t$$

As, $$\cos 2\omega\tau = 2\cos^2\omega t - 1$$

$$\Rightarrow \quad \frac{A - 3a}{A} = 2\left(\frac{A - a}{A}\right)^2 - 1$$

$$\Rightarrow \quad \frac{A - 3a}{A} = \frac{2A^2 + 2a^2 - 4Aa - A^2}{A^2}$$

$$A^2 - 3aA = A^2 + 2a^2 - 4Aa$$

$$a^2 = 2aA \Rightarrow A = 2a$$

Now, $$A - a = A\cos\omega t$$

$$\Rightarrow \quad \cos\omega t = 1/2$$

$$\frac{2\pi}{T} = \frac{\pi}{3}$$

$$T = 6\pi$$

6. (c) Amplitude of damped oscillator is given by

$$A = A_0 e^{-\frac{bt}{2m}}$$

After 5 s, $$0.9 A_0 = A_0 e^{-\frac{b(5)}{2m}}$$

$$\Rightarrow \quad 0.9 = e^{-\frac{b(5)}{2m}} \quad \ldots\text{(i)}$$

After, 10 s, $$A = A_0 e^{-b\frac{(15)}{2m}} \Rightarrow A = A_0 (e^{-\frac{5b}{2m}})^3 \quad \ldots\text{(ii)}$$

From Eqs. (i) and (ii), we get

$$A = 0.729 A_0$$

Hence, $$\alpha = 0.729$$

7. (d) For damped harmonic motion,

$$ma = -kx - mbv$$

or $\quad ma + mbv + kx = 0$

Solution of above equation is

$$x = A_0 e^{-\frac{bt}{2}} \sin \omega t; \quad \text{with } \omega^2 = \frac{k}{m} - \frac{b^2}{4}$$

where, amplitude drops exponentially with time.

i.e. $\quad A_\tau = A_0 e^{-\frac{b\tau}{2}}$

Average time τ is that duration when amplitude drops by 63%, i.e. becomes A_0/e.

Thus, $\quad A_\tau = \frac{A_0}{e} = A_0 e^{-\frac{b\tau}{2}}$

or $\quad \frac{b\tau}{2} = 1$ or $\tau = \frac{2}{b}$

8. (b) Let at any instant, cube be at a depth x from the equilibrium position, then

Net force acting on the cube = Upthrust on the portion of length x

$\therefore \quad F = -\rho l^2 xg = -\rho l^2 g\ x \quad$ ...(i)

$$F \propto -x \text{ (SHM)}$$

Negative sign shows that force is opposite to x. Hence, equation of SHM

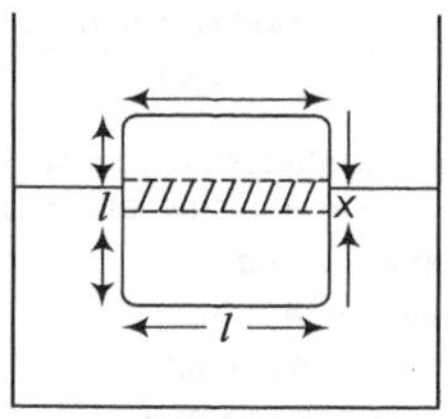

$F = -kx \quad$ ...(ii)

Comparing Eqs. (i) and (ii), we get

$$k = \rho l^2 g$$

$\therefore \quad T = 2\pi\sqrt{\frac{m}{k}} = 2\pi\sqrt{\frac{l^3 d}{\rho l^2 g}} = 2\pi\sqrt{\frac{ld}{\rho g}}$

9. (a) As $\quad F = -kx$

$$k \propto \frac{1}{x} \propto \frac{1}{l}$$

For spring, $\quad k \propto \frac{1}{l}$

$\therefore \quad \frac{k_A}{k} = \frac{l}{l_A}$

$\Rightarrow \quad k_A = \frac{l_A + l_B}{l_A} k = \frac{5}{2} k$

10. (a) Let $\quad x_1 = A\sin(\omega t + \phi_1)$ and $x_2 = A\sin(\omega t + \phi_2)$

$$x_2 - x_1 = A[\sin(\omega t + \phi_2) - \sin(\omega t - \phi_1)]$$

$$= 2A\cos\left(\frac{2\omega t + \phi_1 + \phi_2}{2}\right)\sin\left(\frac{\phi_2 - \phi_1}{2}\right)$$

The resultant motion can be treated as a simple harmonic motion with amplitude $2A\sin\left(\frac{\phi_2 - \phi_1}{2}\right)$.

Given, maximum distance between the particles

$$= X_0 + A$$

$\therefore$ Amplitude of resultant SHM

$$= X_0 + A - X_0 = A$$

$\therefore \quad 2A\sin\left(\frac{\phi_2 - \phi_1}{2}\right) = A$

$\Rightarrow$ Phase difference, $\phi_2 - \phi_1 = \pi/3$

11. (c) At mean position,

$$F_{net} = 0$$

$\therefore$ By conservation of linear momentum,

$$Mv_1 = (M + m)v_2$$

$$M\omega_1 A_1 = (M + m)\,\omega_2 A_2$$

But angular velocity,

$$\omega_1 = \sqrt{\frac{k}{M}}$$

and $\quad \omega_2 = \sqrt{\frac{k}{M + m}}$

On solving, $\quad \frac{A_1}{A_2} = \sqrt{\frac{m + M}{M}}$

12. (c) $\quad U(x) = \frac{a}{x^{12}} - \frac{b}{x^6}$

$$U(x = \infty) = 0$$

As, $\quad F = -\frac{dU}{dx} = -\left[\frac{12a}{x^{13}} + \frac{6b}{x^7}\right]$

At equilibrium, $F = 0$

$\therefore \quad x^6 = \frac{2a}{b}$

$\therefore \quad U_{\text{at equilibrium}} = \frac{a}{\left(\frac{2a}{b}\right)^2} - \frac{b}{\left(\frac{2a}{b}\right)} = \frac{-b^2}{4a}$

$\therefore \quad D = [U(x = \infty) - U_{\text{at equilibrium}}]$

$$= \frac{b^2}{4a}$$

13. (b) $\quad \frac{aT}{x} = \frac{\omega^2 xT}{x}$

$$= \frac{4\pi^2}{T^2} \times T$$

$$= \frac{4\pi^2}{T} = \text{constant}$$

Hence, aT/x does not change with time.

14. (d) $x = x_0 \cos\left(\omega t - \frac{\pi}{4}\right)$

Acceleration, $\quad a = \frac{d^2x}{dt^2}$

$$= -\omega^2 x_0 \cos\left(\omega t - \frac{\pi}{4}\right)$$

$$= \omega^2 x_0 \cos\left(\omega t + \frac{3\pi}{4}\right)$$

So, $\quad A = \omega^2 x_0$

and $\quad \delta = \frac{3\pi}{4}$

15. (d) $\quad f = \frac{1}{2\pi}\sqrt{\frac{k_1 + k_2}{m}}$

and $\quad f' = \frac{1}{2\pi}\sqrt{\frac{4k_1 + 4k_2}{m}}$

$\Rightarrow \quad f' = \frac{1}{2\pi} \cdot 2\sqrt{\frac{k_1 + k_2}{m}}$

$$= 2f$$

16. (a) Average kinetic energy of particle

$$= \frac{1}{4} ma^2\omega^2 \quad [\text{as } \omega = 2\pi\nu]$$

$$= \frac{1}{4} ma^2(2\pi\nu)^2$$

$$= \pi^2\nu^2 ma^2$$

17. (a) $x = (2 \times 10^{-2}) \cos \pi t$

Here, $a = 2 \times 10^{-2}$ m = 2 cm

At $t = 0$, $x = 2$ cm, *i.e.*, the object is at positive extreme, so to acquire maximum speed (*i.e.*, to reach mean position), it takes $\frac{1}{4}$th of time period.

$\therefore$ Required time $= \frac{T}{4}$

As $x = a \cos \omega t$

where, $\omega = \frac{2\pi}{T} = \pi$

$\Rightarrow$ $T = 2$ s

So, required time $= \frac{T}{4} = \frac{2}{4} = 0.5$ s

18. (a) The maximum velocity of a particle performing SHM is given by $v = A\omega$, where A is the amplitude and ω is the angular frequency of oscillation.

$\therefore$ $4.4 = (7 \times 10^{-3}) \times 2\pi/T$

or $T = \frac{7 \times 10^{-3}}{4.4} \times \frac{2 \times 22}{7} = 0.01$ s

19. (b) Here, $y = \sin^2 \omega t$

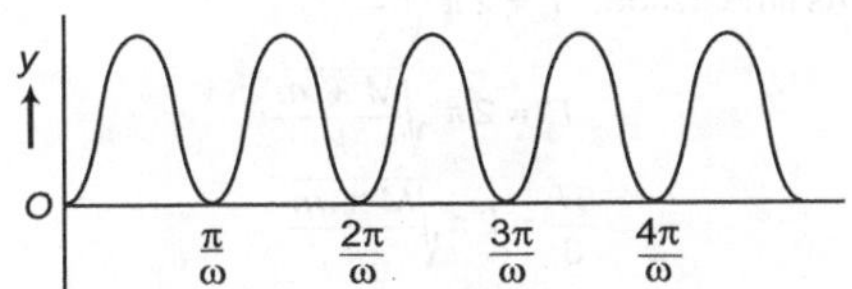

$$\frac{dy}{dt} = 2\omega \sin \omega t \cos \omega t = \omega \sin 2\omega t$$

$$\frac{d^2y}{dt^2} = 2\omega^2 \cos 2\omega t$$

For SHM, $\frac{d^2y}{dt^2} \propto -y$

Hence, function is not SHM but periodic. From the y-t graph, time period is $T = \pi/\omega$. $\left[\because \frac{2\pi}{T} = 2\omega\right]$

20. (a) Given, $y_1 = 0.1 \sin\left(100\pi t + \frac{\pi}{3}\right)$

$\Rightarrow$ Velocity of particle 1,

$$\frac{dy_1}{dt} = v_1 = 0.1 \times 100\pi \cos\left(100\pi t + \frac{\pi}{3}\right)$$

or $v_1 = 10\pi \sin\left(100\pi t + \frac{\pi}{3} + \frac{\pi}{2}\right)$

or $v_1 = 10\pi \sin\left(100\pi t + \frac{5\pi}{6}\right)$

and $y_2 = 0.1 \cos \pi t$

$\Rightarrow$ Velocity of particle 2,

$$\frac{dy_2}{dt} = v_2 = -0.1 \sin \pi t$$

or $v_2 = 0.1 \sin(\pi t + \pi)$

Hence, phase difference

$$\Delta\phi = \phi_1 - \phi_2 = \left(100\pi t + \frac{5\pi}{6}\right) - (\pi t + \pi)$$

$$= \frac{5\pi}{6} - \pi$$

$$= -\frac{\pi}{6} \quad [\text{at } t = 0]$$

21. (b) $\frac{d^2x}{dt^2} = -\alpha x$ ...(i)

We know that, for SHM

$$a = \frac{d^2x}{dt^2} = -\omega^2 x \quad \text{...(ii)}$$

From Eqs. (i) and (ii), we have

$$\omega^2 = \alpha \text{ or } \omega = \sqrt{\alpha}$$

$\Rightarrow$ $\frac{2\pi}{T} = \sqrt{\alpha}$

or $T = \frac{2\pi}{\sqrt{\alpha}}$

22. (a)

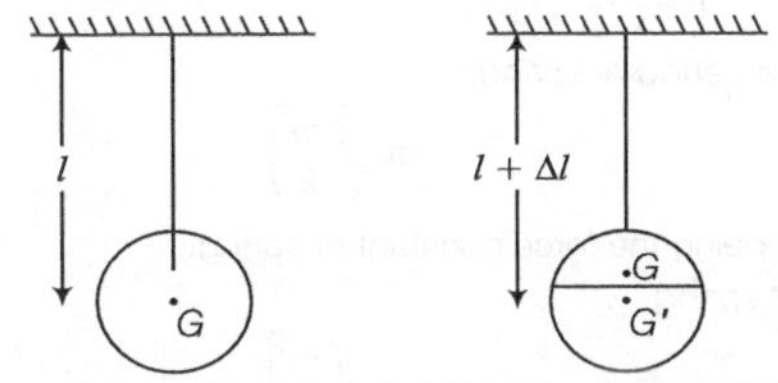

Spherical hollow ball filled with water

Spherical hollow ball half filled with water

$$T = 2\pi\sqrt{\frac{l}{g}}$$

As water is coming out, CG of the system goes down, effective length increase.

$$T_1 = 2\pi\sqrt{\frac{l + \Delta l}{g}}, T_1 > T_2$$

So, time period also increases but upto half empty. As water is coming out after half empty, CG goes up.

Effective length decreases and for complete hollow sphere, CG is also at the initial position, so effective length as well as time period decreases and same as initial at last.

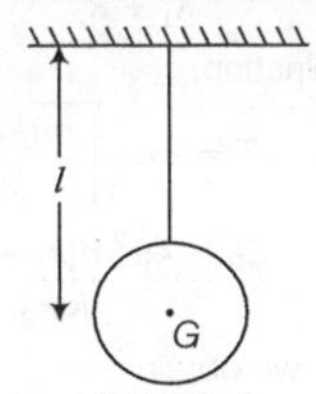

Spherical hollow ball

$$T_2 = 2\pi\sqrt{\frac{l - \Delta l}{g}}$$

and $T_1 < T_2$

Hence, time period first increases and then decreases to the original value.

23. (c) The time period of simple pendulum in air,

$$T = t_0 = 2\pi\sqrt{\left(\frac{l}{g}\right)} \quad \text{...(i)}$$

Here, l being the length of simple pendulum,

In water, effective weight of bob,

w' = Weight of bob in air − Upthrust

$\Rightarrow$ $\rho V g_{eff} = mg - m'g$

$= \rho V g - \rho' V g = (\rho - \rho')Vg$

where, ρ' = density of water,

ρ = density of bob.

$\therefore$ $g_{eff} = \left(\frac{\rho - \rho'}{\rho}\right) g = \left(1 - \frac{\rho'}{\rho}\right) g$

$$\therefore \quad t = 2\pi\sqrt{\left[\frac{l}{\left(1-\frac{\rho'}{\rho}\right)g}\right]} \quad \ldots(\text{ii})$$

$$\text{Thus,} \quad \frac{t}{t_0} = \sqrt{\left[\frac{1}{\left(1-\frac{\rho'}{\rho}\right)}\right]} = \sqrt{\left(\frac{1}{1-\frac{1000}{(4/3)\times 1000}}\right)}$$

$$= \sqrt{\left(\frac{4}{4-3}\right)} = 2$$

or $\quad t = 2t_0$

24. (b) Time period of spring,

$$T = 2\pi\sqrt{\left(\frac{m}{k}\right)}$$

Here, k being the force constant of spring.

For first spring,

$$t_1 = 2\pi\sqrt{\left(\frac{m}{k_1}\right)} \quad \ldots(\text{i})$$

For second spring,

$$t_2 = 2\pi\sqrt{\left(\frac{m}{k_2}\right)} \quad \ldots(\text{ii})$$

In series, f is same.

$$\therefore \quad x = x_1 + x_2$$

$$\frac{-F}{k} = \frac{-F}{k_1} - \frac{F}{k_2}$$

$$\Rightarrow \quad k = \frac{k_1 k_2}{k_1 + k_2}$$

So, the effective force constant in their series combination is

$$k = \frac{k_1 k_2}{k_1 + k_2}$$

Time period of combination,

$$T = 2\pi\sqrt{\left[\frac{m(k_1+k_2)}{k_1 k_2}\right]}$$

$$\Rightarrow \quad T^2 = \frac{4\pi^2 m(k_1+k_2)}{k_1 k_2} \quad \ldots(\text{iii})$$

From Eqs. (i) and (ii), we obtain

$$t_1^2 + t_2^2 = 4\pi^2\left(\frac{m}{k_1} + \frac{m}{k_2}\right)$$

or $$t_1^2 + t_2^2 = 4\pi^2 m\left(\frac{1}{k_1} + \frac{1}{k_2}\right)$$

or $$t_1^2 + t_2^2 = \frac{4\pi^2 m(k_1+k_2)}{k_1 k_2}$$

$$\Rightarrow \quad t_1^2 + t_2^2 = T^2 \quad \text{[from Eq. (iii)]}$$

25. (c) In simple harmonic motion, when a particle is displaced to a position from its mean position, then its kinetic energy gets converted into potential energy.

Hence, total energy of a particle remains constant or the total energy in simple harmonic motion does not depend on displacement x.

26. (b) Initial angular velocity of particle $= \omega_0$

and at any instant t, angular velocity $= \omega$

Therefore, for a displacement x, the resultant acceleration

$$f = (\omega_0^2 - \omega^2)x \quad \ldots(\text{i})$$

External force, $\quad F = m(\omega_0^2 - \omega^2)x \quad \ldots(\text{ii})$

Since, $\quad F \propto \cos \omega t$ [given]

From Eq. (ii), we get

$$m(\omega_0^2 - \omega^2)x \propto \cos \omega t \quad \ldots(\text{iii})$$

Now, from equation of simple harmonic motion,

$$x = A\sin(\omega t + \phi) \quad \ldots(\text{iv})$$

At $t = 0$, $\quad x = A$

$$\therefore \quad A = A\sin(0 + \phi)$$

$$\Rightarrow \quad \phi = \frac{\pi}{2}$$

$$\therefore \quad x = A\sin\left(\omega t + \frac{\pi}{2}\right) = A\cos\omega t \quad \ldots(\text{v})$$

Hence, from Eqs. (iii) and (v), we get

$$m(\omega_0^2 - \omega^2)A\cos\omega t \propto \cos\omega t$$

$$\Rightarrow \quad A \propto \frac{1}{m(\omega_0^2 - \omega^2)}$$

27. (a) For amplitude of oscillation and energy to be maximum, frequency of force must be equal to the initial frequency and this is only possible in case of resonance.

In resonance state, $\omega_1 = \omega_2$.

28. (c) As time period, $T = 2\pi\sqrt{\frac{M}{k}} \quad \ldots(\text{i})$

$$T' = 2\pi\sqrt{\frac{M+m}{k}}$$

$$\Rightarrow \quad \frac{5T}{3} = 2\pi\sqrt{\frac{M+m}{k}} \quad \ldots(\text{ii})$$

Dividing Eq. (i) by Eq. (ii), we have

$$\frac{3}{5} = \sqrt{\frac{M}{M+m}} \quad \text{or} \quad \frac{9}{25} = \frac{M}{M+m}$$

or $9M + 9m = 25M$ or $16M = 9m$

or $$\frac{m}{M} = \frac{16}{9}$$

29. (c) $$v_{max} = a\omega = a\frac{2\pi}{T} = \frac{2\pi a}{2\pi\sqrt{\frac{m}{k}}} = a\sqrt{\frac{k}{m}}$$

Hence, $$\frac{v_{max_1}}{v_{max_2}} = \frac{a_1}{a_2}\sqrt{\frac{k_1}{k_2}}$$

$$\because \quad v_{max_1} = v_{max_2} \quad \text{[given]}$$

$$\Rightarrow \quad \frac{a_1}{a_2} = \sqrt{\frac{k_2}{k_1}}$$

30. (d) $$T = 2\pi\sqrt{\frac{l}{g}}$$

$$T^2 = 4\pi^2\frac{l}{g}$$

$$l = \left(\frac{g}{4\pi^2}\right)T^2$$

$\therefore$ % Change $= x + y + \frac{xy}{100}$

Vaild only for two variables in terms of percentage.

$x \rightarrow$ % change in first variable

$x \rightarrow$ % change in second variable

% Increase in length $= x + x + \frac{x^2}{100}$

$$21 = 2x + \frac{x^2}{100}$$

On solving, $x = 10\%$ [by cross check method]

31. (a) The wire will vibrate with the same frequency as that of source. This can be considered as an example of forced vibration.

$$T = 10 \times 9.8 \text{ N} = 98 \text{ N}, \; m = 9.8 \times 10^{-3} \text{ kg m}^{-1}$$

Frequency of wire, $f = \dfrac{1}{2L}\sqrt{\left(\dfrac{T}{m}\right)}$

$$= \frac{1}{2 \times 1}\sqrt{\left(\frac{98}{9.8 \times 10^{-3}}\right)} = 50 \text{ Hz}$$

32. (a) At $x = 0$, kinetic energy is maximum and potential energy is minimum.

$\because$ Kinetic energy, $K = \dfrac{1}{2} m\omega^2 (a^2 - x^2)$

K_{max} when $x = 0$

$$K_{max} = \frac{1}{2} m\omega^2 a^2$$

and potential energy,

$$U = \frac{1}{2} m\omega^2 x^2$$

at $\qquad x = 0, \; U = 0$

33. (b) As the child stands up, the effective length of pendulum decreases due to the reason that the centre of gravity rises up.

Hence, according to $T = 2\pi\sqrt{\dfrac{l}{g}}$, T will decrease.

34. (c) Kinetic energy of particle of mass m in SHM at any point

$$= \frac{1}{2} m\omega^2 (a^2 - x^2)$$

and potential energy $= \dfrac{1}{2} m\omega^2 x^2$

where, a is amplitude of particle and x is the distance from mean position.

So, at mean position, $x = 0$

$\therefore \qquad KE = \dfrac{1}{2} m\omega^2 a^2 \qquad$ [maximum]

$\qquad PE = 0 \qquad$ [minimum]

35. (b) As we know that spring constant of spring is inversely proportional to length of spring, so new spring constant for each part is given by $k' = nk$, where nk is the spring constant of whole spring.

From the theory of spring pendulum, we know that time period of spring pendulum is inversely proportional to square root of spring constant *i.e.*,

$$T \propto \frac{1}{\sqrt{k}}$$

and $\qquad T' \propto \dfrac{1}{\sqrt{nk}}$

So, $\qquad T' = \dfrac{T}{\sqrt{n}}$

JEE Advanced & IIT JEE Archive

1. (b,d) ***Case I*** For first harmonic oscillator having mass m, angular momentum ω_1 and total energy E_1 as shown in figure.

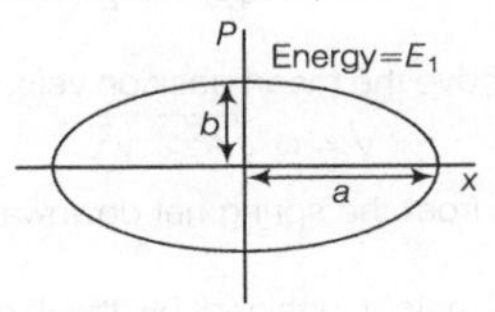

Such that, $E_1 = \dfrac{1}{2} m\omega_1^2 a^2$ and

momentum, $P = mv = m\omega_1 a = b \; [v = \omega_1 a]$

$$\therefore \qquad \frac{a}{b} = \frac{1}{m\omega_1} = n^2 \qquad \ldots(i)$$

Case II For second harmonic oscillator having mass m, angular momentum ω_2 and total energy E_2 as shown in figure.

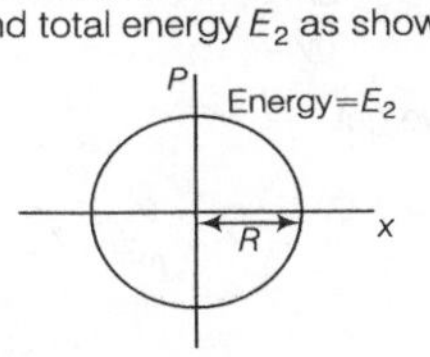

$$E_2 = \frac{1}{2} m\omega_2^2 R^2 = \frac{R^2}{2m} \Rightarrow m\omega_2 = 1 \qquad \ldots(ii)$$

Solving Eqs.(i) and (ii) , we get

$$\frac{\omega_2}{\omega_1} = n^2$$

$$\frac{E_1}{E_2} = \left(\frac{\omega_1}{\omega_2}\right)^2 \times \left(\frac{a}{R}\right)^2 = \frac{1}{n^2} \times \frac{\omega_1}{\omega_2} \times n^2$$

$$\Rightarrow \qquad \frac{E_1}{E_2} = \frac{\omega_1}{\omega_2}$$

$$\Rightarrow \qquad \frac{E_1}{\omega_1} = \frac{E_2}{\omega_2}$$

$\therefore$ Option (b) and (d) are correct

2. (3) Suppose amplitude of individual harmonic wave be A and intensity be I_0.

When they are superimposed such that first and fourth wave produce destructive interference, while 2nd and 3rd form constructive interference only as shown in phaser method.

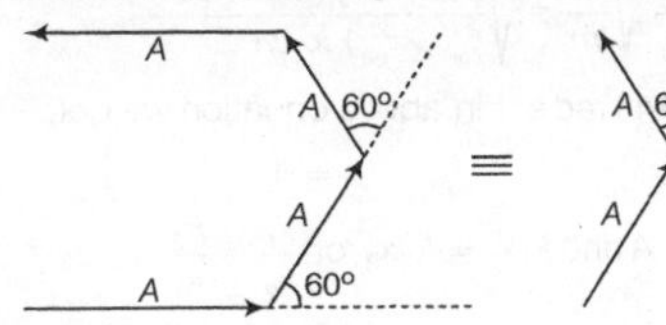

So, resultant amplitude,

$$A_r^2 = A^2 + A^2 + 2.A.A.\cos 60°$$

$$A_r^2 = 3A^2$$

$\therefore$ Intensity of the resulting wave , $I_r = 3I_0$

[$\therefore$ Intensity of a wave, $I = A^2$]

3. (a) t = time of flight of projectile

$$= \frac{2v \sin \theta}{g} \qquad (\theta = 45°)$$

$$\therefore \qquad v = \frac{gt}{2 \sin \theta} = \frac{10 \times 1}{2 \times 1/\sqrt{2}}$$

$$= \sqrt{50} \text{ ms}^{-1}$$

4. (a) Frequency or time period of SHM depends on variable forces. It does not depend on constant external force. Constant external force can only change the mean position. For example, in the given question mean position is at natural length of spring in the absence of electric field. Whereas in the presence of electric field mean position will be obtained after a compression of x_0. Where x_0 is given by

$$Kx_0 = QE$$

or $\qquad x_0 = \dfrac{QE}{K}$

5. (b) If $E > V_0$, particle will escape. But simultaneously for oscillations, $E > 0$.
Hence, the correct answer is $V_0 > E > 0$

6. (b) $[\alpha] = \left[\frac{\text{PE}}{x^4}\right] = \left[\frac{\text{ML}^2\text{T}^{-2}}{\text{L}^4}\right] = [\text{ML}^{-2}\text{T}^{-2}]$

$\therefore \quad \left[\frac{m}{\alpha}\right] = [\text{L}^2\text{T}^2]$

$\therefore \quad \left[\frac{1}{A}\sqrt{\frac{m}{\alpha}}\right] = [\text{T}]$

As dimensions of amplitude A is [L].

7. (d) For $|x| > X_0$, potential energy is constant. Hence, kinetic energy, speed or velocity will also remain constant.
$\therefore$ Acceleration will be zero.

8. (a,d) $\tau_A = \tau_B = \left(mg\,\frac{L}{2}\sin\theta + MgL\sin\theta\right)$

= Restoring torque about point O.

In case A, moment of inertia will be more. Hence, angular acceleration ($\alpha = \tau/I$) will be less. Therefore, angular frequency will be less.

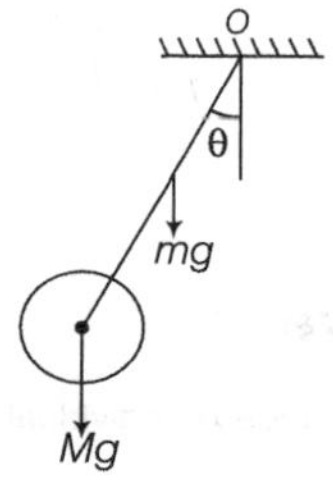

$\therefore$ Correct options are (a) and (d).

9. (4) $\omega = \sqrt{\frac{K}{m}} = \sqrt{\frac{YA}{lm}} = \sqrt{\frac{(n \times 10^9)(4.9 \times 10^{-7})}{1 \times 0.1}}$

Putting, $\omega = 140 \text{ rad s}^{-1}$ in above equation we get,

$$n = 4$$

10. (d) $x_1 + x_2 = A$ and $k_1x_1 = k_2x_2$ or $\frac{x_1}{x_2} = \frac{k_2}{k_1}$

Solving these equations, we get $x_1 = \left(\frac{k_2}{k_1 + k_2}\right)A$

11. (c) $x = \frac{L}{2}\theta$

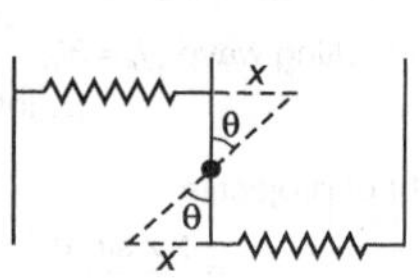

Restoring torque $= -(2kx)\cdot\frac{L}{2}$

$$\alpha = -\frac{kL(L/2\,\theta)}{I} = -\left[\frac{kL^2/2}{ML^2/12}\right]\cdot\theta$$

$$= -\left(\frac{6k}{M}\right)\theta$$

$\therefore \quad f = \frac{1}{2\pi}\sqrt{\left|\frac{\alpha}{\theta}\right|} = \frac{1}{2\pi}\sqrt{\frac{6k}{M}}$

12. (d) $T = 8$ s, $\quad \omega = \frac{2\pi}{T} = \left(\frac{\pi}{4}\right)\text{rads}^{-1}$

$$x = A\sin\omega t$$

$\therefore \quad a = -\omega^2 x = -\left(\frac{\pi^2}{16}\right)\sin\left(\frac{\pi}{4}t\right)$

Substituting $t = \frac{4}{3}$ s, we get

$$a = -\left(\frac{\sqrt{3}}{32}\pi^2\right)\text{cms}^{-2}$$

13. (a) $\quad y = kt^2$

$$\frac{d^2y}{dt^2} = 2k$$

or $\quad a_y = 2 \text{ m/s}^2 \quad$ (as $k = 1 \text{ ms}^{-2}$)

$$T_1 = 2\pi\sqrt{\frac{l}{g}}$$

and $\quad T_2 = 2\pi\sqrt{\frac{l}{g + a_y}}$

$\therefore \quad \frac{T_1^2}{T_2^2} = \frac{g + a_y}{g}$

$$= \frac{10 + 2}{10} = \frac{6}{5}$$

14. At distance y above the mean position velocity of the block.

$$v = \omega\sqrt{A^2 - y^2}$$

After detaching from the spring net downward acceleration of the block will be g.

Therefore, total height attained by the block above the mean position,

$$h = y + \frac{v^2}{2g}$$

$$= y + \frac{\omega^2(A^2 - y^2)}{2g}$$

For h to be maximum $\frac{dh}{dy} = 0$

Putting $\frac{dh}{dy} = 0$, we get

$$y = \frac{g}{\omega^2} = y^*$$

2. Progressive Waves

DPP-1 Basic Parameters and Wave Equation

1. $y = 0.25 \times 10^{-3} \sin(500t - 0.25x)$ is the given equation.

Compare it with standard equation $y = A\sin(\omega t - kx)$

(i) Amplitude is $A = 0.25 \times 10^{-3}$ cm

(ii) Time period is $T = \frac{2\pi}{\omega} = \frac{2\pi}{500}$ s

(iii) Angular frequency is $\omega = 500$ rad s^{-1}

(iv) Angular wave number is $k = 0.25$ m^{-1}

$$\therefore \quad k = \frac{2\pi}{\lambda} = 0.25 \text{ m}^{-1} \Rightarrow \lambda = 8\pi \text{ m}$$

Maximum particle velocity is

$$(v_p)_{max} = A\omega = (0.25 \times 10^{-3})(500)$$

or $(v_p)_{max} = 0.125$ cm s^{-1}

$$(a_p)_{max} = A\omega^2 = 0.25 \times 10^{-3} \times 500 \times 500 = 62.5 \text{ cm s}^{-2}$$

2. (b) $y = 0.5 \sin\left(\frac{\pi}{100}x - 3\pi t\right)$ (given)

Comparing with the equation $y = A\sin(kx - \omega t)$,

we have $k = \frac{\pi}{100}$ and $\omega = 3\pi$ rad s^{-1}

$$\therefore \quad v = \frac{\omega}{k} = \frac{3\pi}{\pi/100} = 300 \text{ ms}^{-1}$$

3. (b) If shape of the string at $t = t_0$ is given by $g(x, t_0)$, then the wave equation for a general x and general time t is given by $g[x - v(t - t_0)]$.

4. (d) Time taken by plate to fall 10 cm freely under gravity,

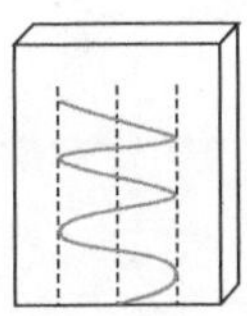

$$t = \sqrt{\frac{2h}{g}} = \sqrt{\frac{2 \times 10}{980}} = \frac{1}{7} \text{ s}$$

In this time 8 oscillations are recorded (given) on the plate. So, the number of oscillations per second i.e. frequency of the tuning fork will be $= 7 \times 8 = 56$ Hz

5. (a) Velocity of pulse $= -\frac{\text{Coefficient of } t}{\text{Coefficient of } x} = -\frac{-4}{1} = +4$ ms^{-1}

6. (c) From the figure,

$$A = 0.06 \text{ m also } \frac{5\lambda}{2} = 0.2 \text{ m}, \lambda = 0.08 \text{ m}$$

The velocity of the left end particle at time $t = 0$ s is downwards i.e. negative,

$\because v_p = -v$ (slope) and slope is positive.

So, equation of wave is of the form $y = A\sin(\omega t - kx + \pi)$

or $\quad y = A\sin(kx - \omega t)$

$$k = \frac{2\pi}{\lambda} = \frac{2\pi}{0.08} = 25\pi$$

$$\omega = kv = 25\pi \times 300 = 7500\pi \text{ rad s}^{-1}$$

$$\therefore \quad y = (0.06 \text{ m})\sin(25\pi x - 7500\pi t)$$

7. (c) At $t = t_0$, $\frac{dy}{dt} = 0$ and $y = A$ (from the first graph)

For $t > t_0$, v is negative and for $t < t_0$, v is positive (from the second graph).

We have to check these conditions in case of each option. Let us do it with the correct option (c).

$$\frac{dy}{dt} = A\frac{2\pi}{T}\cos\left[\frac{2\pi}{T}(t - t_0) + \frac{\pi}{2}\right] = -\frac{2\pi}{T}A\sin\left\{\frac{2\pi}{T}(t - t_0)\right\}$$

$\therefore$ For $t < t_0$, $\frac{dy}{dt}$ is positive and for $t > t_0$, $\frac{dy}{dt}$ is negative.

Alternate Method

The particle is at a distance $x = x_0$ from the origin.

$\therefore$ The displacement was originated at time $t - t_0$ at the origin. And thus the origin has maximum positive displacement at time $t - t_0$.

$\therefore$ Wave equation for particle $x = x_0$ is $y = A\sin\left[\omega(t - t_0) + \frac{\pi}{2}\right]$

or $$y = A\sin\left[\frac{2\pi}{T}(t - t_0) + \frac{\pi}{2}\right]$$

8. (c) $k = \frac{2\pi}{k} = \frac{2\pi}{40} = \frac{\pi}{20}$ rad cm^{-1}, $T = \frac{1}{f} = \frac{1}{8}$ s

$$\omega = 2\pi f = 2\pi(8) = 16\pi \text{ rad s}^{-1}$$

$$v = f\lambda = 8 \times 40 = 320 \text{ cm s}^{-1}$$

At $x = 0$ and $t = 0$, $y = 15$ cm

$$y = A\sin(\omega t - kx + \delta)$$

$$\Rightarrow \quad 15 = 15\sin(0 - 0 + \delta)$$

$$\sin\delta = 1 \Rightarrow \delta = \frac{\pi}{2}$$

Wave equation is $\quad y = 15\sin\left[16\pi t - \frac{\pi}{20}x + \frac{\pi}{2}\right]$

9. (b) $k_{eq} = 2k$

The angular frequency of simple harmonic oscillation is given by

$$\omega = \sqrt{\frac{2k}{m}} = \sqrt{\frac{2 \times 0.5}{0.01}} = 10 \text{ rad s}^{-1}$$

$$\therefore \quad f = \frac{\omega}{2\pi} = \frac{5}{\pi} \text{ Hz}$$

$$A = 0.02 \text{ m}$$

The speed of paper may be assumed as the speed of wave propagation.

$$\therefore \quad \lambda = \frac{v}{f} = \frac{0.1}{5/\pi} = \frac{\pi}{50} \text{ m}$$

10. (c) The curve traced on the paper can be represented by the equation

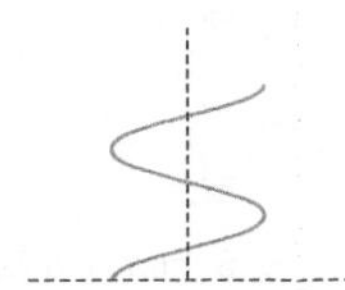

$$y = A\cos kx$$

$$k = \frac{\omega}{v} = \frac{10}{0.1} = 100 \text{ m}^{-1}$$

$$y = 0.02\cos(100x)$$

11. (d) Distance moved by crest (P) in 0.3 s is

$$s = 4 \times 0.3 = 1.2 \text{ cm}$$

i.e. P was at $x = 0$, when $t = 0$, (see figure)

$$y = A\sin(\omega t - kx + \delta)$$

At $t = 0$ and $x = 0$,

$$2 = \sin(0 - 0 + \delta)$$

$$\Rightarrow \quad \sin\delta = 1 \text{ or } \delta = \frac{\pi}{2}$$

$$\therefore \quad y = A\sin\left(\omega t - kx + \frac{\pi}{2}\right) = A\cos(\omega t - kx)$$

$$A = 2 \text{ cm}, \lambda = 7.5 \text{ cm}$$

$$k = \frac{2\pi}{\lambda} = \frac{4\pi}{15} \text{ cm}^{-1}$$

$$\omega = vk = (4)\left(\frac{4\pi}{15}\right) = \frac{16\pi}{15} \text{ rad s}^{-1}$$

$$\therefore \quad y = 2\cos\left(\frac{16\pi}{15}t - \frac{4\pi}{15}x\right)$$

12. (d) From figure, $\lambda = 12\,\text{cm}$, $A = 10$ cm

$\therefore$ $$k = \frac{2\pi}{\lambda} = \frac{\pi}{6}\,\text{cm}^{-1}$$

At $t = 0$, $$y = A\sin(kx + \delta)$$

Let us assume v is positive.

At $x = 0$ (and at $t = 0$),

$$-5 = 10\sin\delta \quad \Rightarrow \quad \sin\delta = -\frac{1}{2}$$

For left end at $t = 0$, $v_p = -v\left(\frac{dy}{dx}\right) = -(+\text{ve})(+\text{ve}) = -\text{ve}$

$\therefore$ $$\delta = \pi + \frac{\pi}{6} = \frac{7\pi}{6}$$

Hence, equation of curve at $t = 0$ is $y = 10\sin\left(\frac{\pi}{6}x + \frac{7\pi}{6}\right)$

Slope is $$\frac{dy}{dx} = (10)\left(\frac{\pi}{6}\right)\cos\left(\frac{\pi}{6}x + \frac{7\pi}{6}\right)$$

$$= \frac{10\pi}{6}\cos\left(\frac{\pi}{6}x + \frac{7\pi}{6}\right)$$

At point P, $x = +6$ cm

$\therefore$ $$\text{Slope} = \tan\theta = \frac{10\pi}{6}\cos\left(\pi + \frac{7\pi}{6}\right)$$

$$= -\frac{10\pi}{6}\cos\left(\frac{\pi}{6}\right)$$

$$\tan\theta = \frac{-10\pi}{6}\left(\frac{\sqrt{3}}{2}\right) = -\frac{5\pi}{2\sqrt{3}}$$

13. (c) The equation should satisfy $\frac{\partial^2 y}{\partial t^2} = v^2\frac{\partial^2 y}{\partial x^2}$

This will be satisfied by II and IV Besides satisfying the equation y should be finite everywhere (i.e. for all values of x) all the time (i.e. for all values of t).

14. (c) Let the equation of the curve is $y = A\sin(\omega t - kx)$

$\therefore$ $$6 = 8\sin(\omega t - 10k)$$

$$4 = 8\sin(\omega t - 25k) \quad \text{(given)}$$

$\therefore$ $$\sin(\omega t - 10k) = \frac{3}{4}$$

$\Rightarrow$ $$\omega t - 10k = 0.8481 \quad \text{...(i)}$$

$$\sin(\omega t - 25k) = \frac{1}{2}$$

$\Rightarrow$ $$\omega t - 25k = \frac{\pi}{6}\,\text{rad} = 0.5236\,\text{rad} \quad \text{...(ii)}$$

Subtracting Eq. (ii) from Eq. (i) we get,

$$k = 0.0216\,\text{cm}^{-1}$$

$\therefore$ $$\lambda = \frac{2\pi}{k} = 290\,\text{cm}$$

15. (a) If $y = f(ax \pm bt)$ represents a wave, then wave speed is $\frac{b}{a}$.

$\therefore$ Here, speed of pulse $= \frac{c}{b}$

16. (c) From the two equations given, the equation of y in terms of t can be written as, $y = \frac{1}{1 + (x - 0.5t)^2}$

$\therefore$ Speed of wave $= \frac{b}{a} = \frac{0.5}{1} = 0.5\,\text{ms}^{-1}$

17. (b,c,d) Velocity of pulse $= -\left(\frac{5}{4}\right)$, so pulse is travelling in negative X-direction.

Distance travelled in 2 s = speed × time $= \frac{5}{4} \times 2 = 2.5$ m

$y(x, t)$ is maximum is when $x = 0$ and $t = 0$

$\therefore$ $$y_{max} = \text{Amplitude} = \frac{0.8}{0 + 5} = 0.16\,\text{m}$$

Pulse will be symmetric (symmetry is checked about y_{max}), if at $t = 0$, $y(x) = y(-x)$.

From the given equation, at $t = 0$,

$$y(x) = \frac{0.8}{16x^2 + 5}$$

and $$y(-x) = \frac{0.8}{16x^2 + 5}$$

So, pulse is symmetric.

18. (a,b,c,d) (a) and (b) are true in case of electromagnetic waves, (c) in transverse waves in a stretched string and (c) and (d) in case of sound waves.

19. (a,b,c,d) $y = 0.05\sin\left(5\pi x - 20\pi t - \frac{\pi}{4}\right)$ (given)

Comparing with $y = A\sin(kx - \omega t + \delta)$, we have

$$k = 5\pi$$

$\Rightarrow$ $$\frac{2\pi}{\lambda} = 5\pi$$

or $$\lambda = \frac{2}{5} = 0.4\,\text{m}$$

Velocity of wave is

$$v = -\frac{\text{Coefficient of } t}{\text{Coefficient of } x}$$

$$= -\frac{-20\pi}{5\pi} = +4\,\text{ms}^{-1}$$

Particle velocity is

$$v_p = \frac{\partial y}{\partial t} = 0.05(-20\pi)\cos\left(5\pi x - 20\pi t - \frac{\pi}{4}\right)$$

$$(v_p)_{\text{at } t = 0.05\,\text{s and } x = 0.5\,\text{m}} = -\pi\cos\left(2.5\pi - \pi - \frac{\pi}{4}\right)$$

$$= -\pi\cos\left(\frac{5\pi}{4}\right)$$

$$= -\pi\cos\left(\pi + \frac{\pi}{4}\right)$$

$$= \pi\left(\frac{1}{\sqrt{2}}\right) = 2.22\,\text{ms}^{-1}$$

Acceleration of particle is

$$a_p = \frac{\partial^2 y}{\partial t^2} = (-\pi)(-1)(-20\pi)\sin\left(5\pi x - 20\pi t - \frac{\pi}{4}\right)$$

$$= -20\pi^2\sin\left(5\pi x - 20\pi t - \frac{\pi}{4}\right)$$

$$(a_p)_{\text{at } t = 0.05\,\text{s and } x = 0.5\,\text{m}} = -20\pi^2\sin\left(2.5\pi - \pi - \frac{\pi}{4}\right)$$

$$= -20\pi^2\sin\left(\pi + \frac{\pi}{4}\right)$$

$$= (-20\pi^2)\left(-\frac{1}{\sqrt{2}}\right)$$

$$= 140\,\text{ms}^{-2}$$

20. (a,c) Wave velocity is

$$v = -\frac{\text{Coefficient of } t}{\text{Coefficient of } x}$$

$$= -\frac{1/0.01}{1/0.05} = -5\,\text{ms}^{-1}$$

Particle velocity is

$$v_p = \frac{\partial y}{\partial t} = (0.02)\left(\frac{1}{0.01}\right)\cos\left[\frac{0.2}{0.05} + \frac{0.3}{0.01}\right]$$

$$= 2\cos 34^\circ$$

$$= 2(-0.85)$$

$$= -1.7\,\text{ms}^{-1}$$

21. (a) Given, $v = 40\ cms^{-1}$

From figure, $\lambda = 4$ cm, $A = 2$ cm

(i) $v = f\lambda \Rightarrow f = \frac{v}{\lambda} = \frac{40}{4} = 10$ Hz

(ii) $\Delta\phi = \frac{2\pi}{\lambda} \cdot \Delta x = \frac{2\pi}{4}(2.5) = 1.25\pi = 3.925$ rad

(iii) $T = \frac{1}{f} = 0.1$ s

$$\Delta t = \frac{T}{2\pi} \cdot \frac{\pi}{3} = \frac{T}{6} = \frac{0.1}{6} = 0.0167\text{s}$$

(iv) In the figure, let $x = 0$ has velocity v' (at the instant shown)

$v' = -v$ (slope) $= -(+40)$ (+ ve)

$\Rightarrow$ v' is negative.

i.e. left end is moving down. Consider the instant shown is at $t = 0$.

$$y = A\sin(\omega t - kx + \delta)$$

If left end is at mean position at moving up, then $\delta = 0$.

In the situation shown,

$$\delta = \frac{\pi}{2} + \frac{\pi}{2} + \frac{\pi}{6} = \frac{7\pi}{6}$$

$$\therefore \quad y = A\sin\left(\omega t - kx + \frac{7\pi}{6}\right)$$

$$= -A\sin\left(\omega t - kx + \frac{\pi}{6}\right)$$

$$k = \frac{2\pi}{\lambda} = \frac{2\pi}{4} = \frac{\pi}{2}\ \text{cm}^{-1}$$

and $$\omega = kv = \frac{\pi}{2} \times 40 = 20\ \pi\text{s}^{-1}$$

$$\therefore \quad y = -2\sin\left(20\ \pi t - \frac{\pi}{2}x + \frac{\pi}{6}\right)$$

$$v_P = \frac{\partial y}{\partial t} = -2(20\pi)\cos\left(20\pi t - \frac{\pi}{2}x + \frac{\pi}{6}\right)$$

At the instant shown, $t = 0$

$$\therefore \quad v_P = -40\pi\cos\left(-\frac{\pi}{2}x + \frac{\pi}{6}\right)$$

At P, $x = 4.5$ cm

$$\therefore \quad v_P = -40\pi\cos\left(-2.25\pi + \frac{\pi}{6}\right)$$

or $$v_P = -121.4\ \text{cm s}^{-1} = -1.214\ \text{ms}^{-1}$$

Hence, (i) $\rightarrow$ (r), (ii) $\rightarrow$ (p), (iii) $\rightarrow$ (q), (iv) $\rightarrow$ (s)

22. (c) Given, $v = 12\text{ms}^{-1}$, $A = 0.05$ m, $\lambda = 0.4$ m and $\delta = 0$

$$\therefore \quad y = A\sin(\omega t - kx) = 0.05\sin(60\pi t - 5\pi x)$$

$$\left(\because k = \frac{2\pi}{\lambda} = \frac{2\pi}{0.4} = 5\pi \text{ and } \omega = kv = 5\pi \times 12 = 60\pi\right)$$

For particle P at $t = 0.15$ s,

$$y = 0.05\sin(60\pi \times 0.15 - 5\pi \times 0.25)$$

$$= -0.0353\ \text{m}$$

$$= -3.53\ \text{cm}$$

For particle P, $\quad y_P = 0.05\sin[60\pi t - 1.25\pi]$

$$v_P = (0.05)(60\pi)\cos[60\pi t - 1.25\pi]$$

$$\therefore \quad (v_P)_{t = 0.15\text{s}} = +6.661\ \text{ms}^{-1}$$

Consider SHM of particle P only. Then, time taken to have a displacement 0.0353 m from mean position is

$$0.0353 = 0.05\sin(60\pi t')$$

$$\Rightarrow \quad 60\pi t' = \frac{\pi}{4}$$

$$t' = \frac{1}{240}\text{s} = 4.167\ \text{ms}$$

Hence, (i) $\rightarrow$ (r), (ii) $\rightarrow$ (t), (iii) $\rightarrow$ (u)

23. (a) Velocity of a particle is $v_P = -v\left(\frac{dy}{dx}\right)$

P is moving up (given). So, v_P is positive. At particle P

Slope i.e. $\frac{dy}{dx}$ is positive. So, v must be negative.

$$20\sqrt{3} = -v(\tan 60^\circ) \Rightarrow v = -20\ \text{cm s}^{-1}$$

24. (a) Let, the equation of wave is $y = A\sin(\omega t + kx + \delta)$

From the figure, $\quad \lambda = 4 \times 10^{-2}\text{m} = 4\text{cm}$

$$\therefore \quad k = \frac{2\pi}{\lambda} = \frac{2\pi}{4} = \frac{\pi}{2}\ \text{cm}^{-1}$$

$$A = 4 \times 10^{-3}\text{m (from figure)}$$

From the figure, at $x = 0$ and $t = 0$, $y = 2\sqrt{2} \times 10^{-3}$m

$$\therefore \quad 2\sqrt{2} \times 10^{-3} = 4 \times 10^{-3}\sin(0 + 0 + \delta)$$

$$\Rightarrow \quad \sin\delta = \frac{1}{\sqrt{2}}$$

$$\therefore \quad \delta = \frac{\pi}{4} \text{ or } \frac{3\pi}{4}$$

For $x = 0$, $v_P = -v\left(\frac{dy}{dx}\right) = (-\text{ve})(+\text{ve}) = +\text{ve}$

So, left end at $t = 0$ is moving up.

$$\therefore \quad \delta = \frac{\pi}{4}$$

$$\therefore \quad \omega = kv = \frac{\pi}{2} \times 20 = 10\pi\ \text{rad s}^{-1}$$

$\therefore$ $y = A\sin(\omega t + kx + \delta)$ gives

$$y = 0.4\sin\left[10\pi t + \frac{\pi}{2}x + \frac{\pi}{4}\right]$$

Note *For finding at y (in cm), we should put x in this equation sink is in* cm^{-1}.

Explanations for Q. Nos. (25-28)

The wave is travelling along positive X-axis (from the figure).

So, let $\quad y = A\sin(\omega t - kx + \delta)$

From the figure, $\quad A = 10$ mm and $\lambda = 2$m

$$\therefore \quad k = \frac{2\pi}{\lambda} = \frac{2\pi}{2} = \pi\text{m}^{-1}$$

At $t = 0$, $x = 0$, $y = 5 \times 10^{-3}$m (from figure)

$$\therefore \quad 5 \times 10^{-3} = 10 \times 10^{-3}\sin(0 + 0 + \delta)$$

$$\Rightarrow \quad \sin\delta = \frac{1}{2}$$

For left end at $x = 0$, $v_P = -v\left(\frac{dy}{dx}\right) = -(+\text{ve})(+\text{ve}) = -\text{ve}$

$$\therefore \quad \delta = \frac{\pi}{2} + \frac{\pi}{3} = \frac{5\pi}{6}$$

Also, at $t = 0.05$ s, $x = 0$, $y = 0$ (from figure)

$$\therefore \quad 0 = 10 \times 10^{-3}\sin\left(\omega \times 0.05 - 0 + \frac{5\pi}{6}\right)$$

$$\therefore \quad \frac{5\omega}{100} + \frac{5\pi}{6} = \pi \Rightarrow \omega = \frac{\pi}{6} \times 20 = \frac{10\pi}{3}$$

$$f = \frac{\omega}{2\pi} = \frac{5}{3}\ \text{Hz}$$

$$(v_P)_{max} = A\omega = 10 \times 10^{-3}\left(\frac{10\pi}{3}\right) = \frac{\pi}{30}\ \text{ms}^{-1}$$

$\therefore$ The equation of the wave is

$$y = 10 \times 10^{-3}\sin\left(\frac{10\pi}{3}t - \pi x + \frac{5\pi}{6}\right)$$

Speed of wave is $v = \lambda f = 2 \times \frac{5}{3} = \frac{10}{3}\ \text{ms}^{-1}$

25. (d) **26.** (c) **27.** (b)

28. (a)

DPP-2 **Speed of Waves**

1. (c) In the figure,

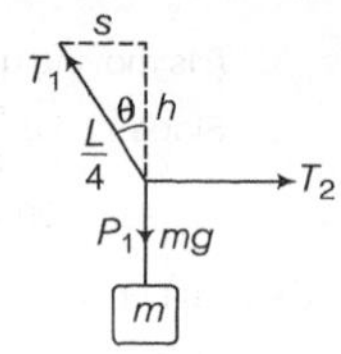

$$s = \frac{0.5}{2} = 0.25 \text{ m}$$

$$\frac{L}{4} = \frac{3}{4} = 0.75 \text{ m}$$

$$h = \sqrt{(0.75)^2 - (0.25)^2} = 0.707 \text{ m}$$

From Lami's theorem,

$$\frac{mg}{\sin(90^\circ + \theta)} = \frac{T_2}{\sin(180^\circ - \theta)}$$

$$T_2 = mg \tan\theta = (2)(9.8)\left(\frac{0.25}{0.707}\right) = 6.93 \text{ N}$$

Mass per unit length of the string is

$$\mu = \frac{10 \times 10^{-3}}{3} = 3.33 \times 10^{-3} \text{ kg m}^{-1}$$

∴ Speed of pulse,

$$v = \sqrt{\frac{T_2}{\mu}} = \sqrt{\frac{6.93}{3.33 \times 10^{-3}}} = 45.62 \text{ ms}^{-1}$$

∴ Time taken $= \frac{P_1P_2}{v} = \frac{L/2}{v} = \frac{1.5}{45.62} = 0.0329 \text{ s} = 32.9 \text{ ms}$

2. (d) Tension at the lower end is $T_1 = 2g$

Tension at the top end is $T_2 = (6 + 2) = 8\,g$

Frequency remains unaltered.

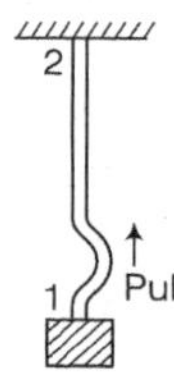

$$v_1 = f\lambda_1 \quad \text{and} \quad v_2 = f\lambda_2$$

∴ $\frac{v_1}{v_2} = \frac{\lambda_1}{\lambda_2}$ Also, $v \propto \sqrt{T}$

∴ $\sqrt{\frac{T_1}{T_2}} = \frac{\lambda_1}{\lambda_2}$

⇒ $\lambda_2 = \lambda_1\sqrt{\frac{T_2}{T_1}}$

$$\lambda_2 = (0.06)\sqrt{\frac{8g}{2g}} = 0.12 \text{ m}$$

3. (a) At a distance x from the bottom tension is $\left(\frac{m_0}{L}xg + mg\right)$

At this position speed of transverse wave

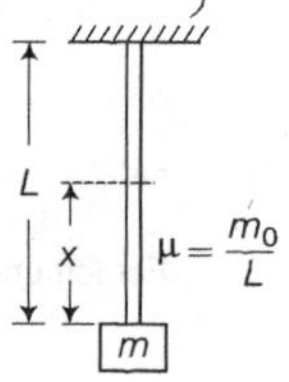

$$= \sqrt{\frac{\left(\frac{m_0}{L}x + m\right)g}{\left(\frac{m_0}{L}\right)}}$$

or $\frac{dx}{dt} = \sqrt{\frac{Lg}{m_0}\left(\frac{m_0}{L}x + m\right)}$

or $\int_0^L \left(\frac{m_0}{L}x + m\right)^{-\frac{1}{2}} dx = \sqrt{\frac{Lg}{m_0}}\int_0^t dt$

$$\frac{\left[\left(\frac{m_0}{L}x + m\right)^{\frac{1}{2}}\right]_0^L}{\frac{1}{2}\cdot\frac{m_0}{L}} = \sqrt{\frac{Lg}{m_0}}\,t$$

$$\frac{2L}{m_0}\left[\left(\frac{m_0}{L}\cdot L + m\right)^{\frac{1}{2}} - (0 + m)^{\frac{1}{2}}\right] = \sqrt{\frac{Lg}{m_0}}\,t$$

$$\sqrt{m_0 + m} - \sqrt{m} = \frac{1}{2}\sqrt{\frac{m_0 g}{L}}\,t$$

$$t = 2\sqrt{\frac{L}{m_0 g}}\left(\sqrt{m_0 + m} - \sqrt{m}\right)$$

Note *if* $m << m_0$, $t = 2\sqrt{\frac{L}{g}}$

4. (a) Tension in the rubber tube P_1P_2 is $T = mg = (5)(9.8) = 49$ N

Linear mass density of the tube is

$$\mu = \frac{0.9}{12} = 0.075 \text{ kg m}^{-1}$$

Speed of wave on tube is

$$v = \sqrt{\frac{T}{\mu}} = \sqrt{\frac{49}{0.075}} = 25.56 \text{ ms}^{-1}$$

So, the required time is $t = \frac{P_1P_2}{v} = \frac{12}{25.56} = 0.47 \text{ s}$

5. (c) $v = \sqrt{\frac{T}{\rho_s}}$, where s is cross-section area.

or $v \propto \frac{1}{\sqrt{\rho}}$

$$\frac{v_1}{v_2} = \sqrt{\frac{\rho_2}{\rho_1}}$$

⇒ $\frac{\rho_2}{\rho_1} = \left(\frac{v_1}{v_2}\right)^2$

or $\frac{\rho_1}{\rho_2} = \left(\frac{v_2}{v_1}\right)^2 = \left(\frac{1}{2}\right)^2 = \frac{1}{4}$

6. (a) $v_P = \sqrt{\frac{4.8}{1.2 \times 10^{-2}}} = 20 \text{ ms}^{-1}$

$$v_Q = \sqrt{\frac{7.5}{1.2 \times 10^{-2}}} = 25 \text{ ms}^{-1}$$

Pulses will meet, when $x_P = x_Q$

or $20t = 25(t - 0.02)$

⇒ $t = 0.1 \text{ s}$

$$x_p \text{ or } x_Q = x = 20 \times 0.1 = 2 \text{ m}$$

7. (b) $v = \sqrt{\frac{T}{\rho s}} = \sqrt{\frac{2 \times 9.8}{8920 \times \pi(1.2 \times 10^{-3})^2}} = 22 \text{ ms}^{-1}$

8. (b) Mass of the differential element is $dm = \mu_0 y dy$

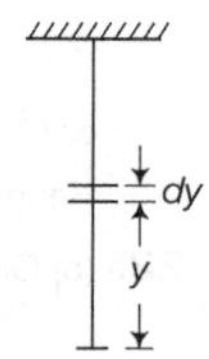

Mass of the string up to the length y is

$$m = \int dm = \int_0^y \mu_0 y\, dy$$

or $m = \frac{\mu_0 y^2}{2}$

Tension at the position of element is

$$T = mg = \frac{\mu_0 y^2}{2} g$$

Speed of wave pulse at this location is $v = \sqrt{\frac{T}{\mu}}$

or $v = \sqrt{\frac{\mu_0 y^2 g}{2\mu_0 y}} = \sqrt{y}\sqrt{\frac{g}{2}}$

$$\frac{dy}{dt} = \sqrt{y}\sqrt{\frac{g}{2}}$$

⇒ $y^{-\frac{1}{2}} dy = \sqrt{\frac{g}{2}}\,dt$

$$\int_0^{L_0} y^{-\frac{1}{2}} dy = \sqrt{\frac{g}{2}}\int_0^t dt$$

⇒ $2\sqrt{L_0} = \sqrt{\frac{g}{2}}\,t$

or $t = \sqrt{\frac{8L_0}{g}}$

9. (d) $g_p = \dfrac{G \cdot 3M}{(3R)^2} = \dfrac{g}{3}$

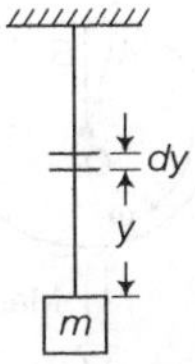

$\mu = \dfrac{m_0}{L_0}$. At the location of differential element tension is

$$F = \left[m + \frac{m_0}{L_0} y\right] g_p$$

and $$v = \sqrt{\frac{F}{\mu}} = \sqrt{\frac{\left(m + \frac{m_0}{L_0} y\right) g}{3\left(\frac{m_0}{L_0}\right)}}$$

$\therefore$ $$\frac{dy}{dt} = \sqrt{\frac{\left(m + \frac{m_0}{L_0} y\right) g}{3\left(\frac{m_0}{L_0}\right)}}$$

$\Rightarrow$ $$\int_0^{L_0} \frac{dy}{\sqrt{mL_0 + m_0 y}} = \sqrt{\frac{g}{3m_0}} \int_0^t dt$$

$\therefore$ $$t = 2\left[\frac{\sqrt{m + m_0} - \sqrt{m}}{\sqrt{m_0}}\right]\sqrt{\frac{3L_0}{g}}$$

10. (a) $v_{min} = \sqrt{\dfrac{mL_0 g}{3m_0}}$ (at the bottom)

$v_{max} = \sqrt{\dfrac{(m + m_0)L_0 g}{3m_0}}$ (at the top)

$$\frac{v_{max}}{v_{min}} = \sqrt{\frac{m + m_0}{m}} = \sqrt{1 + \frac{m_0}{m}}$$

11. (b) Speed of a transverse wave on a wire is

$$v = \sqrt{\frac{F}{\mu}} \quad \text{...(i)}$$

Differentiating w.r.t. tension, we have

$$\frac{dv}{dF} = \frac{1}{2\sqrt{\mu F}} \quad \text{...(ii)}$$

Dividing Eq. (ii) by Eq. (i), we have

$$\frac{dv}{v} = \frac{1}{2}\frac{dF}{F}$$

$\Rightarrow$ $$dF = 2F\frac{dv}{v}$$

or $$dF = 2(500)\frac{(302 - 300)}{300}$$

$$= 6.67 \text{ N}$$

$\therefore$ Tension should be increased by 6.67 N.

12. (a,b) $F_{AB} = 6.4\text{ g} = 64\text{ N}$ (tension in string AB)

$\therefore$ $$v_1 = \sqrt{\frac{F_{AB}}{\mu_{AB}}} = \sqrt{\frac{64}{10 \times 10^{-3}}} = 80 \text{ ms}^{-1}$$

$$F_{CD} = 3.2\text{ g} = 32\text{ N}$$

$$v_2 = \sqrt{\frac{F_{CD}}{\mu_{CD}}} = \sqrt{\frac{32}{8 \times 10^{-3}}} = 63.24 \text{ ms}^{-1}$$

13. (a,b,c) $T = 4t = 20$ ms

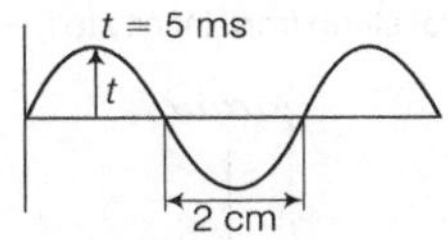

$$= 20 \times 10^{-3}\text{s}$$

$$= 0.02 \text{ s}$$

$$f = \frac{1}{T} = \frac{1}{0.02} = 50 \text{ Hz}$$

$$\lambda = 2 \times 2 = 4\text{cm}$$

$$v = \lambda f = 4 \times 10^{-2} \times 50 = 2 \text{ ms}^{-1}$$

14. (a) Consider a differential element dy of the rope at a distance y from the top. The element is in equilibrium.

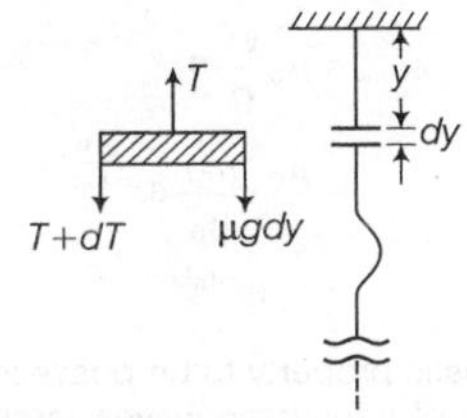

$\therefore$ $$\Sigma F_y = 0$$

or $$[(T + dT) + \mu g\, dy] - T = 0$$

or $$dT = -\mu g\, dy \quad \text{...(i)}$$

Wave speed on a string is

$$v = \sqrt{\frac{T}{\mu}} \Rightarrow T = \mu v^2$$

Differentiating this equation, we get

$$dT = (d\mu)v^2 + 2v \cdot dv \cdot \mu$$

As, v is independent of y, $dv = 0$

$\therefore$ $$dT = v^2 d\mu \quad \text{...(ii)}$$

From Eqs. (i) and (ii), $v^2 d\mu = -\mu g\, dy$

$$\frac{1}{\mu} d\mu = -\frac{g}{v^2} dy$$

$$\int_{\mu_0}^{\mu} \frac{1}{\mu} d\mu = -\frac{g}{v^2}\int_0^y dy \Rightarrow \ln\frac{\mu}{\mu_0} = -\frac{g}{v^2} \cdot y$$

$$\frac{\mu}{\mu_0} = e^{\left(-\frac{g}{v^2} \cdot y\right)}$$

$\Rightarrow$ $$\mu = \mu_0 e^{\left(\frac{-g}{v^2} \cdot y\right)} \quad \text{...(iii)}$$

15. (b) Mass of differential element is $dm = \mu\, dy$

Hence, total mass of the string is $m = \int_0^\infty \mu_0 e^{-\frac{g}{v^2} \cdot y} dy$

or $$m = \mu_0\left(-\frac{v^2}{g}\right)\left[e^{-\frac{g}{v^2} \cdot y}\right]_0^\infty \quad \text{or} \quad m = \frac{\mu_0 v^2}{g}$$

$\Rightarrow$ $$v = \sqrt{\frac{mg}{\mu_0}} \quad \text{...(iv)}$$

16. (c) From Eqs. (iii) and (iv),

$$\mu = \mu_0 e^{-\frac{g\mu_0}{mg} \cdot y} = \mu_0 e^{-\frac{\mu_0 y}{m}}$$

17. (d) If we cut the rope at length L, the tension at the free end must equal the weight of string from infinity to free end. Hence, required mass is

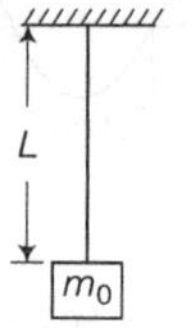

$$m_0 = \int_L^{\infty} \mu dy = \int_L^{\infty} \mu_0 e^{-\frac{g}{v^2}y} dy$$

$$m_0 = \mu_0 \left(-\frac{v^2}{g}\right)\left[e^{-\frac{g}{v^2}\cdot y}\right]_L^{\infty}$$

$$= \mu_0 \frac{v^2}{g} e^{-\frac{g}{v^2}\cdot L}$$

$$= \frac{\mu_0}{g} \cdot \frac{mg}{\mu_0} e^{-\frac{g \cdot \mu_0}{mg}\cdot L} \qquad \left(\because v = \sqrt{\frac{mg}{\mu_0}}\right)$$

$$= m e^{-\mu_0 L/m}$$

18. (a) Required elastic property to be possessed by a medium, for the propagation of transverse waves through it, is modulus of rigidity which is not had by gas media.

19. (a) The free surface of water has the property called surface tension and this plays the role of modulus of rigidity.

20. (d) $v = f\lambda$. For given conditions of a medium, speed of wave remains unaltered. So, if f doubles, λ gets halved, to keep v same.

21. (a) From the figure, it is clear that $\mu_1 < \mu_2$.

We know $v = \sqrt{\frac{T}{\mu}}$. T is same (given).

$\therefore v_1 > v_2$

22. (d) v_P is negative (given). At point P, slope of the curve is negative.

$\therefore v$ i.e. wave velocity should be negative, i.e. wave must be moving along negative X-axis.

23. (7) Thermal stress $= \frac{F}{A} = y\alpha\,\Delta\theta$, where F is tension and A is area of cross-section.

We know, $v = \sqrt{\frac{F}{\rho A}} = \sqrt{\frac{\text{Thermal stress}}{\text{Density}}} = \sqrt{\frac{\gamma\,\alpha\,\Delta\theta}{\rho}}$

$$= \sqrt{\frac{1.3 \times 10^{11} \times 1.7 \times 10^{-5} \times (30 - 10)}{9000}} = 70 \text{ ms}^{-1}$$

$\therefore \qquad D = 7$

24. (5) The string has same shape in figures first and last.

The distance travelled by the pulse (between these two situations)

$= 2 \times 0.3 = 0.6$ m

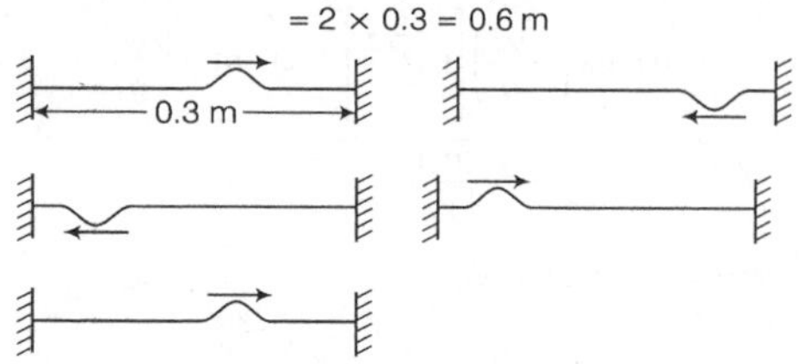

$$v = \sqrt{\frac{T}{\mu}} = \sqrt{\frac{1.2}{10^{-2}/0.3}} = 6 \text{ ms}^{-1}$$

$$\therefore \qquad \Delta t = \frac{0.6}{6} = 0.1 = \frac{1}{2 \times 5} \text{ s}$$

$$\therefore \qquad k = 5$$

25. (i) $dm = \frac{m}{2\pi R} \cdot R\,2\theta = \frac{m}{\pi} \cdot \theta$

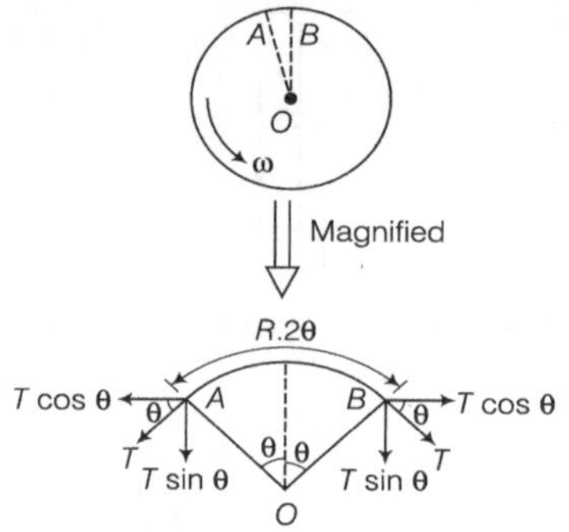

$$2T\sin\theta = dm \cdot \omega^2 R$$

$$\Rightarrow \qquad 2T \cdot \theta = \frac{m}{\pi} \cdot \theta\omega^2 R \qquad (\because \theta \text{ is small})$$

$$\therefore \qquad T = \frac{m\omega^2 R}{2\pi}$$

Linear mass density is

$$\mu = \frac{m}{2\pi R}$$

$$\therefore \qquad v = \sqrt{\frac{T}{\mu}} = \sqrt{\frac{m\omega^2 R}{2\pi} \cdot \frac{2\pi R}{m}} = R\omega$$

Hence proved.

(ii) The kink will be stationary with respect to the ground, if it moves clockwise with respect to the rope.

26. (i) $v = \sqrt{\frac{T}{\rho s}} = \sqrt{\frac{64}{12.5 \times 10^3 \times 0.8 \times 10^{-6}}} = 80 \text{ ms}^{-1}$

(ii) $\omega = 2\pi f = 2\pi(20) = 40\pi$ rad s^{-1}

$$k = \frac{\omega}{v} = \frac{40\pi}{80} = \frac{\pi}{2} \text{ m}^{-1}$$

From the data given $\delta = \pi/2$ rad

$$y = A\sin(\omega t - kx + \delta)$$

$$\therefore \qquad y = (1 \text{ cm}) \cos\left[40\,\pi t - \frac{\pi}{2}x\right]$$

(iii) If we substitute $x = 0.5$ m, $t = 0.05$ s in the above equation, we get $y = \frac{1}{\sqrt{2}}$ cm.

(iv) $v_P = \frac{\partial y}{\partial t} = -(40\pi \text{ ms}^{-1}) \sin\left[40\pi t - \frac{\pi}{2}x\right]$

Substituting $x = 0.5$ m and $t = 0.05$ s, we get

$$v_P = 89 \text{ cm s}^{-1}$$

27. Speed of a wave depends on some special characteristics of the medium. In case of a string, it depends on tension F and linear mass density μ. Wavelength is self adjusted $\left(= \frac{v}{f}\right)$. Let us take an example.

Suppose tension in a string is 100 N. Linear mass density μ of this wire is 1 kgm^{-1}. Then, in such a string speed of the transverse wave is $v = \sqrt{\frac{F}{\mu}} = \sqrt{\frac{100}{1}} = 10 \text{ ms}^{-1}$.

This value of 10 ms^{-1} is now fixed under the given conditions ($F = 100$ N, $\mu = 1$ kgm^{-1}). Now, suppose we keep the frequency of a wave 10 Hz, then wavelength of such a wave will become 1 m $\left(= \frac{v}{f}\right)$. Now, if the frequency is reduced to half i.e. 5 Hz, the wavelength will become two times or 2 m, so that product of λ and f remains constant ($= 10$ ms^{-1}).

DPP-3 Energy, Intensity and Power of Waves

1. The potential energy of a segment AB is the work done by the tension T in stretching the string, which is $T(\Delta l - \Delta x)$. Here, Δl is the length of the stretched string and Δx is its original length.

From the figure, $\Delta l = \sqrt{(\Delta x)^2 + (\Delta y)^2}$

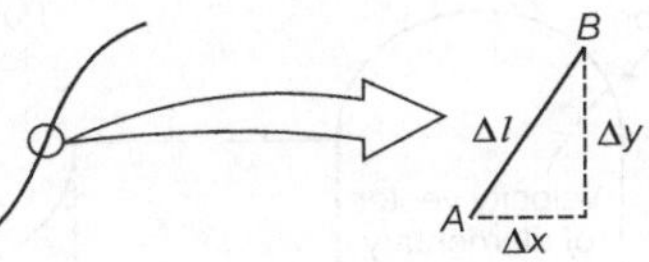

or $$\Delta l = \Delta x\left[1 + \left(\frac{\Delta y}{\Delta x}\right)^2\right]^{1/2} \simeq \Delta x\left[1 + \frac{1}{2}\left(\frac{\Delta y}{\Delta x}\right)^2\right]$$

$\therefore$ $$\Delta l - \Delta x = \frac{1}{2}\left(\frac{\Delta y}{\Delta x}\right)^2 \Delta x$$

$\therefore$ Potential energy is
$$U = T(\Delta l - \Delta x) = T \cdot \frac{1}{2}\left(\frac{\Delta y}{\Delta x}\right)^2 \cdot \Delta x$$

or $$U = \frac{1}{2}T\left(\frac{dy}{dx}\right)^2 dx$$

We know $T = \mu v^2$ and $\mu = \rho s$ (where, s = cross-sectional area of string)

$\therefore$ $$T = \rho s\, v^2$$

$\therefore$ $$U = \frac{1}{2}\rho s v^2\left(\frac{dy}{dx}\right)^2 dx$$

Volume of the element = $s dx$

$\therefore$ u_P = Potential energy unit volume
$$= \frac{U}{\text{volume}} = \frac{1}{2}\rho v^2\left(\frac{dy}{dx}\right)^2$$

$$y = A\sin(kx - \omega t)$$

$\Rightarrow$ $$\frac{dy}{dx} = +Ak\cos(kx - \omega t)$$
$$= A\frac{\omega}{v}\cos(kx - \omega t)$$

$\therefore$ $$u_P = \frac{1}{2}\rho v^2 \cdot A^2\frac{\omega^2}{v^2}\cos^2(kx - \omega t)$$

or $$u_P = \frac{1}{2}\rho A^2\omega^2\cos^2(kx - \omega t)$$

2. Let P be the power of source. Then, $I = \dfrac{P}{4\pi r^2}$

$\therefore$ $I \propto \dfrac{1}{r^2}$. But $I \propto A^2$ (where, A is amplitude)

$\therefore$ $$A \propto \frac{1}{r}$$

The equation in the standard form is $y = A\sin k(r - vt)$

$\therefore$ $y = \dfrac{a_0}{r}\sin k(r - vt)$, simply replacing A by $\dfrac{a_0}{r}$, where a_0 is constant.

Comparing with the usual form $y = A\sin(kr - \omega t)$

$$\omega = kv \quad \text{or} \quad f = \frac{kv}{2\pi}$$

and $$k = \frac{2\pi}{\lambda} \quad \text{or} \quad \lambda = \frac{2\pi}{k}$$

$\therefore$ Speed, $$f\lambda = \frac{kv}{2\pi} \times \frac{2\pi}{k} = v$$

and $$T = \frac{1}{f} = \frac{2\pi}{kv}$$

$$I = \frac{1}{2}\rho\omega^2A^2v = \frac{1}{2}\rho k^2v^2\frac{A_0^2}{r^2}v = \frac{1}{2}\rho A_0^2k^2v^3/r^2$$

3. (i) $$P_{av} = \frac{1}{2}\rho\omega^2A^2sv = \frac{1}{2}\mu(2\pi f)^2A^2v \qquad (\because \mu = \rho s,\ \omega = 2\pi f)$$

or $P_{av} = 2\pi^2\mu v A^2 f^2$,

$$v = \sqrt{\frac{450}{50 \times 10^{-3}}} = \sqrt{9000} = 30\sqrt{10}$$

If $A = 0.2\,\lambda$, then
$$P_{av} = 2\pi^2\mu v\,(0.2\,\lambda)^2f^2$$
$$= 0.8\,\mu v^3$$
$$= 0.8(50 \times 10^{-3})(30\sqrt{10})^3$$
$$= 1080\sqrt{10}\ \text{W}$$

(ii) $$P \propto \mu v$$
$$\propto \mu\sqrt{\frac{T}{\mu}}$$

$\therefore$ $$P \propto \sqrt{\mu}$$

So, if μ is doubled, P_{av} becomes $\sqrt{2}$ times.

4. (c) Intensity at a point 100 m away from source is
$$I = \frac{\text{Power}}{\text{Area}} = \frac{P}{4\pi r^2} = \frac{100}{4 \times 3.141 \times (100)^2} = 7.96 \times 10^{-4}\ \text{Wm}^{-2}$$

Also, $$I = \frac{1}{2}\rho\omega^2A^2v$$

$$A^2 = \frac{2I}{\rho\omega^2v} = \frac{2 \times 7.96 \times 10^{-4}}{(1.29)(2 \times \pi \times 1000)^2 \times 350} = 8.94 \times 10^{-14}$$

$\therefore$ $$A = 2.99 \times 10^{-7}\text{m}$$

5. (b) $y(x = 0, t) = 8\sin 4t$

$\Rightarrow$ $$A = 8\ \text{cm},\ \omega = 4\ \text{rads}^{-1}$$

$$v = \sqrt{\frac{T}{\mu}} = \sqrt{\frac{1}{0.2}} = 2.236\ \text{ms}^{-1}$$

The average power over one time period is $P = \dfrac{1}{2}\rho\omega^2A^2sv$
$$= \frac{1}{2}\omega^2A^2v \cdot \mu$$

($\because \mu = \rho s$, where, s = cross-section area)

$\therefore$ $$P = \frac{1}{2}(4)^2(8 \times 10^{-2})^2\,(2.236)\,(0.2) = 2.29 \times 10^{-2}\ \text{Js}^{-1}$$

Power transferred to the bath is
$$P' = 0.5P = 1.145 \times 10^{-2}\ \text{Js}^{-1}$$

Now, $P't = mC\,\Delta\theta$, where $C = 4.2 \times 10^3\ \text{J kg}^{-1}\text{k}^{-1}$

$$t = \frac{mC\,\Delta\theta}{P'} = \frac{1 \times 4.2 \times 10^3 \times 1}{1.145 \times 10^{-2}} = 3.6 \times 10^5\text{s} \simeq 4.2\ \text{day}$$

6. (d) $v_p = -v\left(\dfrac{dy}{dx}\right) \Rightarrow v_p = -v(\tan\theta)$

$$+20\sqrt{3} = -v\,(\tan 60°) \Rightarrow v = -20\ \text{ms}^{-1}$$

From the figure, $A = 4 \times 10^{-3}$m and $\lambda = 4$m

$$f = \frac{v}{\lambda} = \frac{20}{4} = 5\ \text{Hz}$$

$\therefore$ $$\omega = 2\pi f = 10\pi\ \text{rads}^{-1}$$

$$P = \frac{1}{2}\rho\omega^2A^2sv = \frac{1}{2}\mu\omega^2A^2v \qquad (\because \mu = \rho s)$$
$$= \frac{1}{2}(50 \times 10^{-3})\,(10\pi)^2\,(4 \times 10^{-3})^2(20) = 8 \times 10^{-3}\ \text{W}$$

Energy transferred per cycle is $E = PT = \dfrac{P}{f}$

or $$E = \frac{8 \times 10^{-3}}{5} = 1.6 \times 10^{-3}\ \text{J}$$

7. (a) Power (which is directly proportional to intensity) is given by

$$P = \frac{1}{2}\mu A^2\omega^2 v$$

$\therefore$ $$\frac{P_1}{P_2} = \frac{I_1}{I_2} = \frac{A_1^2}{A_2^2}\cdot\frac{\omega_1^2}{\omega_2^2}$$

$\Rightarrow$ $$\frac{1}{36} = \left(\frac{A_1}{A_2}\right)^2\cdot\left(\frac{1}{2}\right)^2$$

$\therefore$ $$\frac{A_1}{A_2} = 1:3$$

8. (c) $$P = \frac{1}{2}\rho\omega^2A^2sv$$

$$= \frac{1}{2}\mu\omega^2A^2v = \frac{1}{2}\mu\omega^2A^2\sqrt{\frac{T}{\mu}}$$

or $$P = \frac{1}{2}\omega^2A^2\sqrt{\mu T}$$

$$= \frac{1}{2}(2\pi\times 60)^2(6\times10^{-2})^2\sqrt{5\times10^{-2}\times 80}$$

$$\simeq 518 \text{ W}$$

9. (b) After two seconds, both the pulses will have moved 4 cm towards each other. So, by their superposition, the resultant displacement at every point will be zero. Therefore, total energy will be purely in the form of kinetic. Half of the particles will be moving upwards and half downwards.

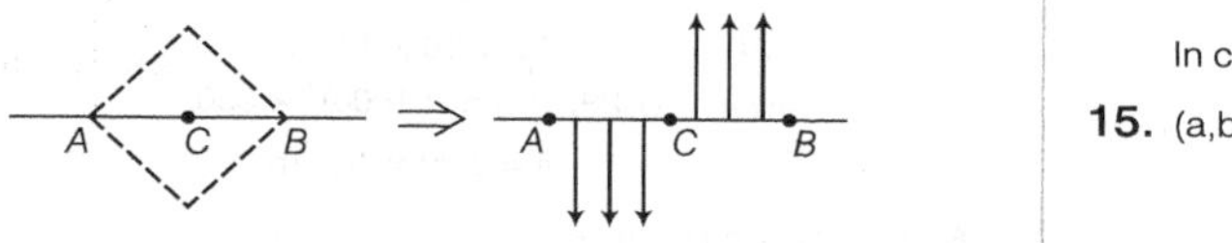

10. (d) Intensity $= \frac{\text{Power}}{4\pi r^2} = \frac{\frac{1}{2}\rho\omega^2A^2sv}{4\pi r^2}$

$$= \frac{1}{2}\rho(2\pi f)^2A^2v \qquad (\because s = 4\pi r^2)$$

or $$I = 2\pi^2f^2A^2\rho v$$

$\Rightarrow$ $$A = \sqrt{\frac{I}{2\pi^2f^2\rho v}}$$

or $$A = \sqrt{\frac{\text{Power}/4\pi r^2}{2\pi^2f^2\rho v}}$$

$$= \sqrt{\frac{10}{4\pi\times10^2\times2\pi^2\times1000\times1.29\times340}}$$

or $$A \simeq 0.96\ \mu\text{m}$$

11. (b) $\mu = \frac{m}{L} = \frac{0.1}{2} = 0.05$ kg m^{-1}

Tension is $T = 60$N

$\therefore$ $$v = \sqrt{\frac{T}{\mu}} = \sqrt{\frac{60}{0.05}} = \sqrt{\frac{6000}{5}} = \sqrt{1200} = 20\sqrt{3}\ \text{ms}^{-1}$$

Given, $A = 10^{-2}$m and $P = 100$W

We know that, $P = \frac{1}{2}\rho\omega^2A^2sv$

$$P = \frac{1}{2}\mu\omega^2A^2v \qquad (\because \mu = \rho s)$$

$$P = \frac{1}{2}\mu\,4\pi^2f^2A^2v \qquad (\because \omega = 2\pi f)$$

$$f^2 = \frac{P}{2\mu\pi^2A^2v} = \frac{100}{2\times0.05\times\pi^2\times(10^{-2})^2\times20\sqrt{3}}$$

$\therefore$ $$f = 171\text{ Hz}$$

12. (a) Intensity is

$$I = \frac{P}{4\pi r^2} = \frac{1}{4\pi(1)^2} = 8\times10^{-2}\text{ Wm}^{-2}$$

13. (c) When both the pulses completely overlap, the velocity, profiles of both the pulses in overlap region are identical. By superposition, the velocity of each elementary section doubles. Therefore, KE of each section becomes four times. Hence, the KE in the complete width (i.e. *d*) of the overlap becomes 4 times.

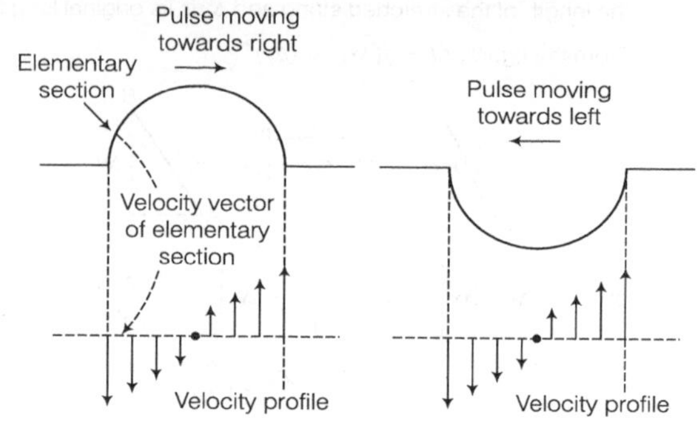

14. (a,c,d) In case of spherical waves (i.e. from a point source), inensity is

$$I = \frac{P}{4\pi r^2}$$

In case of cylindrical waves, (i.e. from a line source), intensity is

$$I = \frac{P}{2\pi rL},\ \text{where } L \text{ is length of line source.}$$

In case of plane waves, *I* is independent of *r*.

15. (a,b) Given, $f = 200$ Hz, $A = 10^{-3}$m, $\mu = 0.006$ kg m^{-1}, $T = 60$ N

$$P_{av} = \frac{1}{2}\mu\omega^2A^2v = \frac{1}{2}(0.006)(2\pi\times200)^2(10^{-3})^2\sqrt{\frac{60}{0.006}}$$

or $$P_{av} = 0.47\text{ W}$$

Now, $$E = \frac{1}{2}m\omega^2A^2 = \frac{1}{2}(0.006\times2)(2\pi\times200)^2(10^{-3})^2$$

or $$E = 9.4\text{ mJ}$$

16. (b,c) $P = \frac{1}{2}\rho\omega^2A^2sv$

$\Rightarrow$ $$A = \frac{1}{\omega}\sqrt{\frac{2P}{\rho sv}}$$

$$\rho_s = \mu = \frac{6\times10^{-3}}{8}\text{ kg m}^{-1} = 7.5\times10^{-4}\text{ kg m}^{-1}$$

$$\omega = 2\pi f = \frac{2\pi v}{\lambda} = 2\times3.14\times\frac{30}{0.2} = 942\text{ rad s}^{-1}$$

$\therefore$ $$A = \frac{1}{942}\sqrt{\frac{2(50)}{7.5\times10^{-4}\times30}} = 0.0708\text{ m} = 7.08\text{ cm}$$

Now, $$P \propto v\omega^2$$

$\Rightarrow$ $$P \propto v(v^2) \qquad (\because \omega = kv)$$

or $$P \propto v^3$$

$\therefore$ $$\frac{P_1}{P_2} = \left(\frac{v_1}{v_2}\right)^3$$

$\Rightarrow$ $$P_2 = \left(\frac{v_2}{v_1}\right)^3 P_1$$

So, if wave speed is doubled, $P_2 = 8P_1$

17. (a,d) v_{max}(velocity amplitude) $= A\omega = 2\times10^{-8}\times(2\pi\times1000)$

$$\simeq 1.3\times10^{-4}\text{ ms}^{-1}$$

$$I = \frac{1}{2}\rho\omega^2A^2v = \frac{1}{2}\rho(v_{max})^2v$$

$$= \frac{1}{2}(1.3)(1.3\times10^{-4})^2(340)$$

$$= 3.7\times10^{-6}\text{ Wm}^{-2}$$

18. (b,c) $P_{av} = \frac{1}{2}\rho\omega^2A^2sv = \frac{1}{2}\mu\sqrt{\frac{T}{\mu}}\,\omega^2A^2 \qquad \left(\because \rho s = \mu \text{ and } v = \sqrt{\frac{T}{\mu}}\right)$

$$P = P_{av} = \frac{1}{2}\sqrt{\mu T}\,\omega^2A^2$$
$$= \frac{1}{2}\sqrt{\left(\frac{0.003}{0.8}\right)(25)}\,(2\pi \times 120)^2\,(1.6 \times 10^{-3})^2$$
$$= 0.223 \text{ W}$$

Halving the amplitude, quarters the average power

$$\therefore \quad P' = \frac{P}{4} = \frac{0.223}{4} = 0.056\text{W}$$

19. (a,b,d) $u = \frac{1}{2}\rho\omega^2A^2$ and $I = \frac{1}{2}\rho\omega^2A^2v$

$$\therefore \quad u = \frac{I}{v}$$
$$\Rightarrow \quad u \propto I$$

We know for a point source $I \propto \frac{I}{r^2}$ $\therefore u \propto \frac{1}{r^2}$

So, option (a) is correct.

If the source is a plane sound source, then intensity at every point in front of the source will be same, if damping does not take place. But if the damping takes place, then the amplitude of oscillation of medium particles, then the amplitude of oscillation of medium particles decreases with distance. In that case, the curve between u and r may have the shape shown. So, option (b) is correct.

For a plane surface, even if the intensity decreases with time, at a particular instant of time the intensity (and thus energy density) is independent of r. So, option (c) is not possible.

$I \propto \rho$. So, if ρ decreases with r, I (and thus u) also decreases with r. So, option (d) is possible.

20. (c) $v = \sqrt{\frac{T}{\mu}}$

$$\therefore \quad v_1 = v_2$$

so, $\quad f_1\lambda_1 = f_2\lambda_2$

$$\frac{f_1}{f_2} = \frac{\lambda_2}{\lambda_1} = \frac{40}{20} = 2$$
$$\Rightarrow \quad \frac{\omega_1}{\omega_2} = 2$$
$$I = \frac{1}{2}\rho\omega^2A^2v \Rightarrow \frac{I_1}{I_2} = \frac{\omega_1^2A_1^2}{\omega_2^2A_2^2}$$

From the figure, $\quad \frac{A_1}{A_2} = \frac{10}{5} = 2$

$$\therefore \quad \frac{I_1}{I_2} = (2)^2(2)^2 = 16$$

21. (a) Let the phase at the origin = 0

$\therefore$ Phase at $P = \frac{\pi}{2}$ and phase at $P' = \pi$

$\therefore$ Phase difference $= \pi - \frac{\pi}{2} = \frac{\pi}{2}$

22. (d) At $x = 20$ cm, for string 1, $\phi_1 = 2\pi$

and for string 2, $\quad \phi_2 = \pi$

$$\therefore \quad \Delta\phi = 2\pi - \pi = \pi$$

23. (b) The equation of wave moving in negative x-direction, assuming origin of position at $x = 2$ m and origin of time (i.e. initial time) at $t = 1$ s, $y = 0.1\sin(2\pi t + 4x)$.

Shifting the origin of position to left by 2 m, that is, to $x = 0$, also shifting the origin of time backwards by 1s, that is, to $t = 0$ s, $y = 0.1\sin[2\pi(t - 1) + 4(x - 2)]$.

24. (c) As, the given particle at $x = 2$m, is at mean position at $t = 1$ s, its velocity is $v = A\omega = (0.1)(2\pi) = 0.2\pi \text{ ms}^{-1}$.

25. (d) Time period is $T = \frac{2\pi}{\omega} = 1$ s

Hence, at $t = 1.25$s, that is $\frac{T}{4}$ seconds after $t = 1$ s, the particle is at rest at extreme position.

Hence, instantaneous power at $x = 2$ m and $t = 1.25$ s is zero.

26. (d) It is the instantaneous displacement that depends on phase, not the amplitude (which is maximum displacement).

27. (a) Since, the wavefronts are planar, the amount of energy passing per unit area per unit time remains same.

28. (b)

(i) Speed of transverse wave on the string is

$$v = \sqrt{\frac{T}{\rho s}} = \sqrt{\frac{100}{(4200)\left(\frac{\pi}{4}\right) \times (0.004)^2}} = 43.53 \text{ ms}^{-1}$$
$$\omega = 2\pi f = 2\pi(10) = 20\,\pi \text{rads}^{-1} = 62.83 \text{ rads}^{-1}$$
$$k = \frac{\omega}{v} = \frac{20\pi}{43.53} = 1.44 \text{ m}^{-1}$$
$$\therefore \quad y = A\sin(kx - \omega t) = 10^{-4}\sin[1.44x - 62.83t]$$

(ii) Energy density is $u = \frac{1}{2}\rho\omega^2A^2 = \frac{1}{2}(4200)(62.83)^2(10^{-4})^2$

$$= 8.29 \times 10^{-2} \text{ Jm}^{-3}$$

(iii) Average energy flow per unit time is power (P),

$$P = \frac{1}{2}\rho\omega^2A^2 \cdot sv = u(sv) = (8.29 \times 10^{-2})\left[\frac{\pi}{4} \times 0.004^2 \times 43.53\right]$$
$$= 4.53 \times 10^{-5} \text{ Js}^{-1}$$

(iv) Power required to drive the oscillator is obviously

$$P = 4.53 \times 10^{-5} \text{ W}$$

Hence, (i) → (p), (ii) → (s), (iii) → (t), (iv) → (t)

29. (3) $v = \sqrt{\frac{T}{\mu}} = \sqrt{\frac{80}{0.05}} = 40 \text{ ms}^{-1}$

$$\omega = 2\pi f = 2 \times 3.14 \times 60 = 377 \text{ rad s}^{-1}$$

Power is $P = \frac{1}{2}\mu\omega^2A^2v = \frac{1}{2}(0.05)(377)^2(6 \times 10^{-2})^2(40)$

$$\approx 513 \text{ W} = 171 \times 3$$
$$\Rightarrow \quad \beta = 3$$

30. (2) Energy in unit volume (i.e. energy density) is

$u = \frac{1}{2}\rho\omega^2A^2 = \frac{1}{2}\frac{m}{sL}\omega^2A^2$ (where, s is cross-section area)

Energy in length L is $E = \frac{1}{2}\frac{m}{sL}\omega^2A^2(sL) = \frac{1}{2}m\omega^2A^2$

or $\quad E = \frac{1}{2}(0.08)(2\pi \times 120)^2(0.16 \times 10^{-3})^2$

$$= 0.58 \text{ mJ} = (0.29)2 \text{ mJ}$$
$$\Rightarrow \quad \beta = 2$$

31. (8) $v = \sqrt{\frac{T}{\mu}} = \sqrt{\frac{80}{0.05}} = 40 \text{ ms}^{-1}$

$P = \frac{1}{2}\rho\omega^2A^2sv = \frac{1}{2}\mu\omega^2A^2v \qquad (\because \mu = s\rho)$

or $P = \frac{1}{2}(0.05)(2\pi \times 60)^2(0.06)^2 \times 40 \approx 512 \text{ W} = 64 \times 8 \text{ W}$

$$\therefore \quad x = 8$$

32. (7) $A = 0.5 \times 10^{-3}$m, $f = 100$ Hz, $T = 100$ N, $v = 100 \text{ ms}^{-1}$ (given)

$$\mu = \frac{T}{v^2} = \frac{100}{100 \times 100} = 0.01 \text{ kg m}^{-1}$$

Now, $P = \frac{1}{2}\mu\omega^2A^2v = \frac{1}{2} \times 0.01(2\pi \times 100)^2(0.5 \times 10^{-3})^2 \times 100$

$0.049\text{W} = (0.007) \times 7$ W

$$\Rightarrow \quad \alpha = 7$$

DPP-4 Reflection and Transmission of Waves

1. (i) $T_2 = T_1$ and $\mu_2 = 4\mu_1$

$$v = \sqrt{\frac{T}{\mu}} \Rightarrow v_2 = \frac{v_1}{2}$$

Frequency does not change in transmission and reflection, since it depends on the source.

$\therefore$ $$\omega_2 = \omega_1$$

$$k = \frac{\omega}{v} \Rightarrow k_1 v_1 = k_2 v_2$$

$$k_2 = k_1 \frac{v_1}{v_2} = k_1 \cdot 2 = 2k_1$$

The amplitude are

$$A_r = \left(\frac{v_2 - v_1}{v_1 + v_2}\right) A_i = A_i \left(\frac{\frac{v_1}{2} - v_1}{v_1 + \frac{v_1}{2}}\right)$$

or $$A_r = -\frac{A_i}{3}$$

$$A_t = \left(\frac{2v_2}{v_1 + v_2}\right) A_i = \frac{2\left(\frac{v_1}{2}\right)}{v_1 + \frac{v_1}{2}} \cdot A_i = \frac{2}{3} A_i$$

Now, the transmitted wave is

$$y_t = A_t \cos(k_2 x - \omega_2 t)$$

or $$y_t = \frac{2}{3} A_i \cos(2k_1 x - \omega_1 t)$$

and the reflected wave is

$$y_r = A_r \cos(k_1 x + \omega_1 t)$$

$$= -\frac{A_i}{3} \cos(k_1 x + \omega_1 t)$$

or $$y_r = \frac{A_i}{3} \cos(k_1 x + \omega_1 t + \pi)$$

(ii) The average power of a harmonic wave on a string is given by

$$P = \frac{1}{2} \rho \omega^2 A^2 s v = \frac{1}{2} \mu \omega^2 A^2 v \qquad (\because \mu = \rho s)$$

$$P_i = \frac{1}{2} \mu_1 v_1 \omega_1^2 A_i^2 \qquad \ldots(i)$$

$$P_t = \frac{1}{2} \mu_2 v_2 \omega_2^2 A_t^2 = \frac{1}{2}(4\mu_1)\left(\frac{v_1}{2}\right)(\omega_1)^2 \left(\frac{2}{3} A_i\right)^2 = \frac{4}{9} \mu_1 v_1 \omega_1^2 A_i^2 \qquad \ldots(ii)$$

$$P_r = \frac{1}{2} \mu_1 v_1 \omega_1^2 A_r^2 = \frac{1}{2} \mu_1 v_1 \omega_1^2 \left(\frac{A_i}{3}\right)^2 = \frac{1}{18} \mu_1 \omega_1^2 v_1 A_i^2 \qquad \ldots(iii)$$

Adding Eq. (ii) and Eq. (iii) gives Eq. (i)

$\therefore$ $$P_i = P_t + P_r$$

2. (i) Reflection of a pulse 'from a free boundary' is really the superposition of two identical waves travelling in opposite direction.

At $t = 0.5$ s,

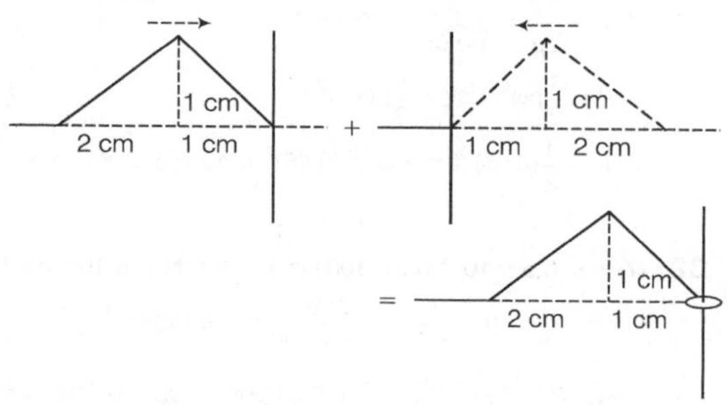

At $t = 1$ s,

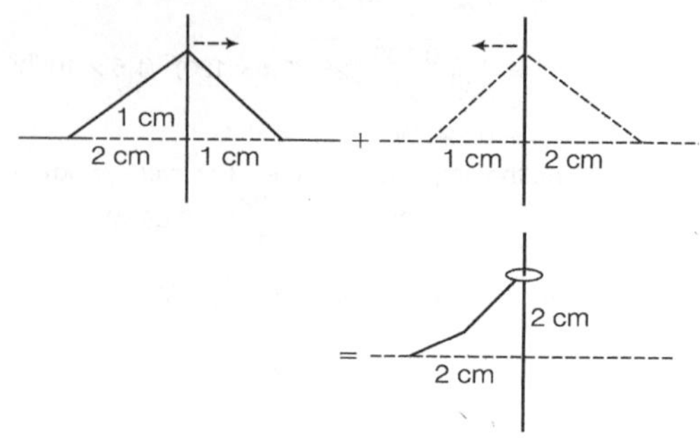

At $t = 1.5$ s,

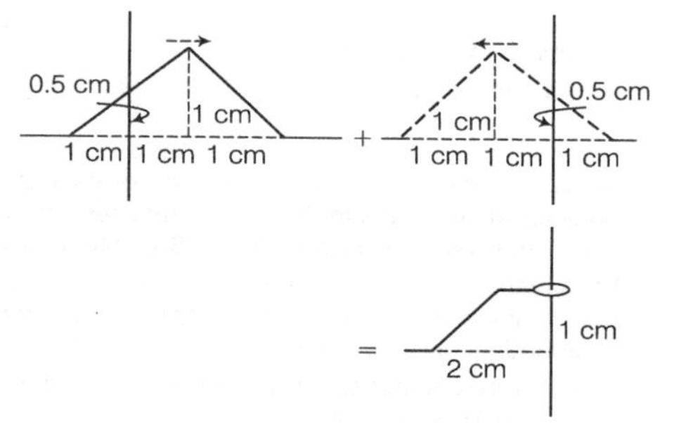

At $t = 2$ s,

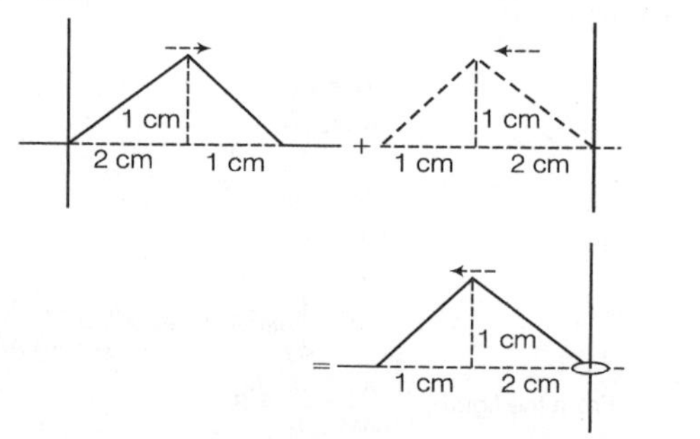

(ii) In the figure depicted in the question, speed of the particle at the trailing edge is

$$v_p = |-v\,(\text{Slope})|$$

$$v = \text{wave speed} = 2 \text{ cm s}^{-1}$$

$$\text{Slope} = \frac{1}{2}$$

$\therefore$ $$v_p \left|-2\left(\frac{1}{2}\right)\right| = 1 \text{ cms}^{-1}$$

3. (i) Reflection of a pulse 'from a fixed boundary' is really the superposition of two similar pulses (which are inverted w.r.t. each other) travelling in opposite directions.

At $t = 0$ s,

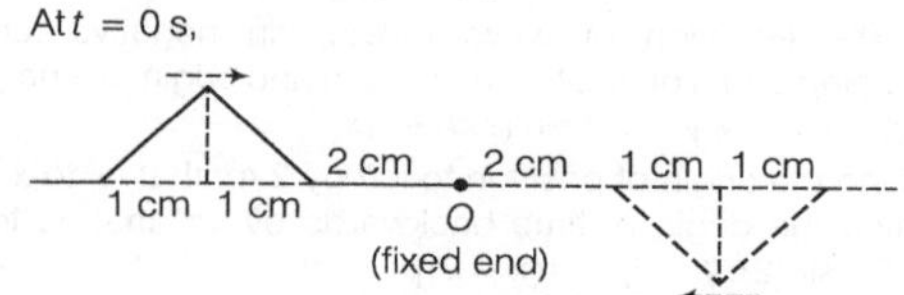

At $t = 3$ s,

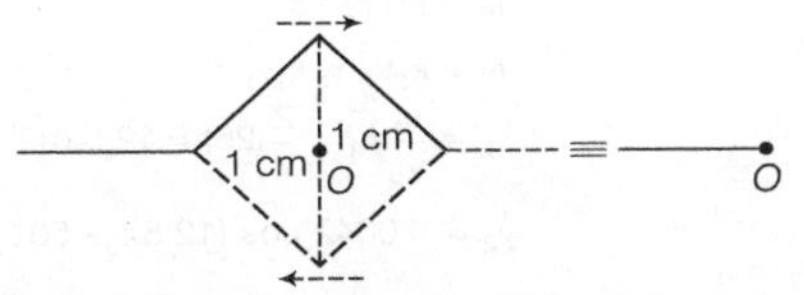

At $t = 4$ s,

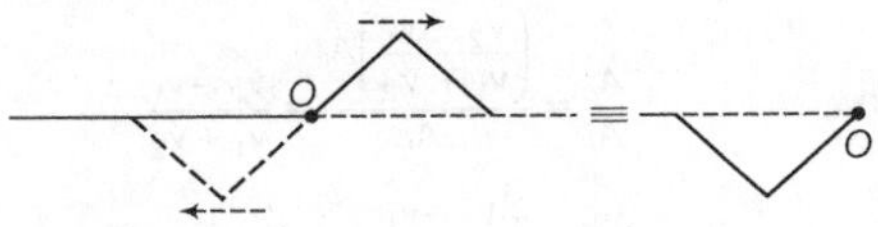

(ii) Reflection of a pulse 'from a free boundary' is equivalent to superposition of two identical pulses travelling in opposite directions.

At $t = 0$ s,

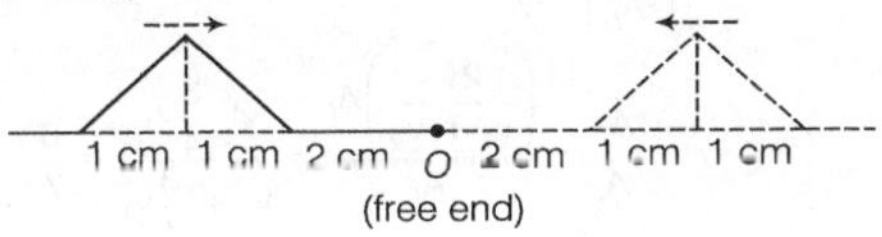

At $t = 3$ s,

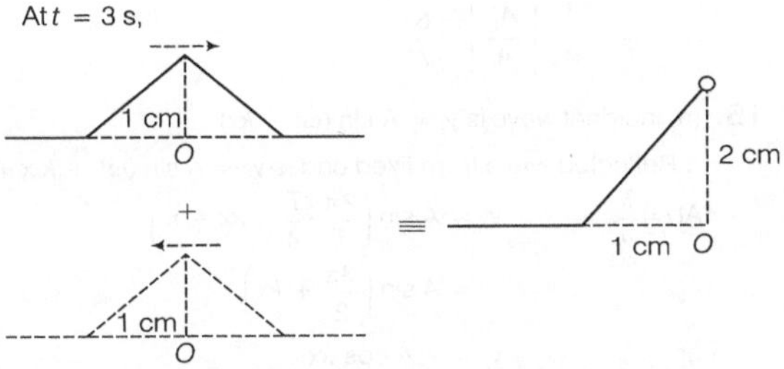

At $t = 4$ s,

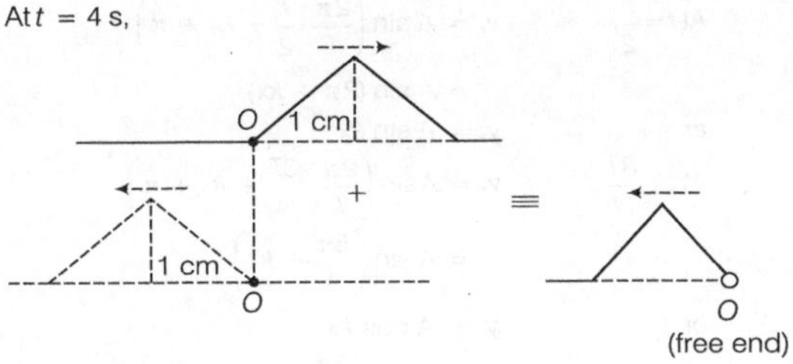

4. (i) O is fixed end

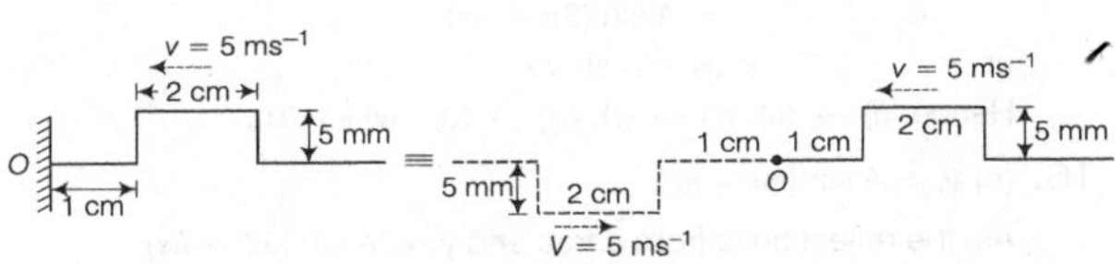

So, the shapes of string will be as follows :

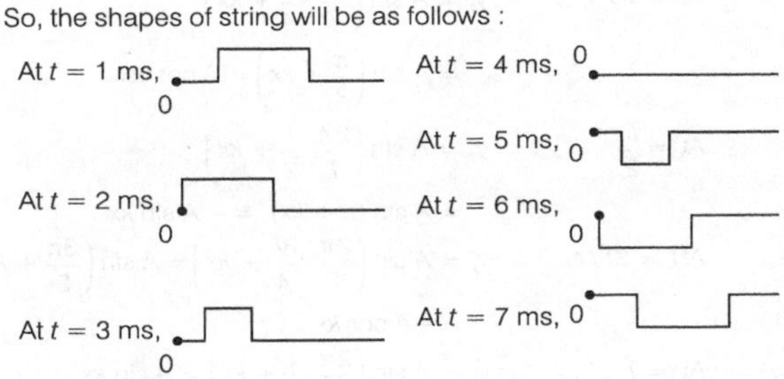

(ii) O is free end

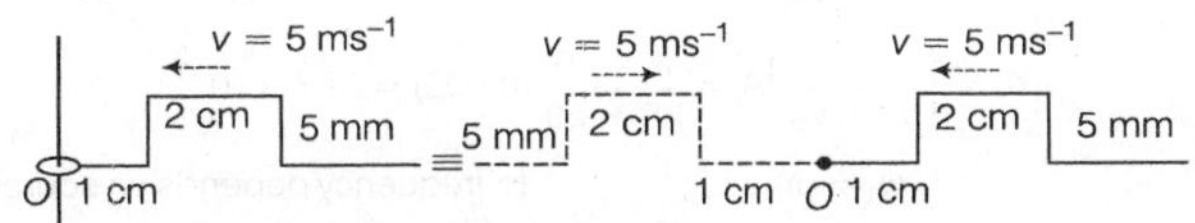

So, the shapes of string will be as fallows :

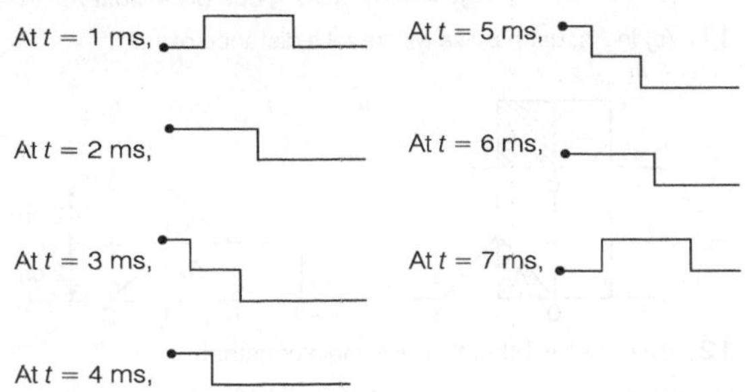

5. (b) $\mu_2 = \frac{\mu_1}{4}$ (given), $T_1 = T_2$

$\therefore \quad \mu_2 v_2^2 = \mu_1 v_1^2$

$\Rightarrow \quad v_2 = \sqrt{\frac{\mu_1}{\mu_2}}\, v_1 = \sqrt{4}\, v_1 = 2v_1$

or $\quad v_2 = 20 \text{ cm s}^{-1}$

6. (d) $A_t = \left(\frac{2v_2}{v_1 + v_2}\right) A_i = \frac{2(20)}{10 + 20} \cdot A_i = \frac{4}{3} A_i$

$$A_r = \left(\frac{v_2 - v_1}{v_1 + v_2}\right) A_i = \frac{20 - 10}{10 + 20} \cdot A_i = \frac{1}{3} A_i$$

$\therefore \quad \frac{A_t}{A_r} = \frac{4/3}{1/3} = 4$

7. (c) The direction of velocity of the pulse reverses after reflection. So, positive sign has to be there between $2x$ and $40t$. And the crest becomes trough after reflection, so $+\pi$ has to be added. So, $y_r = 0.3 \cos(2x + 40t + \pi)$.

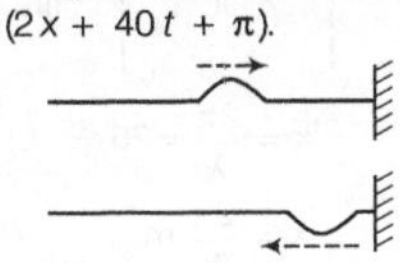

8. (a) After reflection crest remains at crest, but direction of wave is in the negative direction of x. So, positive sign has to be there between $2x$ and $40t$.

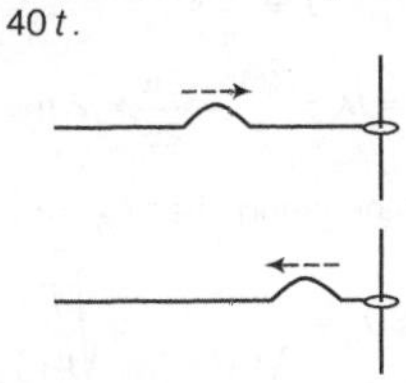

$\therefore \; y_r = 0.3 \cos(2x + 40t)$

9. (b) $v_1 = \frac{\omega_1}{k_1} = \frac{50\pi}{2\pi} = 25 \text{ ms}^{-1}$

$v_2 = 50 \text{ ms}^{-1}$ (given)

$$\therefore \quad A_r = \left(\frac{v_2 - v_1}{v_1 + v_2}\right) A_i = \frac{50 - 25}{25 + 50}(0.002)$$

$$= \frac{0.002}{3} = 0.00067$$

$\therefore \quad y_r = (0.00067) \cos(2\pi x + 50\pi t)$

10. (d) $v_1 = \frac{\omega_1}{k_1} = \frac{50\pi}{2\pi} = 25\ ms^{-1}$ and $v_2 = 50\ ms^{-1}$ (given)

$$A_t = \left(\frac{2v_2}{v_1 + v_2}\right)A_i = \frac{2 \times 50}{25 + 50}(0.002) = 2.67 \times 10^{-3}$$

$\omega_1 = \omega_2$ ($\because$ frequency depends on source)

$\therefore \quad k_2 = \frac{\omega_2}{v_2} = \frac{\omega_1}{v_2} = \frac{50\pi}{50} = \pi$

$\therefore \quad y_t = (2.67 \times 10^{-3}) \cos(\pi x - 50\pi t)$

11. (c) In 2 s, each pulse will travel a distance of 1 cm.

$\therefore$ At $t = 2$ s,

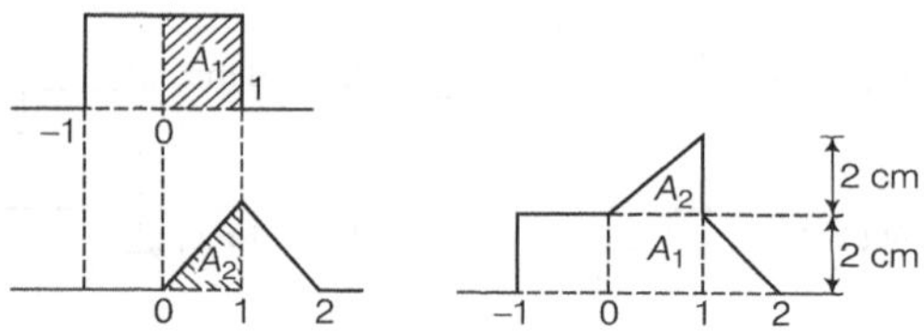

12. (b) $I \propto A^2$ ($\because$ other things remain constant)

$\therefore \quad I_i = kA_i^2,\ I_r = kA_r^2$

Loss in intensity by way of transmission $= I_i - I_r$

So, fractional loss of intensity will be

$$\frac{I_i - I_r}{I_i} = 1 - \frac{I_r}{I_i} = 1 - \left(\frac{A_r}{A_i}\right)^2$$

Now, $\quad \frac{A_r}{A_i} = \frac{1}{2}$ (given)

$\therefore$ The fraction of energy transmitted $= 1 - \left(\frac{1}{2}\right)^2 = \frac{3}{4}$

So, % of energy transmitted $= \frac{3}{4} \times 100 = 75\%$

13. (a,b,c,d) In region I, $y = 0.003 \cos(25x - 50t)$ (given)

$\therefore \quad k = \frac{2\pi}{\lambda} = 25$

$\Rightarrow \quad \lambda = \frac{2\pi}{25}$ m

and $\quad \omega = 2\pi f = 50$

$\Rightarrow \quad f = \frac{25}{\pi} s^{-1}$

Wave velocity is $v_1 = f\lambda = \frac{25}{\pi} = \frac{2\pi}{25} = 2\ ms^{-1}$

If T is the tension in the string, then $v_A = v_1 = \sqrt{\frac{T}{\mu_1}} = 2\ ms^{-1}$

wave velocity at B is $v_B = \sqrt{\frac{T}{\mu_1/4}} = 2\sqrt{\frac{T}{\mu_1}}$

$= 2(2) = 4\ ms^{-1}$

$\therefore \quad v_2 = v_B = 4\ ms^{-1}$

As, the wave is transmitted without reflection at junctions A and B, we can assume that, intensity is constant.

$$I = \frac{1}{2}\mu_1\omega^2 A_i^2 v_1$$

$$= \frac{1}{2}\left(\frac{\mu_1}{4}\right)\omega^2 A_t^2 v_2$$

$\therefore \quad A_t^2 = 4A_i^2 \frac{v_1}{v_2} = 4A_i^2 \left(\frac{2}{4}\right) = 2A_i^2$

$\therefore \quad A_t = \sqrt{2}A_i = \sqrt{2}\ (0.003) = 0.0042$ m

For region III Given, $\quad A = 0.0042$ m,

$\omega = 50\ rad\ s^{-1}$

$\omega = k_2 v_2 = k_1 v_1$

$k_2 = \frac{v_1}{v_2}k_1 = \frac{2}{4}(25) = 12.5\ m^{-1}$

$\therefore \quad y_2 = 0.0042 \cos(12.5x - 50t)$

14. (a,c) $v_2 = \frac{3}{4}v_1$ (given)

Now, $$\frac{A_r}{A_i} = \frac{\left(\frac{v_2 - v_1}{v_1 + v_2}\right)A_i}{A_i} = \frac{v_2 - v_1}{v_1 + v_2}$$

$$\frac{A_r}{A_i} = \frac{\frac{3}{4}v_1 - v_1}{v_1 + \frac{3}{4}v_1} = -\frac{1}{7}$$

$\therefore \quad \left|\frac{A_r}{A_i}\right| = \frac{1}{7}$

$$\frac{A_t}{A_i} = \frac{\left(\frac{2v_2}{v_1 + v_2}\right)A_i}{A_i} = \frac{2 \cdot \frac{3}{4}v_1}{v_1 + \frac{3}{4}v_1} = \frac{6}{7}$$

$\therefore \quad \left|\frac{A_t}{A_i}\right| = \frac{6}{7}$

15. (a) Incident wave is $y_i = A \sin(\omega t - kx)$ (given)

$\therefore$ Reflected wave from fixed end is $y_r = A \sin(\omega t + kx + \pi)$

At $t = \frac{T}{4}$, $\quad y_r = A \sin\left(\frac{2\pi}{T} \cdot \frac{T}{4} + kx + \pi\right)$

$= A \sin\left(\frac{3\pi}{2} + kx\right)$

or $\quad y_r = -A \cos kx$

At $t = \frac{T}{2}$, $\quad y_r = A \sin\left(\frac{2\pi}{T} \cdot \frac{T}{2} + kx + \pi\right)$

$= A \sin(2\pi + kx)$

or $\quad y_r = A \sin kx$

At $t = \frac{3T}{4}$, $\quad y_r = A \sin\left(\frac{2\pi}{T} \cdot \frac{3T}{4} + kx + \pi\right)$

$= A \sin\left(\frac{5\pi}{2} + kx\right)$

or $\quad y_r = A \cos kx$

At $t = T$, $\quad y_r = A \sin\left(\frac{2\pi}{T} \cdot T + kx + \pi\right)$

$= A \sin(3\pi + kx)$

or $\quad y_r = -A \sin kx$

Hence, (i) → (s), (ii) → (r), (iii) → (q), (iv) → (p).

16. (c) $y_i = A \sin(\omega t - kx)$

As, the reflection is from a free end $y_r = A \sin(\omega t + kx)$

At $t = T/4$, $\quad y_r = A \sin\left(\frac{2\pi}{T} \cdot \frac{T}{4} + kx\right)$

$= A \sin\left(\frac{\pi}{2} + kx\right) = A \cos kx$

At $t = \frac{T}{2}$, $\quad y_r = A \sin\left(\frac{2\pi}{T} \cdot \frac{T}{2} + kx\right)$

$= A \sin(\pi + kx) = -A \sin kx$

At $t = 3T/4$, $\quad y_r = A \sin\left(\frac{2\pi}{T} \cdot \frac{3T}{4} + kx\right) = A \sin\left(\frac{3\pi}{2} + kx\right)$

$= -A \cos kx$

At $t = T$, $\quad y_r = A \sin\left(\frac{2\pi}{T} \cdot T + kx\right) = A \sin kx$

Hence, (i) → (s), (ii) → (q), (iii) → (p), (iv) → (r).

17. (b) The equation of reflected wave can be obtained by replacing x by $-x$. Due to reflection from a free boundary, there is no phase change. But reflection from a fixed boundary gives a phase change of π.

Hence, (i) → (r), (ii) → (q), (iii) → (s), (iv) → (p).

18. (a) $\mu_2 = 4\mu_i$ (given)

$$y_i = A_i \sin(\omega t - kx) \quad \text{(given)}$$

$$A_r = \left(\frac{v_2 - v_1}{v_1 + v_2}\right) A_i$$

$$v_1 = \sqrt{\frac{T}{\mu_1}},\ v_2 = \sqrt{\frac{T}{\mu_2}} = \sqrt{\frac{T}{4\mu_1}} = \frac{v_1}{2}$$

$$\therefore \quad A_r = \left(\frac{\frac{v_1}{2} - v_1}{v_1 + \frac{v_1}{2}}\right) A_i = -\frac{A_i}{3}$$

19. (b) From the above discussion, $v_2 = \frac{v_1}{2} = \frac{v}{2}$

20. (a) $y_r = A_r \sin(\omega t + kx) = -\frac{A_i}{3}\sin(\omega t + kx)$

(moving in negative X-direction, so '+' symbol)

$$A_t = \left(\frac{2v_2}{v_1 + v_2}\right) A_i = \frac{2\frac{v}{2}}{v + \frac{v}{2}} A_i = \frac{2}{3} A_i$$

$$\omega_1 = \omega_2 \Rightarrow k_1 v_1 = k_2 v_2$$

$$\therefore \quad k_2 = k_1 \frac{v_1}{v_2} = 2k$$

$$\therefore \quad y_t = \frac{2}{3} A_i \sin(\omega t - 2kx)$$

(moving in positive X-direction, so '–' symbol)

21. (b) $v_1 = \sqrt{\frac{T}{\mu_1}},\ v_2 = \sqrt{\frac{T}{\mu_2}}$

We know, $$A_r = \left(\frac{v_2 - v_1}{v_1 + v_2}\right) A_i$$

$$\therefore \quad A_r = \frac{\sqrt{\frac{T}{\mu_2}} - \sqrt{\frac{T}{\mu_1}}}{\sqrt{\frac{T}{\mu_1}} + \sqrt{\frac{T}{\mu_2}}} A_i$$

$$A_r = \left(\frac{\sqrt{\mu_1} - \sqrt{\mu_2}}{\sqrt{\mu_1} + \sqrt{\mu_2}}\right) \cdot A_i$$

22. (a) $A_t = \left(\frac{2v_2}{v_1 + v_2}\right) A_i$

$$= \frac{2\sqrt{\frac{T}{\mu_2}}}{\sqrt{\frac{T}{\mu_1}} + \sqrt{\frac{T}{\mu_2}}} A_i = \frac{2\sqrt{\mu_1}}{\sqrt{\mu_1} + \sqrt{\mu_2}} A_i$$

23. (9) We know fraction of energy transmitted $= 1 - \left(\frac{A_r}{A_i}\right)^2$

$$\therefore \quad f_r = \text{fraction of energy reflected} = \left(\frac{A_r}{A_i}\right)^2$$

$$A_r = \frac{v_2 - v_1}{v_1 + v_2} \cdot A_i = \frac{v - 2v}{2v + v} \cdot A_i = -\frac{A_i}{3}$$

$$\therefore \quad f_r = \left(\frac{-1/3}{1}\right)^2 = \frac{1}{9}$$

$$\therefore \quad \beta = 9$$

24. (7) $\mu_{AB} = \mu_1 = \frac{0.06}{4.8}\ \text{kg m}^{-1} = \frac{1}{80}\ \text{kg m}^{-1}$

$$\mu_{BC} = \mu_2 = \frac{0.2}{2.56}\ \text{kg m}^{-1} = \frac{20}{256}\ \text{kg m}^{-1} = \frac{10}{128}\ \text{kg m}^{-1}$$

$$v_1 = \sqrt{\frac{T}{\mu_1}} = \sqrt{\frac{80}{1/80}} = 80\ \text{ms}^{-1}$$

$$\therefore \quad t_1 = \frac{4.8}{v_1} = \frac{4.8}{80} = 0.06\ \text{s}$$

$$v_2 = \sqrt{\frac{T}{\mu_2}} = \sqrt{\frac{80}{10} \times 128} = \sqrt{8 \times 2 \times 64} = 32\ \text{ms}^{-1}$$

$$\therefore \quad t_2 = \frac{2.56}{32} = 0.08\ \text{s}$$

$\therefore$ Required time is $t = t_1 + t_2 = 0.06 + 0.08 = 0.14$ s

or $$t = \frac{14}{100} = \frac{2}{100} \times 7$$

$$\Rightarrow \quad \alpha = 7$$

25. (2) Let $v_1 = 3v$ and $v_2 = v$

We know $$A_t = \frac{2v_2}{v_1 + v_2} \cdot A_i = \frac{2v}{3v + v} A_i$$

$$= \frac{A_i}{2} = \frac{\sqrt{13}}{2} \simeq \frac{3.6}{2} = 1.8$$

or $$A_t = 9 \times \frac{2}{10}$$

$$\Rightarrow \quad \Psi = 2$$

Revisal Problems (JEE Main)

1. (b) At the instant shown $v_p = -(v)$ (slope). As, both v and slope of the curve at P are positive, v_p is negative. That is particle P at the instant shown is moving down.

From the equation given,

$$\omega = 2\pi \times 0.1 = 0.2\pi\ \text{rad s}^{-1}$$

$\therefore$ Time period is $T = \frac{2\pi}{\omega} = 10$ s

Total time = 10 min 15 s = 615 s

$\therefore$ In the time duration of given particle P completes 61 oscillations $+\frac{1}{2}$ oscillations, i.e. it is at the mean position and is moving up after this time.

From the given equation, $A = 3$ cm

Distance travelled by particle P in one oscillation $= 4A = 12$ cm

So, distance travelled in 61.5 oscillations = 12(61.5) = 738 cm

2. (a) In the figure, C reaches the position, where A already reached, after $\omega t = \frac{\pi}{2}$ and A reaches the position, where B already reached, after $\omega t = \pi/2$.

3. (d) $y = A\sin(kx - \omega t + \phi)$ (for w_1)...(i)

$$y = A\sin(kx - \omega t) \quad \text{(for } w_2\text{)} \ldots \text{(ii)}$$

For w_1 $$y = A\sin\left[k\left(x + \frac{\phi}{k}\right) - \omega t\right] \quad \ldots\text{(iii)}$$

or $$y = A\sin\left[kx - \omega\left(t + \frac{\phi}{\omega}\right)\right] \quad \ldots\text{(iv)}$$

From Eqs. (ii) and (iii), we can conclude that w_1 is at a distance $\frac{\phi}{k}$ ahead of w_2. From Eqs. (ii) and (iv), we can conclude that wave w_1 is ahead of w_2 by a time difference $\frac{\phi}{\omega}$.

4. (a) According to the given information in the problem, $A = 1$ cm, $T = \frac{1}{125}$ s, $\lambda = 15.6$ cm, $\delta = \pi$ rad

$$\therefore \quad \omega = \frac{2\pi}{T} = 250\,\lambda \approx 786 \text{ rad s}^{-1}$$

and $$k = \frac{2\pi}{\lambda} = 100 \times \frac{2\pi}{15.6} \approx 40.3 \text{ rad m}^{-1}$$

Using this data in the standard equation (for a wave moving in + X-direction) i.e. $y = A \sin(\omega t - kx + \delta)$, we get

$$y = (1 \text{ cm}) \sin[(786 \text{ rad s}^{-1})t - (40.3 \text{ rad m}^{-1})x + \pi]$$

or $$y = (1 \text{ cm}) \sin[(40.3 \text{ rad m}^{-1})x - (786 \text{ rad s}^{-1})t]$$

5. (c) Individual particle executes SHM. So for a particle, we can write $y = A \sin \omega t$ and $v = A\omega \cos \omega t$

When displacement is $\frac{A}{2}$, we can write $\frac{A}{2} = A \sin \omega t$

$$\therefore \quad \omega t = \frac{\pi}{6} \Rightarrow t = \frac{\pi}{6} \cdot \frac{1}{\omega} = \frac{\pi}{6} \cdot \frac{T}{2\pi} = \frac{T}{12}$$

$\therefore$ Then, the velocity is $v = A\omega \cos\left(\omega \cdot \frac{T}{12}\right) = A\omega \cos\left(\frac{2\pi}{T} \cdot \frac{T}{12}\right)$

or $$v = A\omega \cos\frac{\pi}{6} = A\omega \cdot \frac{\sqrt{3}}{2}$$

or $$v = \frac{\sqrt{3}}{2} v_0 \qquad (\because v_{max} = v_0 = A\omega)$$

6. (d) $$\frac{I_1}{I_2} = \frac{\omega_1^2 A_1^2}{\omega_2^2 A_2^2} \qquad \left(\because I = \frac{1}{2}\rho\omega^2 A^2 v\right)$$

or $$\frac{I_1}{I_2} = \frac{f_1^2}{f_2^2} \times \frac{A_1^2}{A_2^2} = \frac{1}{4} \times \frac{36}{9} = 1$$

7. (b) Let in time t, wave gets displaced by x w.r.t. stationary medium. Then, the displacement of wave w.r.t. moving frame is $x' = x + vt$.

$\therefore$ The equation of wave in moving frame is

$$y = A \sin[\omega t - k(x' - vt)] \qquad (\because x = x' - vt)$$

or $$y = A \sin[(\omega + kv)t - kx']$$

Speed of wave is moving frame is $v' = \frac{dx'}{dt} = \frac{d}{dt}\left(x + \frac{\omega}{k}t\right)$

or $$v' = v + \frac{\omega}{k} = \frac{\omega + kv}{k}$$

8. (c) Let the point considered is $x = 0$. Also, let at $t_0 = 0$, its amplitude is maximum. Let our time starts from $t = t_0 = 0$.

Then, $$\delta = \frac{\pi}{2}$$

$$\therefore \quad y = A \sin\left(\omega t - kx + \frac{\pi}{2}\right)$$

or $$y = A \cos(\omega t - kx)$$

Then, its average energy density is $u = \rho\omega^2 A^2 \sin^2(\omega t - kx)$

Now, for $x = 0$ and $t = \frac{T}{6}$, $u = \rho\omega^2 A^2 \sin^2\left(\omega \cdot \frac{T}{6}\right)$

or $u = \rho\omega^2 A^2 \sin^2\left(\frac{2\pi}{T} \cdot \frac{T}{6}\right) = \rho\omega^2 A^2 \sin^2\frac{\pi}{3} = \rho\omega^2 A^2\left(\frac{3}{4}\right)$

But it is given that $u_0 = u$

$$\therefore \quad u_0 = \rho\omega^2 A^2\left(\frac{3}{4}\right) \Rightarrow \rho\omega^2 A^2 = \frac{4}{3}u_0$$

As, average energy density $= \frac{1}{2}\rho\omega^2 A^2 = \frac{2}{3}u_0$

9. (a) The speed of wave pulse is $v = \sqrt{\frac{T}{\mu}}$

$$\therefore \quad \frac{dx}{dt} = \sqrt{\frac{T}{\mu_0 + ax}}$$

or $$\sqrt{\frac{\mu_0 + ax}{T}}\, dx = dt$$

$$\therefore \int_0^L \sqrt{\frac{\mu_0 + ax}{T}}\, dx = \int_0^t dt \Rightarrow t = \frac{2}{3}\frac{1}{a\sqrt{T}}[(\mu_0 + aL)^{3/2} - (\mu_0)^{3/2}]$$

$$t = \frac{2}{3} \cdot \frac{1}{9 \times 10^{-3} \times \sqrt{100}}[(10^{-2} + 9 \times 10^{-3} \times 10)^{3/2} - (10^{-2})^{3/2}]$$

$$= \frac{200}{27}[(10^{-3/2} - (10)^{-3}] = \frac{200}{27}\left[\frac{1}{\sqrt{1000}} - \frac{1}{1000}\right]$$

$$= \frac{200}{27}\left[\frac{1}{31.62} - \frac{1}{1000}\right] = 0.227 \text{ s}$$

10. (d) **Wave 1** $v = \frac{6}{3} = 2 \text{ ms}^{-1}$

and $$(v_p)_{max} = A\omega = (2)(3) = 6 \text{ ms}^{-1}$$

Wave 2 $$v = \frac{12}{4} = 3 \text{ ms}^{-1}, (v_p)_{max} = (3)(12) = 36 \text{ ms}^{-1}$$

Wave 3 $$v = \frac{11}{5} = 2.2 \text{ ms}^{-1},$$

$$(v_p)_{max} = (4)(11) = 44 \text{ ms}^{-1}$$

11. (d) $\omega = \frac{2\pi}{0.04} \text{ rad s}^{-1} = 50\pi \text{ rad s}^{-1}$

$$\therefore \quad f = \frac{50\pi}{2\pi} = 25 \text{ Hz}$$

$$(a_p)_{max} = A\omega^2 = (3)(50\pi)^2$$

$$(a_p)_{max} \approx 75000 \text{ cm s}^{-2} = 750 \text{ ms}^{-2}$$

12. (b) Given, $A = 10^{-10}$ m, $f = 1000 \text{ s}^{-1}$, $\rho = 1.29 \text{ kg m}^{-3}$, $v = 332 \text{ ms}^{-1}$

We know energy density is

$$u = \frac{1}{2}\rho\omega^2 A^2$$

$$= \frac{1}{2}(1.29)(2\pi \times 1000)^2(10^{-10})^2$$

$$= 2.55 \times 10^{-13} \text{ J m}^{-3}$$

13. (c) $y = +\frac{b}{a}\sqrt{a^2 - (x - ct)^2}$

$$v_p = \frac{dy}{dt} = \frac{b}{2a}\frac{1}{\sqrt{a^2 - (x - ct)^2}}[(-2)(x - ct)(-c)]$$

$$= \frac{bc(x - ct)}{a\sqrt{a^2 - (x - ct)^2}}$$

$$(v_p)_{x = \frac{a}{2}, t = 0} = \frac{bc\left(\frac{a}{2} - 0\right)}{a\sqrt{a^2 - \left(\frac{a}{2} - 0\right)^2}}$$

$$= \frac{bc}{2} \cdot \frac{1}{\sqrt{\frac{3a^2}{4}}}$$

$$= \frac{bc}{\sqrt{3}a}$$

14. (b) $\rho_{liquid} = \frac{17}{81}\rho$, where $\rho = \rho$ density of the block

Volume of the block is

$$V = \frac{m}{\rho} = \frac{5}{\rho}$$

Buoyant force, $B_F = V\rho_L g = \frac{5}{\rho} \cdot \frac{17}{81}\rho \cdot g = \frac{85}{81}g$

$\therefore$ Tension in the string is $T = \text{weight} - B_F = 5g - \frac{85}{81}g$

$$= \frac{320\, g}{81}$$

$\therefore$ Speed of transverse wave in the string is $v = \sqrt{\dfrac{T}{\mu}}$

or $$v = \sqrt{\frac{\dfrac{320g}{81}}{2.25 \times 10^{-2}}}$$

$$= \sqrt{1722} = 41.5 \text{ ms}^{-1}$$

15. (a) Given, $y = A \sin (kx - 23562\, t)$

Comparing with $y = A \sin (kx - \omega t)$, we have $\omega = 23562$ rad s^{-1}

$$\therefore \quad f = \frac{23562}{2\pi} = 3750 \text{ Hz}$$

So, $$\lambda = \frac{v}{f} = \frac{300}{3750} = 0.08 \text{ m}$$

For $y = \dfrac{A}{2}$, $\quad t = 0$

$$\frac{A}{2} = A \sin (kx) \Rightarrow kx = \frac{\pi}{6}$$

$$x = \frac{\pi}{6} \cdot \frac{1}{k} = \frac{\pi}{6} \cdot \frac{\lambda}{2\pi} = \frac{\lambda}{12} \text{ m}$$

Required distance $= 2\lambda + \dfrac{\lambda}{12} = \dfrac{25\lambda}{12}$

16. (d) $v = \sqrt{\dfrac{T}{\mu}} \Rightarrow T = \mu v^2 = \mu f^2 \lambda^2$

$$\therefore \quad \mu_1 f_1^2 \lambda_1^2 = \mu_2 f_2^2 \lambda_2^2$$

$$\left(\frac{\lambda_1}{\lambda_2}\right)^2 = \frac{\mu_2}{\mu_1}\left(\frac{f_2}{f_1}\right)^2 = \frac{1}{\eta}\left(\frac{f_2}{f_1}\right)^2 \qquad \left(\because \frac{\mu_1}{\mu_2} = \eta\right)$$

Also given, $$\frac{f_1}{f_2} = \beta$$

$$\therefore \quad \left(\frac{\lambda_1}{\lambda_2}\right)^2 = \frac{1}{\eta}\left(\frac{1}{\beta}\right)^2$$

$$\Rightarrow \quad \frac{\lambda_1}{\lambda_2} = \frac{1}{\beta\sqrt{\eta}}$$

17. (a) Phase velocity is

$$v = \frac{\omega}{k} = \frac{1}{k}\sqrt{\omega_0^2 + \alpha^2 k^2}$$

Group velocity is

$$v_g = \frac{d\omega}{dk} = \frac{1}{2\sqrt{\omega_0^2 + \alpha^2 k^2}} \cdot \alpha^2 \cdot 2k$$

or $$v_g = \frac{\alpha^2 k}{\sqrt{\omega_0^2 + \alpha^2 k^2}} = \alpha^2\left(\frac{k}{\omega}\right) = \frac{\alpha^2}{v}$$

18. (c) The velocity of the particle at $x = 0$, at the given insant $t = 0$ is $v_p = -v$(slope). Here, speed of wave v is positive and slope at $x = 0$ is positive. So, v_p is negative.

That means at $t = 0$, left end is at mean position moving in the downward direction. So in the equation $y = A \sin (\omega t - kx + \delta)$, δ is π.

$$\therefore \quad y = A \sin (\omega t - kx + \pi)$$
$$= -A \sin(\omega t - kx)$$
$$= A \sin (kx - \omega t)$$

19. (c) Given, $$y = 3e^{x - 3t}$$

$$\frac{\partial y}{\partial x} = 3e^{x - 3t},$$

$$\frac{\partial^2 y}{\partial x^2} = 3e^{x - 3t}$$

$$\frac{\partial y}{\partial t} = 3e^{x - 3t}(-3)$$

$$\therefore \quad \frac{\partial^2 y}{\partial t^2} = (-9)(-3)e^{x-3t} = 27e^{x-3t}$$

$$v = \frac{3}{1} = 3$$

$$v^2 \frac{\partial^2 y}{\partial x^2} = 9(3e^{x-3t}) = 27e^{x-3t} \qquad \ldots \text{(i)}$$

$$\frac{\partial^2 y}{\partial t^2} = 27e^{x-3t} \qquad \ldots \text{(ii)}$$

From Eqs. (i) and (ii), it is clear that $\dfrac{\partial^2 y}{\partial t^2} = v^2 \dfrac{\partial^2 y}{\partial x^2}$. But y is not finite for large value of x. So, this cannot represent a travelling wave.

20. (c) A wave travelling in negative X-direction can be written as

$$y = A \cos (\omega t + kx + \delta)$$

At $t = 5$ s, $y = A\cos(5\omega + kx + \delta)$

Comparing with given equation i.e. $y = 0.5 \cos (0.1\, x)$, we have

$$A = 0.5 \text{ cm}, k = 0.1 \text{ cm}^{-1} \text{ and } 5\omega + \delta = 0 \qquad \ldots \text{(i)}$$

$$\therefore \quad \frac{2\pi}{\pi} = 0.1 \Rightarrow \lambda = 20\,\pi \text{ cm}$$

$$\omega = 2\pi f = 200\,\pi \text{ rad s}^{-1} \qquad (\because f = 100 \text{ Hz})$$

$$\therefore \quad v = \frac{\omega}{k} = \frac{200\pi}{0.1} = 2000\,\pi \text{ cm s}^{-1}$$

$$\therefore \quad y = 0.5 \cos (2000\pi t + 0.1x + \delta)$$

From Eq. (i), $\delta = -5\omega = -5(200\,\pi) = -1000\,\pi$ rad

$\therefore$ $\quad y = 0.5 \cos (200\pi t + 0.1x + 1000\pi)$

or $\quad y = 0.5 \cos (200\pi t + 0.1x)$

At $t = 10$ s, $\quad y = 0.5 \cos (2000\pi + 0.1x)$

or $\quad y = 0.5 \cos (0.1\, x)$

21. (d) $t_1 = \dfrac{2x}{v}, t_2 = \dfrac{2y}{v}$

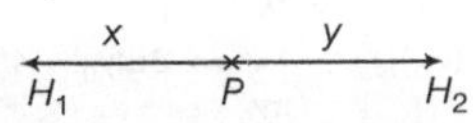

$$t_1 + t_2 = \frac{2}{v}(x + y) \Rightarrow x + y = \frac{v}{2}(t_1 + t_2)$$

or $$x + y = \frac{332}{2}(1.5 + 2.5) = 664 \text{ m}$$

22. (a) $(v_p)_{\max} = A\omega, v = \dfrac{\omega}{k}$

Given, $(v_p)_{\max} < v$ so, $A\omega < \dfrac{\omega}{k}$

$$A < \frac{\lambda}{2\pi}$$

23. (b) At the location of differential element

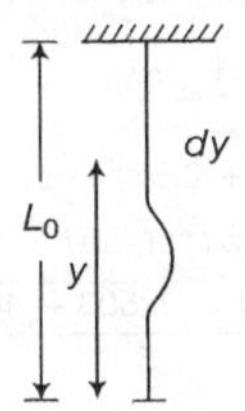

$$T = \frac{m}{L_0} \cdot y\, g$$

$$v = \sqrt{\frac{T}{\mu}} = \sqrt{\frac{m}{L_0} \cdot y \cdot g \cdot \frac{1}{m/L_0}} = \sqrt{g}\sqrt{y}$$

$$v^2 = gy$$

So, the best option is (b).

Revisal Problems (JEE Advanced)

1. (d) If the echo comes back, when the next is on, then it won't be clear. Sound travels $(x + x) = 2x$ distance

Time gap $= \frac{60}{40}$ s

Speed of sound × time gap $= 2x$

$\therefore \quad v = \frac{2x}{60/40} = \frac{4}{3}x$

Similarly for second case, $v \times$ time gap $= (2x - 180)$

$$\frac{4}{3}x \times \frac{60}{60} = 2x - 180$$

$$\Rightarrow \quad 2x - \frac{4}{3}x = 180$$

$$2x = 540$$

$$\Rightarrow \quad x = 270 \text{ m}$$

2. (c) For a general time t, the given expression can be written as $y = \frac{6}{(x - 2t)^2}$ ($\because$ $v = 2 \text{ ms}^{-1}$ and the pulse is moving in + x-direction)

$$v_p = \frac{\partial y}{\partial t} = 6(-2)(-2)(x - 2t)^{-3} = + \frac{24}{(x - 2t)^3}$$

$$(v_p)_{\text{at } x = 2\text{m}, t = 2\text{s}} = + \frac{24}{(2 - 2 \times 2)^3} = \frac{24}{-8} = -3 \text{ ms}^{-1}$$

3. (b) When incident ray is reflected back from a yielding end, then the last end is treated as free end. So, option (b) is correct.

4. (b) $v_p = \frac{\partial y}{\partial t} = 2\pi f \, y_0 \cos 2\pi\left(ft - \frac{x}{\lambda}\right)$

$$(v_p)_{\max} = 2\pi f y_0$$

Wave velocity is $v = f\lambda$

$$(v_p)_{\max} = 4v \Rightarrow 2\pi f y_0 = 4f\lambda$$

$$\Rightarrow \quad \lambda = \frac{\pi y_0}{2}$$

5. (c) Let the acceleration of the system is a. Then,

$$m_1 g - T = m_1 a$$

$$\Rightarrow \quad 100 - T = 10a \quad \ldots \text{(i)}$$

and $\quad T = m_2 a \Rightarrow T = 10a \quad \ldots \text{(ii)}$

From Eqs. (i) and (ii), $\quad a = 5 \text{ ms}^{-2}, T = 50 \text{ N}$

$$v = \sqrt{\frac{T}{\mu}} = \sqrt{\frac{50}{0.02}} = 50 \text{ ms}^{-1}$$

In ground frame, speed of transverse pulse $= v + at$ (at time t)

$$\therefore \quad \frac{dl}{dt} = v + at \Rightarrow \int_0^t (v + at)\, dt = L_0$$

$$L_0 = vt + \frac{1}{2}at^2$$

$$\Rightarrow \quad 4 = 50t + 2.5t^2$$

$$2.5t^2 + 50t - 4 = 0$$

$$\Rightarrow \quad t = \frac{-50 \pm \sqrt{2500 + 40}}{5} = 0.08 \text{ s}$$

6. (b) From the given information, $\lambda = 2 \times 10 = 20$ m

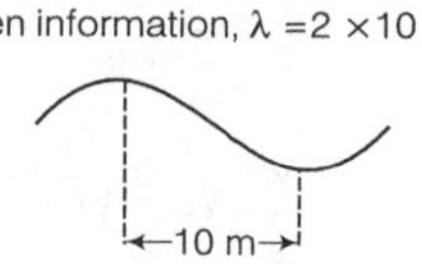

Time period is $T = 4$ s

$$\therefore \quad f = \frac{1}{4}$$

So, speed of wave is $v = f\lambda = \frac{1}{4} \times 20 = 5 \text{ ms}^{-1}$

7. (b) From the given information, $\omega = 3\pi$

$$\Rightarrow \quad f = \frac{3}{2} \text{ Hz}$$

At $x = 1$ cm, $\quad kx = \frac{\pi}{8}$

$$\Rightarrow \quad k \cdot 1 \text{ cm} = \frac{\pi}{8} \Rightarrow k = \frac{\pi}{8} \text{ cm}^{-1}$$

$$\therefore \quad \frac{2\pi}{\lambda} = \frac{\pi}{8} \Rightarrow \lambda = 16 \text{ cm}$$

So, $\quad v = f\lambda = 16 \times \frac{3}{2} = 24 \text{ cm s}^{-1}$

8. (a) Velocity of pulse is $v = \sqrt{\frac{T}{\mu}} = \sqrt{\frac{100}{0.25}} = 20 \text{ ms}^{-1}$

For the pulse to remain stationary w.r.t. ground, the velocity of cart should be 20 ms^{-1} towards left.

9. (c) $v = \frac{\Delta y}{\Delta t} = \frac{15}{0.25} = 60 \text{ ms}^{-1}$

The tension in the cable is $F = \mu v^2 = \frac{8}{15} \times 60^2$

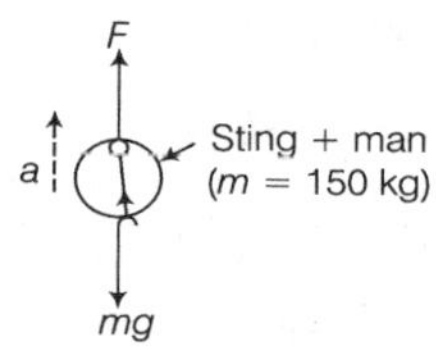

or $\quad F = 1920$ N

$$F - mg = ma$$

$$a = \frac{F - mg}{m} = \frac{1920 - 150 \times 9.8}{150}$$

or $\quad a = 3 \text{ ms}^{-2}$

so, acceleration of the helicopter is 3 ms^{-2}.

A real cable has stiffness in addition to tension. Stiffness tends to return a wire to its original straight line shape, even when it is not under tension. For example, piano wire straightens, if released from a curved shape, package wrapping string does not.

Stiffness represents a restoring force in addition to tension and increase the wave speed. Consequently, for a real cable, the speed of 60 ms^{-1}, that we determined, is most likely associated with a smaller acceleration of the helicopter.

10. (c) $I_r = I_i (0.64) \Rightarrow \frac{I_r}{I_i} = 0.64$

$$I \propto A^2$$

$$\therefore \quad \frac{A_r}{A_i} = \sqrt{0.64} = 0.8$$

So, $\quad A_r = 0.8 A_i = 0.8 A$

The reflected wave moves opposite to the direction of incident wave, so positive sign appears between kx and ωt. As, the wave is reflected from denser medium, an additional phase 180° adds. So, correct option is (c).

11. (b) Given, $s = 1 \text{ mm}^2 = 10^{-6} \text{m}^2$ and $\rho = 8000 \text{ kg m}^{-3}$

$\therefore$ Linear mass density is $\mu = \rho s = 8000 \times 10^{-6} = 0.008 \text{ kgm}^{-1}$

From the equation given, $k = 1 \text{ m}^{-1}$ and $\omega = 30 \text{ rad s}^{-1}$

$$\therefore \quad \lambda = \frac{2\pi}{k} = 2\pi m, \; f = \frac{30}{2\pi} \text{s}^{-1}$$

Speed of wave is $\quad v = f\lambda = \frac{30}{2\pi} \times 2\pi = 30 \text{ ms}^{-1}$

Tension in the string is

$$F = \mu v^2 = (0.008)(30)^2 = 7.2 \text{ N}$$

12. (a,b,c,d) Let the wave is

$$y = A\sin(\omega t - kx)$$

$$\lambda = \frac{v}{f} = \frac{700}{1000} = 0.7 \text{ m}$$

$$\omega = 2\pi f = 2000\pi \text{ rad s}^{-1}$$

and $$k = \frac{2\pi}{\lambda} = \frac{2\pi}{0.7} = 8.971 \text{ rad m}^{-1}$$

Phase $= \omega t - kx$

Phase at the origin (i.e. $x = 0$) at time $t = 2000\,\pi t$

Let $s = 0.117 + 0.7$ m

When, $n = 0$, $s = 0.117$ m,

Then, $\phi = 2000\pi t - 8.971 \times 0.117$

or $\phi = 2000\pi t - 1.04$

or $\phi = 2000\pi t - \frac{\pi}{3}$

When $n = 1$ $s = 0.817$ m. then $\phi = 2000\pi t - 8.971 \times 0.817$

or $\phi = 2000\pi t - (2\pi + 1.04) = 2000\pi t - \left(2\pi + \frac{\pi}{3}\right)$

When $n = 2$, $s = 1.517$ m, then

$\phi = 2000\pi t - 8.971 \times 1.517$

or $\phi = 2000\pi t - 13.53 = 2000\pi t - \left(4\pi + \frac{\pi}{3}\right)$

Time period is $T = \frac{2\pi}{\omega} = \frac{2\pi}{2000\pi} = 10^{-3}$ s

$\therefore$ Gain in phase is 10^{-3}s is 2π.

So, gain in phase in 0.5×10^{-3} s is π

and in 0.1×10^{-3} s is $\frac{\pi}{5}$ i.e. 0.628 rad.

13. (a,d) $v = f\lambda = 50 \times 0.5 = 25 \text{ ms}^{-1}$

$A = 1$ cm (given)

$$t_1 = \frac{8}{v} = \frac{8}{25} = 0.32 \text{ s}$$

Time period is $T = \frac{1}{50} = 0.02$ s

Distance travelled by a particle in time $T = 4A = 4(1) = 4$ cm

$\therefore$ Time taken for completing a distance 8 m is

$$t_2 = \frac{T}{4} \times 800 = 200\,T = 200 \times 0.02 = 4 \text{ s}$$

14. (b,c) Let the equation of wave is $y = A(\omega t - kx + \delta)$

$\therefore$ $v_p = \frac{\partial y}{\partial t} = A\omega\cos(\omega t - kx + \delta)$

It is given that, when $x = 0$, $t = \frac{T}{4}$, y is equal to a.

$\therefore$ $a = A\sin\left(\frac{2\pi}{T}\cdot\frac{T}{4} - k(0) + \delta\right)$

or $a = A\cos\delta$, At $x = 0, t = \frac{T}{4}$

$v_p = 0$ (given)

$\therefore$ $0 = A\,\omega\cos\left(\frac{2\pi}{T}\cdot\frac{T}{4} - k(0) + \delta\right)$

So, $\frac{\pi}{2} + \delta = \pi/2 \Rightarrow \delta = 0$

Also, it is given that, $v_p = v'$ at $x = \frac{\lambda}{4}$ and $t = \frac{T}{4}$

$\therefore$ $v' = a\omega\cos\left(\frac{2\pi}{T}\cdot\frac{T}{4} - \frac{2\pi}{\lambda}\cdot\frac{\lambda}{4}\right)$

$v' = a\omega\cos 0^\circ$

$\Rightarrow$ $v' = a\omega$

or $\omega = \frac{v'}{a}$

As, $\delta = 0, A = a$

Now, the wave equation is $y = a\sin\omega\left(t - \frac{x}{v}\right)$

or $$y = a\sin\frac{v'}{a}\left(t - \frac{x}{v}\right)$$

15. (a,b,d) It is a known, as well as experimentally and analytically verified fact that wave speed depends on the properties of the medium and is same for the entire wave. The particle velocity is given by $v_p = \frac{\partial y}{dt} = A\omega\cos(\omega t - kx)$, where symbols have their usual meaning. It is clear from above expression that v_p depends on amplitude and frequency of wave which are wave properties.

16. (a) Let $y = A\sin(\omega t - kx)$. Then,

$v_p = A\omega\cos(\omega t - kx)$

$dU = F(dl - dx)$ (= work done to stretch the element from dx to dl by tension F)

$$= F[\sqrt{(dx)^2 + (dy)^2} - dx]$$

$$= Fdx\left[\left(1 + \frac{dy}{dx}\right)^{\frac{1}{2}} - 1\right]$$

$$\approx \frac{1}{2}Fdx\left(\frac{\partial y}{\partial x}\right)^2$$

$$d(\text{KE}) = \frac{1}{2}\mu dx\left(\frac{\partial y}{\partial t}\right)^2$$

Now, $\frac{\partial y}{\partial x} = -Ak\cos(\omega t - kx)$

and $\frac{\partial y}{\partial t} = v_p = A\omega\cos(\omega t - kx)$

By using these in equation of dU and d(KE). You will find both of them are same.

17. (d) $P = \frac{F\omega^2A^2}{2v} = \frac{F\omega^2A^2}{2\sqrt{F/\mu}}$ $(\because v = \sqrt{F/\mu})$

or $P = \frac{1}{2}\omega^2A^2\sqrt{\mu}\sqrt{F}$

$\Rightarrow P \propto \sqrt{F}$

18. (a) $v = \sqrt{\frac{F}{\mu}}$

$\Rightarrow$ $v_1 = \sqrt{\frac{F}{\mu}}$ and $v_2 = \sqrt{\frac{F}{9\mu}} = \frac{v_1}{3}$

$\therefore$ $\frac{v_1}{v_2} = 3$

$$A_1 = A_r = \left(\frac{v_2 - v_1}{v_1 + v_2}\right)A_i = \left(\frac{v_2 - 3v_2}{3v_2 + v_2}\right)A_i = -\frac{A_i}{2}$$

$$A_2 = A_t = \left(\frac{2v_2}{v_1 + v_2}\right)A_i = \frac{2v_2}{3v_2 + v_2}\cdot A_i = \frac{A_i}{2}$$

$\therefore$ $\left|\frac{A_1}{A_2}\right| = 1$

$$I = \frac{1}{2}\rho\omega^2A^2v = \frac{1}{2}\frac{\mu}{s}\omega^2A^2v$$

We do not have the information about density (and cross-sectional area s of two ropes). So, I_1/I_2 cannot be calculated.

Power is $P = \frac{1}{2}\rho\omega^2A^2sv$

As $\mu = \rho s$, $P = \frac{1}{2}\mu\omega^2A^2v \Rightarrow P \propto \mu vA^2$

$\therefore$ $\frac{P_1}{P_2} = \frac{\mu_1}{\mu_2}\cdot\frac{v_1}{v_2}\left(\frac{A_1}{A_2}\right)^2 = \frac{1}{9}\cdot 3(1) = \frac{1}{3}$

$\therefore$ $\frac{P_2}{P_1} = 3$

Hence, (i) $\rightarrow$ (q), (ii) $\rightarrow$ (r), (iii) $\rightarrow$ (s), (iv) $\rightarrow$ (r)

19. (a) $P_{av} = \frac{1}{2}\rho\omega^2 A^2 sv$

P_{inst} or $P = \rho\omega^2 A^2 sv \cos^2(kx - \omega t)$

Hence, (i) → (q,s), (ii) → (q,s), (iii) → (p,r), (iv) → (p,r)

20. (a) From wave 1, $k = 3 \Rightarrow \lambda = \frac{2\pi}{3}$ m

$$\omega = 6 \Rightarrow f = \frac{6}{2\pi}$$

$$v = f\lambda = \frac{6}{2\pi} \cdot \frac{2\pi}{3} = 2 \text{ ms}^{-1}$$

Speed of wave is same for all the waves given

($\because \mu$ and F are same)

$$T_1 = \frac{2\pi}{6} = \frac{\pi}{3} \Rightarrow f_1 = \frac{3}{\pi}$$

$$T_2 = \frac{2\pi}{8} = \frac{\pi}{4} \Rightarrow f_2 = \frac{4}{\pi}$$

$$T_3 = \frac{2\pi}{12} = \frac{\pi}{6} \Rightarrow f_3 = \frac{6}{\pi}$$

$\therefore$ $f_1 : f_2 : f_3 = 3 : 4 : 6$

Hence, (i) → (p), (ii) → (q,p), (iii) → (r,p), (iv) → (s,p)

21. (a)

(i) $y = A\sin(kx + \omega t)$

Slope is $\frac{\partial y}{\partial x} = Ak\cos(kx + \omega t)$

Slope at $x = 0 = Ak\cos\omega t$.

(ii) $y = A\sin(kx - \omega t)$

$$v_p = \frac{\partial y}{\partial t} = -A\omega\cos(kx - \omega t)$$

$\frac{\partial y}{\partial t}$ at $x = 0 = -A\omega\cos\omega t$

(iii) $y = A\sin(\omega t - kx)$ at $x = 0$, $y = A\sin\omega t$

Hecne, (i) → (r), (ii) → (q), (iii) → (p)

22. (c) $v_1 = \sqrt{\frac{F}{\mu}}$, $v_2 = \sqrt{\frac{F}{4\mu}} = \frac{1}{2}\sqrt{\frac{F}{\mu}}$

$$v_3 = \sqrt{\frac{F}{9\mu}} = \frac{1}{3}\sqrt{\frac{F}{\mu}}, \quad v_4 = \sqrt{\frac{F}{16\mu}} = \frac{1}{4}\sqrt{\frac{F}{\mu}}$$

Total time taken $= \frac{L}{v_1} + \frac{L}{v_2} + \frac{L}{v_3} + \frac{L}{v_4}$

$$= \frac{L}{\sqrt{F/\mu}}[1 + 2 + 3 + 4] = \frac{10L}{\sqrt{F/\mu}}$$

23. (c) Frequency of wave is same on all the four strings. So,

$$\lambda_1 = \frac{v_1}{f}, \quad \lambda_2 = \frac{v_2}{f} = \frac{v_1}{2f}$$

$$\lambda_3 = \frac{v_3}{f} = \frac{v_1}{3f}, \quad \lambda_4 = \frac{v_1}{4f}$$

$$\lambda_1 : \lambda_2 : \lambda_3 : \lambda_4 = 1 : \frac{1}{2} : \frac{1}{3} : \frac{1}{4}$$

$$= \frac{12 : 6 : 4 : 3}{12}$$

Required rate is $\lambda_4 : \lambda_3 : \lambda_2 : \lambda_1 = 3 : 4 : 6 : 12$

24. (c) $\omega = 2\pi f = 6\pi$ rad s^{-1},

$$k = \frac{\omega}{v} = \frac{6\pi}{15} = 1.26 \text{ m}^{-1}$$

$$\delta = \frac{\pi}{2}$$

$$y = A\sin(\omega t - kx + \delta)$$

$\Rightarrow$ $y = (0.1 \text{ m})\sin\left(6\pi t - 1.26x + \frac{\pi}{2}\right)$

$$y = (0.1 \text{ m})\cos(18.8t - 1.26x)$$

25. (a) At $x = 2.5$ m,

$$y = (0.1 \text{ m})\cos(18.8t - 1.26 \times 2.5)$$

or $y = (0.1 \text{ m})\cos(18.8t - \pi)$

$$= -(0.1 \text{ m})\cos(18.8t)$$

26. (d) $\lambda = \frac{v}{f} = \frac{15}{3} = 5$ m

$\therefore$ Time taken to reach 2.5 m is $T/2$.

$\therefore$ $\Delta\phi = \frac{2\pi}{T} \cdot \Delta t = \frac{2\pi}{T} \cdot \frac{T}{2} = \pi$

27. (4) The linear mass density is

$$\mu = \frac{5 \times 10^{-3}}{50 \times 10^{-2}} = 0.01 \text{ kg m}^{-1}$$

The wave sped is $v = \sqrt{F/\mu}$, where F is the tension in the string.

$\therefore$ $F = \mu v^2 = (0.01)(80)^2 = 64$ N

We know that, Young's modulus is $Y = \frac{F/A}{\Delta L/L}$

So, the extension is $\Delta L = \frac{FL}{Ay} = \frac{64 \times 0.5}{10^{-6} \times 8 \times 10^{11}}$

$$= 4 \times 10^{-5} \text{ m}$$

or $\Delta L = 4 \times 10^{-2}$ mm

28. (2) Particle velocity is $v_p = \frac{\partial x}{\partial t} = 40[-\sin(50\pi t - 0.02\pi y)](50\pi)$

or $v_p = -2000\pi\sin[(50\pi t - 0.02\pi y]$

$[v_p]_{\text{at } y = 25 \text{ cm when } t = 0.005 \text{ s}}$

$$= -2000\pi\sin[50\pi(0.005) - (0.02\pi)(25)]$$

$$= -2000\pi\sin(0.25\pi - 0.5\pi) = -2000\pi\sin\left(\frac{-\pi}{4}\right)$$

$$= 2000\pi\left(\frac{1}{\sqrt{2}}\right) = 1000\pi\sqrt{2} \text{ cm s}^{-1}$$

$$= 10\pi\sqrt{2} \text{ ms}^{-1}$$

$\therefore$ $\alpha = 2$

29. (4) Tension is the string is $F = m(g + a) = 4(10 + 2)$ 48 N

Relative to the string speed of wave is

$$v = \sqrt{\frac{F}{\mu}} = \sqrt{\frac{48}{19.2 \times 10^{-3}}}$$

or $v = 50 \text{ ms}^{-1} = 12.5 \times 4 \text{ ms}^{-1}$

$\Rightarrow$ $\beta = 4$

30. (4) $y = \frac{0.8}{3[x^2 + 8xt + 16t^2] + 4}$

$$= \frac{0.8}{3(x + 4t)^2 + 4}$$

Speed of wave $= \frac{\text{Coefficient of } t}{\text{Coefficient of } x} = \frac{4}{1} = 4 \text{ ms}^{-1}$

31. (2) $a_{max} = \omega^2 A = g$

$$\omega = \frac{2\pi v}{\lambda}, \quad v = \sqrt{\frac{F}{\mu}}$$

$\therefore$ $\omega^2 = \frac{4\pi^2}{\lambda^2} \times \frac{F}{\mu}$

$\therefore$ $a_{max} = \frac{4\pi^2 F}{\lambda^2 \mu} \cdot A_{min} = g$

$$A_{min} = \frac{g\lambda^2\mu}{4\pi^2 F} = \frac{\lambda^2\mu}{4F} \quad (\because \pi^2 = g)$$

$$A_{min} = \frac{(0.5)^2(0.1)}{4(3.125)}$$

$$= 2 \times 10^{-3} \text{ m} = 2 \text{ mm}$$

JEE Main & AIEEE Archive

1. (a) By superposition principle,

$$y = y_1 + y_2$$
$$= A\sin(\omega t - kx) + A\sin(\omega t + kx)$$
$$y = 2A\sin\omega t\cos kx$$

Clearly, it is equation of standing wave for position of nodes $y = 0$.

i.e. $x = (2n+1)\dfrac{\lambda}{4}$

$\Rightarrow \left(n + \dfrac{1}{2}\right)\dfrac{\lambda}{2}, n = 0, 1, 2, 3$

2. (d) As $v = \sqrt{\dfrac{T}{\mu}}$

So, $T = \mu v^2 = \mu\dfrac{\omega^2}{k^2}$

$$= 0.04\frac{(2\pi/0.004)^2}{(2\pi/0.50)^2} = 6.25\text{ N}$$

3. (a) As general equation is $y(x, t) = a\cos(kx - \omega t)$

where, $k = \dfrac{2\pi}{\lambda}$ and $\omega = \dfrac{2\pi}{T}$

$\therefore \quad y(x, t) = 0.005\cos(\alpha x - \beta t)$

$\Rightarrow \quad \dfrac{2\pi}{\lambda} = \alpha$ and $\dfrac{2\pi}{T} = \beta$

So, $\alpha = \dfrac{2\pi}{0.08} = 25\pi\text{ m}^{-1}$

and $\beta = \dfrac{2\pi}{2} = \pi\text{ s}^{-1}$

4. (b) As given $y = 10^{-6}\sin\left(100t + 20x + \dfrac{\pi}{4}\right)$ ...(i)

Comparing it with

$y = a\sin(\omega t + kx + \phi)$, we obtain ...(ii)

$$\omega = 100\text{ rad s}^{-1}, k = 20\text{ m}^{-1}$$

$\therefore \quad v = \dfrac{\omega}{k} = \dfrac{100}{20} = 5\text{ ms}^{-1}$

5. (a) The given equation of wave,

$$y = 10^{-4}\sin\left(600t - 2x + \frac{\pi}{3}\right) \quad \text{...(i)}$$

Standard equation of wave

$$y = a\sin(\omega t - kx + \phi) \quad \text{...(ii)}$$

Now, comparing Eqs. (i) and (ii), we get

$$\omega = 600\text{ rad s}^{-1} \text{ and } k = 2\text{ m}^{-1}$$

$\therefore$ Velocity of wave $= \dfrac{\omega}{k} = \dfrac{600}{2} = 300\text{ ms}^{-1}$

6. (c) $x = 4(\cos\pi t + \sin\pi t)$

$$= \frac{4}{\sqrt{2}} \times \sqrt{2}[\cos\pi t + \sin\pi t]$$
$$x = 4\sqrt{2}\left[\sin\pi t\cdot\frac{1}{\sqrt{2}} + \frac{1}{\sqrt{2}}\cos\pi t\right]$$
$$x = 4\sqrt{2}\left[\sin\pi t + \cos\frac{\pi}{4} + \cos\pi t\sin\frac{\pi}{4}\right]$$
$$x = 4\sqrt{2}\sin\left[\pi t + \frac{\pi}{4}\right]$$

So, amplitude $= 4\sqrt{2}$

JEE Advanced & IIT JEE Archive

1. (b)

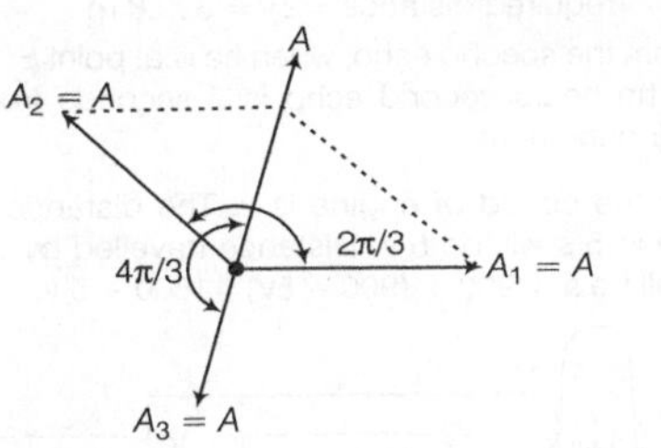

Resultant amplitude of x_1 and x_2 is A at angle $\left(\dfrac{\pi}{3}\right)$ from A_1. To make resultant of x_1, x_2 and x_3 to be zero. A_3 should be equal to A at angle $\phi = \dfrac{4\pi}{3}$ as shown in figure.

Alternate Method

It we substitute, $x_1 + x_2 + x_3 = 0$

or $A\sin\omega t + A\sin\left(\omega t + \dfrac{2\pi}{3}\right) + B\sin(\omega t + \phi) = 0$

Then by applying simple mathematics we can prove that

$$B = A$$

and $\phi = \dfrac{4\pi}{3}$.

2. (a) $f \propto v \propto \sqrt{T}$

$$f_{AB} = 2f_{CD}$$

$\therefore \quad T_{AB} = 4T_{CD}$ ...(i)

Further $\quad \Sigma\tau_p = 0$

$\therefore \quad T_{AB}(x) = T_{CD}(l - x)$

or $\quad 4x = l - x$ (as $T_{AB} = 4T_{CD}$)

or $\quad x = l/5$

3. (a) Particle velocity $v_p = -v$ (slope of y-x graph)

Here, $v = +$ ve, as the wave is travelling in positive x-direction.

Slope at P is negative.

$\therefore$ Velocity of particle is in positive y(or $\hat{j}$) direction.

3. Sound

DPP-1 Basic Parameters of Sound Waves

1. (i) Sound waves from the different instruments of performers on stage reach the audience simultaneously. This shows that the speed of sound is same for all wavelengths.

(ii) Sound travels faster in iron than in air. So, there is a difference between the time taken by sound to travel through iron and air. This is why the observer hears two sounds.

(iii) During the day, the temperature of air is maximum near the ground and it progressively diminishes upwards. Therefore, the velocity of sound is greater near the ground and it decreases as we move upwards ($v \propto \sqrt{T}$). The plane wavefronts, initially vertical, are turned upwards. So, the sound rays curl up during the day. At night the conditions are reversed. So, sound rays curl down, producing better, audilibility of distant sounds.

2. (i) Given, $y = 50 \cos (1800t - 5.3x)$

Note *Here, y and x are along the same line.*

Displacement amplitude, $A = 50\ \mu m$

$$k = 5.3 \text{ rad m}^{-1} \text{ (from the equation)}$$

$$\therefore \quad \frac{2\pi}{\lambda} = 5.3$$

$$\Rightarrow \quad \lambda = \frac{2\pi}{5.3} \text{ m} = 1.185 \text{ m}$$

$$\text{Required ratio} = \frac{A}{\lambda} = \frac{50 \times 10^{-6}}{1.185} = 4.22 \times 10^{-5}$$

(ii) $$v_p = \frac{\partial y}{\partial t} = -50 \times 10^{-6}(1800) \sin [1800t - 5.3x]$$

$$= -(0.09 \text{ ms}^{-1}) \sin [1800t - 5.3x]$$

Wave speed is

$$v = \frac{\omega}{k} = \frac{1800}{5.3} = 339.6 \text{ ms}^{-1}$$

Required ratio is

$$\frac{(v_p)_{max}}{v} = \frac{0.09}{339.6} = 2.65 \times 10^{-4}$$

3. Given, $f = 100$ Hz, $v = 350 \text{ ms}^{-1}$

$$T = \frac{1}{f} = 0.01 \text{ s}$$

(i) $$\Delta\phi = \frac{2\pi}{T} \cdot \Delta t = \frac{2\pi}{0.01}(2.5 \times 10^{-3}) = \frac{\pi}{2} \text{ rad}$$

(ii) $$\lambda = \frac{v}{f} = \frac{350}{100} = 3.5 \text{ m}$$

$$\Delta\phi = \frac{2\pi}{\lambda} \cdot \Delta x = \frac{2\pi}{3.5} \times 10 \times 10^{-2} = \frac{2\pi}{35} \text{ rad}$$

4. (i) The standard form of equation is

$$\Delta p = (\Delta p)_{max} \sin (\omega t - kx)$$

Given, $\Delta p = 0.02 \sin (3000t - 9x)$

$$\therefore \quad \omega = 3000 \text{ rads}^{-1}$$

$$\Rightarrow \quad f = \frac{\omega}{2\pi} = \frac{3000}{2\pi} = \frac{1500}{\pi} \text{ Hz}$$

$$\lambda = \frac{2\pi}{k} = \frac{2\pi}{9} \text{ m}$$

$$v = f\lambda = \frac{1500}{\pi} \times \frac{2\pi}{9} = \frac{1000}{3} \text{ ms}^{-1}$$

(ii) Maximum variation in pressure (from the equation) is

$$(\Delta p)_{max} = 0.02 \text{ Pa}$$

$\therefore$ Maximum pressure $= (1.01 \times 10^5 + 0.02)$ Pa

Minimum pressure $= (1.01 \times 10^5 - 0.02)$ Pa

5. (a) The situation is shown in the figure. The sound reaching the ground observer O, was emitted by the plane, when it was at the point P vertically above his head. The time taken by the sound to reach the observer is $t = \frac{2000}{340} = \frac{100}{17}$ s.

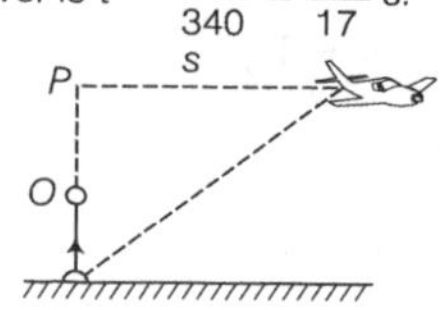

The distance moved by the plane during this period is

$$s = vt = \left(510 \times \frac{5}{18}\right)\left(\frac{100}{17}\right) = 833.33 \text{ m}$$

6. (b) Motorist is travelling with a velocity, $v = 36 \times \frac{5}{18} = 10 \text{ ms}^{-1}$.

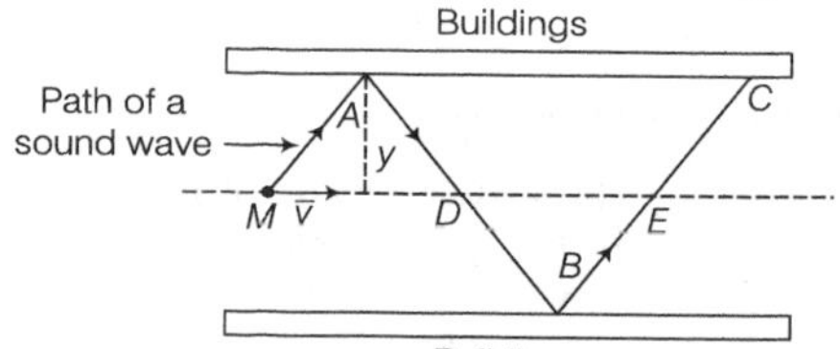

So, $MD = 10$ m

He hears first echo after 1 s. So, when he travels a distance MD, he hears first echo (i.e. at D).

In this time, distance travelled by sound $= MA + AD = 330$ m

$$\therefore \quad AD = 165 \text{ m}$$

$$\therefore \quad y^2 + 5^2 = 165^2 \Rightarrow y = 164.9 \text{ m}$$

So, required distance $= 2y = 329.8$ m

He hears the second echo, when he is at point E. So, the time after which he hears second echo is 2 seconds from the instant of sounding the horn.

7. (c) Let the speed of engine is v. The distance travelled by the engine in 5 s will be $5v$. Distance travelled by the sound in this time will be $s = 900 + (900 - 5v) = 1800 - 5v$

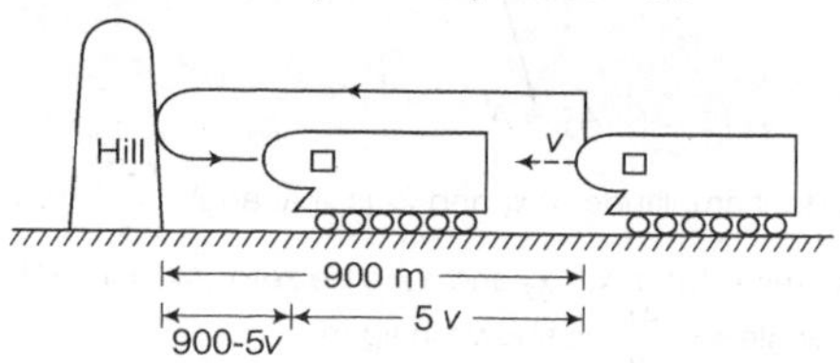

Time interval between the original sound and its echo is

$$t = \frac{s}{v_{sound}} \Rightarrow 5 = \frac{1800 - 5v}{330}$$

or $$v = 30 \text{ ms}^{-1} = 30 \times \frac{18}{5} \text{ km h}^{-1} = 108 \text{ km h}^{-1}$$

8. (d) Distance s travelled by sound in 8 s is $s = 8 \times 330 = 2640$ m

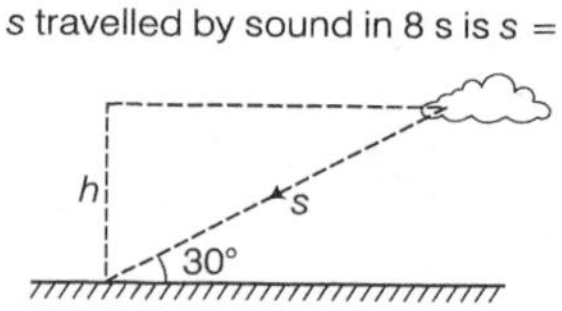

$$\sin 30° = \frac{h}{s} \Rightarrow h = s (\sin 30°) = 2640 \times \frac{1}{2}$$

or $$h = 1320 \text{ m} = 1.32 \text{ km}$$

9. (b) From the given equation,

$$(\Delta p)_m = 12 \text{ Pa}$$

$$k = 8.18 \text{ m}^{-1} \text{ and } \omega = 2700 \text{ s}^{-1}$$

$$\therefore \quad \lambda = \frac{2\pi}{k} = \frac{2\pi}{8.18} \text{ and } f = \frac{\omega}{2\pi} = \frac{2700}{2\pi}$$

So, speed of wave is

$$v = f\lambda = \frac{2700}{2\pi} \times \frac{2\pi}{8.18} = 330 \text{ ms}^{-1}$$

$$v = \sqrt{\frac{B}{\rho}}$$

$$B = \rho v^2 = 1.29 \times (330)^2 = 140.5 \times 10^3 \text{ Nm}^{-2}$$

We know

$$(\Delta p)_m = BAk$$

$$\therefore \quad A = \frac{(\Delta p)_m}{Bk}$$

$$= \frac{12}{140.5 \times 10^3 \times 8.18} = 1.044 \times 10^{-5} \text{ m}$$

10. (c) Audible range frequency = 20 Hz to 20000 Hz

$$\therefore \quad f_{min} = 20 \text{ Hz and } f_{max} = 20000 \text{ Hz}$$

Speed of sound in water is $v = 1450 \text{ ms}^{-1}$ (given)

$$\therefore \quad \lambda_{min} = \frac{v}{f_{max}} = \frac{1450}{20000} = 0.0725\text{m} = 72.5 \text{ mm}$$

and $$\lambda_{max} = \frac{v}{f_{min}} = \frac{1450}{20} = 72.5 \text{ m}$$

11. (d) Given, $\lambda = 35 \times 10^{-2}$ m, $(\Delta p)_m = 14$ Pa, $A = 5.5 \times 10^{-6}$ m

$$\therefore \quad k = \frac{2\pi}{\lambda} = \frac{2\pi}{0.35} = 17.94 \text{ m}^{-1}$$

$$(\Delta p)_m = BAk \Rightarrow B = \frac{(\Delta p)_m}{Ak} = \frac{14}{5.5 \times 10^{-6} \times 17.94}$$

$$= 1.42 \times 10^5 \text{ Pa}$$

12. (c) $$(\Delta p)_{max} = BAk = BA \cdot \frac{2\pi}{\lambda} = 5 \times 10^5 \times 10^{-4} \times \frac{2\pi}{0.2}$$

$$\approx 1571 \text{ Pa}$$

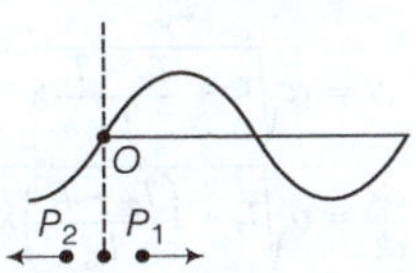

In the figure, particles P_1 has positive displacement and particle P_2 negative displacement. So, at O pressure should be minimum. And we know pressure is 90° out of phase with displacement wave.

13. (d) Pressure variation in the medium is

$$\Delta p = -B\frac{\Delta V}{V} = -B\frac{\Delta s}{\Delta x}$$

where B is bulk modulus of the gas.

14. (a,c) Pressure amplitude is $(\Delta p)_m = BAk$

where, $k = \frac{\omega}{v} = \frac{2\pi f}{v}$ and $B = \rho v^2$ $\quad \left(\because v = \sqrt{\frac{B}{\rho}}\right)$

$$\therefore \quad (\Delta p)_m = \rho v^2 \cdot A \cdot \frac{2\pi f}{v}$$

$$A = \frac{(\Delta p)_m}{2\pi \rho v f}$$

So, $$A_1 = \frac{10}{2 \times 3.14 \times 1.29 \times 340 \times 100} = 3.63 \times 10^{-5} \text{ m}$$

Now, $$(\Delta p)_m = 2\pi \rho v f A$$

$$\therefore \quad (\Delta p)_{m2} = 2 \times 3.14 \times 1.29 \times 340 \times 300 \times 10^{-7}$$

$$= 8.26 \times 10^{-2} \text{ Nm}^{-2}$$

15. (b,d) Let displacement at particle x is given by

$$y = A \sin(\omega t - kx)$$

then pressure at the same point is given by

$$\Delta p = (\Delta p)_m \cos(\omega t - kx)$$

where $(\Delta p)_m = BAk$

$$B = \rho v^2 = 1.29 \times (340)^2 = 149124 \text{ Pa}$$

$$\omega = 2\pi f = 2\pi(1000) = 2000\pi \text{ s}^{-1}$$

$$k = \frac{\omega}{v} = \frac{2000\pi}{340} = 18.47 \text{ m}^{-1}$$

Given, $$(\Delta p)_m = 10 \text{ Pa}$$

$$\therefore \quad A = \frac{(\Delta p)_m}{Bk} = \frac{10}{149124 \times 18.47}$$

$$= 3.63 \times 10^{-6} \text{ m}$$

When ΔP is maximum, y is minimum,

$$\therefore \quad y_1 = 0$$

16. (a,c,d) At A, particles are crowded, so maximum pressure occurs. At C, particles are displaced away from it. So, it corresponds to minimum pressure. At B, right side particles have negative displacemnet and left side particles also have same negative displacement. So, at B pressure is normal i.e. atmospheric.

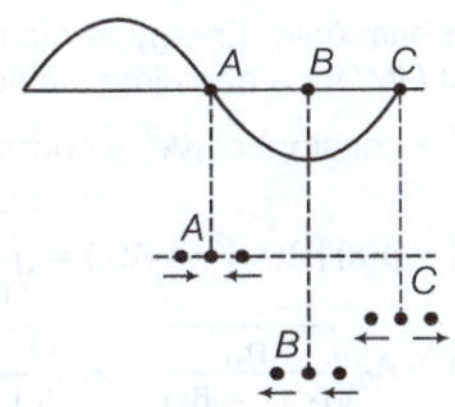

Note $\Delta p = -B\frac{dy}{dx} = B\left(-\frac{dy}{dx}\right)$

When the slope is negative, Δp is positive and maximum pressure occurs. When slope positive, Δp is negative and minimum pressure occurs. When slope is zero, Δp is zero and pressure is normal.

17. (c) When the pulse is incident on the wall, pressure gets doubled and the detector can record only, when the pressure is 16 Pa and above.

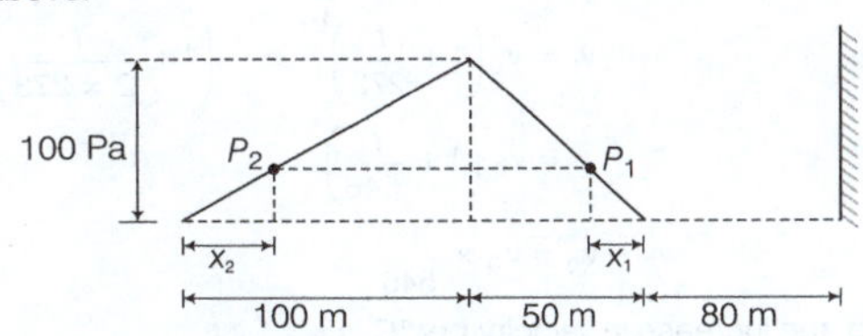

Let at point p_1 pressure is 8 Pa. When this point reaches the wall, detector starts showing pressure.

Now, $$x_1 = \frac{50}{100} \times 8 = 4 \text{ m}$$

$\therefore$ Required time is

$$t_1 = \frac{80 + x_1}{400} = \frac{84}{400} \text{ s} = 0.21 \text{ s}$$

18. (a) The detector will record the pulse till point P_2 (where, pressure is 8 Pa) reaches the wall. Distance between P_2 and the wall is $s = (230 - x_2)$ m

Now, $$x_2 = \frac{100}{100} \times 8 = 8 \text{ m}$$

$$\therefore \quad s = 230 - 8 = 222 \text{ m}$$

$$t_2 = \frac{222}{400} = 0.555 \text{ s}$$

Required time interval $= t_2 - t_1$

$$= 0.555 - 0.210$$

$$= 0.345 \text{ s}$$

19. (c) Area of the wall, $A = 2 \times 2 = 4\ m^2$

Maximum pressure on the wall $= 100 \times 2 = 200$ Pa

$\therefore \quad F_{max} = 200 \times 4 = 800$ N

20. (a) Total time that the pressure acts on the wall is

$$\Delta t = \Delta t_1 + \Delta t_2 = \frac{50}{400} + \frac{100}{400}$$
$$= 0.125 + 0.25 \text{ s}$$
$$= 0.375 \text{ s}$$

Average force is

$$F_{av} = \frac{0 + 800}{2} = 400 \text{ N}$$

$\therefore$ Impulse is $\quad J = F_{av} \times \Delta t_1 + F_{av} \times \Delta t_2$

$$= 400 \times 0.125 + 400 \times 0.25$$
$$= 150 \text{ Ns}$$

21. (d) At the cross-section MM', the density is $\rho = \rho_0(1 + \beta x)$

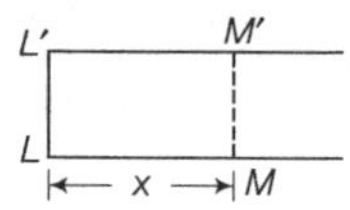

Source is simple harmonic. Energy at cross-section LL' must be equal to enegy at MM' for a one dimensional wave.

$\therefore \quad \frac{1}{2}\rho\omega^2 A^2 = \text{constant or } \rho A^2 = \text{constant}$

$\therefore \quad \rho_0 A_0^2 = \rho(x)\,[A(x)]^2 \Rightarrow A(x) = \sqrt{\frac{\rho_0}{\rho(x)} A_0^2}$

or $\quad A(x) = A_0 \sqrt{\frac{\rho_0}{\rho_0(1 + \beta x)}} = A_0 \sqrt{\frac{1}{1 + \beta x}}$

22. (c) $f = \frac{\omega}{2\pi}$ remains constant at all cross-sections. Velocity at cross- section MM' is

$$v(x) = \sqrt{\frac{Y}{\rho(x)}} = \sqrt{\frac{Y}{\rho_0(1 + \beta x)}}$$
$$\lambda(x) = \frac{v(x)}{f} = \frac{2\pi}{\omega}\sqrt{\frac{Y}{\rho_0(1 + \beta x)}}$$

23. (b) We know, $\quad v(x) = \sqrt{\frac{Y}{\rho_0(1 + \beta x)}}$

$\therefore \quad \frac{dx}{dt} = \sqrt{\frac{Y}{\rho_0}} \frac{1}{(1 + \beta x)^{1/2}} \Rightarrow (1 + \beta x)^{\frac{1}{2}} dx = \sqrt{\frac{Y}{\rho_0}}\, dt$

$$\int_0^x (1 + \beta x)^{1/2} dx = \sqrt{\frac{Y}{\rho_0}} \int_0^t dt \Rightarrow \frac{[(1 + \beta x)^{3/2}]_0^x}{(3/2)\,\beta} = \sqrt{\frac{Y}{\rho_0}}\, t$$

$\Rightarrow \quad t = \frac{2}{3\beta}\sqrt{\frac{\rho_0}{Y}}(1 + \beta x)^{3/2}$

24. (a) Standard equation of the wave is $y = A \sin(\omega t - kx)$

Here, $\quad A = A(x) = A_0\sqrt{\frac{1}{1 + \beta x}}$

and $\quad k = k(x) = \frac{2\pi}{\lambda(x)} = \omega\sqrt{\frac{\rho_0(1 + \beta x)}{Y}}$

$\therefore \quad y = A_0\sqrt{\frac{1}{1 + \beta x}} \sin\left[\omega t - \omega\sqrt{\frac{\rho_0(1 + \beta x)}{Y}} \cdot x\right]$

$$= A_0\sqrt{\frac{1}{1 + \beta x}} \sin\omega\left[t - \sqrt{\frac{\rho_0(1 + \beta x)}{Y}} \cdot x\right]$$

Note *y and x are along the same line.*

DPP-2 **Speed of Sound**

1. (c) We know, $v \propto \sqrt{T}$

$\therefore \quad \frac{v_t}{v_0} = \sqrt{\frac{T_t}{T_0}} = \sqrt{\frac{273 + t}{273}}$

or $\quad v_t = v_0\left(1 + \frac{t}{273}\right)^{1/2} \simeq v_0\left(1 + \frac{t}{2 \times 273}\right)$

$$= v_0\left(1 + \frac{t}{546}\right)$$

$\therefore \quad v_t - v_0 = v_0 \times \frac{t}{546}$

So, the increase in velocity per °C is

$$\frac{v_0 \times 1}{546} = \frac{332}{546} \simeq 0.61 \text{ ms}^{-1}$$

2. (b) $\frac{v_t}{v_0} = \sqrt{\frac{T_t}{T_0}}$, now, $v_t = 2v_0$

$$\frac{2v_0}{v_0} = \sqrt{\frac{T_t}{273}}$$

$\Rightarrow \quad T_t = 4 \times 273 = 1092$ K

or $\quad t = 1092 - 273 = 819°C$

3. (a) $v = \sqrt{\frac{\gamma p}{\rho}} = \sqrt{\frac{\gamma RT}{M_0}}$

Given, $\quad \gamma = 1.3,\ R = 8.3\ \text{J mol}^{-1}\,°\text{C}^{-1}$

$T = -130\ °C = -130 + 273 = 143$ K

M_0 = Molecular weight of methane (CH_4)

$= 12 + 4 = 16$ g

$\therefore \quad v = \sqrt{\frac{1.3 \times 8.3 \times 143}{16 \times 10^{-3}}} \simeq 311 \text{ ms}^{-1}$

4. (d) The temperature at a distance x from point I is

$$T_x = T_1 + \frac{T_2 - T_1}{L_0} x \qquad (\because \text{Linear variation})$$

At this point, $\quad v = \alpha\sqrt{T_1 + \frac{T_2 - T_1}{L_0} x}$

or $\quad \frac{dx}{dt} = \alpha\sqrt{T_1 + \left(\frac{T_2 - T_1}{L_0}\right) x}$

$$\frac{1}{\sqrt{T_1 + \left(\frac{T_2 - T_1}{L_0}\right) x}} dx = \alpha\, dt$$

$$\int_0^{L_0} \frac{1}{\sqrt{T_1 + \left(\frac{T_2 - T_1}{L_0}\right) x}} dx = \alpha \int_0^t dt$$

$\Rightarrow \quad \frac{2L_0}{\alpha(T_2 - T_1)}\left[\sqrt{T_1 + \frac{T_2 - T_1}{L_0} x}\right]_0^{L_0} = t$

$$t = \frac{2L_0}{\alpha(T_2 - T_1)}(\sqrt{T_2} - \sqrt{T_1})$$

$\Rightarrow \quad t = \frac{2L_0}{\alpha(\sqrt{T_2} + \sqrt{T_1})}$

Alternate Method

Given that $\quad v = \alpha\sqrt{T}$

$\therefore$ Velocity at I is $\quad v_1 = \alpha\sqrt{T_1}$ and

Velocity of F is $\quad v_2 = \alpha\sqrt{T_2}$

Average velocity is $v = \frac{v_1 + v_2}{2} = \frac{\alpha}{2}(\sqrt{T_1} + T_2)$

Time taken from A to B is $t = \frac{L_0}{v} = \frac{2L_0}{\alpha(\sqrt{T_1} + \sqrt{T_2})}$

5. (a) $\frac{v_1}{v_2} = \sqrt{\frac{T_2}{T_2}}$

$\Rightarrow \quad v_2 = \sqrt{\frac{T_2}{T_1}} \cdot v_1$

$$v_2 = \sqrt{\frac{273 + 32}{273 + 17}} \times 340 = 349 \text{ ms}^{-1}$$

6. (b) $v = \sqrt{\frac{\gamma RT}{M_0}}$, Given $v_{H_2} = v_{O_2}$

$\therefore \quad \sqrt{\frac{\gamma_{H_2} RT_{H_2}}{(M_0)_{H_2}}} = \sqrt{\frac{\gamma_{O_2} RT_{O_2}}{(M_0)_{O_2}}}$

As O_2 and H_2 are diatomic,

$$\gamma_{H_2} = \gamma_{O_2}$$

$\therefore \quad T_{H_2} = \frac{(M_0)_{H_2}}{(M_0)_{O_2}} \cdot T_{O_2}$

$$= \frac{1}{16}(100 + 273) = 23.31 \text{ K}$$

$\therefore \quad t_{H_2} = 23.31 - 273 = -249.7°\text{C}$

7. (a) Let the densities of two gases are ρ_1 and ρ_2 respectively.

$$m_1 + m_2 = m_{mix}$$

$$\rho_1 V_1 + \rho_2 V_2 = \rho_{mix}(V_1 + V_2)$$

$\therefore \quad \rho_{mix} = \frac{\rho_1 V_1 + \rho_2 V_2}{V_1 + V_2}$

Speeds of sound in respective gases are

$$v_1 = \sqrt{\frac{\gamma p}{\rho_1}} \text{ and } v_2 = \sqrt{\frac{\gamma p}{\rho_2}}$$

$$v_{mix} = \sqrt{\frac{\gamma p}{\rho_{mix}}}$$

$$= \sqrt{\frac{\gamma p (V_1 + V_2)}{(\rho_1 V_1 + \rho_2 V_2)}}$$

or $\quad v_{mix} = \sqrt{\frac{\gamma p (V_1 + V_2)}{\frac{\gamma p}{v_1^2} \cdot V_1 + \frac{\gamma p}{v_2^2} \cdot V_2}}$

$$= \sqrt{\frac{v_1^2 v_2^2 (V_1 + V_2)}{V_1 v_2^2 + V_2 v_1^2}}$$

or $\quad v_{mix} = v_1 v_2 \sqrt{\frac{V_1 + V_2}{V_1 v_2^2 + V_2 v_1^2}}$

8. (d) Number of moles of first gas is

$$n_1 = \frac{m_1}{M_{01}}$$

Number of moles of second gas is

$$n_2 = \frac{m_2}{M_{02}}$$

where, M_{01}, M_{02} are molecular masses of gases. Let M_{0mix} is the molecular mass of mixture.

Then, $\quad m_1 + m_2 = \left(\frac{m_1}{M_{01}} + \frac{m_2}{M_{02}}\right) M_{0mix}$

$\therefore \quad M_{0mix} = \frac{m_1 + m_2}{\frac{m_1}{M_{01}} + \frac{m_2}{M_{02}}} \quad \ldots \text{(i)}$

We know that $\quad v_1 = \sqrt{\frac{\gamma RT}{M_{01}}}, v_2 = \sqrt{\frac{\gamma RT}{M_{02}}}$

and $\quad v_{mix} = \sqrt{\frac{\gamma RT}{M_{0mix}}}$

$\therefore \quad M_{01} = \frac{\gamma RT}{v_1^2}, M_{02} = \frac{\gamma RT}{v_2^2} \text{ and } M_{0mix} = \frac{\gamma RT}{v_{mix}^2}$

Using these in Eq. (i), we get

$$\frac{\gamma RT}{v_{mix}^2} = \frac{m_1 + m_2}{\frac{m_1 v_1^2}{\gamma RT} + \frac{m_2 v_2^2}{\gamma RT}}$$

$\Rightarrow \quad v_{mix}^2 = \frac{m_1 v_1^2 + m_2 v_2^2}{m_1 + m_2}$

$\therefore \quad v_{mix} = \sqrt{\frac{m_1 v_1^2 + m_2 v_2^2}{m_1 + m_2}}$

9. (c) $v = f\lambda = 250 \times 8 = 2000 \text{ ms}^{-1}$

Also, $\quad v = \sqrt{\frac{B}{\rho}} \Rightarrow B = \rho v^2$

$\therefore \quad B = 900 (2000)^2 = 36 \times 10^8 \text{ Pa}$

$$= 3.6 \times 10^9 \text{ Pa}$$

10. (d) Speed of sound is

$$v = \frac{\text{Distance}}{\text{Time}} = \frac{1.5 \text{ m}}{3.9 \times 10^{-4} \text{ s}}$$

$$= 3846 \text{ ms}^{-1}$$

Density of metal is $\rho = 6400 \text{ kg m}^{-3}$ (given)

We know, $\quad v = \sqrt{\frac{Y}{\rho}}$

$\Rightarrow \quad Y = \rho v^2$

or $\quad Y = 6400 (3846)^2 = 9.47 \times 10^{10} \text{ Pa}$

11. (a) This concept is same as that of mirage i.e. total internal reflection of light wave due to varying refractive index of air of atmosphere.

Here, as we move up, if temperature increases, speed of sound increases ($v \propto \sqrt{T}$). So, the atmosphere becomes rarer and rarer. So, a sound wave from P_1 may keep bending away from normal, at one point total internal reflection may take place and ultimately sound will reach point P_2.

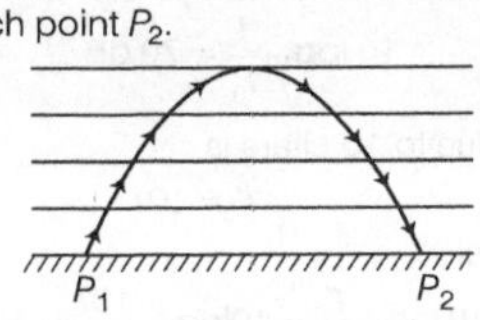

12. (a,b) $v_{O_2} = \sqrt{\frac{B}{\rho}} = \sqrt{\frac{1.41 \times 10^5}{1.43}} = 314 \text{ ms}^{-1}$

$$v_{He} = \sqrt{\frac{B}{\rho}} = \sqrt{\frac{1.7 \times 10^5}{0.18}} = 972 \text{ ms}^{-1}$$

13. (c) $v = \sqrt{\frac{\gamma p}{\rho}}$. But change in pressure (without change in temperature) has no effect on speed of sound because density also changes accordingly

$$v = \sqrt{\frac{\gamma RT}{M_0}}$$

$\Rightarrow \quad v \propto \sqrt{T}$

Hence, (i) $\rightarrow$ (q), (ii) $\rightarrow$ (r), (iii) $\rightarrow$ (q), (iv) $\rightarrow$ (s)

14. (b) Frequency depends on the source. So, it does not change in reflection and refraction.

$$v_{rarer} > v_{denser}$$

so, $\quad \lambda_{rarer} < \lambda_{denser}$

Thus, in refraction from rarer to denser medium wavelength decreases. When a wave is reflected from a denser medium, a phase change of π takes place.

In reflection, as the medium is same v does not change.

Hence, (i) $\rightarrow$ (r), (ii) $\rightarrow$ (p,r), (iii) $\rightarrow$ (q,r), (iv) $\rightarrow$ (p,r,s)

15. (9) Given, $v_L = 30 v_T$

Speed of longitudinal waves is $v_L = \sqrt{\dfrac{Y}{\rho}}$

Speed of transverse waves is $v_T = \sqrt{\dfrac{F}{\mu}} = \sqrt{\dfrac{F}{\rho A}}$

where A is cross- sectional area.

$$\therefore \quad 900\sqrt{\frac{F}{\rho A}} = \sqrt{\frac{Y}{\rho}} \Rightarrow \frac{F}{A} = \frac{Y}{900} = \frac{Y}{100 \times 9}$$

$$\therefore \quad x = 9$$

16. (3) Let A_i and A_t are the amplitudes of incident and transmitted waves respectively. Then,

$$A_t = \frac{2v_2}{v_1 + v_2} = \frac{2(100)}{200 + 100} = \frac{2}{3} \Rightarrow k = 3$$

17. (5) In a liquid, speed of sound is $v = \sqrt{\dfrac{B}{\rho}}$

$$\therefore \quad B = \rho v^2$$

By definition, $$B = -V\frac{\Delta p}{\Delta V}$$

$$\therefore \quad -V\frac{\Delta p}{\Delta V} = v^2 \rho \Rightarrow \Delta V = \frac{V(-\Delta p)}{\rho v^2}$$

$$\Delta p = \rho g H_2 - \rho g H_1 = 13.6 \times 981\,(75 - 200)$$
$$= -1.667 \times 10^6 \text{dyne cm}^{-2}$$

$$\therefore \quad \Delta V = \frac{(6 \times 10^3)(1.667 \times 10^6)}{(0.81)(1.28 \times 10^5)^2} = 0.75 \text{ cc}$$

or $$\Delta V = 15 \times 5 \text{ cc} \Rightarrow \beta = 5$$

18. (4) Source should be located midway $x = -8$ m and $x = 2$ m i.e. at $x = -3$. For the same wavefront to reach point (3), distance between S and (3) must be 5 m.

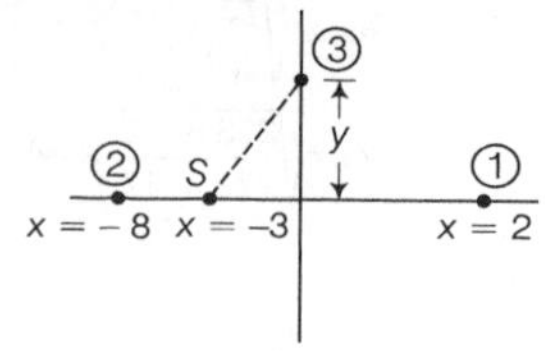

$$\therefore \quad y^2 + 3^2 = 5^2$$

$$\Rightarrow \quad y = 4 \text{ m}$$

19. (2) $1200 = \sqrt{\dfrac{\gamma p}{\rho_{H_2}}}$ ...(i)

Let x is volume of H_2 and y volume of O_2. Then,

$$(x + y)\rho_{mix} = x\rho_{H_2} + y\rho_{O_2} = x\rho_{H_2} + 16y\rho_{H_2}$$

$$\therefore \quad \rho_{mix} = \frac{x + 16y}{x + y} \cdot \rho_{H_2}$$

$$\therefore \quad 500 = \sqrt{\frac{\gamma p (x + y)}{(x + 16y)\rho_{H_2}}} \quad \text{...(ii)}$$

Dividing Eq. (i) by Eq. (ii), we get

$$\frac{12}{5} = \sqrt{\frac{x + 16y}{x + y}}$$

or $$\frac{x}{y} = \frac{2.2}{1} = \frac{11}{10} \times 2 \Rightarrow \alpha = 2$$

DPP-3 Intensity of Sound

1. (a) Let the intensity due to each sitar = I

Then, $$10 \log_{10}\frac{I}{I_0} = 70 \text{ dB} \quad \text{(given) ...(i)}$$

Now, intensity due to 10 sitars is

$$I' = 10I$$

Now, sound level

$$10 \log_{10}\frac{I'}{I_0} = 10\log_{10}\frac{10I}{I_0} = 10\log_{10} 10 + 10\log_{10}\frac{I}{I_0}$$

$$\therefore \quad \text{Sound level} = 10 \text{ dB} + 70 \text{ dB} \quad \text{(using Eq. (i))}$$
$$= 80 \text{ dB}$$

2. (a) Given, $I_2 = 32 I_1$

$$\beta_2 - \beta_1 = 10 \log_{10}\frac{I_2}{I_0} - 10\log_{10}\frac{I_1}{I_0}$$
$$= 10\log_{10}\frac{I_2}{I_1} = 10\log_{10} 32$$
$$= 15 \text{ dB}$$

3. (c) The solid angle subtended by right circular cone is $\Omega = 2\pi\,(1 - \cos\theta)$

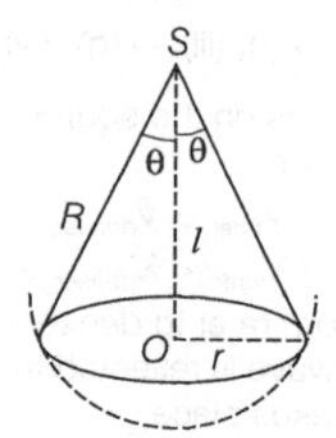

In case of a spherical wavefront, the power per unit solid angle is constant.

$\therefore$ Power per unit solid angle is

$$P_\Omega = \frac{4\pi l^2 I_0}{4\pi} = l^2 I_0$$

$\therefore$ Mean energy flow across the area enclosed by the ring is

$$P' = P_\Omega\, 2\pi(1 - \cos\theta)$$
$$= I_0 l^2 2\pi(1 - \cos\theta)$$
$$= 30 \times 1^2 \times 2\pi\left(1 - \frac{l}{\sqrt{l^2 + r^2}}\right)$$
$$= 60\pi\left(1 - \frac{1}{\sqrt{1 + 0.25}}\right) \simeq 20\ \mu\text{Js}^{-1} \text{ or } \mu\text{W}$$

4. (b)

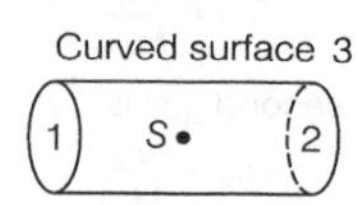

$$P = P_1 + P_2 + P_3$$
$$P_3 = P - (P_1 + P_2)$$
$$P_1 = P_2 = \frac{P}{2}(1 - \cos\theta)$$

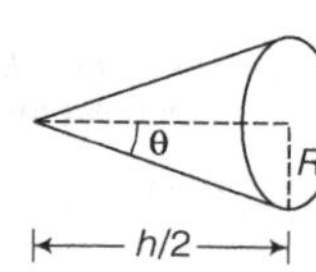

$$= \frac{P}{2}\left(1 - \frac{h/2}{\sqrt{R^2 + \dfrac{h^2}{4}}}\right)$$

$$= \frac{P}{2}\left[1 - \frac{h}{2\sqrt{R^2 + \dfrac{h^2}{4}}}\right]$$

$$P_1 = P_2 = \frac{P}{2}\left[1 - \frac{h}{\sqrt{4R^2 + h^2}}\right]$$

$$\therefore \quad P_3 = P - 2P_1 = P - P\left[1 - \frac{h}{\sqrt{4R^2 + h^2}}\right]$$

$$= \frac{Ph}{\sqrt{4R^2 + h^2}} = \frac{(0.1)\,(2)}{\sqrt{4(1)^2 + (2)^2}} = \frac{0.1}{\sqrt{2}}$$

or $P_3 = 0.07$ W

Note

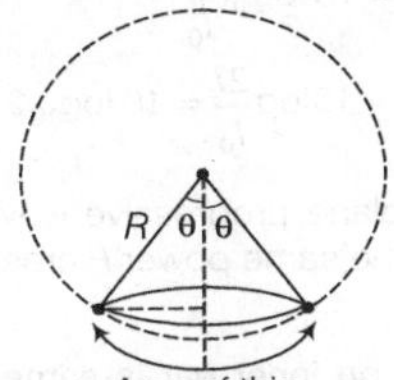

Area of this part of sphere $= A$

$A = 2\pi R^2(1 - \cos\theta)$

5. (b) Due to one dog, $I_1 = \dfrac{P}{2\pi r^2} = \dfrac{10^{-3}}{2\pi(5)^2}$

or $I_1 = 6.37 \times 10^{-6}\,\text{Wm}^{-2}$

Due to 5 dogs, $I_R = 5I_1 = 31.85 \times 10^{-6}\,\text{Wm}^{-2}$

Now, sound level is

$$\beta = 10\log_{10}\frac{I_R}{I_0} = 10\log_{10}\left(\frac{31.85 \times 10^{-6}}{10^{-12}}\right)$$

or $\beta = 10\log_{10}(31.85 \times 10^6)$

$= 10[\log_{10}(6.37 \times 5) + 6\log_{10}10\,]$

or $\beta = 10[\log_{10}6.37 + \log_{10}5 + 6\,]$

$= 10\,[0.8 + 0.7 + 6] = 75$ dB

6. (d) For maximum possible sound intensity, maximum pressure amplitude of wave can be at most equal to atmosphere pressure. Thus,

$(\Delta p)_{max} = 1.01 \times 10^5\,\text{Nm}^{-2}$

$$I = \frac{1}{2}\rho\omega^2 A^2 v = \frac{1}{2}\rho\left[\frac{(\Delta p)_{max} v}{BA}\right]^2 A^2 v$$

$$\left(\because (\Delta p)_{max} = BAk = BA\,\frac{\omega}{v}\right)$$

$$= \frac{1}{2}\rho\,\frac{(\Delta p)^2_{max} v^2}{(\rho v^2)^2}\cdot v \qquad (\because B = \rho v^2)$$

$$= \frac{1}{2}\cdot\frac{1}{\rho}\cdot\frac{1}{v}\cdot(\Delta p)^2_{max} = \frac{(\Delta p)^2_{max}}{2\rho v}$$

So, here $I = \dfrac{(1.01 \times 10^5)^2}{2\,(1.3)\,(332)} = 1.18 \times 10^7\ \text{Wm}^{-2}$

Corresponding sound level is

$$\beta = 10\log_{10}\frac{I}{I_0} \simeq 10\log\frac{10^7}{10^{-12}} = 190\ \text{dB}$$

7. (b) $\beta = 10\log_{10}\dfrac{I}{I_0}$

$\Rightarrow \quad 10\log_{10}\dfrac{I}{I_0} = 60$

$\dfrac{I}{I_0} = 10^6$

$\Rightarrow \quad I = I_0 \times 10^6 = 10^{-12} \times 10^6 = 10^{-6}\,\text{Wm}^{-2}$

Power of sound $= I \times$ area $= 10^{-6} \times 2 = 2\ \mu\omega$

Thus, energy absorbed in 10 h is

$E = P \times t = 2 \times 10^{-6} \times (10 \times 60 \times 60) = 72$ mJ

8. (d) β_1 (for signal) $= 10\log_{10}\dfrac{I_1}{I_0}$, β_2 (for noise) $= 10\log_{10}\dfrac{I_2}{I_0}$

Given, $\beta_1 - \beta_2 = 65$ dB

$\therefore \quad 10\log_{10}\dfrac{I_1}{I_0} - 10\log_{10}\dfrac{I_2}{I_0} = 65$

$10\log_{10}\dfrac{I_1}{I_2} = 65$

$\Rightarrow \quad \dfrac{I_1}{I_2} = 10^{6.5} = 10^6\sqrt{10} = 3.16 \times 10^6$

9. (d) $\beta = 10\log_{10}\left(\dfrac{I}{10^{-12}}\right)$

$\log_{10}\dfrac{I}{10^{-12}} = \dfrac{\beta}{10} \Rightarrow \dfrac{I}{10^{-12}} = 10^{\beta/10}$

$\Rightarrow \quad I = [10^{\beta/10} \times 10^{-12}]\ \text{Wm}^{-2}$

For your baby, $I_b = 10^{75/10}\cdot 10^{-12} = \sqrt{10} \times 10^{-5}\,\text{Wm}^{-2}$

For the music, $I_m = 10^{80/10}\cdot 10^{-12} = 10 \times 10^{-5}\,\text{Wm}^{-2}$

The combined intensity is

$I_{total} = I_b + I_m = 3.2 \times 10^{-5} + 10 \times 10^{-5}$

or $I_{total} = 13.2 \times 10^{-5}\,\text{Wm}^{-2}$

10. (b) $\beta_{total} = 10\log_{10}\left(\dfrac{I_{total}}{10^{-12}}\right) = 10\log_{10}\left(\dfrac{13.2 \times 10^{-5}}{10^{-12}}\right)$

$= 10\log_{10}(13.2 \times 10^7) = 10(\log_{10}13.2 + \log_{10}10^7)$

$= 10[1.12 + 7] = 81.2$ dB

11. (b,d) $I_{at\ 4m} = \dfrac{P_{av}}{4\pi r^2} = \dfrac{20}{4 \times 3.14 \times 4^2} = 9.95 \times 10^{-2}\,\text{Wm}^{-2}$

$\therefore \quad \beta = 10\log_{10}\left(\dfrac{9.95 \times 10^{-2}}{10^{-12}}\right) \simeq 10\log_{10}10^{11} = 110$ dB

$\beta' = \dfrac{\beta}{2} = 55$ dB

$\beta' = 10\log_{10}\dfrac{I'}{I_0} \Rightarrow \log_{10}\dfrac{I'}{I_0} = 5.5$

$\dfrac{I'}{I_0} = 10^{5.5} \Rightarrow I' = 10^{-12}(10^{5.5}) = 3.16 \times 10^{-7}\,\text{Wm}^{-2}$

$\therefore \quad \dfrac{P}{4\pi(r')^2} = 3.16 \times 10^{-7} \Rightarrow (r')^2 = \dfrac{20}{4\pi \times 3.16 \times 10^{-7}}$

$\Rightarrow \quad r' = 2.24$ km

12. (a,b) Sound level β is given by

$$\beta = 10\log_{10}\frac{I}{I_0}$$

$\log_{10}\dfrac{I}{I_0} = \dfrac{\beta}{10} \Rightarrow \dfrac{I}{I_0} = 10^{\beta/10}$

$\ln\dfrac{I}{I_0} = \dfrac{\beta}{10}\ln 10 \Rightarrow I = I_0 e^{\frac{(\ln 10)\beta}{10}}$

Differentiating w.r.t. β gives, $\dfrac{dI}{d\beta} = I_0\,e^{\frac{(\ln 10)\beta}{10}}\cdot\dfrac{(\ln 10)\beta}{10}$

or $\dfrac{dI}{d\beta} = \dfrac{\ln 10}{10}I \Rightarrow \dfrac{\Delta I}{I} = \dfrac{\ln 10}{10}\Delta\beta = 0.23$

$\therefore \quad x = 23\%$

Now, from the above equation, we get.

$\dfrac{dI}{I} = \dfrac{\ln 10}{10}d\beta = \dfrac{\ln 10}{10}\dfrac{d\beta}{dt}\cdot dt$

Given, $\dfrac{d\beta}{dt} = 1\ \text{dB/year}$

$\therefore \quad \dfrac{dI}{I} = (0.23 \times 1)\,dt \Rightarrow \displaystyle\int_I^{2I}\frac{1}{I}dI = (0.23)\int_0^t dt$

$\ln 2 = 0.23 \times t \Rightarrow t = \dfrac{0.693}{0.23} \simeq 3$ yr

$\therefore \quad y = 3$ yr

13. (a,b,d) From the wave equation, $k = 15\pi \text{m}^{-1}$, $\omega = 6000\pi \text{rads}^{-1}$

$$v = \frac{\omega}{k} = \frac{6000\pi}{15\pi} = 400 \text{ ms}^{-1}$$

Also, $$v = \sqrt{\frac{B}{\rho}} \Rightarrow \rho = \frac{B}{v^2} = \frac{1.6 \times 10^5}{(400)^2} = 1 \text{ kg m}^{-3}$$

Pressure amplitude is

$$(\Delta p)_{max} = BAk$$

$$\therefore \quad A = \frac{(\Delta p)_{max}}{Bk} = \frac{24\pi}{1.6 \times 10^5 \times 15\pi} = 10^{-5} \text{m}$$

Intensity at a distance r is $I = \frac{\text{Power}}{4\pi r^2}$

Also $$I = \frac{(\Delta p)_{max}}{2\rho v}$$

$$\therefore \quad \frac{\text{Power}}{4\pi r^2} = \frac{(\Delta p)_{max}}{2\rho v}$$

$$\Rightarrow \quad \text{Power} = \frac{(24\pi)^2 \times 4\pi\,(10)^2}{2(1)(400)} = 8.93 \times 10^3 \text{ W}$$

14. (b,d) $20 = 10\log_{10}\frac{I}{10^{-12}} \Rightarrow I = 10^{-10} \text{ Wm}^{-2}$

(This is constant for a plane wave)

$$v = \sqrt{\frac{B}{\rho}} = \sqrt{\frac{4 \times 10^4}{1}} = 200 \text{ ms}^{-1}$$

$$I = \frac{(\Delta p)^2_{max}}{2\rho v} \Rightarrow (\Delta p)_{max} = \sqrt{2I\rho v}$$

or $$(\Delta p)_{max} = \sqrt{2 \times 10^{-10} \times 1 \times 200} = 2 \times 10^{-4} \text{ Nm}^{-2}$$

Also, $$I = \frac{1}{2}\rho\omega^2 A^2 v$$

$$\Rightarrow \quad A = \sqrt{\frac{2I}{\rho\omega^2 v}} = \sqrt{\frac{2 \times 10^{-10}}{1 \times (500)^2 \times 200}}$$

or $$A = 2 \times 10^{-9} \text{m}$$

$$k = \frac{\omega}{v} = \frac{500}{200} = \frac{5}{2} \text{m}^{-1}$$

So, equation of wave is

$$y = 2 \times 10^{-9} \sin\left[500t - \frac{5}{2}x\right]$$

15. (a,d) $I = \frac{\text{Power}}{4\pi r^2} = \frac{0.8}{4\pi(1.5)^2} \Rightarrow I = 2.83 \times 10^{-2} \text{ Wm}^{-2}$

Also, $$I = \frac{(\Delta p)^2_{max}}{2\rho v} \Rightarrow (\Delta p)_{max} = \sqrt{2I\rho v}$$

or $$(\Delta p)_{max} = \sqrt{2 \times 2.83 \times 10^{-2} \times 1.29 \times 340} = 4.98 \text{ Nm}^{-2}$$

$$v = \sqrt{\frac{B}{\rho}}$$

$$\Rightarrow \quad B = \rho v^2 = 1.29 \times (340)^2 \text{Nm}^{-2} = 149124 \text{ Nm}^{-2}$$

$$k = \frac{\omega}{v} = \frac{2\pi \times 600}{340} = 11.08 \text{ m}^{-1}$$

$$(\Delta p)_{max} = BAk \Rightarrow A = \frac{(\Delta p)_{max}}{Bk}$$

$$\therefore \quad A = \frac{4.98}{149124 \times 11.08} = 3 \times 10^{-6} \text{m}$$

16. (c)

(i) Pressure amplitude is $(\Delta p)_m = BAk = \rho v^2 \cdot A \cdot \frac{\omega}{v}$

or $(\Delta p)_m = \rho v A\omega = (1.29)(340)(10^{-5}) \cdot (2\pi \times 1000)$

$= 27.54$ Pa

(ii) Intensity is $I = \frac{1}{2}\rho\omega^2 A^2 v = \frac{1}{2}(1.29)(2\pi \times 1000)^2 (10^{-5})^2 (340)$

$= 0.865 \text{ Wm}^{-2}$

(iii) Power is $P = \frac{1}{2}\rho\omega^2 A^2 \cdot Sv = \left(\frac{1}{2}\rho\omega^2 A^2 v\right)S$ (where, S = area)

or $P = (0.865)(\pi \times 0.1)^2 = 2.72 \times 10^{-2}$ W

(iv) $I_{\text{at } r = 10\text{m}} = \frac{P_{av}}{\text{Area}} = \frac{2.7 \times 10^{-2}}{4\pi\,(10)^2} = 2.15 \times 10^{-5} \text{ Wm}^{-2}$

Hence, (i) → (r), (ii) → (s), (iii) → (p), (iv) → (q)

17. (d) Let the intensity due to one person is I.

Then, $$60 = 10\log_{10}\frac{I}{I_0}$$

$$\beta_2 = 10\log\frac{2I}{I_0} = 10\log_{10}2 + 10\log_{10}\frac{I}{I_0} = 63.01 \text{ dB}$$

18. (a) In case of a plane progressive wave, if a total power of P crosses an area, the same power P crosses the same area at all locations.

19. (d) Force exerted on inner ear is same as that exerted on the eardrum due to negligible mass of ossicles.

$$\therefore \quad F_{\text{max, eardrum}} = F_{\text{max, stirrup}}$$

$$\Rightarrow \quad (p_{max} \times A)_{\text{eardrum}} = (p_{max} \times A)_{\text{stirrup}}$$

$$\therefore \quad (\Delta p)_{\text{max, stirrup}} \times A_{\text{stirrup}} = (\Delta p)_{\text{max, eardrum}} \times A_{\text{eardrum}}$$

$$\therefore \quad (\Delta p)_{\text{max, stirrup}} = \frac{4 \times 10^{-2} \times 50 \times 10^{-6}}{5 \times 10^{-6}} = 0.4 \text{ Pa}$$

[$= \Delta p_0$ (say)]

20. (d) Displacement amplitude in fluid is

$$A = \left(\frac{\Delta p_0}{Bk}\right)$$

$$A = \frac{0.4}{2.18 \times 10^9 \times \left(\frac{\omega}{v_{\text{fluid}}}\right)} = \frac{0.4 \times 1500}{2.18 \times 10^9 \times 2\pi \times 1200}$$

$$= 3.65 \times 10^{-11} \text{m}$$

21. (a) $\beta = 10\log\frac{I}{I_0}$, where β is sound level.

$$\beta_2 - \beta_1 = 10\log_{10}\frac{I_2}{I_0} - 10\log_{10}\frac{I_1}{I_0} = 10\log_{10}\frac{I_2}{I_1}$$

$$30 = 10\log_{10}\frac{I_2}{I_1} \Rightarrow \frac{I_2}{I_1} = 10^3$$

$$\Rightarrow \quad I_2 = 1000\,I_1$$

22. (a) Energy taken up by the person in 1 hour is

$[3.2 \times 10^{-5} \times 50 \times 10^{-6} \times 3600] \times 2$ (through two ears)

$= 1.152 \times 10^{-5} \text{ J} \approx 1.2 \times 10^{-5}$ J

23. (4) Given, $I_1 = 10^{-8} \text{ Wm}^{-2}$, $r_1 = 5$ m, $r_2 = 25$ m $I_2 = ?$,

We know, $$I \propto \frac{1}{r^2}$$

$$\Rightarrow \quad I_1 r_1^2 = I_2 r_2^2$$

$$I_2 = I_1\left(\frac{r_1}{r_2}\right)^2 = 10^{-8}\left(\frac{5}{25}\right)^2 = \frac{1}{25} \times 10^{-8}$$

$$= 4 \times 10^{-10} \text{ Wm}^{-2}$$

24. (6) $I_1 = Ka^2 = I_2$; resultant amplitude at A is $a + a = 2a$

$$\therefore \quad I = K(2a)^2 = 4Ka^2 = 4I_1$$

$$10\log_{10}\frac{I}{I_0} = \left[10\log_{10}\frac{I_1}{I_0}\right] + n \text{ dB}$$

$$10\log_{10}\frac{I}{I_1} = n\text{dB}$$

$$\Rightarrow \quad 10\log_{10}\frac{4I_1}{I_1} = n$$

$$n = 10\log_{10}4 = 10(0.6) = 6$$

25. (7) Let at distance $r_1 = 0.7\,m$, intensity is I_1

$$\therefore \quad 20 = 10\log_{10}\frac{I_1}{I_0} \quad \ldots(i)$$

Let at distance r_2, sound level is zero. Then, intensity at that point must be equal to I_0. (where, $I_0 = 10^{-12}\,Wm^{-2}$)

$$I \propto \frac{1}{r^2} \Rightarrow \frac{I_1}{I_0} = \left(\frac{r_2}{r_1}\right)^2$$

Using this in Eq. (i), $20 = 10\log_{10}\left(\frac{r_2}{r_1}\right)^2$

or $\left(\frac{r_2}{r_1}\right)^2 = 10^2 \Rightarrow r_2 = \sqrt{100}\, r_1 = 10(0.7) = 7\,m$

26. (1230) Since, aeroplane is producing intensity level 100 dB at a distance of 200 m from it and this is also the maximum allowable sound level, so as not to violate the regulations, the minimum distance of the plane from microphone has to be 200 m.

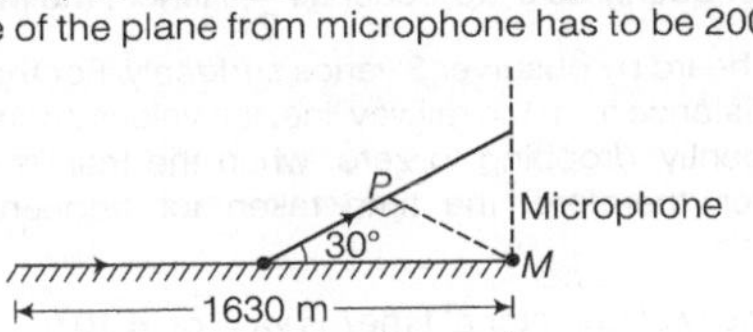

The diagram shows the situation exactly. Let the aeroplane leaves the runway, before x metre of the location of microphone. Then, PM is the shortest distance between the microphone and the aeroplane.

$$PM = x\sin 30° = \frac{x}{2}$$

$$\Rightarrow \quad \frac{x}{2} = 200 \Rightarrow x = 400\,m$$

$\therefore$ Required distance $= 1630 - 400 = 1230\,m$

27. (i) As intensity is power per unit area, $P \propto I$

$$\beta_2 - \beta_1 = 10\log_{10}\frac{I_2}{I_1} = 10\log\frac{P_2}{P_1}$$

$$= 10\log\frac{400}{20}$$

or $\beta_2 - \beta_1 = 10\log_{10}20 = 13\,dB$

(ii) Increase in loudness level of sound is

$$\beta_2 - \beta_1 = 10\log_{10}\frac{I_2}{I_1}$$

$$80 - 20 = 10\log_{10}\frac{I_2}{I_1}$$

$$\Rightarrow \quad \log_{10}\frac{I_2}{I_1} = 6$$

$$I_2 = 10^6 I_1$$

DPP-4 **Doppler Effect-I**

1. (a) Speed of source i.e. car $= v_s = ?$

Speed of sound is $v = 330\,ms^{-1}$

Speed of observer is $v_O = 0$

Apparent frequency is $f' = 0.9f_0$ (given)

As source moves away from stationary observer, apparent frequency is

$$f' = \left(\frac{v}{v + v_s}\right)f_0 \Rightarrow 0.9\,f_0\left(\frac{330}{330 + v_s}\right)f_0$$

$$330 + v_s = 366.67 \Rightarrow v_s = 36.67\,ms^{-1}$$

2. (a) When the locomotive is approaching the stationary observer,

$$f' = \left(\frac{v}{v - v_s}\right)f_0 = \frac{v}{v - \frac{v}{20}} \times 256 = \frac{20}{19} \times 256 = 269.5\,Hz$$

When the locomotive is moving away from the stationary observer :

$$f' = \left(\frac{v}{v + v_s}\right)f_0 = \frac{v}{v + \frac{v}{20}} \times 256 = \frac{20}{21} \times 256 = 243.8\,Hz$$

3. (b) When the sound reaches the wall, apparent frequency heard by the wall is $f = \left(\frac{v - v_O}{v}\right)f_0$

or $$f = \frac{340 - 0.06}{340} \times 1700 = 1699.7\,Hz$$

The wall reflects the same frequency i.e. 1699.7 Hz. When this sound is received by the receiver, apparent frequency is

$$f' = \left(\frac{v}{v + v_s}\right)f = \frac{340}{340 + 0.06} \times 1699.7 = 1699.4\,Hz$$

4. (b) Number of revolutions per second = 2

Radius of the circle = 1.988 m

Linear velocity of the tuning fork is

$$v = (2\pi R) \times 2 = 4 \times \frac{22}{7} \times 1.988 = 25\,ms^{-1}$$

Let the listner is located on the left side at a large distance in the diagram.

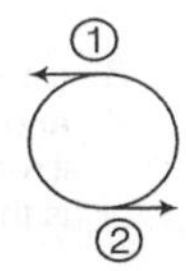

(1) $\therefore$ Apparent frequency, when the tuning fork is approaching the listener is

$$f_1 = \left(\frac{v}{v - v_s}\right)f_0 = \frac{350}{350 - 25}f_0 = 1.077\,f_0$$

(2) Apparent frequency, when the tuning fork is moving away from the listener is $f_2 = \left(\frac{v}{v + v_s}\right)f_0$

$$f_2 = \frac{350}{350 + 25} \cdot f_0 = 0.933f_0$$

f_1 is the highest note and f_2 is the lowest note.

Now, $$\frac{f_1}{f_2} = \frac{1.077}{0.933} = 1.154$$

5. (d)

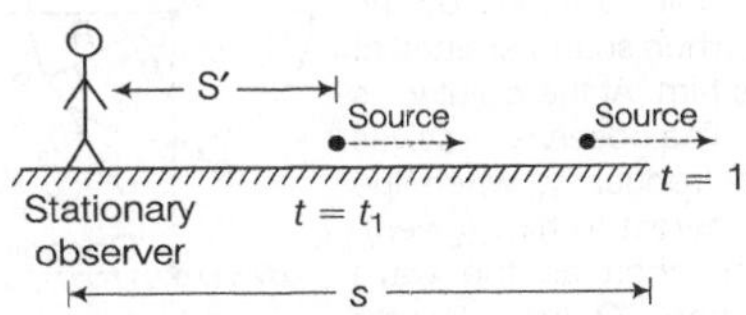

In 10 s, the distance travelled by source is $s = \frac{1}{2}at^2$

or $$s = \frac{1}{2}(10)(10)^2 = 500\,m$$

At this instant, observer will receive that sound which source had emitted at time $t = t_1$ (say). At time t_1

source was at a distance, $s' = \frac{1}{2}(10)t_1^2 = 5t_1^2$

Now in a time $(10 - t_1)$, sound travels a distance s' from source to observer. Thus, we have,

$$5t_1^2 = 340(10 - t_1)$$

$$\Rightarrow \quad 5t_1^2 + 340t_1 - 3400t_1 = 0$$

$$t_1^2 + 68t_1 - 680 = 0$$

$$\Rightarrow \quad t_1 = \frac{-68 \pm \sqrt{(68)^2 - 4(1)(-680)}}{2(1)}$$

$$t_1 = 8.85\,s \text{ (neglecting negative root)}$$

Thus, velocity of source at instant, t_1 is

$$v = at_1 = (10)(8.85)$$

or $\quad v = 88.5 \text{ ms}^{-1}$

Thus, frequency observed by the observer at 10 s is given by

$$f = \left(\frac{340}{340 + 88.5}\right) 1700 = 1349 \approx 1350 \text{ Hz}$$

6. (b) The situation is shown in the figure.

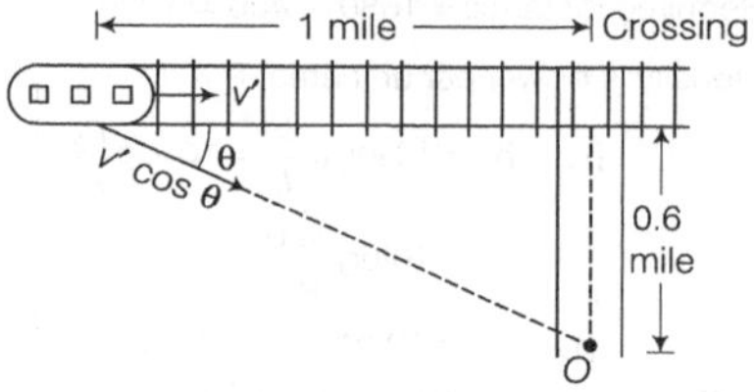

$$\cos\theta = \frac{1}{\sqrt{1^2 + 0.6^2}} = 0.8575 \text{ and } v' = 80 \text{ mile h}^{-1} \quad \text{(given)}$$

$$v' = 80 \times \frac{1}{3600} \text{ mile s}^{-1} = 0.0222 \text{ mile s}^{-1}$$

$$v_s = v'\cos\theta = (0.0222)(0.8575) = 0.019 \text{ mile s}^{-1}$$

The apparent frequency observed by the observer is

$$f' = \left(\frac{v}{v - v_s}\right) f_0 = \left(\frac{0.2}{0.2 - 0.019}\right) 400 = 442 \text{ Hz}$$

7. (c) Here, $f' = \left(\frac{v - v_O - v_w}{v + v_s - v_w}\right) f_0 = \left(\frac{343 - 20 - 5}{343 + 10 - 5}\right)(261.6)$

$= 239 \text{Hz} \qquad$ (v_w is velocity of wind)

8. (b) First, sound emitted by the source passes through the water and strikes the submarine, which acts as a listener. Second, when it reflects the sound, the submarine acts like a source and warship is the listener.

$$f_1' = \left(\frac{v + v_O - v_w}{v - v_s - v_w}\right) f_0 = \left(\frac{1470 + 14 - 1.3}{1470 - 9.5 - 1.3}\right) 8000 = 8131 \text{ Hz}$$

$$f_2' = \left(\frac{v + v_O + v_w}{v - v_s + v_w}\right) f_1' = \left(\frac{1470 + 9.5 + 1.3}{1470 - 14 + 1.3}\right) 8131 = 8262.2 \text{ Hz}$$

Here, v_w is velocity of medium i.e. ocean current.

9. (a) Let the speed of the plane (source) is v_s. Maximum frequency will be heard by the observer, when sound emitted at S reaches him. At the point v_s is along SO. The observer receives maximum frequency, when the plane is nearest to him (given). That is as soon as the wave pulse reaches O from S with speed v, the plane reaches S' from S with speed v_s. Hence,

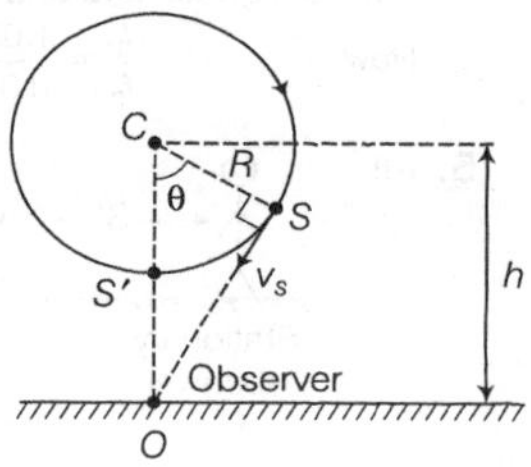

$$t = \frac{SO}{v} = \frac{SS'}{v_s} \Rightarrow v_s = \left(\frac{SS'}{SO}\right) v$$

or $\quad v_s = \left[\frac{R\theta}{\sqrt{h^2 - R^2}}\right] v$

Here, $\quad \cos\theta = \frac{R}{h}$ or $\theta = \cos^{-1}\left(\frac{R}{h}\right)$

So, $\quad \theta = \cos^{-1}\left(\frac{100}{200}\right) = \cos^{-1}\left(\frac{1}{2}\right) = \frac{\pi}{3}$

Thus, $\quad v_s = \frac{(100)\left(\frac{\pi}{3}\right)}{\sqrt{200^2 - 100^2}}(340) = 205 \text{ units}$

10. (b) $f' = \left(\frac{v}{v - v_s\cos\theta}\right) f_0$

As, θ increases, $\cos\theta$ decreases. Thus, $v - v_s\cos\theta$ increases. So f' decreases.

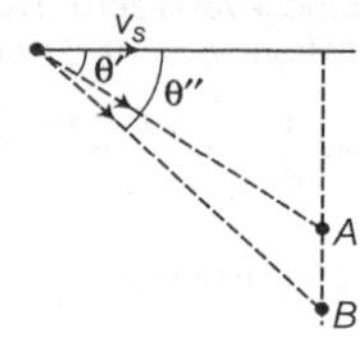

In the figure, $\theta'' > \theta'$. So, listener at B hears smaller frequency. So, he must be observer 1, because in the graph 1 corresponds to smaller frequency. So, observer 2 is closer to the railway track.

For the listener near to moving train, component of velocity toward listener $v_s\cos\theta'$ is almost same as the velocity of train as it varies suddenly as θ approaches $\frac{\pi}{2}$. Hence, the frequency of the sound heard by observer 2 varies suddenly. For the observer at a large distance from the railway line, the velocity component varies more gently, dropping to zero, when the train is closest to the observer, therefore, the time taken for frequency change is greater.

11. (d) Velocity of detector at time t is $v_O = gt = 10t$

Apparent frequency detected by the detector at time t is

$$f' = \left(\frac{v + v_O}{v}\right) f_0 \Rightarrow f'\left(1 + \frac{10t}{v}\right) 1000$$

$$f' = 1000 + \frac{1000}{v} t$$

From the graph slope $= \frac{100}{3}$

$\therefore \quad \frac{10000}{v} = \frac{100}{3} \Rightarrow v = 300 \text{ ms}^{-1}$

12. (c,d) **For f_1** $v_s = 10 \text{ ms}^{-1}$, $v_O = 20 \text{ ms}^{-1}$

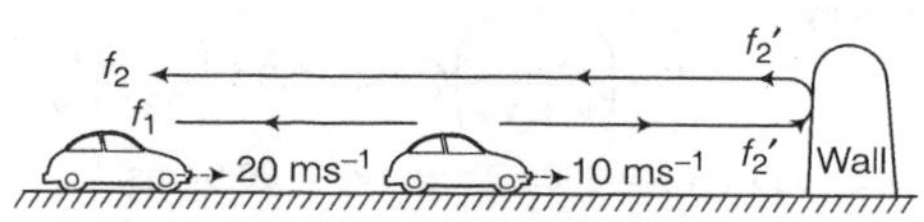

$$f_1 = \left(\frac{v + v_O}{v + v_s}\right) f_0 = \left(\frac{340 + 20}{340 + 10}\right) 500 = 514.3 \text{ Hz}$$

For f_2 Frequency received by wall is $f_2' = \left(\frac{v}{v - v_s}\right) f_0$

$$f_2' = \left(\frac{340}{340 - 10}\right) 500 = 515.2 \text{ Hz}$$

Wall reflects the same frequency f_2' and the wall will be the stationary source for second car. So, now

$$f_2 = \left(\frac{v + v_O}{v}\right) f_2' = \left(\frac{340 + 20}{340}\right)(515.2) = 545.5 \text{ Hz}$$

Alternate Method

Note : *This method of images should not be used, when velocities are large and reflector is in motion.*

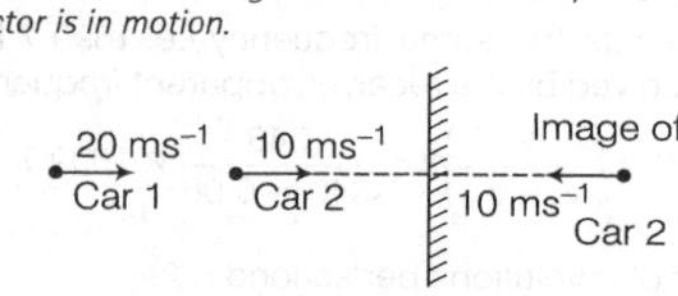

Assume the image of car 2, treating the wall as a mirror. Now for f_2, car 1 is the observer and image of car 2 is the source with original frequency 500 Hz.

$$\therefore \quad f_2 = \left(\frac{v + v_O}{v - v_s}\right) f_0 = \left(\frac{340 + 20}{340 - 10}\right) 500 = 545.5 \text{ Hz}$$

13. (b,d) The situation is shown in the figure. When the source is at point Q it is closest to the observer. At this instant, the observer listens a sound, which was emitted when the source was at P (say).

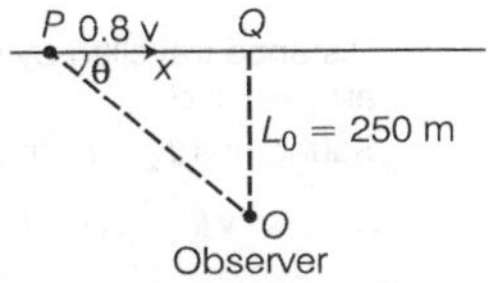

The velocity of the source along the line joining P and observer is

$$v_s = (0.8v)\cos\theta \quad \ldots\text{(i)}$$

When the source is at Q, the sound emitted at P is heard.

Let $x = (0.8v)t_1$

(Time taken by the source to travel a distance x is t_1)

$\therefore$ PO = speed of sound $\times t_1 = vt_1$

So, $\cos\theta = \dfrac{x}{PO} = \dfrac{0.8vt_1}{vt_1} = 0.8$

$\therefore$ $v_s = (0.8v)(0.8)$ (from Eq. (i))

$= 0.64\,v$

Now, $f_1 = \left(\dfrac{v}{v - v_s}\right)f_0 = \left(\dfrac{v}{v - 0.64v}\right)(1800) = 5000$ Hz

Observer receives a frequency equal to original frequency of source corresponding to the waves emitted by source from position Q, as in this case velocity component of source is in the direction of observer is zero.

In this case, the time taken by sound to reach observer from Q is $t_2 = \dfrac{L_0}{v}$.

In this time distance travelled by source is $x' = (0.8v)t_2$.

or $x' = 0.8L_0$

At this instant, the distance between observer and source is

$$d = \sqrt{(x')^2 + L_0^2} = \sqrt{0.64\,L_0^2 + L_0^2} = \sqrt{1.64}\,L_0$$

or $d = \sqrt{1.64}\,(250) = 320$ m

14. (a,c) Frequency as heard (i.e. received) by the wall is

$$f_1' = \left(\frac{v + v_O}{v}\right)f_0 = \left(\frac{340 + 3.3}{340}\right)1000 = 1009.7\text{ Hz}$$

Frequency of reflected sound as heard by the driver is

$$f_2' = \left(\frac{v}{v - v_s}\right)f_1' = \left(\frac{340}{340 - 3.3}\right)(1009.7) = 1020\text{ Hz}$$

$$\%\text{ increase} = \frac{f_2' - f_0}{f_0} \times 100 = \frac{1020 - 1000}{1000} \times 100 = 2\%$$

15. (b) (i) $f' = \left(\dfrac{v + v_O}{v}\right)f_0 = \left(\dfrac{343 + 15}{343}\right)(800) = 835$ Hz

(ii) $f' = \left(\dfrac{v}{v - v_s}\right)f_0 = \left(\dfrac{343}{343 - 15}\right)(800) = 836.6$ Hz

(iii) $f' = \left(\dfrac{v + v_O}{v - v_s}\right)f_0 = \left(\dfrac{343 + 15}{343 - 15}\right)800 = 873$ Hz

(iv) $f' = \left(\dfrac{v - v_O}{v + v_s}\right)f_0 = \left(\dfrac{343 - 15}{343 + 15}\right)800 = 733$ Hz

Here, v = velocity of sound, v_s = velocity of source, v_O = velocity of observer, f' = apparent frequency and f_0 = original frequency.

Hence, (i) → (r), (ii) → (s), (iii) → (p), (iv) → (q)

Explanations for Q. Nos. (16-19)

Frequency detected by D_1 will be maximum and that detected by D_2 will be minimum. Where as original frequency (i.e. as observed by D_3) will be intermittent.

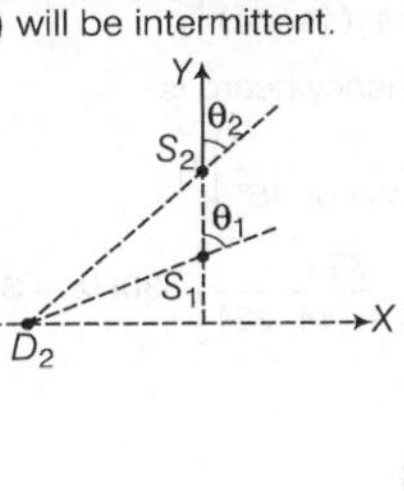

When the source moves along Y-axis with constant velocity. Due to the angle indicated (as θ_1 and θ_2) keeps changing with time, the velocity of source along the line joining S and D_2 keeps changing. Thus, recorded frequency cannot be constant.

16. (b) **17.** (c) **18.** (a) **19.** (d)

Explanations for Q. Nos. (20-22)

+ ve -direction

Velocity of man w.r.t plank is $v_{mp} = -8\text{ ms}^{-1}$

$$v_{mp} = v_m - v_p$$

$\Rightarrow$ $v_m = v_{mp} + v_p = -8 + v_p$

Momentum is conserved.

$\therefore$ $(50 + 150)(0) = (50)v_m + (150)v_p$

or $-50(-8 + v_p) = 150\,v_p$

$\Rightarrow$ $200v_p = 400$

or $v_p = 2\text{ ms}^{-1}$

$\therefore$ $v_m = -8 + 2 = -6\text{ ms}^{-1}$

$\therefore$ Speed of observer = 2 ms^{-1} and speed of source = 6 ms^{-1}

When the man is running on the plank,

$$f_1' = \left(\frac{v + v_O}{v - v_s}\right)f_0$$

or $f_1' = \dfrac{330 + 2}{330 - 6} \times 500 = 512.35$ Hz

When the man is running on the ground,

$$f_2' = \left(\frac{v - v_O}{v + v_s}\right)f_0$$

or $f_2' = \left(\dfrac{330 - 2}{330 + 6}\right)500 = 488.1$ Hz

Apparent frequency is independent of the distance. So, the best graph is (a).

20. (b) **21.** (c) **22.** (a)

23. Speed of sound in water is

$$v_1 = \sqrt{\frac{B}{\rho}} = \sqrt{\frac{2.088 \times 10^9}{1000}}$$

or $v_1 = 1445\text{ ms}^{-1}$

Frequency of sound in still water

$$f_0 = \frac{v_1}{\lambda_w} = \frac{1445}{14.45 \times 10^{-3}} = 10^5\text{Hz}$$

Calculating f_1'

$$f_1' = \left(\frac{v_1 + v_w}{v_1 - v_s + v_w}\right)f_0$$

where, v_w is velocity of stream and v_s velocity of source (i.e. the boat.)

$\therefore$ $f_1' = \left[\dfrac{1445 + 2}{1445 - 10 + 2}\right]10^5 = 1.00696 \times 10^5$ Hz

Calculating f_2' Speed of sound in air is

$$v_2 = \sqrt{\frac{\gamma RT}{M_0}}$$

or $v_2 = \sqrt{\dfrac{1.4 \times 8.31 \times 293}{28.8 \times 10^{-3}}} = 344\text{ ms}^{-1}$

Wind is blowing from detector to source with velocity 5 ms^{-1}

Now, $f_2' = \left(\dfrac{v_2 - v_w}{v_2 - v_s - v_w}\right)f_0$,

where v_w = velocity of wind

$\therefore$ $f_2' = \left[\dfrac{344 - 5}{344 - 10 - 5}\right](10^5) = 1.0304 \times 10^5$ Hz

24. Time taken by the source to cross the origin $= \dfrac{100}{5} = 20$ s

Distance moved by detector (O') in this time $= 10 \times 20 = 200$ m
The singnals received by detector at $t = 20$ s are not produced, when the source is at the origin but from some previous instant t' (say), shown in the figure. The time in which source travels $S'O$, the signal travels $S'O'$.

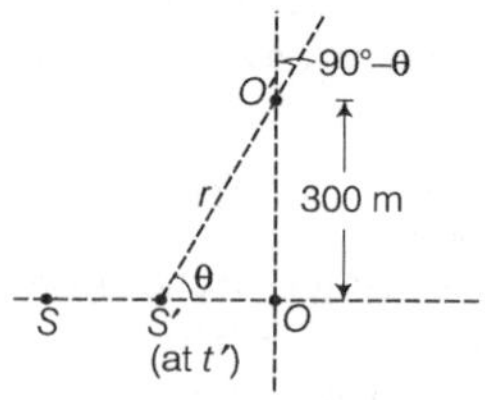

$\therefore \quad t = \dfrac{S'O}{v_s} = \dfrac{S'O'}{v}$, where v = speed of sound

Now, $\cos\theta = \dfrac{S'O}{S'O'}$

$\therefore \quad \cos\theta = \dfrac{v_s}{v} = \dfrac{5}{330} = \dfrac{1}{66}$

$\therefore$

$\therefore \quad \sin\theta = \sqrt{1 - \left(\dfrac{1}{66}\right)^2} \simeq 1$

Frequency observed by detector is $f' = \left(\dfrac{v - v_O \cdot \cos(90 - \theta)}{v - v_s \cos\theta}\right) f_0$

or $\quad f' = \left[\dfrac{v - v_O \cdot \sin\theta}{v - v_s \cos\theta}\right] f_0 = \left[\dfrac{330 - 10(1)}{330 - 5\left(\dfrac{1}{66}\right)}\right](100)$

or $\quad f' = 97$ Hz

25. Acceleration of the system, $a = \left(\dfrac{m_O - m_s}{m_O + m_s}\right) g$

or $\quad a = \left(\dfrac{8-2}{8+2}\right)(10) = 6\ \text{ms}^{-2}$

The signal received by the detector, at the instant the source strikes the pulley, is produced by the source some time earlier. Let this signal be produced at instant t_1.

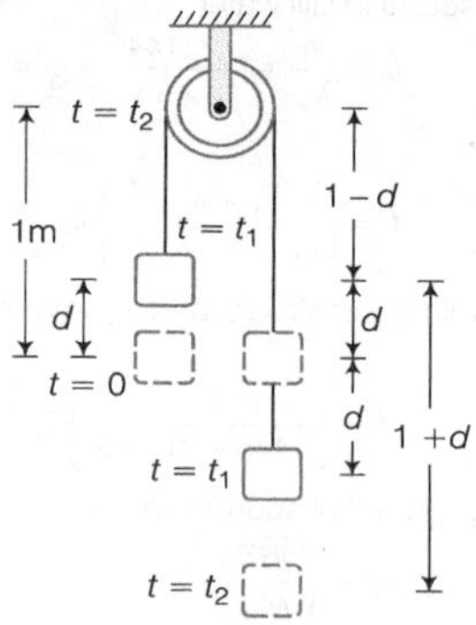

Time taken by the source to reach the pulley is

$$t_2 = \sqrt{\frac{2 \times 1}{a}}$$

or $\quad t_2 = \sqrt{\dfrac{2}{6}} = \dfrac{1}{\sqrt{3}}$ s

Distance travelled by source in the time t_1 is $d = \dfrac{1}{2} a t_1^2$

or $\quad d = \dfrac{1}{2} \times 6 t_1^2 = 3t_1^2$

Distance travelled by signal produced at t_1 to reach the detector at $t_2 = 1 + d$

Sound has $(t_2 - t_1)$ time to travel this distance.

$\therefore \quad v(t_2 - t_1) = 1 + d \Rightarrow 330\left(\dfrac{1}{\sqrt{3}} - t_1\right) = 1 + 3t_1^2$

$190.53 - 330t_1 = 1 + 3t_1^2 \Rightarrow 3t_1^2 + 330t_1 - 189.5 = 0$

Solving, we get $\quad t_1 = 0.571$ s

Velocity of detector, when it receives signal is

$$v_O = at_2 = 6\left(\frac{1}{\sqrt{3}}\right)$$

or $\quad v_O = 2\sqrt{3} = 3.464\ \text{ms}^{-1}$

Velocity of source at time t_1 is $v_s = at_1 = 3.426\ \text{ms}^{-1}$.

As, source and detector are moving away from each other.

So, $\quad f' = \left(\dfrac{v - v_O}{v + v_s}\right) f_0 = \left(\dfrac{330 - 3.464}{330 + 3.426}\right) 500 = 489.6$ Hz

26. When the boy has displaced by an angle θ (i.e. from P_1 to P_1'), then the train has moved from P_2 to P_2'.

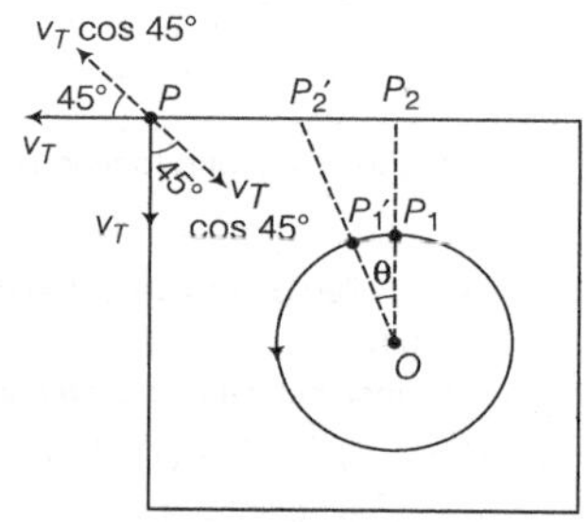

Displacement of train, $P_2 P_2' = \dfrac{L}{2} \tan\theta$

Velocity of train is $v_T = \dfrac{d}{dt}\left(\dfrac{L}{2} \tan\theta\right)$

or $\quad v_T = \dfrac{L}{2} \sec^2\theta \left(\dfrac{d\theta}{dt}\right) = \dfrac{L}{2} \omega \sec^2\theta$

Angular velocity of boy is $\omega = \dfrac{v}{R} = \dfrac{v}{L/4} \quad \left(\because R = \dfrac{L}{4}\right)$

or $\quad \omega = \dfrac{4v}{L} = \dfrac{4 \times 11}{L} = \dfrac{44}{L}$

Velocity of toy train at $P_2 = \dfrac{L}{2} \omega \sec^2 0 = \dfrac{L}{2} \omega = \dfrac{L}{2} \times \dfrac{44}{L} = 22\ \text{ms}^{-1}$

Angular velocity is same for both boy and toy train.
As, θ changes from 0 to $\pi/4$, as the train moves from P_2 to P, the velocity of train increases. Just before the turn, the component of velocity of train along the line joining boy and train is away from the boy. Similarly just after the turn, the component of velocity of the train is towards the boy. Thus, in the former case, the frequency heard by the boy is minimum and in the latter case it is maximum.

Velocity of train at P is $v_T = \dfrac{L}{2} \omega \sec^2 45° = (22)(2) = 44\ \text{ms}^{-1}$

$\therefore$ Minimum frequency heard is

$$f_1' = \left(\frac{v}{v + v_T \cos 45°}\right) f_0$$

or $\quad f_1' = \left[\dfrac{330}{330 + (44/\sqrt{2})}\right] 330\ \text{Hz} = 301.6$ Hz

and maximum frequency heard is

$$f_2' = \left(\frac{v}{v - v_T \cos 45°}\right) f_0$$

or $\quad f_2' = \left[\dfrac{330}{330 - (44/\sqrt{2})}\right] 330\ \text{Hz} = 364.4$ Hz

DPP-5 Doppler Effect-II

1. (c) Source will be stationary when it hits the ground. So, the required distance is

$$d = \sqrt{440^2 + 55^2} = \sqrt{55^2 \times 8^2 + 55^2}$$

or $$d = 55\sqrt{64 + 1} = 55\sqrt{65}\text{ m}$$

Note *The sound produced by the source, when it is at a height* 55 m *from the ground, also will reach the observer after the source becomes stationary, in the situation of the problem.*

2. (d) $$f_1' = \left(\frac{v - v_O \cos\theta}{v}\right) f_0$$

$$f_2' = \left(\frac{v - v_O}{v}\right) f_0$$

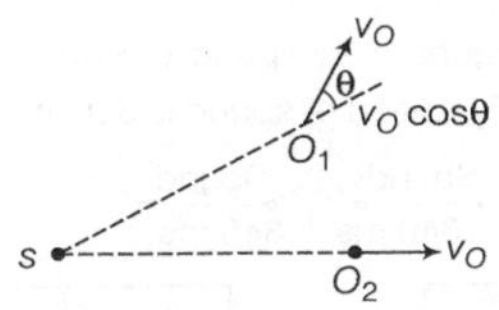

$\therefore \quad f_1' > f_2'$

3. (a) Consider the wave received by R_1. Say at $t = 0$, a wavefront comes from S. At $t = T$ another wavefront is emitted from S. The distance travelled by S in this time is $\frac{u}{2}T$ and the distance travelled by first wavefront is $10uT$.

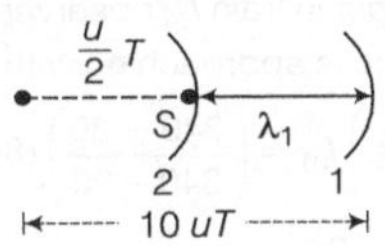

$$\because \quad \lambda_1 = 10uT - \frac{u}{2}T = \frac{19}{2}uT$$

Similarly, for waves coming in the backward direction

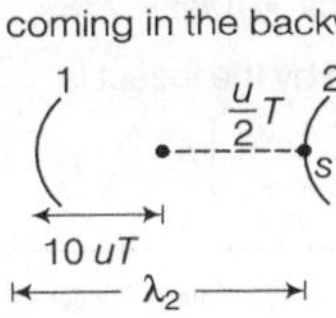

$$\lambda_2 = 10uT + \frac{u}{2}T = \frac{21}{2}uT$$

$$\therefore \quad \frac{\lambda_1}{\lambda_2} = \frac{19}{21}$$

4. (d) The velocity of source (c_2) here is constant and $v_s = 2\text{ ms}^{-1}$ At the time of receiving signal, the velocity of observer (c_1) is

$$v_O = 2 + 2 \times 10 = 22\text{ ms}^{-1}$$

Apparent frequency is

$f' = 352$ Hz (given)

$$f' = \left(\frac{v - v_O}{v - v_s}\right) f_0$$

$$\Rightarrow \quad f_0 = (352)\left(\frac{330 - 2}{330 - 22}\right) = 374.86\text{ Hz}$$

5. (b) $$f' = \left(\frac{v + v_w}{v - v_s + v_w}\right) f_0$$

$$f_0 = \left(\frac{1200 + 40}{1200 - 40 + 40}\right)(580) \approx 599\text{ Hz}$$

6. (a) Let the initial separation i.e. at $t = 0$ between source and detector be x.

The first wavefront arrives at detector at time $t_1 = \frac{x}{v}$

The second similar wavefront is emitted after a time period $T = \frac{1}{f_0}$

The distance moved by source in this time $= \frac{1}{2}aT^2$

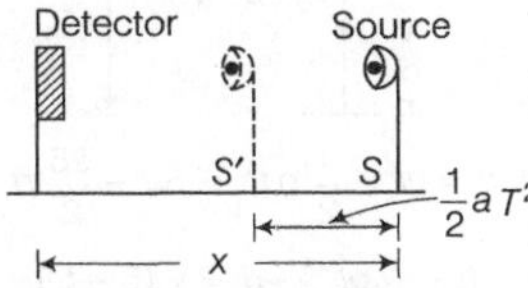

The separation between second wavefront and detector is now

$$x - \frac{1}{2}aT^2$$

Time taken by second wavefront to reach detector $= \dfrac{x - \frac{1}{2}aT^2}{v}$

The time when the second wavefront reaches the detector is

$$t_2 = T + \frac{x - \frac{1}{2}aT^2}{v}$$

Time interval between successive wavefronts (in reaching the detector) is

$$\Delta t = t_2 - t_1 = T + \frac{x - \frac{1}{2}aT^2}{v} - \frac{x}{v} = T - \frac{1}{2v}aT^2$$

Hence, frequency detected by detector $= \frac{1}{\Delta t}$

$$= \frac{1}{T - \frac{1}{2}\frac{aT^2}{v}} = \frac{1}{\frac{1}{f_0} - \frac{a}{2vf_0^2}} = \frac{2vf_0^2}{2vf_0 - a}$$

7. (b) Sound takes a finite time to travel. Hence, the frequency emitted at t' reaches the observer at time $(t' < t)$.

Distance moved by source in time $t' = \frac{1}{2}at'^2$. Sound has to travel this distance in time $(t - t')$. So, we have

$$\frac{1}{2}at'^2 = v(t - t')$$

$$\Rightarrow \quad at'^2 + 2vt' - 2vt = 0$$

$$\therefore \quad t' = \frac{-2v + \sqrt{4v^2 - 4(a)(-2vt)}}{2(a)}$$

$$= \frac{-2v + \sqrt{4v^2 + 8avt}}{2a}$$

Velocity of source at t' is

$$v_s = at' = a\left[\frac{-2v + \sqrt{4v^2 + 8avt}}{2a}\right]$$

or $$v_s = -v + \sqrt{v^2 + 2avt}$$

$$\therefore \quad f' = \left(\frac{v}{v + v_s}\right) f_0 = \left[\frac{v}{v - v + \sqrt{v^2 + 2avt}}\right] f_0$$

$$\therefore \quad f' = \frac{vf_0}{\sqrt{v^2 + 2avt}}$$

8. (d) The sound wave received at $t = 5$ s must have been emitted earlier. Let t' be the time at which the source emits a sound wave which is detected at t = 5s. So, the time for which this sound travels $t - t'$ i.e. $5 - t'$. Distance travelled by waves is $v\,(5 - t')$. Let the distance moved by the detector be d, then

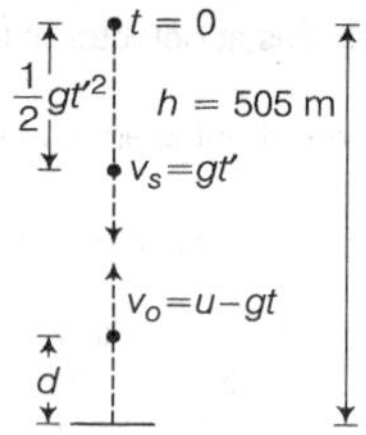

$$d = ut - \frac{1}{2}gt^2 = 5u - \frac{25}{2}g$$

and so, $$h - \frac{1}{2}gt'^2 - d = v\,(5 - t')$$

or $$505 - \frac{1}{2}gt'^2 - 5u + \frac{25}{2}g = 300\,(5 - t')$$

$$505 - 5t'^2 - 5(50) + \frac{25}{2}(10) = 1500 - 300t'$$

$$5t'^2 - 300t' + 1500 - 505 + 250 - 125 = 0$$

$$5t'^2 - 300t' + 1120 = 0$$

$$t'^2 - 60t' + 224 = 0 \quad \Rightarrow \quad t' = 4\text{ s}$$

$\therefore$ $$\text{Apparent frequency} = \left(\frac{v + v_O}{v - v_S}\right) f_0$$

$$v_O = u - gt = 50 - 10(5) = 0$$
$$v_S = gt' = 10(4) = 40\text{ ms}^{-1}$$

$\therefore$ $$f' = \left(\frac{300 + 0}{300 - 40}\right) 1300 = 1500\text{ Hz}$$

9. (a,d) When there is no relative motion between source and observer, wind does not affect the frequency received by the observer and Doppler effect cannot be observed. Where as when there is relative motion between O and S, apparent frequency is different from original frequency.

10. (c,d) For region PO',

$$f' = f_0\left(\frac{v}{v - v_S\cos\theta}\right) \qquad \ldots\text{(i)}$$

For region $O'Q$,

$$f'' = f_0\left(\frac{v}{v + v_S\cos\theta}\right) \qquad \ldots\text{(ii)}$$

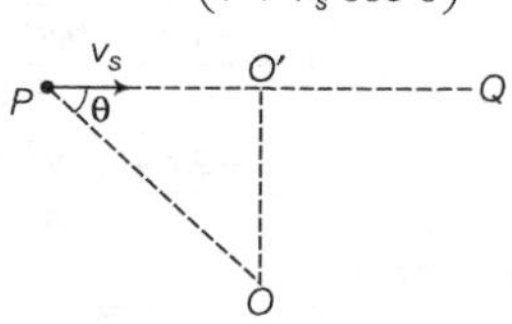

Minimum value of f'' is when $\cos\theta = 1$

i.e. $$f''_{min} = f_0\left(\frac{v}{v + v_S}\right)$$

$$1800 = 2000\left(\frac{300}{300 + v_S}\right)$$

$\Rightarrow$ $$v_S = 33.33\text{ ms}^{-1}$$

Maximum value of f' can be $f'_{max} = f_0\left(\frac{v}{v - v_S}\right)$, when $\cos\theta = 1$

$$f'_{max} = 2000\left(\frac{300}{300 - 33.33}\right)$$

$$= 2250\text{ Hz}$$

11. (d) Initial vertical components of velocity for both are same. So, there is no relative motion in the vertical direction (Note that relative acceleration is also zero)

$$f' = \left(\frac{v}{v + v_S}\right) f_0$$

$$= \left(\frac{310}{310 + 10\sqrt{2}\cos 45^\circ}\right) 960 = 930\text{ Hz}$$

And this apparent frequency is constant as both of them reach the ground simultaneously at t = 2 s.

12. (d) The best option is (d).

13. (a) In the vertical direction, there is no relative motion. And in the horizontal direction, relative velocity is constant. So, apparent frequency heard by the detector will be constant.

14. (c) For passengers in T_1, speed of sound is 360 ms^{-1} and for passengers in T_2, speed of sound is 310 ms^{-1}.

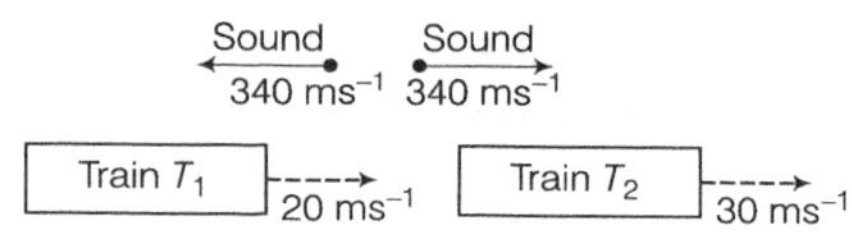

15. (a) For the passengers in train T_1, there is no relative motion between source and observer, as both are moving with velocity 20 ms^{-1}. Thus, there is no change in observed frequencies and correspondingly there is no change in their intensities.

16. (b) For the passenger in train T_2, observer is receding with a speed 30 ms^{-1} and source is approaching with speed 20 ms^{-1}.

$\therefore$ $$f_1' = \left(\frac{v - v_O}{v - v_S}\right) f_{01} = \left(\frac{340 - 30}{340 - 20}\right)(800) = 775\text{ Hz}$$

and $$f_2' = \left(\frac{340 - 30}{340 - 20}\right) 1120 = 1085\text{ Hz}$$

$\therefore$ Spread of frequency = 1085 − 775 = 310 Hz

17. The frequency heard by the insect is

$$f_1' = \left(\frac{v + v_O}{v - v_S}\right) f_0$$

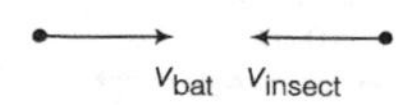

or $$f_1' = \left(\frac{v + v_{insect}}{v - v_{bat}}\right) f_{bat}$$

The frequency received by bat (i.e. f_{refl}) is

$$f_2' = \left(\frac{v + v_O}{v - v_S}\right) f_1' = \left(\frac{v + v_{bat}}{v - v_{insect}}\right)\left(\frac{v + v_{insect}}{v - v_{bat}}\right) f_{bat}$$

Thus $$f_{refl} = \left(\frac{v + v_{bat}}{v - v_{insect}}\right)\left(\frac{v + v_{insect}}{v - v_{bat}}\right) f_{bat}$$

If we make v_{insect} the subject, we get

$$v_{insect} = v\left[\frac{f_{refl}\,(v - v_{bat}) - f_{bat}\,(v + v_{bat})}{f_{refl}\,(v - v_{bat}) + f_{bat}\,(v + v_{bat})}\right]$$

18. Wavelength in still water is $\lambda_0 = \frac{v}{f_0}$

When the medium moves with a velocity v_m, the wavelength downstream (DS) is increased by $v_m T$ and decreased upstream (US) by $v_m T$.

$\because$ $$\lambda_{DS} = \lambda_0 + v_m T \text{ and } \lambda_{US} = \lambda_0 - v_m T$$

When particles of medium are in motion and through these particles a wave travels, the velocity of the received wavefronts is the sum of the velocities of waves in still water and the velocity of medium.

$\therefore$ $$v_{DS} = v + v_m \text{ and } v_{US} = v - v_m$$

The frequency f' registered by a receiver is the ratio of the speed of sound to the wavelength.

For downstream receiver,

$$f_1' = \frac{v + v_m}{\lambda_0 + v_m T} = \frac{v + v_m}{\frac{v}{f_0} + \frac{v_m}{f_0}} = f_0$$

Similarly for upstream receiver,

$$f_2' = \frac{v - v_m}{\lambda_0 - v_m T} = \frac{v - v_m}{\frac{v}{f_0} - \frac{v_m}{f_0}} = f_0$$

Frequency received by both the receivers is same. Thus, we conclude that there is no Doppler shift due to the motion of medium only.

19. Time period of circular motion of source is

$$T = \frac{2\pi}{\omega} = \frac{2\pi}{10} = \frac{\pi}{5}\text{ s}$$

Time period of oscillation of detector is

$$T = \frac{1}{f} = \frac{\pi}{5}\text{ s}$$

Time periods of both are same.

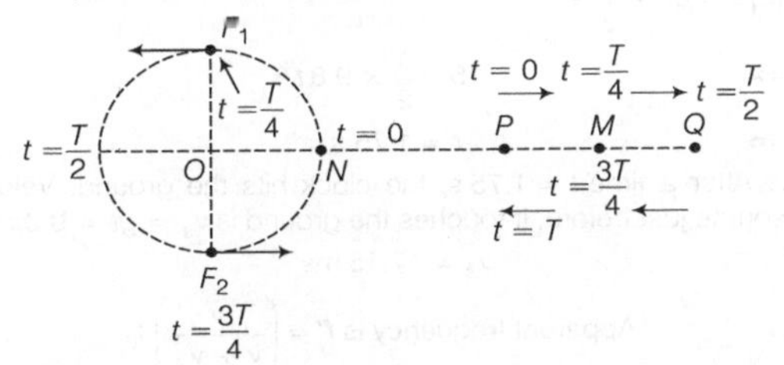

At $t = \frac{T}{4}$, source is at F_1 and detector is at M and both are receding away from each other. So, the frequency recorded would be minimum.

Speed of source is

$$v_s = R\omega = 3 \times 10 = 30\text{ ms}^{-1}$$

Speed of detector at mean position M is

$$v_O = A \cdot 2\pi f$$

or $$v_O = 6 \times 2\pi \times \frac{5}{\pi} = 60\text{ ms}^{-1}$$

$\therefore$ $$f'_{min} = \left(\frac{v - v_O}{v + v_s}\right) f_0$$

$$= \frac{330 - 60}{330 + 30}(340) = 255\text{ Hz}$$

At $t = \frac{3T}{4}$, source is at F_2 and detector is at M and both are approaching each other. So, the frequency recorded would be maximum.

$$f'_{max} = \left(\frac{v + v_O}{v - v_s}\right) f_0$$

$$= \frac{330 + 60}{330 - 30} \times 440 = 442\text{ Hz}$$

Note *θ is very small as distance is large. So, we have used $\cos\theta \approx 1$*

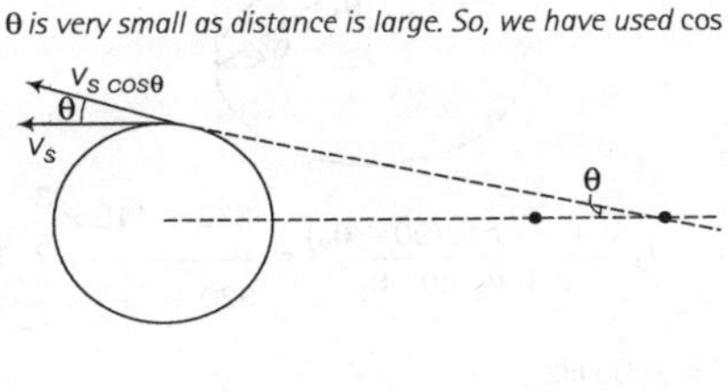

20. In a binary system, two stars revolve about the centre of mass of system, with the same frequency. When a star approaches the observer, minimum wavelength (λ_{min}) is observed and when it recedes, maximum wavelength (λ_{max}). In one time period of revolution, the λ_{min} and λ_{max} are observed twice. So the time period of revolution of stars is $2T$, (where T = 18 hr, according to given problem)

$$f_A' = f_0\left[1 + \frac{v_r}{c}\right]$$

and $$f_R' = f_0\left[1 - \frac{v_r}{c}\right] \quad \text{(where, } v_r << c\text{)}$$

Here, v_r is relative speed of approach (A) or relative speed of recession (R).

$\therefore$ $$\frac{c}{\lambda_A'} = \frac{c}{\lambda_0}\left[1 + \frac{v_r}{c}\right]$$

$\Rightarrow$ $$\lambda_A' = \lambda_0\left[1 + \frac{v_r}{c}\right]^{-1} = \lambda_0\left[1 - \frac{v_r}{c}\right]$$

and $$\frac{c}{\lambda_R'} = \frac{c}{\lambda_0}\left[1 - \frac{v_r}{c}\right]$$

$\Rightarrow$ $$\lambda_R' = \lambda_0\left[1 - \frac{v_r}{c}\right]^{-1} = \lambda_0\left[1 + \frac{v_r}{c}\right]$$

Here, $\lambda_A' = \lambda_{min}$ and $\lambda_R' = \lambda_{max}$

Let the distance between two stars is $2d$, then speed of each star

$$v = \frac{2\pi d}{2T} = \frac{\pi d}{T}$$

Now, $$\lambda_{max} - \lambda_{min} = \lambda_0\left[1 + \frac{v_r}{c}\right] - \lambda_0\left[1 - \frac{v_r}{c}\right]$$

$$= \lambda_0 \frac{2v_r}{c}$$

(As, observer is stationary $v_r = v$)

We can also use the approximation,

$$\lambda_0 = \frac{\lambda_{max} + \lambda_{min}}{2}$$

$$\lambda_{max} - \lambda_{min} = \frac{\lambda_{max} + \lambda_{min}}{2} \cdot \frac{2v}{c}$$

$$v = c\left[\frac{\lambda_{max} - \lambda_{min}}{\lambda_{max} + \lambda_{min}}\right]$$

$\Rightarrow$ $$\frac{\pi d}{T} = c\left[\frac{\lambda_{max} - \lambda_{min}}{\lambda_{max} + \lambda_{min}}\right]$$

$$d = \frac{cT}{\pi}\left[\frac{\lambda_{max} - \lambda_{min}}{\lambda_{max} + \lambda_{min}}\right]$$

$$= \frac{3 \times 10^8 \times (18 \times 60 \times 60)}{\pi}\left[\frac{563 - 539}{563 + 539}\right]$$

or $$d = 1.35 \times 10^{11}\text{ m}$$

The centripetal force of revolution is provided by the gravitational attraction between the stars.

$\therefore$ $$\frac{Gm^2}{(2d)^2} = \frac{mv^2}{d} \quad \text{(where, } d = \text{radius)}$$

$$\frac{Gm}{4d} = v^2$$

$\Rightarrow$ $$m = \frac{4dv^2}{G} = \frac{4d}{G}\left(\frac{\pi d}{T}\right)^2$$

or $$m = \frac{4d}{G}\frac{\pi^2 d^2}{T^2} = \frac{4\pi^2}{G}\frac{d^3}{T^2}$$

$$= \frac{4 \times \pi^2}{6.67 \times 10^{-11}} \frac{(1.35 \times 10^{11})^3}{(18 \times 60 \times 60)^2}$$

or $$m = 3.46 \times 10^{35}\text{ kg}$$

Revisal Problems (JEE Main)

1. (a) Nitrogen and Oxygen are both diatomic.

$\therefore \quad \gamma = 1.4$

$$\frac{M_1 + M_2}{M_{0\text{mix}}} = \frac{M_1}{M_{01}} + \frac{M_2}{M_{02}}$$

($\because$ number of moles remains constant)

$$\frac{7+12}{M_{0\text{mix}}} = \frac{7}{28} + \frac{12}{32}$$

$\Rightarrow \quad M_{0\text{mix}} = 30.4$

$$\therefore \quad v = \sqrt{\frac{\gamma RT}{M_{0\text{mix}}}} = \sqrt{\frac{1.4 \times 8.3 \times 300}{30.4 \times 10^{-3}}} = 338.6\ \text{ms}^{-1}$$

2. (b) $\beta = 10 \log_{10} \frac{I}{I_0} \Rightarrow \log_{10} \frac{I}{I_0} = \frac{\beta}{10} \Rightarrow I = I_0\, 10^{\beta/10}$

$\therefore \quad I_1 = I_0\,(10^5)$ and $I_2 = I_0\,(10^6)$

$\therefore \quad \frac{I_2}{I_1} = 10$

The relation between pressure amplitude and intensity is

$$I = (\Delta p)^2_{\max}/2\rho v \quad \Rightarrow \quad I \propto (\Delta p)^2_{\max}$$

3. (d) At a distance of 400 m from the explosion $(\Delta p)_{\max} = 10$ Pa.

At this point, the intensity is

$$I = \frac{(\Delta p)^2_{\max}}{2\rho v} = \frac{(10)^2}{2\,(1.2)\,(343)}$$

$\Rightarrow \quad I = 0.121\ \text{Wm}^{-2}$

From the inverse square law, the intensity (due to distance alone) 4 km away is $I' = \frac{I\,(400)^2}{(4000)^2} = 1.21 \times 10^{-3}\ \text{Wm}^{-2}$ and thus,

$$\beta' = 10 \log\left(\frac{I'}{I_0}\right) = 10 \log\left[\frac{1.21 \times 10^{-3}}{10^{-12}}\right] = 90.8\ \text{dB}$$

At a distance of 4 km from the explosion, absorption from the air will decrease the sound level by an additional amount,

$$\Delta\beta' = 7\ \frac{\text{dB}}{\text{km}}\ (3.6\ \text{km}) = 25.2\ \text{dB}$$

So, at 4 km the sound level will be $\beta = \beta' - \Delta\beta' = 90.8 - 25.2$

$\Rightarrow \quad \beta = 65.6$ dB

4. (b) $I = \frac{(\Delta p)^2_{\max}}{2\rho V}$

$\Rightarrow \quad v = \frac{(\Delta p)^2_{\max}}{2\rho I} = (2 \times 10^{-4})^2/2 \times 1 \times 10^{-10} = 200\ \text{ms}^{-1}$

Amplitude of wave is $A = \frac{(\Delta p)_{\max}}{\omega \rho v} = \frac{2 \times 10^{-4}}{10^3 \times 1 \times 200} = 10^{-9}$ m

$\therefore \quad k = \frac{\omega}{v} = \frac{1000}{200} = 5\ \text{m}^{-1}$

As, the wave is travelling in $-x$-direction, the only equation with positive sign between ωt and kx, with suitable ω and k values is option (b).

5. (c) $\Delta\beta = 10 \log \frac{I_2}{I_0} - 10 \log \frac{I_1}{I_0} = 10 \log \frac{I_2}{I_1}$

$\therefore \quad 30 = 10 \log \frac{I_2}{I_1} \Rightarrow \frac{I_2}{I_1} = 10^3 = 1000$

6. (a) Here, source and observer are approaching each other.

$$\therefore \quad f' = \left(\frac{v + v_O}{v - v_s}\right) f_0 = \frac{v + \frac{v}{15}}{v - \frac{v}{10}} \times 600 = \frac{16}{15} \times \frac{10}{9} \times 600$$

or $\quad f' = \frac{32}{27} \times 600 = 711$ Hz

7. (a) 1 $\rightarrow$ wavefront 1

2 $\rightarrow$ wavefront 2

$i \rightarrow$ initial position of source

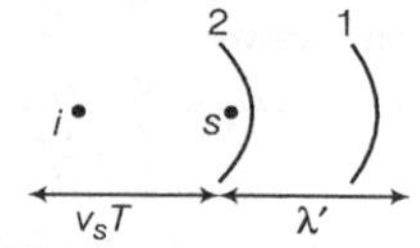

Time gap between 1 and 2 wavefronts is one time period (T)

$$v_s T + \lambda' = \lambda_0$$

$$\Rightarrow \quad v_s \frac{\lambda_0}{v} + \lambda' = \lambda_0$$

$$\lambda' = \lambda_0 \left(1 - \frac{v_s}{v}\right)$$

$$= 60\left(1 - \frac{20}{320}\right) = 56.25\ \text{cm}$$

8. (c) $s = ut + \frac{1}{2} at^2$

$\Rightarrow \quad 15 = \frac{1}{2} \times 9.8 t^2$

$\Rightarrow \quad t = 1.75$ s

$\therefore$ After a time $t = 1.75$ s, the clock hits the ground. Velocity of source just before, it touches the ground is $v_s = gt = 9.8x\,1.75$

$$v_s = 17.15\ \text{ms}^{-1}$$

$\therefore \quad$ Apparent frequency is $f' = \left(\frac{v}{v + v_s}\right) f_0$

$$= \left(\frac{343}{343 + 17.15}\right) 600$$

$$= 571.4\ \text{Hz}$$

9. (b) The motions of A and B are synchronised. Maximum speed of each block is $A\omega = 1 \times 2\pi \times 2 = 12.56\ \text{ms}^{-1}$

$$\therefore \quad f'_{\max} = \left(\frac{v + v_O}{v - v_s}\right) f_0 = \left(\frac{340 + 12.56}{340 - 12.56}\right)(300) = 323\ \text{Hz}$$

and $\quad f'_{\min} = \left(\frac{v - v_O}{v + v_s}\right) f_0 = \left(\frac{340 - 12.56}{340 + 12.56}\right)(300) = 278.6$ Hz

10. (c) $\quad \frac{v}{v - v_s} : \frac{v}{v + v_s} = 6 : 5$

$$\Rightarrow \quad \frac{v + v_s}{v - v_s} = \frac{6}{5}$$

$$5v + 5v_s = 6v - 6v_s$$

$$\Rightarrow \quad 11 v_s = v$$

$$\Rightarrow \quad v_s = \frac{v}{11} = \frac{330}{11} = 30\ \text{ms}^{-1}$$

11. (a) Frequency received by guard is

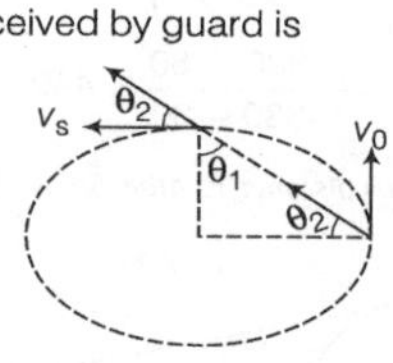

$$f' = f_0\, \frac{v + v_O \cos(90 - \theta_2)}{v + v_s \cos \theta_2} = \frac{330 + 110 \times \frac{3}{5}}{330 + 110 \times \frac{4}{5}} \cdot f_0$$

$\Rightarrow \quad f' = 1800$ Hz

12. (c) As apparent frequency in case of 2nd car is more, its speed must be more and it must have won the race. I is initial point and F is final point.

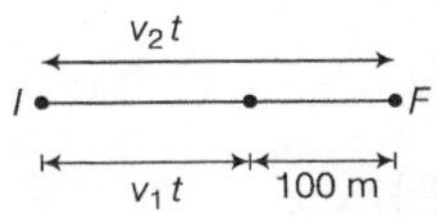

$$f_1' = \left(\frac{v}{v - v_1}\right) f_0$$

$$\Rightarrow \quad 330 = (300)\left(\frac{330}{330 - v_1}\right)$$

$$f_2' = \left(\frac{v}{v - v_2}\right) f_0$$

$$\Rightarrow \quad 360 = 300\left(\frac{330}{330 - v_2}\right)$$

By solving, we get $v_1 = 30 \text{ ms}^{-1}$ and $v_2 = 55 \text{ ms}^{-1}$

From the figure, $v_1 t + 100 = v_2 t$

$\therefore \quad t(v_2 - v_1) = 100 \Rightarrow t(55 - 30) = 100$

$\Rightarrow \quad t = 4 \text{ s}$

13. (b) Intensity after passing through one slab is

$$I' = \left(I_0 - \frac{20}{100} \times I_0\right) = I_0 - \frac{I_0}{5} = \frac{4I_0}{5}$$

So, intensity after passing through two slabs is

$$I'' = \left(I' - \frac{20}{100} \times I'\right)$$

$$= \frac{4I_0}{5} - \frac{1}{5} \cdot \frac{4I_0}{5} = \frac{16}{25} I_0$$

$$\% \text{ decrease} = \frac{I_0 - \frac{16}{25} I_0}{I_0} \times 100 = \frac{9}{25} \times 100 = 36\%$$

14. (b) The time recorded in summer is more accurate. The velocity of sound is directly proportional to the square root of absolute temperature. Hence, the sound of the gun fired at the starting point will reach the finishing point quicker in summer than in winter. The lapse of time due to the time taken by the sound in reaching the finish point will be less in summer and hence, the time recorded will be more accurate in summer than in winter.

15. (d) $T = \dfrac{2\pi}{\omega_1 - \omega_2} = \dfrac{1}{f_1 - f_2} = \dfrac{1}{1000} \text{ s}$

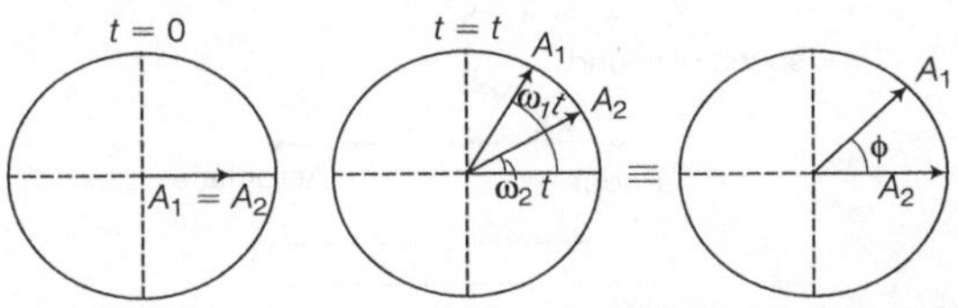

where, $\phi = (\omega_1 - \omega_2)t$

Now, it is given that the detector can detect signals only, when resultant intensity is $\geq 2I_0$ or resultant amplitude $\geq \sqrt{2}\, A_0$, because $I \propto A^2$. We can see that resultant amplitude is less than $\sqrt{2}\, A_0$ from $\phi = \dfrac{\pi}{2}$ to $\phi = \dfrac{3\pi}{2}$ i.e. for each half cycle. Hence, required time for which the detector remains idle is $t = \dfrac{T}{2} = \dfrac{\frac{1}{1000}}{2}$ s

$$\Rightarrow \quad t = 5 \times 10^{-4} \text{ s} = 500 \ \mu\text{s}$$

16. (a) $\beta_2 - \beta_1 = 10 \log \dfrac{I_2}{I_2}$

$$I \propto \frac{1}{r^2} \Rightarrow \frac{I_2}{I_1} = \left(\frac{r_1}{r_2}\right)^2 = \frac{1}{4}$$

$$\therefore \quad \beta_2 - \beta_1 = 10 \log\left(\frac{1}{4}\right) = -6 \text{ dB}$$

17. (c) $\cos\theta = \dfrac{80}{100} = \dfrac{4}{5}$

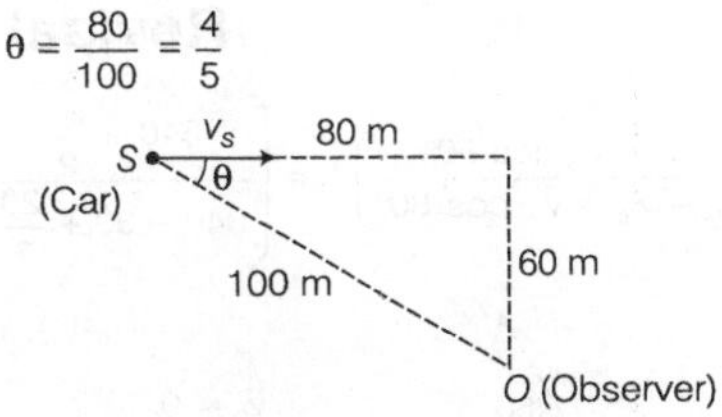

$\therefore$ Apparent frequency is $f' = \left(\dfrac{v}{v - v_s \cos\theta}\right) f_0$

or

$$f' = \left(\frac{330}{330 - 20 \times \frac{4}{5}}\right) 500 = 525.5 \text{ Hz}$$

18. (c) $\beta_2 - \beta_1 = 10 \log_{10} \dfrac{I_2}{I_1}$

$$\Rightarrow \quad 10 \log_{10} \frac{I_2}{I_1} = 13 \Rightarrow \log_{10} \frac{I_2}{I_1} = 1.3$$

$$\frac{I_2}{I_1} = 10^{1.3} \Rightarrow I_2 = 10^{1.3} I_1$$

19. (d) $I = \dfrac{1}{2}\rho\omega^2 A^2 v = \dfrac{1}{2}\rho(2\pi f)^2 A^2 v = 2\pi^2 f^2 A^2 \rho v$

$\therefore \quad I = 2\pi^2 \times (300)^2 \times (0.2 \times 10^{-3})^2 \times 1.29 \times 330 \text{ Wm}^{-2}$

$\approx 30.2 \text{ Wm}^{-2}$

$$\beta = 10\log_{10}\frac{30.2}{10^{-12}} \approx 10\log_{10}(3 \times 10^{13})$$

$$= 10\log_{10} 3 + (13\log_{10} 10)10 = 134.7 \text{ dB}$$

20. (a) Time period is $T = 1.6$ s, speed of waves,

$$v = 0.32 \text{ ms}^{-1}$$

$$\therefore \quad \lambda_0 = vT = 0.32 \times 1.6 = 0.512 \text{ m}$$

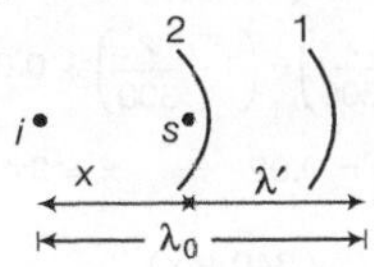

From initial position (i) of the duck distance moved by duck in time $T = x = v_s \times T = 1.6\, v_s\, \lambda' = 0.12$ m (given).

From the figure,

$$x + \lambda' = \lambda_0$$

$$\therefore \quad 1.6\, v_s + 0.12 = 0.512$$

$$\Rightarrow \quad v_s = 0.245 \text{ ms}^{-1}$$

21. (b) $\beta = 10 \log_{10} \dfrac{I}{I_0}$

$$\Rightarrow \quad 60 = 10 \log_{10} \frac{I}{10^{-12}}$$

$$\therefore \quad I = 10^6 \times 10^{-12} = 10^{-6} \text{ Wm}^{-2}$$

But $\quad I = \dfrac{\text{Power}}{\text{Area}} = \dfrac{P}{4\pi r^2}$

$$\Rightarrow \quad P = (4\pi r^2)\, I = (4\pi)(500)^2 (10^{-6})$$

or $\quad P = 3.14$ W

$$\therefore \quad \text{Time, } t = \frac{1000 \times \frac{30}{100}}{3.14} = 95.5 \text{ s}$$

22. (d) Due to increase in relative speed of sound w.r.t. observer, apparent frequency is larger than real frequency. So, Statement I is incorrect.

23. (d) Pressure variation is maximum, when displacement is zero i.e. if $y = 0$, pressure may be maximum or minimum.

Revisal Problems (JEE Advanced)

1. (a) $f' = \left(\dfrac{v + v_w \cos 60°}{v - v_s + v_w \cos 60°}\right) f_0 = \left(\dfrac{340 + \dfrac{20}{2}}{340 - 30 + \dfrac{20}{2}}\right) 1000$

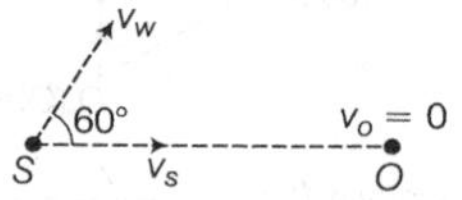

or $\quad f' \approx 1094$ Hz

2. (d) $\beta = 10 \log \dfrac{I}{I_0} = 10 \log I - 10 \log I_0 = 10 \log \left(\dfrac{P}{4\pi r^2}\right) - 10 \log I_0$

$$= 10 \log P - 10 \log 4\pi r^2 - 10 \log I_0$$

$$= 10 \log \frac{P}{I_0} - 10 \log 4\pi - 10 \log r^2$$

$$= \left(10 \log \frac{P}{I_0} - 10 \log 4\pi\right) - 20 \log r$$

or $\quad \beta = k_1 - k_2 \log r$

3. (a) $f_1' = \left(\dfrac{v}{v - v_s}\right) f_0$ and $f_2' = \left(\dfrac{v}{v + v_s}\right) f_0$

Given, $f_1' - f_2' = 0.02\, f_0$

$$\frac{v}{v - v_s} - \frac{v}{v + v_s} = 0.02$$

$$\Rightarrow \quad \frac{300}{300 - x} - \frac{300}{300 + x} = 0.02$$

$$\frac{300}{300\left[1 - \dfrac{x}{300}\right]} - \frac{300}{300\left[1 + \dfrac{x}{300}\right]} = 0.02$$

$$\left(1 - \frac{x}{300}\right)^{-1} - \left(1 + \frac{x}{300}\right)^{-1} = 0.02$$

$$\Rightarrow \quad \left(1 + \frac{x}{300}\right) - \left(1 - \frac{x}{300}\right) \approx 0.02$$

$$\frac{x}{150} = 0.02 \quad \Rightarrow \quad x = 3 \text{ ms}^{-1}$$

4. (b) $f' = \left(\dfrac{v + v_O}{v - v_s}\right) f_0 \Rightarrow \left(\dfrac{340 + x}{340 - x}\right) 400 = 450$

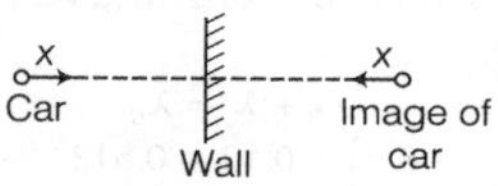

$$\frac{340 + x}{340 - x} = \frac{9}{8} \Rightarrow 2720 + 8x = 3060 - 9x \Rightarrow x = 20 \text{ ms}^{-1}$$

5. (c) $\lambda' = \dfrac{\text{Wave speed relative to observer}}{f_0} \Rightarrow \lambda' = \dfrac{v + v_w}{f_0}$

or $\lambda' = \left(\dfrac{v + v_w}{v}\right) \lambda_0$

6. (c) The displacements of the neighbouring particles are shown in the figure. Clearly, point 3 corresponds to maximum pressure.

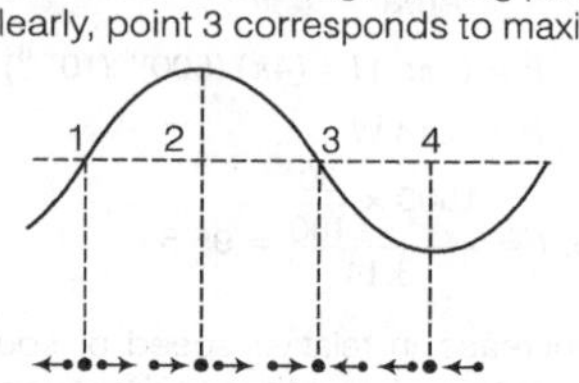

or

$p = -B \dfrac{dy}{dx}$. At point 3, slope is most negative.

7. (c) $f_1 = \left(\dfrac{v + v_O}{v}\right) f_0$

$$f_2 = \left(\frac{v}{v - v_O}\right) f_0 \qquad (\because v_s = v_O)$$

$$f_3 = \left(\frac{v + v_O/2}{v - v_O/2}\right) f_0$$

($\because$ relative speed is same in all cases)

It is clear from the above that, $f_1 \neq f_2 \neq f_3$, $f_3 > f_0$, $f_1 < f_2$, we can also prove that, $f_2 > f_3 > f_1$.

8. (d) Let t' is the time, when the sound is emitted which is received at $t = 4$ s. Then,

v = speed of sound

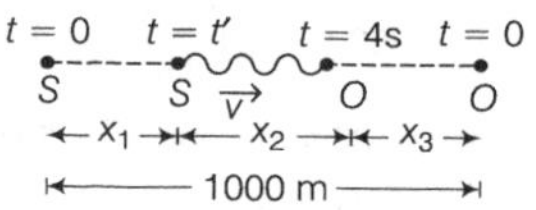

$$x_1 + x_2 + x_3 = 1000 \text{ m}$$

$$\Rightarrow \quad \frac{1}{2} a_0 t'^2 + v(4 - t') + \frac{1}{2} a_0 (4)^2 = 1000$$

or $\quad \dfrac{1}{2}(10) t'^2 + 300(4 - t') + \dfrac{1}{2}(10)(4)^2 = 1000$

$$5t'^2 - 300t' + 280 = 0$$

$$\Rightarrow \quad t'^2 - 60t' + 56 = 0$$

or $\quad t' = 0.95$ s and 59 s

But here t' must be 0.95 s.

So, the pulse emitted at $t' = 0.95$ s is received by the detector at $t = 4$ s.

Speed of source at $t = 0.95$ s is

$$v_s = a_0 (0.95) = (10)(0.95) = 9.5 \text{ ms}^{-1}$$

Speed of observer at $t = 4$ s is

$$v_O = a_0 (4) = (10)(4) = 40 \text{ ms}^{-1}$$

Both of them are approaching each other

$$\therefore \quad f' = \left(\frac{v + v_O}{v - v_s}\right) f_0 = \left(\frac{300 + 40}{300 - 9.5}\right)(2000)$$

$$= 2340.74 \text{ Hz} \approx 2341 \text{ Hz}$$

9. (b) $f_1 = \left(\dfrac{v + v_O}{v}\right) f_0 = \left(\dfrac{v + x}{v}\right) f_0 \qquad \ldots\text{(i)}$

v = speed of sound

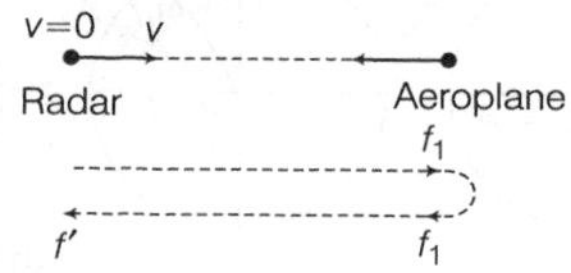

$$f' = \left(\frac{v}{v - v_s}\right) f_1 = \left(\frac{v}{v - x}\right) f_1 \qquad \ldots\text{(ii)}$$

Given, $\quad f' = f_0 + 2.6$ kHz $= 780$ MHz $+ 2.6$ KHz

or $\quad f' = 780.0026$ MHz

From Eqs. (i) and (ii). we get

$$f' = \left(\frac{v + x}{v - x}\right) f_0$$

Here, radar emits light.

$\therefore \quad v = c$

$$780.0026 = \frac{3 \times 10^8 + x}{3 \times 10^8 - x} \times 780$$

$$\Rightarrow \quad x = 500 \text{ ms}^{-1}$$

10. (a,d) 123 has negative slope. So, it represents compression and 345 has positive slope. So, it represents rarefaction.

11. (b,c,d) $v \propto \sqrt{T}$

$$60\text{ dB} = 10\log\frac{I_1}{I_0}$$

and $$30\text{ dB} = 10\log\frac{I_2}{I_0}$$

$$60 - 30 = 10\log\frac{I_1}{I_2} \Rightarrow \frac{I_1}{I_2} \neq 2$$

12. (d) When they recede from each other, apparent frequency is smaller than f_0, but does not decrease with time.

Time taken for the line SO to become perpendicular to their lines of motion is $t_0 = \frac{L_0 \cot\theta_0}{2v_0}$. The sound emitted by source at this instant (i.e. t_0), when reaches the observer will have apparent frequency = f_0. But it takes some finite time to reach observer.

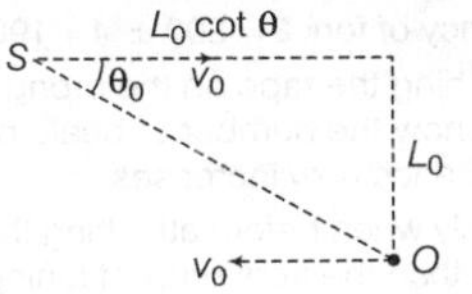

13. (a,c,d) $u = \frac{I}{v}$, where v is the wave propagation velocity i.e. $u \propto I$.

So, option (a) correct but option (b) is incorrect.

$$I = 2\pi^2 f_0^2 A^2 \rho v$$

$$\Rightarrow \quad u = \frac{I}{v} = 2\pi^2 f_0^2 A^2 \rho$$

$u \propto f_0^2$, so option (d) is correct

$$v_{max} = A\omega = A \cdot 2\pi f_0$$

$$\pi f_0 A = \frac{v_{max}}{2}$$

$$\therefore \quad u = 2\left(\frac{v_{max}}{2}\right)^2 \rho = \frac{\rho}{2}(v_{max})^2$$

So, option (c) is correct.

14. (a,c,d) At the rigid boundary, displacement is zero, but pressure variation is maximum i.e. reflected wave is non-inverted w.r.t. the incident pressure waves. At the free end, reflected and incident pressure waves are out of phase by π.

Note *Open end of a tube is equivalent to a rigid boundary in case of string and closed end of a tube is equivalent to a free end in case of a string.*

15. (b,c) $v = \sqrt{\frac{\gamma RT}{M_0}}$. Change in temperature affects the velocity of sound in air and as long as temperature remains same, change in pressure has no effect.

$$v = \sqrt{\frac{B}{\rho}}$$

Bulk modulus of water is very high, so velocity of sound in water is higher than that in air.

16. (b) Velocity of sound is $v = \sqrt{\frac{E}{\rho}}$

Modulus of elasticity (E) for solids is larger. So, speed of sound in solids is greater.

17. (a) Though ρ is larger for solids, the coefficient of elasticity is much larger as compared to liquid and gases i.e. v is more.

18. (a) The wood offers high damping to the sound waves. In the manufacture of bells, low damping is required. The metals are most suitable for low damping. So, the bells are made of metals.

19. (c) Imagine an image of submarine, consider the fixed island as a plane mirror.

Given, $$f' = 1.1 f_0$$

Also, $$f' = \left(\frac{v + x}{v - x}\right) f_0$$

$$\therefore \quad 1.1 f_0 = \frac{1050 + x}{1050 - x} \cdot f_0$$

$$\Rightarrow \quad 1155 - 1.1x = 1050 + x$$

or $$x = 50\text{ ms}^{-1}$$

20. (d) Because the apparent frequency is larger than initial case, the enemy ship should be moving towards submarine. Let the speed of enemy ship is y.

The frequency received by enemy ship is f_1' (say),

$$\therefore \quad f_1' = \left(\frac{v + v_O}{v - v_s}\right) f_0 = \left(\frac{1050 + y}{1050 - 50}\right) f_0 = \left(\frac{1050 + y}{1000}\right) f_0$$

The enemy ship reflects same frequency i.e. f_1', when this is received by Indian submarine. Let the apparent frequency is f'. Now,

$$f' = \left(\frac{v + v_O}{v - v_s}\right) f_1' = \frac{1050 + 50}{1050 - y} \cdot \left(\frac{1050 + y}{1000}\right) f_0$$

Given, $f' = 1.21 f_0$

$$\therefore \quad 1.21\, f_0 = \frac{1100}{1000} \cdot \frac{1050 + y}{1050 - y} \cdot f_0$$

$$\Rightarrow \quad 1.1 = \frac{1050 + y}{1050 - y}$$

$$1155 - 1.1y = 1050 + y \Rightarrow y = 50\text{ ms}^{-1}$$

21. (c) $\lambda' = \frac{\text{Speed w.r.t. observer}}{f_1'} = \frac{v + y}{\left(\frac{1050 + y}{1000}\right) f_0}$

or $\lambda' = \frac{1050 + 50}{1050 + 50} \cdot \frac{1000}{f_0} = \frac{1000}{f_0}$

$\lambda'' = \frac{\text{Speed w.r.t. observer}}{f'} = \frac{v + y}{1.21 f_0} = \frac{1100}{1.21 f_0}$ or $\lambda'' = \frac{1000}{1.1 f_0}$

$$\therefore \quad \frac{\lambda'}{\lambda''} = 1.1$$

22. (d) $v = \sqrt{\frac{B}{\rho}} \Rightarrow B = \rho v^2 = 1000 \times (1050)^2$

$$\therefore \quad B \approx 10^9\text{ Nm}^{-2}$$

23. (a) This problem is an analogy of Doppler effect.

$f_0 = 20\text{ min}^{-1}$, $v = 300\text{ m min}^{-1}$, $v_s = 0$ and $v_O = 0$

Spacing between the pies $= \frac{300}{20} = 15\text{ m}$

and $$f' = f_0 = 20\text{ min}^{-1}$$

24. (b) $v_s = 30\text{ m min}^{-1}$

Spacing between the pies $= \frac{300 - 30}{20} = 13.5\text{ m}$

25. (c) $f' = \left(\frac{v}{v - v_s}\right) f_0 = \left(\frac{300}{300 - 30}\right)(20) = 22.22\text{ min}^{-1}$

26. (1) Intensity is $I = \frac{(\Delta p)_m^2}{2\rho v}$. Here, ρ and v are same for both. Also, it is given that intensity (I) is same for both.

So, ratio of their pressure amplitudes is 1 : 1.

JEE Main & AIEEE Archive

1. (b) Apparent frequency heard by the person before crossing the train.

$$f_1 = \left(\frac{c}{c - v_s}\right) f_0 = \left(\frac{320}{320 - 20}\right) 1000$$

Similarly, apparent frequency heard, after crossing the trains

$$f_2 = \left(\frac{c}{c + v_s}\right) f_0 = \left(\frac{320}{320 + 20}\right) 1000 \qquad [c = \text{speed of sound}]$$

$$\Delta f = f_1 - f_2 = \left(\frac{2cv_s}{c^2 - v_s^2}\right) f_0 \quad \text{or} \quad \frac{\Delta f}{f_0} \times 100 = \left(\frac{2cv_s}{c^2 - v_s^2}\right) \times 100$$

$$= \frac{2 \times 320 \times 20}{300 \times 340} \times 100 = \frac{2 \times 32 \times 20}{3 \times 34} = 12.54\% = 12\%$$

2. (b) For motor cycle, $u = 0, a = 2\text{ ms}^{-2}$

Observer is in motion and source is at rest, then apparent frequency,

$$n' = n\,\frac{v - v_o}{v + v_s}$$

$$\Rightarrow \quad \frac{94}{100} n = n\,\frac{330 - v_o}{330} \quad \Rightarrow 330 - v_o = \frac{330 \times 94}{100}$$

$$\Rightarrow \quad v_o = 330 - \frac{94 \times 33}{10} = \frac{33 \times 6}{10}\text{ ms}^{-1}$$

$$s = \frac{v^2 - u^2}{2a} = \frac{9 \times 33 \times 33}{100} = \frac{9 \times 1089}{100} \approx 98\text{ m}$$

3. (c) Maximum number of beats $= v + 1 - (v - 1) = 2$

4. (d) Speed of sound is given by

$$v = \sqrt{\frac{\gamma RT}{M}}$$

$$v_{O_2} = \sqrt{\frac{\frac{7}{5}RT}{32}} \quad \text{and} \quad v_{He} = \sqrt{\frac{\frac{5}{3}RT}{4}}$$

$$\therefore \quad \frac{v_{O_2}}{v_{He}} = \sqrt{\frac{7 \times 3 \times 4}{5 \times 32 \times 5}}$$

or $$v_{He} = 460 \times \sqrt{\frac{5 \times 32 \times 5}{7 \times 3 \times 4}} \approx 1420\text{ ms}^{-1}$$

5. (b) $$l_1 = 18\text{cm}$$

$$f = \frac{v_1}{4l_1} \qquad \text{[for first resonances]}$$

$$f = \frac{3v_2}{4l_2} \qquad \text{[for second resonances]}$$

where, $l_2 = x$ according to given situation and also $v_1 < v_2$ as during summer temperature would be higher.

$$\Rightarrow \quad \frac{3v_2}{4l_2} = \frac{v_1}{4l_1} \quad \text{or} \quad l_2 = 3l_1 \times \frac{v_2}{v_1}$$

$\Rightarrow$ $x = 54 \times$ (A quantity greater than 1)

So, $x > 54$

6. (d) Let intensity of sound be I and I'.

Loudness of sound initially, $\beta_1 = 10 \log\left(\frac{I}{I_0}\right)$

Later, $\beta_2 = 10 \log\left(\frac{I'}{I_0}\right)$

Given, $\beta_1 - \beta_2 = 20$

$\therefore$ $20 = 10 \log\left(\frac{I}{I'}\right)$ or $I' = \frac{I}{100}$

Therefore, intensity decreases by a factor of 100.

7. (c) Given, velocity of sound in air = 300 m/s

If a source of sound is moving towards a stationary listener, the frequency heard by the listener would be different from the actual frequency of the source, this apparent frequency is given by $f_{app} = \left(\frac{v_{\text{sound in air}}}{v_{\text{sound in air}} \pm v_{\text{source}}}\right)$, where symbols have their usual meanings.

In the denominator, +ve sign would be taken when source is receding away from the listener, while –ve sign would be taken when source is approaching the listener.

Let v be the maximum value of source velocity for which the person is able to hear the sound, then

$$10000 = f_{app} = \left(\frac{300}{300 - v}\right) \times 9500 \quad \text{or} \quad v = 15\text{ ms}^{-1}$$

8. (b) The frequency of fork 2 = 200 ± 4 = 196 or 204 Hz

Since, on attaching the tape on the prong of fork 2, its frequency decreases but now the number of beats per second is 6 *i.e.*, the frequency difference now increases.

It is possible only when before attaching the tape, the frequency of fork 2 is less than the frequency of tuning fork 1. Hence, the frequency of fork 2 is 196 Hz.

9. (d) $$v_o = \frac{v}{5} \quad \Rightarrow \quad v_o = \frac{320}{5} = 64\text{ ms}^{-1}$$

When observer moves towards the stationary source, then

$$n' = \left(\frac{v + v_o}{v}\right) n \quad \Rightarrow \quad n' = \left(\frac{320 + 64}{320}\right) n$$

or $$n' = \left(\frac{384}{320}\right) n \quad \text{or} \quad \frac{n'}{n} = \frac{384}{320}$$

Hence, percentage increase

$$\left(\frac{n' - n}{n}\right) = \left(\frac{384 - 320}{320} \times 100\right) = \left(\frac{64}{320} \times 100\right) = 20\%$$

10. (c) $f_1 = 256$ Hz

For tuning fork, where $f_2 - f_1 = \pm 5$

f_2 = frequency of piano

$f_2 = (256 + 5)$ Hz or $(256 - 5)$ Hz

When tension is increased, the beat frequency decreases to 2 beats per second.

If we assume that the frequency of piano string is 261 Hz, then on increasing tension, frequency, more than 261 Hz. But it is given that beat frequency decreases to 2, therefore 261 Hz is not possible.

Hence, 251 Hz i.e., (256 – 5) was the frequency of piano string before increasing tension.

11. (b) As, $f = \frac{1}{2l}\sqrt{\frac{T}{m}}$

When temperature increases, then length increases and frequency $\left(f \propto \frac{1}{l}\right)$ decreases.

12. (b) The tuning fork of frequency 288 Hz is producing 4 beats per second with the unknown tuning fork *i.e.*, the frequency difference between them is 4. Therefore, the frequency of unknown tuning fork

$$= 288 \pm 4 = 292 \text{ or } 284$$

On placing a little wax on unknown tuning fork, its frequency decreases but now the number of beats produced per second is 2 i.e. the frequency difference now decreases.

It is possible only when before placing the wax, the frequency of unknown fork is greater than the frequency of given tuning fork. Hence, the frequency of unknown tuning fork = 292 Hz.

JEE Advanced & IIT JEE Archive

1. (a,b) When wind blows from S to 0

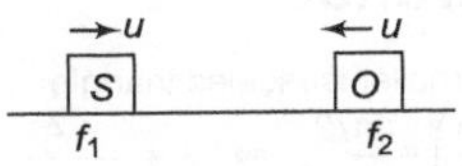

$$f_2 = f_1\left(\frac{v + w + u}{v + w - u}\right)$$

or $\quad f_2 > f_1$

when wind blows from O to S

$$f_2 = f_1\left(\frac{v - w + u}{v - w - u}\right)$$

$\therefore \quad f_2 > f_1$

2. (b,d) At open end phase of pressure wave changes by 180°. So, compression returns as rarefaction. At closed end there is no phase change. So, compression returns as compression and rarefaction as rarefaction.

3. (b) With end correction,

$$f = n\left[\frac{v}{4(l + e)}\right], \quad \text{where } n = 1, 3, \ldots$$

$$= n\left[\frac{v}{4(l + 0.6\,r)}\right]$$

because, $e = 0.6\,r$, where r is radius of pipe.

For first resonance, $\quad n = 1$

$$\therefore \quad f = \frac{v}{4(l + 0.6\,r)}$$

or $l = \frac{v}{4f} - 0.6r = \left[\left(\frac{336 \times 100}{4 \times 512}\right) - 0.6 \times 2\right] \text{cm} = 15.2 \text{ cm}$

4. (a) $36 \text{ km/h} = 36 \times \frac{5}{18} = 10 \text{ ms}^{-1}$

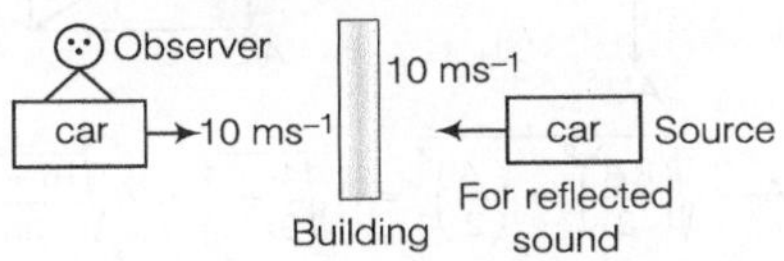

Apparent frequency of sound heard by car driver (observer) reflected from the building will be

$$f' = f\left(\frac{v + v_o}{v - v_s}\right) = 8\left(\frac{320 + 10}{320 - 10}\right) = 8.5 \text{kHz}$$

5. (a,d) $l < \frac{\lambda}{4}$

Further, larger the length of air column, feebler is the intensity.

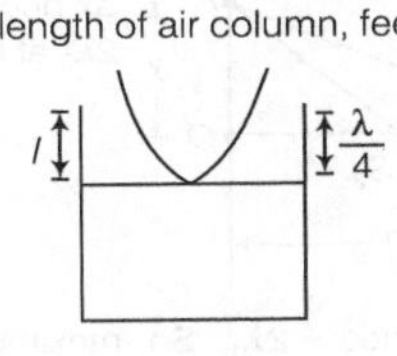

Explanations for Q. Nos. (6-8)

$$v_{SA} = 340 + 20 = 360 \text{ms}^{-1}$$

$$v_{SB} = 340 - 30 = 310 \text{ms}^{-1}$$

For the passengers in train A, there is no relative motion between source and observer, as both are moving with velocity 20 ms^{-1}. Therefore, there is no change in observed frequencies and correspondingly there is no change in their intensities. Therefore, the correct option is (a).

For the passengers in train B, observer is receding with velocity 30 m/s and source is approaching with velocity 20 ms^{-1}.

$$\therefore \quad f_1' = 800\left(\frac{340 - 30}{340 - 20}\right) = 775 \text{Hz}$$

and $$f_2' = 1120\left(\frac{340 - 30}{340 - 20}\right) = 1085 \text{ Hz}$$

$\therefore$ Spread of frequency $= f_2' - f_1' = 310$ Hz

6. (b) **7.** (a) **8.** (a)

Explanations for Q. Nos. (9-11)

In one second number of maximas is called the beat frequency. Hence,

$$f_b = f_1 - f_2 = \frac{100\pi}{2\pi} - \frac{92\pi}{2\pi} = 4 \text{ Hz}$$

Speed of wave $v = \frac{\omega}{k}$

or $$v = \frac{100\pi}{0.5\,\pi}$$

or $$\frac{92\,\pi}{0.46\,\pi} = 200 \text{ m/s}$$

At $x = 0$, $$y = y_1 + y_2$$
$$= 2A \cos 96\pi t \cos 4\pi t$$

Frequency of $\cos(96\pi t)$ function is 48 Hz and that of $\cos(4\pi t)$ function is 2 Hz.

In one second, cos function becomes zero at $2f$ times, where f is the frequency. Therefore, first function will become zero at 96 times and the second at 4 times. But second will not overlap with first. Hence, net y will become zero 100 times in 1 s.

9. (a)

10. (a)

11. (c)

12. (d) The question is incomplete, as speed of sound is not given. Let us assume speed of sound as 330 ms^{-1}. Then, method will be as under.

$$\frac{\lambda}{2} = (632 - 30.7) \text{cm} \quad \text{or} \quad \lambda = 0.65 \text{ m}$$

$\therefore$ Speed of sound observed,

$$v_0 = f\lambda = 512 \times 0.65 = 332.8 \text{ms}^{-1}$$

$\therefore$ Error in calculating velocity of sound $= 2.8 \text{ ms}^{-1}$

$$= 280 \text{cms}^{-1}$$

4. Superposition and Standing Waves

DPP-1 Superposition and Interference

1. (a) $\because \quad \dfrac{A_1}{A_2} = \dfrac{3}{5}$

So, $\quad \sqrt{\dfrac{I_1}{I_2}} = \dfrac{3}{5} \qquad (\because I \propto A^2)$

$$I = I_1 + I_2 + 2\sqrt{I_1 I_2}\cos\phi$$

I is maximum, when $\cos\phi = 1$ and I is minimum when $\cos\phi = -1$

$\therefore \quad I_{max} = (\sqrt{I_1} + \sqrt{I_2})^2$ and $I_{min} = (\sqrt{I_1} - \sqrt{I_2})^2$

$$\frac{I_{min}}{I_{max}} = \left(\frac{\sqrt{I_1} - \sqrt{I_2}}{\sqrt{I_1} + \sqrt{I_2}}\right)^2 = \left[\frac{\sqrt{\dfrac{I_1}{I_2}} - 1}{\sqrt{\dfrac{I_1}{I_2}} + 1}\right]^2 = \left[\frac{\dfrac{3}{5} - 1}{\dfrac{3}{5} + 1}\right]^2 = \frac{1}{16}$$

2. (c) Path difference is $\Delta x = TRQ - TPQ = 11.5$ cm

There is a complete silence at P, so $\Delta x = (2n + 1)\dfrac{\lambda}{2}$

For f_{min}, $\quad \dfrac{\lambda}{2} = 11.5$ cm or $\lambda = 23$ cm

$\therefore \quad f_{min} = \dfrac{v}{\lambda} = \dfrac{331.2}{0.23} = 1440$ Hz

3. (b) $A_1 = 10, A_2 = 20, \delta = \dfrac{\pi}{3}$

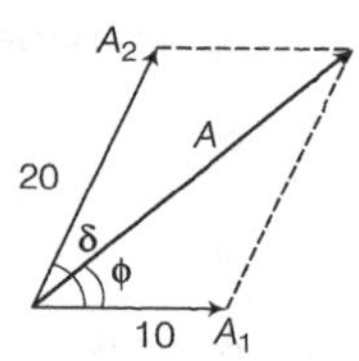

The amplitude of resultant wave is

$$A = \sqrt{A_1^2 + A_2^2 + 2A_1A_2\cos\delta}$$

$$= \sqrt{10^2 + 20^2 + 2(10)(20)\left(\frac{1}{2}\right)}$$

$= 10\sqrt{7}$ units

$$\phi = \tan^{-1}\left[\frac{A_2\sin\delta}{A_1 + A_2\cos\delta}\right] = \tan^{-1}\left[\frac{20\left(\dfrac{\sqrt{3}}{2}\right)}{10 + 20\left(\dfrac{1}{2}\right)}\right] = \tan^{-1}\left(\frac{\sqrt{3}}{2}\right)$$

So, the resultant wave is

$$y = 10\sqrt{7}\cos\left[5x + 25t + \tan^{-1}\left(\frac{\sqrt{3}}{2}\right)\right]$$

4. (b) The diagram for $n = 6$ is shown. Let each side of isosceles triangle is r.

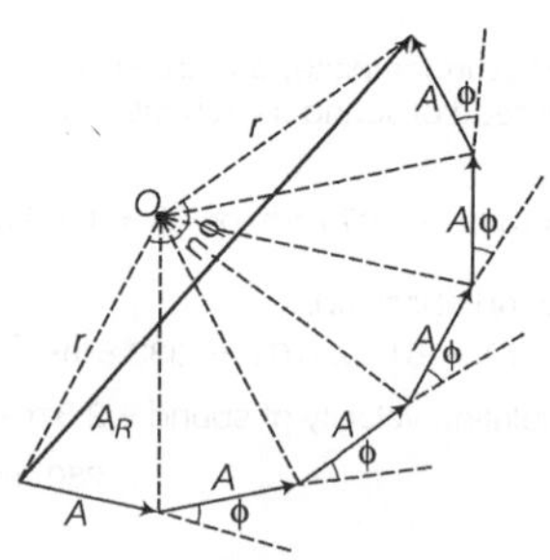

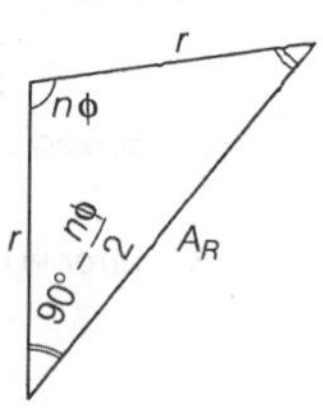

From sine rule

$$\frac{A_R}{\sin(n\phi)} = \frac{r}{\sin\left(90° - \dfrac{n\phi}{2}\right)}$$

$$A_R = \frac{r\,2\sin\left(\dfrac{n\phi}{2}\right)\cos\left(\dfrac{n\phi}{2}\right)}{\cos\left(\dfrac{n\phi}{2}\right)} = 2r\sin\left(\frac{n\phi}{2}\right) \qquad \ldots(i)$$

Now, consider smaller isosceles triangle

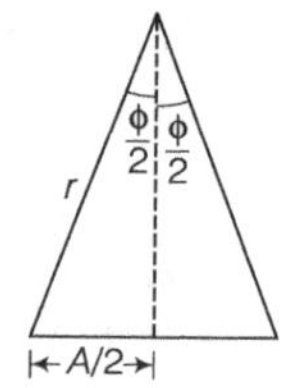

$$\sin\left(\frac{\phi}{2}\right) = \frac{A/2}{r} \Rightarrow r = \frac{A}{2\sin\left(\dfrac{\phi}{2}\right)}$$

Using this result in Eq. (i), we get

$$A_R = 2\frac{A}{2\sin\left(\dfrac{\phi}{2}\right)}\sin\left(\frac{n\phi}{2}\right)$$

or $$A_R = \frac{A\sin\left(\dfrac{n\phi}{2}\right)}{\sin\left(\dfrac{\phi}{2}\right)}$$

5. (c) In Fig. (i), path difference is $\Delta x = S_1m - S_2m = 0$

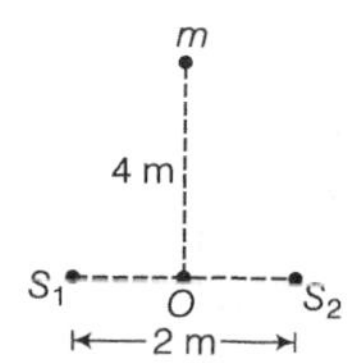

Initially Fig. (i)

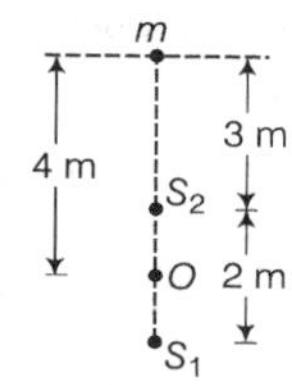

Finally Fig. (ii)

In Fig. (ii) path difference is $\Delta x = S_1m - S_2m = 5 - 3 = 2$ m

As, 5 maxima are observed $5\lambda = 2 \Rightarrow \lambda = \dfrac{2}{5} = 0.4$ m

6. (d) The phasor diagram is

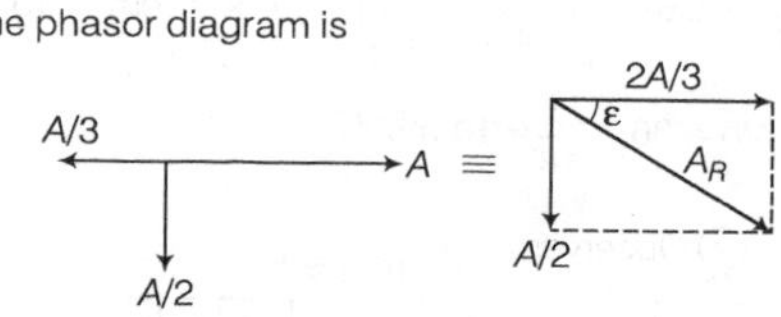

$$A_R = \sqrt{\left(\frac{2A}{3}\right)^2 + \left(\frac{A}{2}\right)^2} = A\sqrt{\frac{4}{9} + \frac{1}{4}} = A\sqrt{\frac{16 + 9}{36}}$$

or $\quad A_R = \dfrac{5}{6}A$

$$\tan(-\varepsilon) = \frac{A/2}{2A/3} = \frac{3}{4}$$

$$-\tan\varepsilon = \frac{3}{4} \Rightarrow \tan\varepsilon = \frac{-3}{4} \Rightarrow \varepsilon = -\tan^{-1}\left(\frac{3}{4}\right)$$

7. (d)

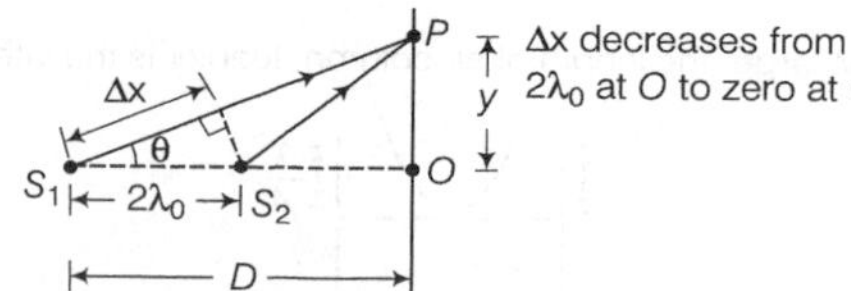

At O path difference $= 2\lambda_0$. So, maximum occurs at O. If at P, maximum has to occur, $\Delta x = \lambda_0 \qquad \ldots(i)$

From figure, $\quad \Delta x = 2\lambda_0\cos\theta \qquad \ldots(ii)$

From Eqs. (i) and (ii), we get

$$2\lambda_0\cos\theta = \lambda_0 \Rightarrow \theta = \frac{\pi}{3}\text{ rad}$$

Now, $\quad \tan\dfrac{\pi}{3} = \dfrac{y}{D} \qquad$ (from the figure)

$$\sqrt{3} = \frac{y}{D} \Rightarrow y = \sqrt{3}D$$

8. (b) Wave 1 directly coming from S reaches D and wave 2 coming after reflection from the obstacle reaches D. As 2nd wave has undergone reflection from wall, a phase change of π has occurred. So, condition for maximum at D is

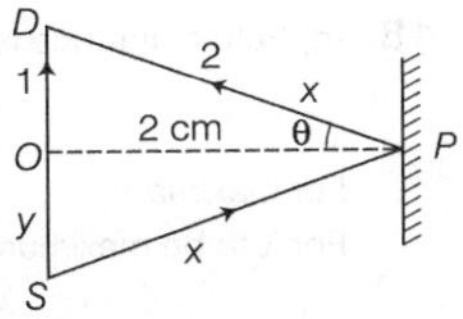

$$(SP + PD) - SD = (2n+1)\frac{\lambda_0}{2} \quad (n = 0, 1, \ldots)$$

For SD to be minimum, $\quad n = 0 \quad$ (Here, $\lambda_0 = \frac{v}{f_0} = \frac{360}{180} = 2\,\text{m}$)

$$x + x - y = \frac{\lambda_0}{2} = 1\,\text{m}$$

$$2x - y = 1$$

$$2\sqrt{\frac{y^2}{4} + 4} - y = 1 \Rightarrow 2\sqrt{\frac{y^2}{4} + 4} = 1 + y$$

Squaring on both the sides

$$4\left(\frac{y^2}{4} + 4\right) = 1 + y^2 + 2y$$

$$y^2 + 16 = 1 + y^2 + 2y$$

$$\Rightarrow \quad 2y = 15 \Rightarrow y = 7.5\,\text{m}$$

9. (c) Say at a general point $P(x, y)$, intensity is minimum. So, path difference is

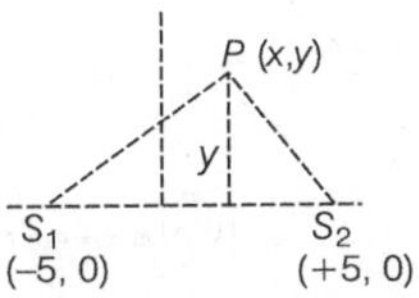

$$\therefore \quad S_1P - S_2P = (2n-1)\frac{\lambda}{2}$$

$$(n = 1, 2, 3, \cdots)$$

$$\sqrt{(x+5)^2 + y^2} - \sqrt{(x-5)^2 + y^2} = (2n-1)\frac{\lambda}{2}$$

Here, $\lambda = \frac{v}{f} = \frac{343}{21.44} \simeq 16\,\text{m}$

$$\therefore \quad \sqrt{(x+5)^2 + y^2} - \sqrt{(x-5)^2 + y^2} = (16n - 8)$$

10. (d) Path difference is

$$\Delta x = L_2D - L_1D = \sqrt{40^2 + 9^2} - 40 = 41 - 40 = 1\,\text{m}$$

For first maximum, $\Delta x = \lambda \Rightarrow \lambda = 1\,\text{m}$

Corresponding frequency is $f = \frac{v}{\lambda} = \frac{330}{1} = 330\,\text{Hz}$

11. (a) $r_1 = \sqrt{1.15^2 + 8^2} = 8.08\,\text{m}$,

$r_2 = \sqrt{1.85^2 + 8^2} = 8.21\,\text{m}$

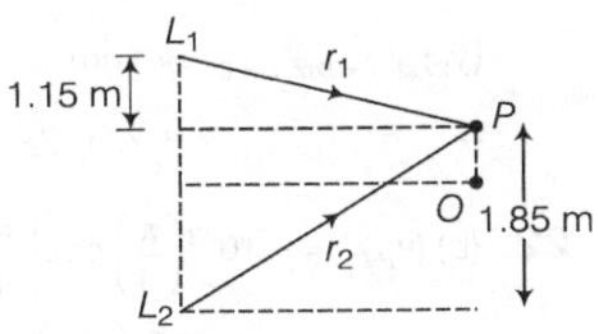

Path difference is

$\Delta x = 8.21 - 8.08 = 0.13\,\text{m}$

This path difference corresponds to first minimum

$$\therefore \quad \Delta x = \frac{\lambda}{2} = 0.13\,\text{m} \Rightarrow \lambda = 0.26\,\text{m}$$

So, $\quad f = \frac{v}{\lambda} = \frac{343}{0.26} \simeq 1320\,\text{Hz}$

12. (b) $(SA + AD) - SD = n\lambda$ for reinforcement at D

$(SB + BD) - SD = n\lambda + \frac{\lambda}{2}$ for total annulment at D

Subtracting $(SB + BD) - (SA + AD) = \frac{\lambda}{2}$

$$(SB - SA) + (BD - AD) = \frac{\lambda}{2} \Rightarrow 2(SB - SA) = \frac{\lambda}{2}$$

$$\lambda = 4(SB - SA)$$

Now, $\quad SB = \sqrt{(H+h)^2 + \left(\frac{d}{2}\right)^2} = \frac{\sqrt{4(H+h)^2 + d^2}}{2}$

and $\quad SA = \sqrt{H^2 + \left(\frac{d}{2}\right)^2} = \frac{\sqrt{4H^2 + d^2}}{2}$

$$\therefore \quad \lambda = 2\sqrt{4(H+h)^2 + d^2} - 2\sqrt{4H^2 + d^2}$$

13. (c) The path difference is

$$\Delta x = ABCD - AED$$

$$= (10 + 20 + 10) - (\sqrt{500} + 10) = 7.64\,\text{cm}$$

Wavelength of sound is $\quad \lambda = \frac{v}{f} = \frac{343}{f}\,\text{m}$

For maximum intensity, $\quad \Delta x = n\lambda$

$$\therefore \quad 7.64 \times 10^{-2} = n\frac{343}{f} \Rightarrow f \simeq (4490)n\,\text{Hz}$$

Here, $n \geq 1$. For $n = 1$, $f = 4490$ Hz and for $n = 2$, $f = 8980$ Hz. Thus, the frequency within the specified range is 4490 Hz.

14. (b) Path difference between two routes is

$$\Delta x = 2\sqrt{H^2 + \frac{D^2}{4}} - D \quad \ldots\text{(i)}$$

The fluctuation in intensity takes place due to interference between the two signals arriving at the receiver. Each time path difference, i.e. Δx changes by λ (i.e. wavelength of radiation), the received signal strength will vary through one cycle, thus the frequency of observed fluctuations is

$$f = \frac{d(\Delta x)}{dt} \cdot \frac{1}{\lambda} = \frac{d(\Delta x)}{dt} \cdot \frac{f_0}{c} \quad \ldots\text{(ii)}$$

where, f_0 = frequency of radiation and c = speed of light

Differentiating Eq. (i), we get

$$\frac{d(\Delta x)}{dt} = 2 \cdot \frac{1}{2}\left(H^2 + \frac{D^2}{4}\right)^{-1/2} 2H\frac{dH}{dt} \quad \ldots\text{(iii)}$$

From Eqs. (ii) and (iii), we get

$$\frac{fc}{f_0} = 2H\left(H^2 + \frac{D^2}{4}\right)^{-1/2}\frac{dH}{dt} \quad \text{or} \quad \frac{dH}{dt} = \frac{fc}{f_0} \cdot \frac{1}{2H}\left(H^2 + \frac{D^2}{4}\right)^{1/2}$$

or

$$v = \frac{\left(\frac{8}{60}\right) \times 3 \times 10^8}{(10 \times 10^6)} \cdot \frac{1}{2(200 \times 10^3)}\left[(200 \times 10^3)^2 + \frac{(5 \times 10^2)^2}{4}\right]^{1/2}$$

or $\quad v = 2.0\,\text{ms}^{-1}$

15. (a) $S_1P = \sqrt{16 + 4} = \sqrt{20}\,\text{m}$

and $\quad S_2P = \sqrt{16 + 1} = \sqrt{17}\,\text{m}$

Path difference is $\Delta x = S_1P - S_2P = \sqrt{20} - \sqrt{17} = 0.35\,\text{m}$

Also, $\quad \phi = \frac{2\pi}{\lambda} \cdot \Delta x = \frac{2\pi}{v}f(\Delta x) = \frac{2\pi}{350}f(0.35) = \frac{\pi}{500}f$

For constructive interference at P, $\phi = 2\pi, 4\pi, 6\pi, \cdots$

or $\quad f = 1000\,\text{Hz}, 2000\,\text{Hz}, 3000\,\text{Hz}$, etc. $\quad \left[\because \phi = \frac{\pi}{500}f\right]$

For destructive interference at P, $\phi = \pi, 3\pi, 5\pi, \cdots$

or $f = 500\,\text{Hz}, 1500\,\text{Hz}, 2500\,\text{Hz}$, etc. $\left[\because \phi = \frac{\pi}{500}f\right]$

Now, when the intensity at P is $\frac{P_0}{4\pi}$. Let intensity due to S_1 (at P) is I_1 and due to S_2 (at P) is I_2. Then,

$$I_1 = \frac{P_0}{4\pi(\sqrt{20})^2} = \frac{P_0}{80\pi} \text{ and } I_2 = \frac{P_0}{4\pi(\sqrt{17})^2} = \frac{P}{68\pi}$$

$$I = I_1 + I_2 + 2\sqrt{I_1I_2}\cos\phi$$

$$12.98 \times 10^{-3}P_0 = \frac{P_0}{80\pi} + \frac{P_0}{68\pi} + 2\frac{P_0}{\pi}\sqrt{\frac{1}{80} \cdot \frac{1}{68}}\cos\phi$$

$$\Rightarrow \quad \cos\phi = \frac{1}{2}$$

$$\therefore \quad \phi = \frac{\pi}{3}$$

Also, $\quad \phi = \frac{\pi}{500}f$

$$\therefore \quad \frac{\pi}{500}f = \frac{\pi}{3} \Rightarrow f \simeq 167\,\text{Hz}$$

Hence, (i) → (p,r) (ii) → (q), (iii) → (s)

16. (a) Like mechanical waves, electromagnetic waves too obey the principle of superposition. As ($D >> d$) the angular position of a point P, on the line of motion of detector, is approximately same for each source. The path difference between any two consecutive sources is $d \sin\theta$. The phase difference is

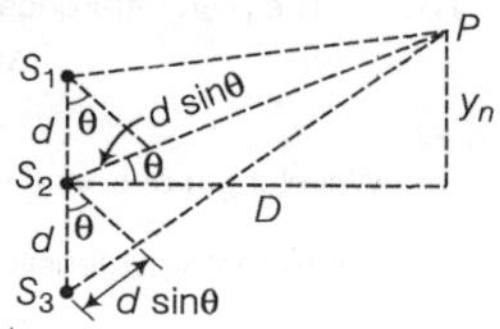

$$\phi = \frac{2\pi}{\lambda} d \sin\theta = kd \sin\theta \qquad \left[\because k = \frac{2\pi}{\lambda}\right]$$

So, the equations of the three waves arriving at P can be written as

$$y_1 = A\sin(kx - \phi),\ y_2 = A\sin kx,\ y_3 = A\sin(kx + \phi)$$

The resultant amplitude of the three waves is

$$y = y_1 + y_2 + y_3$$
$$= A[\sin(kx - \phi) + \sin kx + \sin(kx + \phi)]$$
$$= 2A\sin kx \cos\phi + A\sin kx = A\sin kx(1 + 2\cos\phi)$$
$$= A(1 + 2\cos\phi)\sin kx$$

Thus, resultant amplitude is $A_R = A[1 + 2\cos(kd\sin\theta)]$

$$I \propto A_R^2,\ I_0 \propto A^2$$

$$\therefore \quad \frac{A_R^2}{A^2} = \frac{I}{I_0} \Rightarrow I = I_0[1 + 2\cos(kd\sin\theta)]^2$$

For maxima $\cos(kd\sin\theta) = 1 \Rightarrow kd\sin\theta = 2n\pi$

$$\therefore \quad \sin\theta = \frac{2n\pi}{kd} = \frac{n\lambda}{d}$$

From the figure, $\sin\theta \approx \tan\theta = \frac{y_n}{D}$, where y_n is the distance of nth maxima from point O.

$$\therefore \quad \frac{y_n}{D} = \frac{n\lambda}{d} \Rightarrow y_n = \frac{n\lambda D}{d}$$

Separation between consecutive maxima is

$$y_n - y_{n-1} = \frac{n\lambda D}{d} - \frac{(n-1)\lambda D}{d} = \frac{\lambda D}{d}$$

(i) $\lambda = \frac{c}{f} = 3 \times 10^8/10^8 = 3\text{ m}$

(ii) Now, $\frac{\lambda D}{d} = 10\text{ km} \Rightarrow D = \frac{10000 \times 1000}{3} = \frac{10^7}{3}\text{ m}$

(iii) Resultant intensity at maxima is

$$I_{max} = I_0[1 + 2(1)]^2 = 9I_0$$

At minima $\quad I_{min} = I_0[1 + 2(-1)]^2 = I_0$

$$\therefore \quad \frac{I_{max}}{I_{min}} = \frac{9I_0}{I_0} = 9$$

(iv) Let d' be the new distance between the adjacent sources

$$y_1 = A\sin(kx - \phi')$$
$$y_2 = A\sin(kx + \phi'),\ \text{where } \phi' = kd'\sin\theta$$

Resultant amplitude is $A_R = 2A\cos(kd'\sin\theta)$

Hence, the resultant intensity is $I = 4I_0^2\cos^2(kd'\sin\theta)$

For maxima $\cos^2(kd'\sin\theta) = 1$

$$kd'\sin\theta = n\pi$$

or $$\frac{2\pi}{\lambda} d'\left(\frac{y_n'}{D}\right) = n\pi \Rightarrow y_n' = \frac{n\lambda D}{2d'}$$

As position of maxima is unchanged, $y_n = y_n'$

$$\frac{n\lambda D}{d} = \frac{n\lambda D}{2d'}$$

$$\Rightarrow \quad d' = \frac{d}{2} = 0.5\text{ km} = 500\text{ m}$$

Hence, (i) → (q), (ii) → (r), (iii) → (s), (iv) → (t)

17. (c) $I = I_1 + I_2 + 2\sqrt{I_1 I_2}\cos\phi$, I is maximum, when $\cos\phi = 1$

$$\therefore \quad I_{max} = \frac{I_0}{2} + \frac{I_0}{2} + 2\sqrt{\frac{I_0}{2}\cdot\frac{I_0}{2}} = 2I_0$$

$$\left(I_1 = I_2 = \frac{I_0}{2},\ \text{same wave is equally divided}\right)$$

18. (d) Path difference is

$$\Delta x = (2\pi R)\frac{3}{4} - (2\pi R)\frac{1}{4} = (2\pi R)\frac{1}{2} = \pi R$$

For maxima, $\quad \Delta x = n\lambda$

For λ to be maximum, $n = 1$

$$\therefore \quad \lambda_{max} = \pi R$$

19. (c) For minima, $\quad \Delta x = (2n - 1)\frac{\lambda}{2} \qquad [n = 1, 2, 3 \cdots]$

Also, $\quad \Delta x = \pi R$

$$\therefore \quad (2n - 1)\frac{\lambda}{2} = \pi R,$$

For maximum λ, $\quad n = 1$

$$\therefore \quad \frac{\lambda_{max}}{2} = \pi R$$

$$\Rightarrow \quad \lambda_{max} = 2\pi R$$

20. (d) $y = -A\sin(\omega t - kx)$ [At $t = 0$, $x = 0$ is moving down]

From the figure, $\lambda = 8$ m and $A = 10$ mm

$$\therefore\ k = \frac{2\pi}{k} = \frac{\pi}{4}\text{ m}^{-1},\ \omega = kv = \frac{\pi}{4} \times 1 = \frac{\pi}{4}\text{ s}^{-1}$$

$$\therefore \quad y = -10 \times 10^{-3}\sin\left[\frac{\pi}{4}t - \frac{\pi}{4}x\right]$$

$$\frac{\partial y}{\partial t} = v_P = -10 \times 10^{-3}\left(\frac{\pi}{4}\right)\cos\left[\frac{\pi}{4}t - \frac{\pi}{4}x\right]$$

$$|(v_P)_{\text{at } x = 4\text{m}, t = 0\text{s}}| = \left|\frac{\pi}{4} \times 10^{-2}\cos\left[\frac{\pi}{4} \times 0 - \frac{\pi}{4} \times 4\right]\right|$$

$$= \frac{\pi}{4}\text{ cm s}^{-1}$$

21. (d) Distance moved by each pulse in $t = 8$ s is $s = 8 \times 1 = 8$ m

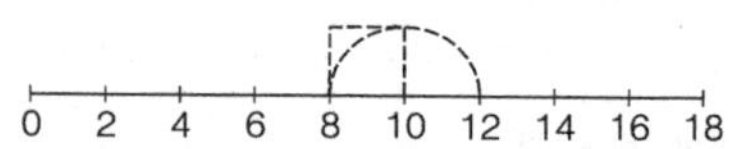

Due to sinusoidal pulse

$$(y_1)_{\text{at } x = 9\text{m}\ t = 8\text{s}} = -10 \times 10^{-3}\sin\left[\frac{\pi}{4} \times 8 - \frac{\pi}{4} \times 9\right]$$

or $$y_1 = -10^{-2}\sin\left(-\frac{\pi}{4}\right) = \frac{10^{-2}}{\sqrt{2}}\text{ m} = \frac{1}{\sqrt{2}}\text{ cm}$$

$$(y_2)_{\text{at } x = 9\text{m}, t = 8\text{s}} = 1\text{ cm}$$

$$\therefore \quad y = y_1 + y_2 = \frac{1}{\sqrt{2}} + 1 = \frac{\sqrt{2} + 1}{\sqrt{2}}\text{ cm}$$

22. (b) $(v_P)_1 = -10^{-2}\left(\frac{\pi}{4}\right)\cos\left[\frac{\pi}{4} \times 8 - \frac{\pi}{4} \times 9\right]$

$$= -10^{-2}\frac{\pi}{4}\left(\frac{1}{\sqrt{2}}\right)\text{ms}^{-1} = -\frac{\pi}{4\sqrt{2}}\text{ cm s}^{-1}$$

Due to square pulse $(v_P)_2$ at $x = 9$ m, $t = 1$ s in y-direction is zero.

$$\therefore \quad v_P = (v_P)_1 + 0 = -\frac{\pi}{4\sqrt{2}}\text{ cms}^{-1}$$

23. (6) $v = 330\text{ ms}^{-1}$, $\quad f = 220$ Hz (given)

$$\therefore \quad \lambda = \frac{v}{f} = \frac{330}{220} = 1.5\text{ m}$$

Path difference is $\Delta x = S_2Q - S_1Q = 3 - 0.75 = 2.25$ m

Phase difference at point Q is

$$\phi = \frac{2\pi}{\lambda}\cdot\Delta x = \frac{2\pi}{1.5}(2.25) = 3\pi$$

So, they are out of phase at point Q. So, destructive interference happens at Q. So, resultant power is

$$P = 1.8\text{ mW} - 1.2\text{ mW} = 0.6\text{ mW} = \frac{6}{10}\text{ mW}$$

$$\therefore \quad k = 6$$

DPP-2 Standing Waves in Strings

1. (d) $f_0 = \frac{1}{2L}\sqrt{\frac{F}{\mu}} \Rightarrow 4f_0^2L^2 = \frac{F}{\mu}$

$$L^2 = \frac{1}{4f_0^2\mu}\cdot F$$

Now taking logarithm on both sides, we give

$$2\log L = \log\frac{1}{4f_0^2\mu} + \log F$$

or $$\log L = \left(\frac{1}{2}\log\frac{1}{4f_0^2\mu}\right) + \left(\frac{1}{2}\right)\log F$$

So, the $\log L$ - $\log F$ graph is a straight line with slope $= \frac{1}{2}$

[Compare it with $y = mx + c$]

2. (a) Fourth overtone corresponds to the fifth harmonic, and so the wire vibrates in five segments.

$\therefore$ Length of each segment $= \frac{100\text{ cm}}{5} = 20\text{ cm}$

3. (b) The frequency corresponding to nth harmonic is given by

$$f_n = \frac{n}{2L}\sqrt{\frac{F}{\mu}} \qquad (n = 1, 2, 3, \cdots)$$

$\therefore$ The third harmonic and the seventh harmonic are given by

$f_3 = \frac{3}{2L}\sqrt{\frac{F}{\mu}}$ and $f_7 = \frac{7}{2L}\sqrt{\frac{F}{\mu}}$, respectively.

$$f_7 - f_3 = 1600\text{ Hz} \qquad \text{(given)}$$

$$\frac{(7-3)}{2L}\sqrt{\frac{F}{\mu}} = 1600 \Rightarrow \frac{1}{2L}\sqrt{\frac{F}{\mu}} = 400\text{ Hz}$$

$\therefore$ The fundamental frequency is

$$f_1 = \frac{1}{2L}\sqrt{\frac{F}{\mu}} = 400\text{ Hz}$$

4. (c) $f_4 = \frac{4}{2L}\sqrt{\frac{F}{\mu}} = \frac{4}{2L}v$

$\therefore$ $$\frac{4}{2L}\cdot v = 2000$$

$\Rightarrow$ $$v = 1000 \times L = 1000(0.5) = 500\text{ ms}^{-1}$$

5. (d) Initially $F_1 = v\rho_s g$

and $F_2 = v(\rho_s - \rho_w)g$,

$F_3 = v(\rho_s - \rho_k)g$

Now, $f \propto \frac{\sqrt{F}}{L}$ (other things remaining constant) $\left[\because f = \frac{n}{2L}\sqrt{\frac{F}{\mu}}\right]$

The wire resonates with the same tuning fork, so

$$\frac{\sqrt{F}}{L} = \text{constant or } \frac{F}{L^2} = \text{constant}$$

$\therefore$ $$\frac{F_1}{L^2} = \frac{F_2}{L_w^2} = \frac{F_3}{L_k^2}$$

$\therefore$ $$\frac{v\rho_s g}{L^2} = \frac{v(\rho_s - \rho_w)g}{L_w^2} \qquad \ldots\text{(i)}$$

and $$\frac{V\rho_s g}{L^2} = \frac{V(\rho_s - \rho_k)g}{L_k^2} \qquad \ldots\text{(ii)}$$

From Eq. (i), $\frac{\rho_s}{L^2} = \frac{\rho_s - \rho_w}{L_w^2}$

$$1 - \frac{\rho_w}{\rho_s} = \left(\frac{L_w}{L}\right)^2 \Rightarrow \frac{\rho_w}{\rho_s} = 1 - \left(\frac{L_w}{L}\right)^2$$

$$\frac{\rho_s}{\rho_w} = \left[1 - \left(\frac{L_w}{L}\right)^2\right]^{-1}$$, which is specific gravity of solid.

6. (b) Equating right hand sides of Eqs. (i) and (ii) of previous solution, we have

$$\frac{\rho_s - \rho_w}{L_w^2} = \frac{\rho_s - \rho_k}{L_k^2} \Rightarrow \frac{\rho_s}{L_w^2} - \frac{\rho_s}{L_k^2} - \frac{\rho_w}{L_w^2} = \frac{-\rho_k}{L_k^2}$$

$$-\rho_k = \frac{\rho_s}{L_w^2}\cdot L_k^2 - \frac{\rho_s}{L_k^2}\cdot L_k^2 - \frac{\rho_w}{L_w^2}\cdot L_k^2 = \rho_s\left[\frac{L_k^2}{L_w^2} - 1\right] - \frac{\rho_w}{L_w^2}\cdot L_k^2$$

$$-\frac{\rho_k}{\rho_w} = \frac{\rho_s}{\rho_w}\left[\frac{L_k^2}{L_w^2} - 1\right] - \frac{1}{L_w^2}\cdot L_k^2$$

$$-\frac{\rho_k}{\rho_w} = \left(\frac{1}{1 - \frac{L_w^2}{L^2}}\right)\left(\frac{L_k^2}{L_w^2} - 1\right) - \frac{L_k^2}{L_w^2} \quad \left[\because \frac{\rho_s}{\rho_w} = \left[1 - \left(\frac{L_w}{L}\right)^2\right]^{-1}\right]$$

$$= \left(\frac{L^2}{L^2 - L_w^2}\right)\left(\frac{L_k^2 - L_w^2}{L_w^2}\right) - \frac{L_k^2}{L_w^2}$$

$$= \frac{L^2L_k^2 - L^2L_w^2 - L_k^2L^2 + L_k^2L_w^2}{L_w^2(L^2 - L_w^2)} = \frac{L_w^2(L_k^2 - L^2)}{L_w^2(L^2 - L_w^2)}$$

$\therefore$ $$\frac{\rho_k}{\rho_w} = \left(\frac{L^2 - L_k^2}{L^2 - L_w^2}\right)$$

7. (c) Fundamental frequency is given by $f = \frac{1}{2L}\sqrt{\frac{F}{\mu}} \Rightarrow f \propto \sqrt{F}$

$\therefore$ $$\frac{f_1'}{f_1} = \sqrt{\frac{F'}{F}}$$

$$\frac{F'}{F} = 1.44 \qquad \text{(given)}$$

$\therefore$ $$\frac{f_1'}{f_1} = 1.2 \text{ or } f_1' = 1.2f_1$$

$$f_1' - f_1 = 15 \text{ (given)} \Rightarrow 1.2f_1 - f_1 = 15$$

or $$f_1 = 75\text{ Hz}$$

If the length of the wire is changed, other things remaining constant, then $f \propto \frac{1}{L}$

or $\frac{f_1''}{f_1} = \frac{L}{L''}$, Given $\frac{\Delta L}{L} = 40\%$

$\therefore$ $$\frac{L'' - L}{L} = 0.4 \text{ or } L'' = 1.4L$$

$\therefore$ $$\frac{f_1''}{75} = \frac{1}{1.4} \Rightarrow f_1'' = 53.6\text{ Hz}$$

So, the required change in frequency

$$f_1 - f_1'' = 75 - 53.6 = 21.4\text{ Hz}$$

8. (d) $\frac{n_1}{2(4L)}\sqrt{\frac{F}{(1\mu)}} = \frac{n_2}{2(1L)}\sqrt{\frac{F}{(4\mu)}} \Rightarrow \frac{n_1}{n_2} = \frac{4}{2} = 2:1$

9. (a) $f = \frac{1}{2L}\sqrt{\frac{F}{\rho S}}$, where S is cross-sectional area. When the wire is stretched, the volume remains same.

$\therefore$ $$S_1L_1 = S_2L_2$$

$$S_2 = \frac{L_1}{L_2}\cdot S_1 = \frac{S}{2}$$

$\therefore$ $$f' = \frac{1}{2(2L)}\sqrt{\frac{F}{\rho\cdot(S/2)}} = \frac{1}{\sqrt{2}}\left(\frac{1}{2L}\sqrt{\frac{F}{\rho S}}\right) = \frac{f}{\sqrt{2}}$$

10. (c) $\Delta L = L_0 \propto \Delta\theta = -(1)(1.21 \times 10^{-5})(20)$

$$= -2.42 \times 10^{-4}\text{ m}$$

$\therefore$ Strain $= 2.42 \times 10^{-4}$

So, stress $= Y \times \text{strain} = (2 \times 10^{11}) \times (2.42 \times 10^{-4})$

So, tension is $F = \text{stress} \times \text{area} = (2 \times 10^{11})(2.42 \times 10^{-4})(10^{-6})$

or $F = 48.4\text{ N}$

Point of plucking would be an antinode. For minimum number of loops, the wire has to vibrate in its second harmonic.

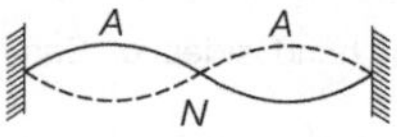

$\therefore \quad f_2 = \frac{2}{2L}\sqrt{\frac{F}{\mu}} = \frac{1}{1}\sqrt{\frac{48.4}{0.1/1}} = \sqrt{484} = 22$ Hz

11. (a, d) In this problem two cases are possible. In case I, points A and B have same displacement amplitude. In case II, C and D have the same amplitude *viz* 3.5 mm.

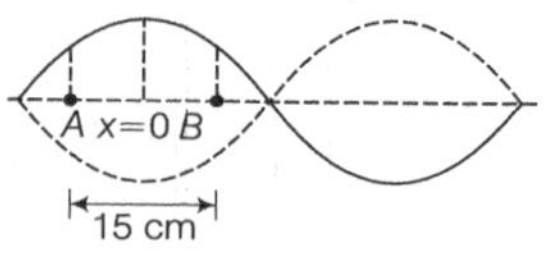

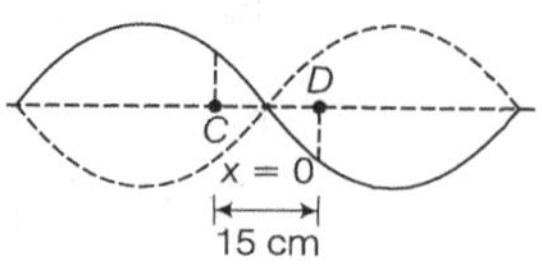

Case I If $x = 0$ is taken at antinode, then amplitude as a function of x is $A = a \cos kx$

Case II If $x = 0$ is taken at node, then amplitude as a function of x is $A = a \sin kx$

But nothing, regarding this, is given in the question. Hence, from both the cases, results should be same. This is possible only when

$$a \cos kx = a \sin kx \text{ or } kx = \frac{\pi}{4}$$

$$\therefore \quad a = \frac{A}{\cos kx} = \frac{3.5 \text{ mm}}{\cos(\pi/4)} = (3.5)(\sqrt{2}) \text{ mm}$$

$$= 4.95 \text{ mm} \simeq 5 \text{ mm}$$

$$kx = \frac{\pi}{4} \Rightarrow \left(\frac{2\pi}{\lambda}\right)\left(\frac{15}{2}\right) = \frac{\pi}{4} \text{ or } \frac{\lambda}{2} = 30 \text{ cm}$$

$$\therefore \text{ Number of loops} = \frac{120 \text{ cm}}{\lambda/2} = \frac{120}{30} = 4$$

Hence, it corresponds to 'third overtone'.

12. (a,b,c,d) The displacement equation of resultant wave is $y = y_1 + y_2$

or $\quad y = 10 \text{ cm} \sin(100\pi t - \pi x) - (10 \text{ cm}) \sin(100\pi t + \pi x)$

$$= 10 \text{ cm}\left[2 \sin\left(\frac{100\pi - \pi x - 100\pi t - \pi x}{2}\right) \cos\left(\frac{100\pi t - \pi x + 100\pi t + \pi x}{2}\right)\right]$$

$$y = 20 \text{ cm} \sin(-\pi x)\cos(100\pi t)$$

or $\quad y = 20 \text{ cm} \sin[\pi(x+1)\cos(100\pi t]$

$[\because \sin(-\pi x) = \sin(\pi x + \pi)]$

From the above equation, the amplitude of particle x is

$$A = (20 \text{ cm})\sin[\pi(x+1)] \quad \ldots \text{(i)}$$

Now, nodes are points of permanent rest

$$A = 0 \text{ or } \sin[\pi(x+1)] = 0$$

or $\quad \pi(x+1) = n\pi$

$\Rightarrow \quad x = n - 1 \quad [n = 1, 2, 3, 4, \cdots]$

Putting $n = 1, 2, 3, 4 \ldots\ldots$ etc, we get

$$x = 0, 1 \text{ m}, 2 \text{ m}, 3 \text{ m}, \cdots \text{ etc.}$$

Antinodes are the points having maximum amplitude.

$\therefore \quad \sin[\pi(x+1)] = \pm 1$

or $\quad \pi(x+1) = (2n+1)\frac{\pi}{2} \quad [n = 1, 2, 3, 4, \cdots]$

$$x = \frac{2n+1}{2} - 1 = \frac{2n-1}{2}$$

Putting $n = 1, 2, 3, 4, \ldots\ldots$ etc, we get

$$x = \frac{1}{2} \text{ m}, \frac{3}{2} \text{ m}, \frac{5}{2} \text{ m}, \frac{7}{2} \text{ m}, \cdots \text{ etc.}$$

Now, amplitude of the particle at $x = \frac{1}{4}$ m is

$$A = \left|(20 \text{ cm})\sin\left[\pi\left(\frac{1}{4}+1\right)\right]\right| \quad \text{[Using Eq. (i)]}$$

$$= \left|(20 \text{ cm})\sin\frac{5\pi}{4}\right|$$

$$= (20 \text{ cm})\left|-\frac{1}{\sqrt{2}}\right| = 10\sqrt{2} \text{ cm}$$

13. (a,b,d) $y = 5\sin\left(\frac{\pi x}{20}\right)\cos(100\pi t)$

At $x = 10$ cm, displacement is

$$y = 5\sin\left(\frac{\pi}{2}\right)\cos(100\pi t) = 5\cos(100\pi t)$$

At $x = 10$ cm, maximum displacement is $y_{max} = 5$ cm

[by taking $\cos(100\pi t) = 1$]

Now, for antinodes, displacement amplitude should be maximum.

$$\therefore \quad \left|5\sin\left(\frac{\pi x}{20}\right)\right| = 5 \text{ or } \sin\left(\frac{\pi x}{20}\right) = \pm 1$$

or $\quad \frac{\pi x}{20} = (2n+1)\frac{\pi}{2}$, where $n = 0, 1, 2, 3 \cdots$

$$\frac{\pi x}{20} = \frac{\pi}{2} \Rightarrow x = 10 \text{ cm}$$

$$\frac{\pi x}{20} = \frac{3\pi}{2} \Rightarrow x = 30 \text{ cm}$$

$$\frac{\pi x}{20} = \frac{5\pi}{2} \Rightarrow x = 50 \text{ cm}$$

$$\frac{\pi x}{20} = \frac{7\pi}{2} \Rightarrow x = 70 \text{ cm, here } x \ngtr 60 \text{ cm.}$$

So, practical values of x are 10 cm, 30 cm, 50 cm.

For constituent waves, use

$$2\sin A\cos B = \sin(A+B) + \sin(A-B)$$

$$\therefore \quad y = 5\sin\left(\frac{\pi x}{20}\right)\cos(100\pi t)$$

$$\Rightarrow y = \frac{5}{2}\left[\sin 2\pi\left(\frac{x}{10} + 50t\right) + \sin 2x\left(\frac{x}{10} - 50t\right)\right]$$

So, individual waves are

$$y_1 = \frac{5}{2}\sin 2\pi\left(\frac{x}{10} - 50t\right)$$

and $\quad y_2 = \frac{5}{2}\sin 2\pi\left(\frac{x}{10} + 50t\right)$

14. (a,d) Distance between two nodes is $\frac{\lambda}{2}$ or $\frac{\pi}{k}$. The volume of the string between two nodes, therefore, is $V = \frac{\pi}{k} \cdot s$. Energy density, i.e. energy per unit volume of each wave will be

$$u_1 = \frac{1}{2}\rho\omega^2(8)^2 = 32\rho\omega^2$$

and $\quad u_2 = \frac{1}{2}\rho\omega^2(6)^2 = 18\rho\omega^2$

$\therefore$ Total mechanical energy between consecutive nodes will be

$$E = (u_1 + u_2) \text{ volume} = (32 + 18)\rho\omega^2\left(\frac{\pi}{k}s\right)$$

or $\quad E = \frac{50\pi}{k}\rho\omega^2 s$

Now, $y = y_1 + y_2 = 8\sin(\omega t - kx) + 6\sin(\omega t + kx)$

or $\quad y = 2\sin(\omega t - kx) + \{6\sin(\omega t - kx) + 6\sin(\omega t + kx)\}$

$= 2\sin(\omega t - kx) + 12\cos kx \sin\omega t$

Thus, the resultant wave is a sum of standing wave and travelling wave.

Energy crossing through a node per second

= Power of travelling wave.

$$\therefore \quad P = \frac{1}{2}\rho\omega^2(2)^2 sv = 2\rho\omega^2 s\left(\frac{\omega}{k}\right) = \frac{2\rho\omega^3 s}{k}$$

15. (b) In stationary waves, the displacement of nodes is zero. So, Statement I is correct. Statement II is also correct but it is not the correct explanation of Statement I.

16. (a) From the figure, we observe that, at nodes the amplitude is zero and velocity of particle is also zero and at antinodes the amplitude is maximum. Hence, the velocity of particle is also maximum as all particles between two successive nodes cross the mean position at the same time.

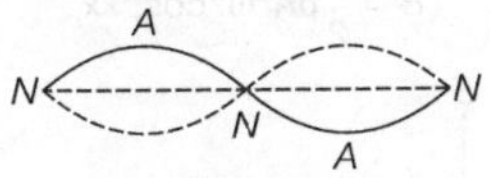

Standing wave

17. (a) In a travelling wave at mean position M velocity is maximum and stretching of string is maximum, so KE and PE are maximum. At E, KE and PE are minimum.

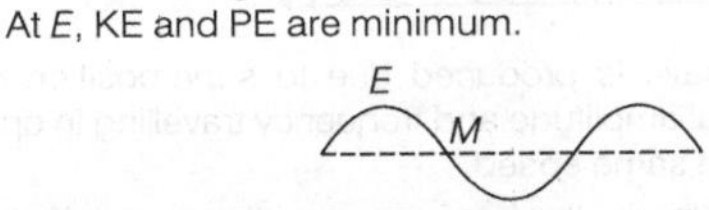

In case of a standing wave, all the particles between node and antinode pass through position simultaneously. So, at this moment, all energy is KE.

Hence, (i) $\rightarrow$ (q,r), (ii) $\rightarrow$ (p,s), (iii) $\rightarrow$ (r,s), (iv) $\rightarrow$ (p,q)

18. (c) $y = 4\sin\frac{\pi x}{15}\cos(96\pi t)$

Amplitude of particle x is $A = 4\sin\left(\frac{\pi x}{15}\right)$

At $x = 5$ cm, y_{max} or $A = 4\sin\frac{5\pi}{15} = 4\sin\frac{\pi}{3} = 2\sqrt{3}$ cm

19. (d) At nodes amplitude is zero.

$\therefore \quad 4\sin\left(\frac{\pi x}{15}\right) = 0$ or $\frac{\pi x}{15}$, $0, \pi, 2\pi \cdots$

or $\quad x = 0$, 15 cm, 30 cm, 45 cm

20. (a) Velocity of particle x is

$$v_P = \frac{\partial y}{\partial t} = (-4\times 96\pi)\sin\frac{\pi x}{15}\sin(96\pi t)$$

or $\quad v_P = -384\pi\sin\left(\frac{\pi x}{15}\right)\sin(96\pi t)$

Now, velocity of particle $x = 7.5$ cm at $t = 0.25$ s is

$$V_{P\text{ at }x=7.5\text{cm},\,t=0.25\text{s}} = -384\,\pi\sin\left(\frac{\pi}{2}\right)\sin(24\pi) = 0$$

21. (a) Amplitude of each component wave $= \frac{4}{2}$ cm $= 2$ cm

$$\omega = 96\pi \text{ and } k = \frac{\pi}{15}$$

$\therefore$ Component waves are $y_1 = 2\sin\left[\frac{\pi}{15}x - 96\pi t\right]$ and

$$y_2 = 2\sin\left[\frac{\pi}{15}x + 96\pi t\right]$$

22. Let n_A loops are formed in aluminium wire and n_s in steel. Then,

$$f_A = f_S$$

$\Rightarrow \quad n_A\frac{v_A}{2L_A} = n_S\frac{v_S}{2L_S}$

$\Rightarrow \quad \frac{n_A}{n_S} = \left(\frac{v_S}{v_A}\right)\left(\frac{L_A}{L_S}\right)$

We know $\quad v = \sqrt{\frac{F}{\mu}} = \sqrt{\frac{F}{\rho s}}$

where, F is tension, ρ is density and s is cross-sectional area.

$\therefore \quad \frac{v_S}{v_A} = \sqrt{\frac{\rho_A}{\rho_S}}$

$\therefore \quad \frac{n_A}{n_S} = \sqrt{\frac{\rho_A}{\rho_S}}\cdot\frac{L_A}{L_S} = \sqrt{\frac{2600}{10400}}\,\frac{0.6}{0.9} = \frac{1}{3}$

So, at the lowest frequency, one loop is formed in aluminium wire and 3 in steel wire

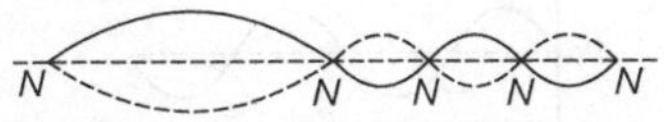

So, total number of nodes is 5.

$$f_{min} = \frac{n_A}{2L_A}\sqrt{\frac{F}{\rho_A S_A}} = \frac{1}{2(0.6)}\sqrt{\frac{10\times 10}{2600\times 10^{-6}}} = 163.4 \text{ Hz}$$

[Here, $F = mg$]

23. (i) $y = A\cos kx\cos\omega t$ ...(i)

At $t = 0$, $\quad y = A\cos kx$

$\therefore$ Y-X plot would be

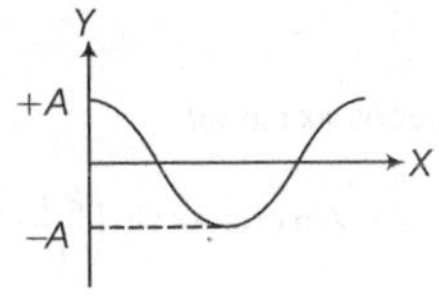

At $t = \frac{T}{2}$,

$$y = A\cos kx\cos\left(\frac{2\pi}{T}\cdot\frac{T}{2}\right) = -A\cos kx$$

So Y-X plot would be

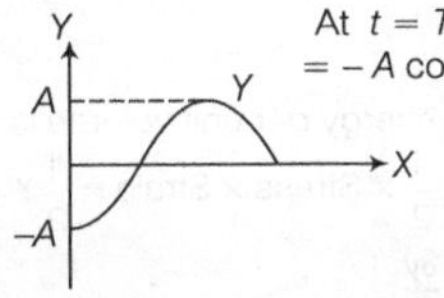

Differentiating Eq. (i) $\frac{\partial y}{\partial x} = -Ak\sin kx\cos\omega t$

At $t = 0$, $\frac{\partial y}{\partial x} = -Ak\sin kx \quad \therefore \left(\frac{\partial y}{\partial x} - X\right)$ plot would be

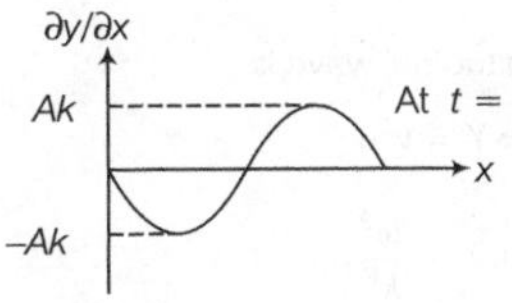

At $t = T/2$,

$$\frac{\partial y}{\partial x} = -Ak\sin kx\cos\left(\frac{2\pi}{T}\cdot\frac{T}{2}\right) = Ak\sin kx.$$

So, $\left(\frac{\partial x}{\partial y} - x\right)$ graph would be

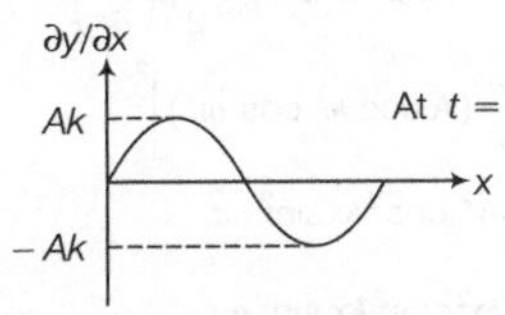

(ii) $dP = -B\frac{dV}{V} = -B\frac{\partial y}{\partial x}$

$$= -B(-Ak\sin kx\cos\omega t)$$

or $\Delta p = BAk\sin kx\cos\omega t = (\Delta p)_{max}\sin kx\cos\omega t$

The change in density ($\Delta\rho$) of medium is proportional to the change in pressure.

$\therefore \quad \Delta\rho = (\Delta\rho)_{max}\sin kx\cos\omega t$

At $t = 0$,

$$\Delta\rho = (\Delta\rho)_{max} \sin kx, \text{ so } \rho = \rho_0 + \Delta\rho$$

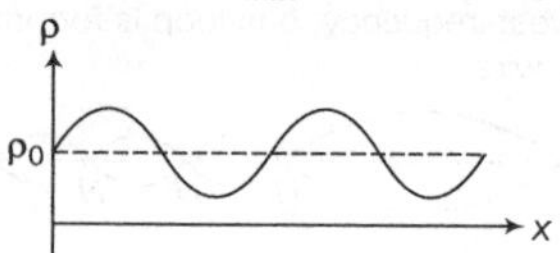

At $t = \frac{T}{2}$, $\quad \Delta\rho = (\Delta\rho)_{max} \sin kx(-1) = -(\Delta\rho)_{max} \sin kx$

So, $\quad \rho = \rho_0 + \Delta\rho$

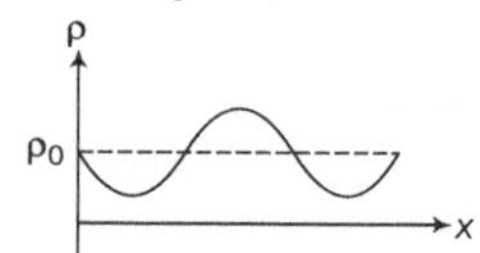

(iii) $v = \frac{\partial y}{\partial t} = -A\omega\cos kx \sin \omega t$

At $t = \frac{T}{4}$, $\quad v = -A\omega \cos kx \sin\left(\frac{2\pi}{T}\cdot\frac{T}{4}\right)$

$$v = -A\omega \cos kx$$

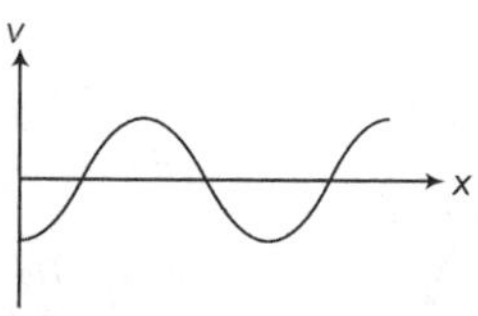

24. (i) Elastic potential energy per unit volume is defined as

$$U = \frac{1}{2} \times \text{Stress} \times \text{Strain} = \frac{1}{2} Y\,(\text{Strain})^2$$

$$\text{Strain} = \frac{\partial y}{\partial x}$$

$\therefore \quad U = \frac{1}{2} Y \left[\frac{\partial}{\partial x}(A \cos kx \cos \omega t)\right]^2$

or $\quad U = \frac{1}{2} YA^2k^2 \sin^2 kx \cos^2 \omega t$

The velocity of a longitudinal wave is

$$v = \sqrt{\frac{Y}{\rho}} \Rightarrow Y = v^2\rho$$

$$v = \frac{\omega}{k}. \text{ So, } Y = \frac{\omega^2}{k^2}\rho$$

$\therefore \quad U = \frac{1}{2}\frac{\omega^2}{k^2}\rho A^2k^2 \sin^2 kx \cos^2 \omega t$

or $\quad U = \frac{1}{2}\rho A^2\omega^2 \sin^2 kx \cos^2 \omega t \quad \ldots$(i)

Kinetic energy per unit volume is $K = \frac{1}{2}\rho\left(\frac{\partial y}{\partial t}\right)^2$

or $\quad K = \frac{1}{2}\rho\left[\frac{\partial}{\partial t}(A \cos kx \cos \omega t)\right]^2$

$$= \frac{1}{2}\rho\omega^2A^2 \cos^2 kx \sin^2 \omega t$$

or $\quad K = \frac{1}{2}\rho A^2\omega^2 \cos^2 kx \sin^2 \omega t \quad \ldots$(ii)

(ii) Now, total energy density of the medium is

$$E = U + K = \frac{1}{2}\rho A^2\omega^2[\sin^2 kx \cos^2 \omega t + \cos^2 kx \sin^2 \omega t]$$

At $t = 0$,

$$E = \frac{1}{2}\rho A^2\omega^2 \sin^2 kx$$

At $t = T/4$,

$$E = \frac{1}{2}\rho A^2\omega^2 \cos^2 kx$$

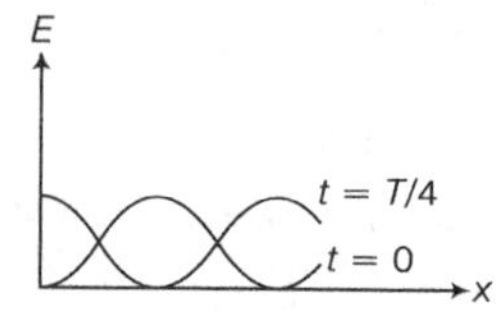

25. (i) A standing wave is produced due to superposition of two waves of equal amplitude and frequency travelling in opposite directions with same speed.

Here, the amplitude, the frequency and the speed of all the waves are equal. Z_1 is travelling in + x-direction, Z_2 in − x-direction and Z_3 in + y-direction.

So, superposition of Z_1 and Z_2 will produce stationary waves

$$Z = Z_1 + Z_2 = A \cos(\omega t - kx) + A \cos(\omega t + kx)$$

Note $\cos(-\theta) = \cos\theta$]

or $Z = 2A \cos kx \cos \omega t$. *This is a standing wave of amplitude* ($2A \cos kx$).

$$I \propto \text{Amplitude}^2$$

$$\Rightarrow \quad I \propto 4A^2 \cos^2 kx$$

I will be zero, if $\cos^2 kx = 0 \Rightarrow kx = \frac{\pi}{2}, \frac{3\pi}{2}, \frac{5\pi}{2}$

or $\quad x = \frac{\pi}{2k}, \frac{3\pi}{2k}, \frac{5\pi}{2k}, \ldots$

As $k = \frac{2\pi}{\lambda}$, $\quad x = \frac{\lambda}{4}, \frac{3\lambda}{4}, \frac{5\lambda}{4}, \ldots$

(ii) The waves Z_1 and Z_3 are travelling in + x-direction and + y-direction, respectively with same speed,

$$v = \frac{\omega}{k} = \frac{2\pi f}{2\pi/\lambda} = f\lambda.$$

So, the superposition of Z_1 and Z_3 will produce a wave travelling in a direction making an angle of 45° with positive X-axis and positive Y-axis. The resulting wave is

$$Z = Z_1 + Z_3$$
$$= A \cos(\omega t - kx) + A \cos(\omega t - ky)$$

or $\quad Z = 2A \cos\left\{k\frac{(x-y)}{2}\right\} \cos\left\{\omega t - k\frac{(x+y)}{2}\right\}$

Amplitude of this wave is

$$2A \cos\left\{\frac{k(x-y)}{2}\right\}$$

$\therefore \quad I \propto \cos^2\left\{\frac{k(x-y)}{2}\right\}$

Therefore, the intensity will be zero,

if $\frac{k(x \sim y)}{2} = \frac{\pi}{2}, \frac{3\pi}{2}, \frac{5\pi}{2}, \ldots$

or $\quad x \sim y = \frac{\pi}{k}, \frac{3\pi}{k}, \frac{5\pi}{k}$

As, $\quad k = \frac{2\pi}{\lambda}, x \sim y = \frac{\lambda}{2}, \frac{3\lambda}{2}, \frac{5\lambda}{2}, \ldots$

DPP-3 Resonating Air Columns

1. Length of the pipe is 3.6 m and speed of sound is $v = 340\ \text{ms}^{-1}$. Let L_0 be the length of air column corresponding to the fundamental frequency. Then,

$$\frac{v}{4L_0} = 212.5 \Rightarrow L_0 = \frac{340}{4(212.5)} = 0.4\ \text{m}$$

In closed pipe, only odd harmonics are obtained. Now, let L_1, L_2, L_3, L_4 etc. be the lengths corresponding to the 3rd harmonic, 5th harmonic, 7th harmonic, etc. Then,

$$3\left(\frac{v}{4L_1}\right) = 212.5 \Rightarrow L_1 = 1.2\text{m}$$

$$5\left(\frac{v}{4L_2}\right) = 212.5 \Rightarrow L_2 = 2\text{m}$$

$$7\left(\frac{v}{4L_3}\right) = 212.5 \Rightarrow L_3 = 2.8\ \text{m}$$

and $$9\left(\frac{v}{4L_4}\right) = 212.5 \Rightarrow L_4 = 3.6\ \text{m}$$

The heights of corresponding water levels are (3.6 – 0.4) m, (3.6 – 1.2) m, (3.6 – 2) and (3.6 – 2.8) m, or 3.2 m, 2.4 m, 1.6 m and 0.8 m.

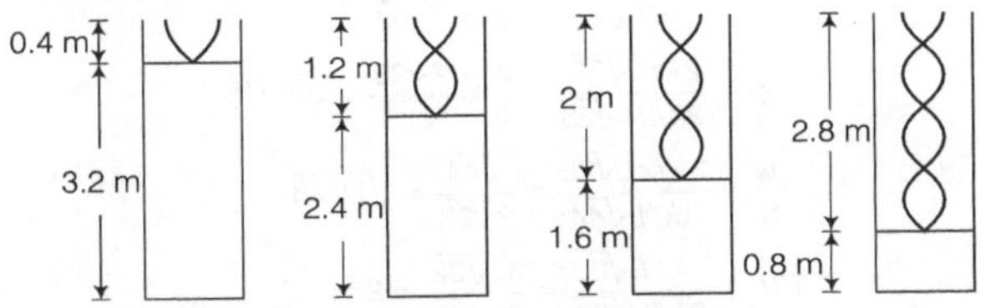

Now, let A and a are the areas of cross-section of the pipe and the hole respectively. Then,

$$A = \pi(2\times10^{-2})^2 = 1.26\times10^{-3}\text{m}^2$$

and $$a = \pi(10^{-3})^2 = 3.14\times10^{-6}\ \text{m}^2$$

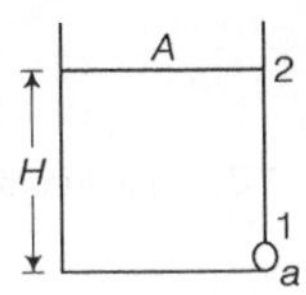

Since, velocity of flux, $v = \sqrt{2gH}$

Applying equation of continuity at (1) and (2), we get

$$a\sqrt{2gH} = A\left(-\frac{dH}{dt}\right)$$

So, rate of fall of water level in the pipe is given by

$$-\frac{dH}{dt} = \frac{a}{A}\sqrt{2gH}$$

$\therefore$ $$-\frac{dH}{dt} = \frac{3.14\times10^{-6}}{1.26\times10^{-3}}\sqrt{2\times10\times H} = 1.11\times10^{-2}\sqrt{H}$$

$\Rightarrow$ $$-\frac{dH}{dt} = 1.11\times10^{-2}\sqrt{H}$$

$$H^{-1/2}\,dH = -(1.11\times10^{-2})\,dt$$

Between first two resonances, the water level falls from 3.2 m to 2.4m.

$$\int_{3.2}^{2.4} H^{-1/2}\,dH = -(1.11\times10^{-2})\int_0^t dt$$

$\Rightarrow$ $$2(\sqrt{2.4} - \sqrt{3.2}) = -(1.11\times10^{-2})t \quad \text{or} \quad t = 43\ \text{s}$$

2. (i) Frequency of 2nd harmonic in P = frequency of 3rd harmonic in Q

$$2\left(\frac{v_1}{2L}\right) = 3\left(\frac{v_2}{4L}\right) \Rightarrow \frac{v_1}{v_2} = \frac{3}{4}$$

$$\frac{\sqrt{\dfrac{\gamma_1 RT}{M_1}}}{\sqrt{\dfrac{\gamma_2 RT}{M_2}}} = \frac{3}{4} \Rightarrow \sqrt{\frac{\gamma_1}{\gamma_2}}\sqrt{\frac{M_2}{M_1}} = \frac{3}{4}$$

$$\sqrt{\frac{M_2}{M_1}} = \frac{3}{4}\sqrt{\frac{\gamma_2}{\gamma_1}}$$

$\Rightarrow$ $$\frac{M_1}{M_2} = \frac{16}{9}\left(\frac{5/3}{7/5}\right) = \frac{16}{9}\times\frac{25}{21}$$

or $$\frac{M_1}{M_2} = \frac{400}{189}$$

(ii) Ratio of fundamental frequency in pipe P and in pipe Q is

$$\frac{f_1}{f_2} = \frac{v_1/2L}{v_2/2L} = \frac{v_1}{v_2}$$

or $$\frac{f_1}{f_2} = \frac{\sqrt{\dfrac{\gamma_1 RT}{M_1}}}{\sqrt{\dfrac{\gamma_2 RT}{M_2}}} = \sqrt{\frac{\gamma_1}{\gamma_2}}\sqrt{\frac{M_2}{M_1}} = \sqrt{\frac{5/3}{7/5}\times\frac{189}{400}}$$

or $$\frac{f_1}{f_2} = \sqrt{\frac{25}{21}\times\frac{189}{400}} = \sqrt{\frac{9}{16}} = \frac{3}{4}$$

3. (c) Fundamental frequency for a closed pipe is

$$f = \frac{1}{4L}v = \frac{v}{4}L^{-1}$$

$$\frac{df}{dt} = \frac{v}{4}(-1)L^{-2}\frac{dL}{dt}, \quad \text{But } \frac{dL}{dt} = -u \quad \text{(given)}$$

$\therefore$ $$\frac{df}{dt} = \frac{vu}{4L^2}$$

4. (b) $$3\left(\frac{v}{4L_c}\right) = 2\left(\frac{v}{2L_0}\right) \quad \text{(given)}$$

$\therefore$ $$\frac{L_c}{L_0} = \frac{3}{2}$$

Also, $$n\left(\frac{v}{4L_c}\right) = m\left(\frac{v}{2L_0}\right) \quad \text{(given)}$$

$$\frac{n}{m} = 2\left(\frac{L_c}{L_0}\right) = \frac{6}{4} = \frac{3}{2}$$

$$\frac{n}{m} = \frac{3}{2} = \frac{9}{6}$$

$\therefore$ $$n = 9 \quad \text{and} \quad m = 6$$

5. (c) $L_1 + e = \dfrac{\lambda}{4}$

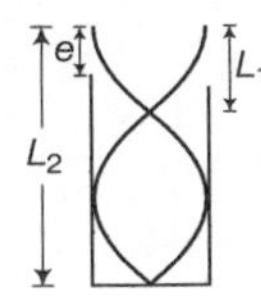

$$L_2 + e = \frac{3\lambda}{4}$$

$\therefore$ $$3(L_1 + e) = L_2 + e$$

$$2e = L_2 - 3L_1$$

$\Rightarrow$ $$e = \frac{L_2 - 3L_1}{2}$$

6. (a) For an open pipe, fundamental frequency is

$$f_0 = \frac{v}{2L} = \frac{360}{2\times1.5}$$

or $f_0 = 120$ Hz (open tube with air)

Let the speed of sound in the gas at 30°C is v'

For a closed pipe, fundamental frequency is

$$f_c = \frac{v'}{4L} = \frac{v'}{4\times1.5} \quad \text{or} \quad f_c = \frac{v'}{6}$$

Given $$f_c = f_0$$

$\therefore$ $$120 = \frac{v'}{6} \Rightarrow v' = 720\ \text{ms}^{-1}$$

Let the speed of sound in the gas at 0°C is v_0, then $\dfrac{v_0}{v'} = \sqrt{\dfrac{T_0}{T'}}$,

Here, $T' = 30 + 273 = 303$ K, $T_0 = 273$ K and $v' = 720\ \text{ms}^{-1}$.

$\therefore$ $$v_0 = 720\left(\sqrt{\frac{273}{303}}\right) = 683.43\ \text{ms}^{-1}$$

7. (b) When the two positions of the resonance are obtained at distances L_1 and L_2 respectively, then the velocity v of sound is obtained as below.

$$L_2 - L_1 = \frac{\lambda}{2}$$

$$\lambda = 2(L_2 - L_1)$$

$$v = f\lambda = 2f(L_2 - L_1)$$

Here, $L_2 - L_1 = 8.5 \text{ cm} = 8.5 \times 10^{-2} \text{ m}$

$$v = 2(2000)(8.5 \times 10^{-2}) = 340 \text{ ms}^{-1}$$

8. (b) Let h' be the length of second liquid poured in A.

Now let the first liquid comes down a level y in arm A and rises up the same level y in arm B.

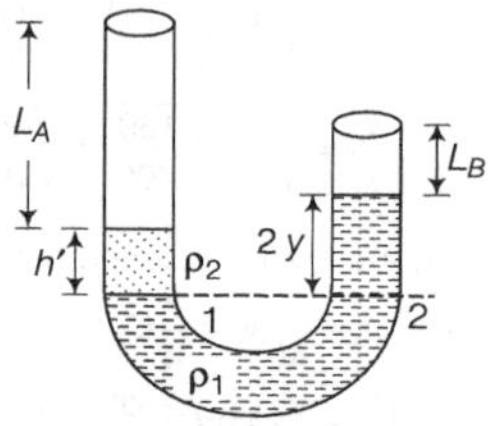

Pressure at 1 = Pressure at 2

$$\rho_2 gh' = \rho_1 g(2y)$$

$$y = \frac{\rho_2}{2\rho_1} h' = \frac{h'}{4} \qquad \left[\because \frac{\rho_2}{\rho_1} = \frac{1}{2}\right]$$

Length of air column in A is

$$L_A = (L_1 - h) - (h' - y)$$

or $$L_A = L_1 - h - \left(h' - \frac{h'}{4}\right) = L_1 - h - \frac{3h'}{4}$$

Length of air column B is

$$L_B = L_2 - h + y = L_2 - h + \frac{h'}{4}$$

It is given that first overtone of A is in unison with fundamental tone of B.

Thus, $$3\left[\frac{v}{4L_A}\right] = \frac{v}{4L_B} \quad \text{or} \quad L_A = 3L_B$$

$\therefore$ $$L_1 - h - \frac{3h'}{h} = 3\left[L_2 - h + \frac{h'}{4}\right]$$

$\Rightarrow$ $$L_1 - h - \frac{3h'}{4} = 3L_2 - 3h + \frac{3h'}{4}$$

$$\frac{3h'}{2} = L_1 - 3L_2 + 2h$$

$\Rightarrow$ $$h' = \frac{2(L_1 - 3L_2 + 2h)}{3}$$

9. (d) Fundamental frequency of an organ pipe is $f = \dfrac{v}{2L}$.

Let L be the length of organ pipe at temperature T.

$\therefore$ $$L = L_0[1 + \alpha(T - T_0)]$$

Speed of sound is $v = \sqrt{\dfrac{\gamma RT}{M_0}}$

We have to find the temperature at which

$$f(T) = f(T_0) \text{ for small } T - T_0$$

$\therefore$ $$\frac{\sqrt{\dfrac{\gamma RT}{M_0}}}{2 \cdot L_0[1 + \alpha(T - T_0)]} = \frac{\sqrt{\dfrac{\gamma RT_0}{M_0}}}{2L_0}$$

$\Rightarrow$ $$\sqrt{\frac{T}{T_0}} = 1 + \alpha(T - T_0)$$

$$\frac{T}{T_0} = [1 + \alpha(T - T_0)]^2$$

For small $T - T_0$, we can use Binomial approximation.

$\therefore$ $$\frac{T}{T_0} = 1 + 2\alpha(T - T_0)$$

$\Rightarrow$ $$\frac{T - T_0}{T_0} = 2\alpha(T - T_0) \quad \text{or} \quad T_0 = \frac{1}{2\alpha}$$

10. (a) Here, $\dfrac{\lambda}{4} = L_1$ and $\dfrac{3\lambda}{4} = L_2$

$$L_3 = \frac{5\lambda}{4}$$

$\therefore$ $$L_3 = 2L_2 - L_1$$

11. (c) We know $\gamma_{H_2} = \gamma_{O_2} = \gamma_{N_2} = \dfrac{7}{5}$ and given $\gamma_{CO_2} = \dfrac{7}{5}$. So, γ is same for all.

$$(M_0)_{H_2} = 2, (M_0)_{O_2} = 32, (M_0)_{N_2} = 28, (M_0)_{CO_2} = 44$$

$$v = \sqrt{\frac{\gamma RT}{M_0}} \Rightarrow v \propto \frac{1}{\sqrt{M_0}}$$

$$f = \frac{v}{2L} \Rightarrow f \propto \frac{1}{L\sqrt{M_0}} \Rightarrow f = \frac{k}{L\sqrt{M_0}}$$

$\therefore$ $$f_1 = \frac{k}{L\sqrt{2}}, f_2 = \frac{2k}{L\sqrt{32}}, f_3 = \frac{3k}{2L\sqrt{28}}, f_4 = \frac{3k}{L\sqrt{44}}$$

$$\frac{f_4}{f_1} = \frac{3k/L\sqrt{44}}{k/L\sqrt{2}} = \frac{3\sqrt{2}}{\sqrt{44}} = \frac{3}{\sqrt{22}}$$

$$\frac{f_2}{f_3} = \frac{2k/L\sqrt{32}}{3k/2L\sqrt{28}} = \frac{2}{3} \times \frac{2\sqrt{28}}{\sqrt{32}} = \sqrt{28/18}$$

$$\frac{f_3}{f_4} = \frac{3k/2L\sqrt{28}}{3k/L\sqrt{44}} = \frac{\sqrt{44}}{2\sqrt{28}} = \sqrt{11/28}$$

$$f_1/f_2 = \frac{k/L\sqrt{2}}{2k/L\sqrt{32}} = \frac{1}{2}\frac{\sqrt{32}}{\sqrt{2}} = 2$$

12. (a) For a closed organ pipe, $f_1 = \dfrac{v_1}{4L}$

For a stretched string, $f_2 = \dfrac{v_2}{2(0.8\,L)}$

$$\frac{f_2}{f_1} = \frac{v_2}{1.6\,L} \times \frac{4L}{v_1} \Rightarrow \frac{v_2}{v_1} = 0.4$$

13. (c) In fundamental mode, in a closed pipe,

$$\frac{\lambda}{4} = L + e = L + 0.3D$$

$$\frac{\lambda}{4} = 16 + 0.3 \times 5 = 17.5 \Rightarrow \lambda = 70 \text{ cm}$$

$$v = f\lambda = 480 \times 70 = 33600 \text{ cms}^{-1} = 336 \text{ ms}^{-1}$$

14. (a,b) As, this is third harmonic, $3\left(\dfrac{\lambda}{2}\right) = L$

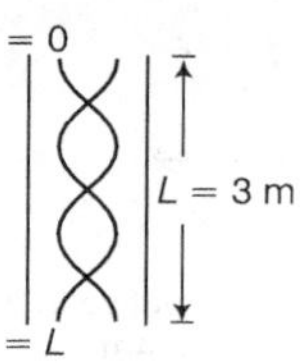

$$\lambda = \frac{2L}{3} = \frac{2 \times 3}{3} = 2\text{m}$$

Angular frequency is $\omega = 2\pi f = \dfrac{2\pi v}{\lambda}$

or $$\omega = \frac{2\pi \times 332}{2} = 332\pi \text{ rad s}^{-1}$$

$$k = \frac{2\pi}{\lambda} = \frac{2\pi}{2} = \pi \text{ m}^{-1}$$

$$(\Delta p)_{max} = 1\% \text{ of } p_0 \quad \text{(given)}$$

So, $$(\Delta p)_{max} = \frac{1}{100} \times 10^5 = 10^3 \text{ Pa}$$

Also, $$(\Delta p)_{max} = BAk$$

$\Rightarrow$ $$1000 = \rho v^2 Ak \qquad \left[\because v = \sqrt{\frac{B}{\rho}}\right]$$

$$1000 = (1.03)(332)^2 A(\pi)$$

$\Rightarrow$ $$A = 0.0028 \text{ m} = 0.28 \text{ cm}$$

$$(\Delta\rho)_{max} = \frac{(\Delta p)_{max}}{v^2} = \frac{1000}{(332)^2} = 9 \times 10^{-3} \text{ kgm}^{-3}$$

15. (c,d) Frequency of second overtone of the closed pipe $= 5\left(\frac{v}{4L}\right) = 440\text{ Hz}$

$$L = \frac{5v}{4 \times 440} = \frac{5 \times 330}{4 \times 440} = \frac{15}{16}\text{ m}$$

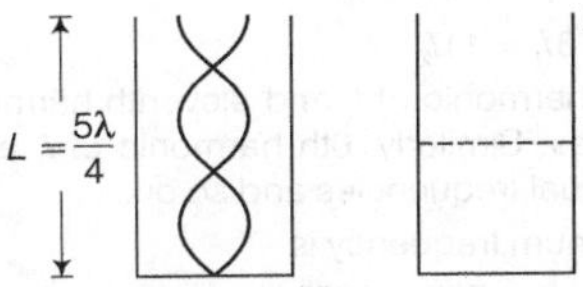

$x = 0, \Delta p = 0$

$x = x, \Delta p = \pm (\Delta p)_{max} \sin kx$

Open end is displacement antinode. So, it would be a pressure node. Thus, at $x = 0$, $\Delta p = 0$. Hence, pressure amplitude at $x = x$ can be written as $\Delta p = \pm (\Delta p)_{max} \sin kx$.

$$\lambda = \frac{4L}{5} = \frac{4\left(\frac{15}{16}\right)}{5} = \frac{3}{4}\text{ m}$$

So, $$k = \frac{2\pi}{\lambda} = \frac{8\pi}{3}\text{ m}^{-1}$$

So, pressure amplitude at $x = \frac{L}{2}\left(\text{i.e. } x = \frac{15}{32}\text{ m}\right)$

$$\Delta p = \pm (\Delta p)_{max} \cdot \sin\left(\frac{8\pi}{3} \cdot \frac{15}{32}\right)$$

$$= \pm \frac{(\Delta p)_{max}}{\sqrt{2}}$$

16. (b,c,d) $5 \times 85 = 425$ Hz, $7 \times 85 = 595$ Hz and $9 \times 85 = 765$ Hz As, only add harmonics are present, the pipe is closed at one end and fundamental frequency is 85 cm.

$\therefore$ $$\frac{\lambda_1}{4} = L = \frac{v}{f_1 \cdot 4} = L$$

or $$\frac{340}{85 \times 4} = L$$

$\Rightarrow$ $$L = 1\text{ m}$$

17. (a, b, c) $f = 340$ Hz,

$$v = 340\text{ ms}^{-1}$$

$\therefore$ $$\lambda = \frac{v}{f} = \frac{340}{340} = 1\text{ m}$$

Let end correction is e.

Here, the pipe is closed. So, resonating lengths of air column can be

$$\frac{\lambda}{4}, \frac{3\lambda}{4}, \frac{5\lambda}{4}, \frac{7\lambda}{4} \ldots$$

$\frac{5\lambda}{4} = \frac{5 \times 1}{4} = 1.25$ m is close to length of pipe, i.e. 1.2 m

$\therefore$ $$L + e = 1.25\text{ m}$$

$\Rightarrow$ $$e = 5\text{ cm}$$

Now, water is poured, let next resonating air column length is L'.

$\therefore$ $$L' + e = \frac{3\lambda}{4}$$

$\Rightarrow$ $$L' = \frac{3(100)}{4} - 5 = 70\text{ cm}$$

$\therefore$ Water length $= 120 - 70 = 50$ cm

Now, when the maximum water column is present, let the length of air column L''.

$\therefore$ $$L'' + e = \frac{\lambda}{4}$$

$\Rightarrow$ $$L'' = \frac{100}{4} - 5 = 20\text{ cm}$$

$\therefore$ Maximum water column length = 100 cm

Distance between the successive nodes $= \frac{\lambda}{2} = 50$ cm

18. (b,d) In case string, $$f_1 = \frac{v}{2L}$$

Fundamental frequencies in four cases respectively are

$$\frac{v}{4L}, \frac{v}{2(2L)}\left(= \frac{v}{2L}\right), \frac{v}{2\left(\frac{L}{2}\right)}\left(= \frac{v}{L}\right) \text{ and } \frac{v}{4\left(\frac{L}{2}\right)}\left(= \frac{v}{2L}\right)$$

Harmonics in case (a) : $\frac{v}{4L}, \frac{3v}{4L}, \frac{5v}{4L}, \frac{7v}{4L} \ldots$

Harmonics in case (b) : $\frac{v}{4L}, \frac{2v}{4L}, \frac{3v}{4L}, \ldots$

Harmonics in case (c) : $\frac{v}{L}, \frac{2v}{L}, \frac{3v}{L}, \ldots$

Harmonics in case (d) : $\frac{v}{2L}, \frac{3v}{2L}, \frac{5v}{2L}, \ldots$

19. (a) Fundamental frequency when the pipe is open at both ends is

$$f_1 = \frac{v}{2L} = \frac{340}{2(0.6)} = 283.33\text{ Hz}$$

20. (d) Opening a hole in the side effectively shortens the length of the resonating column.

Let the hole is uncovered at a distance L' from the mouth piece, then the fundamental frequency will be

$$f_1' = \frac{v}{2L'}$$

$\Rightarrow$ $$L' = \frac{v}{2f_1'} = \frac{340}{2(330)} = 0.515\text{ m}$$

or $$L' = 51.5\text{ cm}$$

21. (a) When the frequency is 2000 Hz, the distance between two adjacent nodes is 8 cm (given).

$\therefore$ $$\lambda_1 = 2 \times 8 = 16\text{ m}$$

$\therefore$ $$v = f_1\lambda_1 = (2000)(0.16) = 320\text{ ms}^{-1}$$

22. (c) $320 = f_2\lambda_2$

$\Rightarrow$ $$\lambda_2 = \frac{320}{1600} = 0.2\text{ m} = 20\text{ cm}$$

$\therefore$ Distance between two adjacent nodes, now is $\frac{20}{2} = 10$ cm.

23. (b) Let there are x_1 loops in 1st case and x_2 loops in 2nd case. Then, $x_1 - x_2 = 1$.

If the distance between the diaphragm and the closed end is L, then

$$8x_1 = 10x_2 = L$$

$\Rightarrow$ $$x_1 = \frac{5}{4}x_2$$

$\therefore$ $$\frac{5}{4}x_2 - x_2 = 1 \Rightarrow x_2 = 4$$

$\therefore$ $$L = 10 \times 4 = 40\text{ cm}$$

24. (c) $y = (4\text{ mm}) \sin \pi x \cos \pi t$ (given)

Comparing it with $y = 2A \sin kx \cos \omega t$, we get

$$k = \pi\text{ m}^{-1}, \frac{2\pi}{\lambda} = \pi \Rightarrow \lambda = 2\text{m}$$

It is second overtone in closed pipe.

$\therefore$ $$\frac{5\lambda}{4} = L \Rightarrow L = 2.5\text{ m}$$

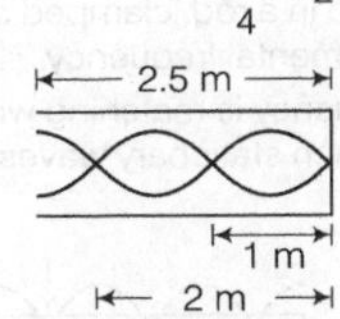

Displacement variation

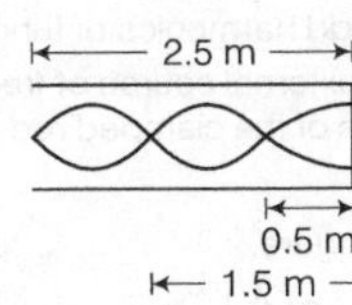

Pressure variation

Hence, (i) → (s), (ii) → (q), (iii) → (p,r) (iv) → (q)

25. (3) P and Q are set in oscillations, hence antinodes are situated at these points. The portions of the cylinders i.e. PR and RQ behave as a closed cylinder at R. Thus, there is immediate node at R.

The fundamental frequency of each pipe corresponds to just one node and one antinode. Let f_1 and f_2 be the fundamental frequencies of gases in PR and QR respectively, then

$$f_1 = \frac{v_1}{4L} = \frac{1100}{4 \times 0.5} = 550 \text{ Hz}$$

and $$f_2 = \frac{v_2}{4L} = \frac{300}{4 \times 0.5} = 150 \text{ Hz}$$

Here, v_1 and v_2 are the velocities of sound in hydrogen and oxygen respectively.

The two frequencies are different and hence the two gas columns are not vibrating in the fundamental mode. In case of a closed pipe only odd harmonics with frequencies 3, 5, 7, 9, ... etc. times the fundamental frequency are observed. Now, the problem is to find out that which harmonics of f_1 and f_2 have the same frequency. We notice that

$$\frac{f_1}{f_2} = \frac{500}{150} = \frac{11}{3}$$

or $$3f_1 = 11f_2$$

This shows that third harmonic of f_1 and eleventh harmonic of f_2 have equal frequencies. Similarly, 6th harmonic of f_1 and 22nd harmonic of f_2 have equal frequencies and so on.

So, the common minimum frequency is

$$3f_1 = 3 \times 550 = 1650 \text{ Hz}$$

and $$11f_2 = 11 \times 150 = 1650 \text{ Hz}$$

$\therefore$ $$p = 3.$$

DPP-4 Standing Waves in Rods

1. The given figure (i) shows a rod AB clamped at its middle point. If we gently hit the rod at its one end, it begins to oscillate. In the natural vibrations, the rod vibrates at its lowest frequency and maximum wavelength, which we call fundamental mode of oscillations. With maximum wavelength, when stationary waves are set up in the rod, the free ends vibrate as antinodes and the clamped end as a node as shown in the figure (i).

The oscillations of the particles of the rod are longitudinal and so the broken lines in the figure represent 'longitudinal' displacements of various parts of the rod. For clarity we have drawn them in transverse direction.

In fig (i),

$$l = \frac{\lambda}{2} \Rightarrow \lambda = 2l$$

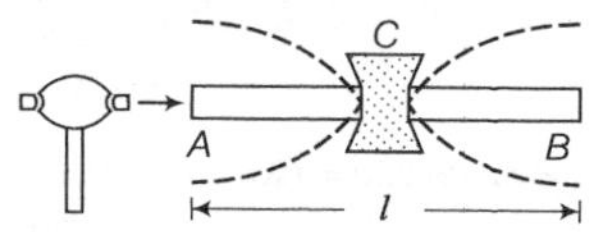

Fig (i)

Thus, the frequency of fundamental oscillations of a rod clamped at mid-point can be given as

$$f_0 = \frac{v}{\lambda} = \frac{1}{2l}\sqrt{\frac{Y}{\rho}}$$

where Y is the Young's modulus of the material of the rod and ρ is the density of the material of rod.

Higher frequencies

(i) First overtone or 3rd harmonic

$$l = \frac{3\lambda}{2} \text{ or } \lambda = \frac{2l}{3}$$

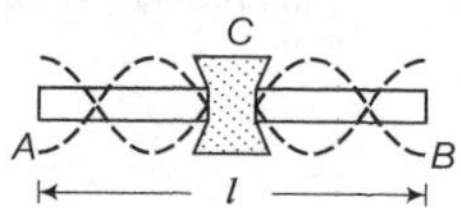

$$f_1 = \frac{v}{\lambda} = \frac{3}{2l}\sqrt{\frac{Y}{\rho}} = 3f_0$$

(ii) Second overtone or fifth harmonic

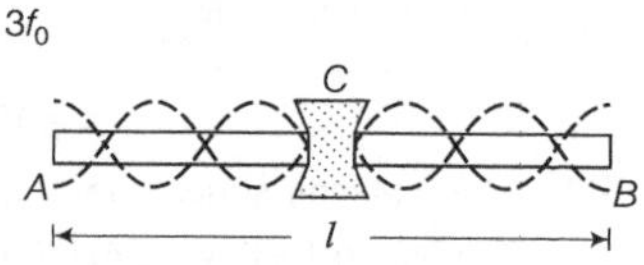

$$l = \frac{5\lambda}{2} \text{ or } \lambda = \frac{2l}{5}$$

$$f_2 = \frac{v}{\lambda} = \frac{5}{2l}\sqrt{\frac{Y}{\rho}} = 5 f_0$$

We see from the above analysis that, the resonant frequencies at which stationary waves are set up in a rod, clamped at the centre, are only odd harmonics of fundamental frequency.

When an external source of frequency is matching with any of the harmonics of the clamped rod, then stationary waves are setup in the rod.

2. From the figure,

$$4\left(\frac{\lambda}{2}\right) + 2\left(\frac{\lambda}{4}\right) = 1 \text{ m}$$

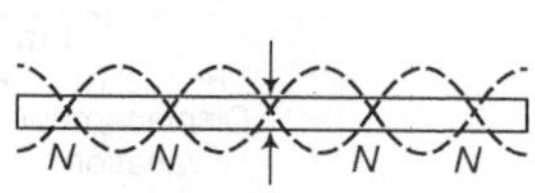

$$\lambda = \frac{2}{5} = 0.4 \text{m}$$

Velocity of longitudinal wave in the rod is

$$v = \sqrt{\frac{Y}{\rho}} = \sqrt{\frac{2 \times 10^{11}}{8000}}$$

or $$v = 5000 \text{ ms}^{-1}$$

$$v = f\lambda \Rightarrow f = \frac{v}{\lambda} = \frac{5000}{0.4} = 12500 \text{ Hz}$$

Let the incident wave is $y_i = A\sin(\omega t - kx)$ and reflected wave is

$$y_r = A\sin(\omega t + kx + \delta)$$

From principle of superposition, $y = y_i + y_r$

i.e. $$y = A\sin(\omega t - kx) + A\sin(\omega t + kx + \delta)$$

$$= 2A\sin\left[\frac{\omega t - kx + \omega t + kx + \delta}{2}\right]\cos\left[\frac{\omega t - kx - (\omega t + kx + \delta)}{2}\right]$$

$$= 2A\cos\left(kx + \frac{\delta}{2}\right)\sin\left(\omega t + \frac{\delta}{2}\right)$$

The free end of the rod is an antinode, here amplitude is maximum.

$\therefore$ $\cos\left(kx + \frac{\delta}{2}\right) = 1$, Now at $x = 0$, $\frac{\delta}{2} = 0$ or $\delta = 0$

$$A_{max} = 2A = 2 \times 10^{-6} \text{ m}$$

The equation of resulting standing wave is

$$y = 2 \times 10^{-6}\cos\left(\frac{2\pi}{\lambda}x\right)\sin(2\pi ft)$$

$$= (2 \times 10^{-6} \text{ m})\cos(5\pi x)\sin(25000\pi t)$$

For a point 2 cm from the mid-point, $x = (0.5 + 0.02)$ m

$\therefore$ $$y = (2 \times 10^{-6}\text{m})\cos[5\pi(0.5 \pm 0.02)]\sin(25000\pi t)$$

The equations of component waves can be obtained by applying

$$2\cos A\sin B = \sin(A + B) - \sin(A - B)$$

Thus, $y = 10^{-6}\sin(5\pi x + 25000\pi t) - 10^{-6}\sin(5\pi x - 25000\pi t)$

or $y = 10^{-6}\sin(25000\pi t + 5\pi x) + 10^{-6}\sin(25000\pi t - 5\pi x)$

The two terms in the expression of y represents the two component waves.

3. $y = 10[\sin 2\pi(100t - 0.02x) + \sin 2\pi(100t + 0.02x)]$

$$y = 20\cos\frac{2\pi x}{50}\sin\left(\frac{2\pi}{50} \cdot 5000t\right)$$

Comparing it with standard equation $y = 2A\cos kx\sin\omega t$, we get

$$k = \frac{2\pi}{50} \Rightarrow \frac{2\pi}{\lambda} = \frac{2\pi}{50} \Rightarrow \lambda = 50 \text{ units}$$

$$\omega = 200\pi \Rightarrow 2\pi f = 200\pi \Rightarrow f = 100 \text{ units}$$

Maximum amplitude $= 2A = 20$ units

Loop length $= \frac{\lambda}{2} = 25$ units and velocity, $v = f\lambda = 5000$ units

4. (d) Speed of longitudinal waves in the rod is

$$v = \sqrt{\frac{Y}{\rho}} = \sqrt{\frac{144 \times 10^9}{9000}} \quad \text{or} \quad v = 4000 \text{ ms}^{-1}$$

Points P and Q are clamped. So nodes are formed at these points or rod oscillates with integer number of loops in the middle part. Let the number of these loops is m. Length of each loop is $\frac{\lambda}{2}$.

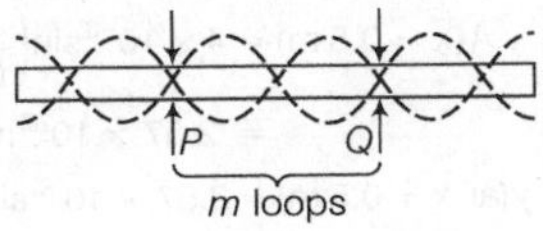

$\therefore \quad m\frac{\lambda}{2} = 1 - 2d$

or $\quad m\lambda = 2 - 4d = 2 - 4(0.3) = 0.8 \text{ m}$

or $\quad m\lambda = 80 \text{ cm} \quad \ldots\text{(i)}$

The ends of the rod are free. Therefore, antinodes are formed at each free end of the rod. That means at one end of each end part is an antinode and at the other end is a node. Thus, number of loops in each end part will be an odd multiple of half.

Let these be $\frac{(2n-1)}{2}$, where n is an integer.

Thus, $\left(\frac{2n-1}{2}\right)\frac{\lambda}{2} = d$ or $(2n-1)\lambda = 120 \text{ cm} \quad \ldots\text{(ii)}$

Dividing Eq. (i) by Eq. (ii), we get $\frac{m}{2n-1} = \frac{2}{3} \quad \ldots\text{(iii)}$

Minimum possible frequency corresponds to maximum possible wavelength and hence minimum number of loops.

For minimum frequency, from Eq. (iii),

$m = 2$ and $2n - 1 = 3$ (or $n = 2$)

Substituting $m = 2$ in Eq. (i), $\lambda_0 = 40$ cm

$\therefore$ Minimum frequency $f_0 = \frac{v}{\lambda_0} = \frac{4000}{0.4} = 10 \text{ kHz}$

5. (b) Next higher frequency corresponds to next higher integer values of m and n which satisfy Eq. (iii) of above solution.

Hence, for this case $m = 6$ and $(2n - 1) = 9$ or $n = 5$

Substituting $m = 6$ in Eq. (i) $m\lambda = 80$ cm

We have, $\quad \lambda = \frac{80}{6} \text{ cm} = \frac{40}{3} \text{ cm}$

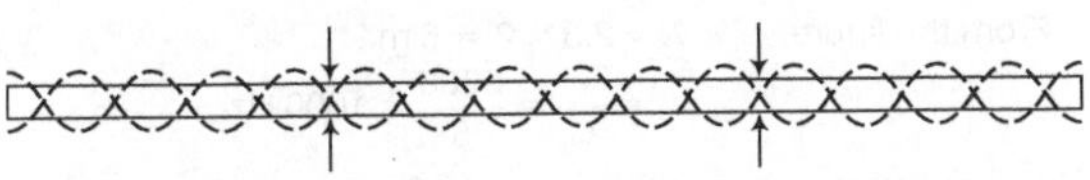

$\therefore$ Next higher frequency is $f_1 = \frac{v}{\lambda} = \frac{4000}{\frac{40}{3} \times 10^{-2}} = 30 \text{ kHz}$

(It means rod oscillates with odd harmonics.)

6. (a) For section AB of the rod, $a = (2n-1)\frac{\lambda}{4}$, where $n = 1, 2, 3, \cdots$

$$f_1 = \frac{v}{\lambda} = \frac{(2n-1)v}{4a}$$

For section BC of the rod,

$2a = m\frac{\lambda}{2}$, where $m = 1, 2, 3, \cdots$

$$f_2 = \frac{v}{\lambda} = \frac{mv}{4a}$$

But $\quad f_1 = f_2$

So, $\quad \frac{2n-1}{4a} = \frac{m}{4a}$ or $2n - m = 1$

For fundamental, $m = 1$ and $n = 1$ and

For first overtone, $m = 3$ and $n = 2$

$\therefore$ Frequency of the rod in I[st] overtone is $f = \frac{3v}{4a} \quad \ldots\text{(i)}$

Frequency of the pipe in 5th harmonic is $f' = \frac{5v_s}{4L_0} \quad \ldots\text{(ii)}$

Given, $\quad f = f'$

$\therefore \quad \frac{3v}{4a} = \frac{5v_s}{4L_0} \Rightarrow 4a = \frac{12vL_0}{5v_s}$

or Length of the rod $= \frac{12L_0}{5v_s}\sqrt{\frac{Y}{\rho}} \quad \left(\because v = \sqrt{\frac{Y}{\rho}}\right)$

7. (b) Particle velocity is

$$v_p = \frac{\partial y}{\partial t} = (4)(40\pi)\cos\frac{\pi x}{3}\cos(40\pi t)$$

Now, the velocity of particle $x = 3$ cm at time $t = \frac{1}{8}$ s is

$$v_p = 160\pi\cos\left(\frac{\pi \times 3}{3}\right)\cos\left(40\pi\frac{1}{8}\right) = 160\pi \text{ cms}^{-1}$$

8. (c) $L_0 = \frac{5\lambda_{air}}{2} \Rightarrow \lambda_{air} = \frac{2L_0}{5}$

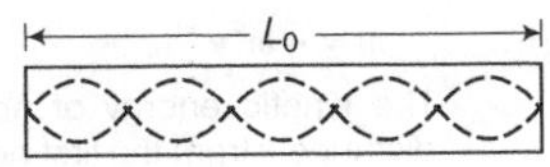

$$f = v_{air}/\lambda_{air} = \frac{343 \times 3}{2L_0} = \frac{1029}{2L_0}$$

$\frac{\lambda}{4} = 1 \text{ m} \quad \Rightarrow \lambda = 4 \text{ m}$

$\therefore \quad f = v_{steel}/\lambda = \frac{5250}{4} = 1312.5 \text{ Hz}$

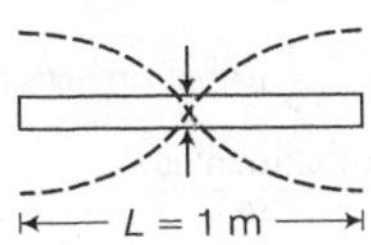

As both of them are in unison, $\frac{1029}{2L_0} = 1312.5 \text{ Hz}$

$\therefore \quad L_0 = 0.392 \text{ m} \simeq 39 \text{ cm}$

9. (d) $\frac{\lambda}{2} = L \Rightarrow \lambda = 2L$

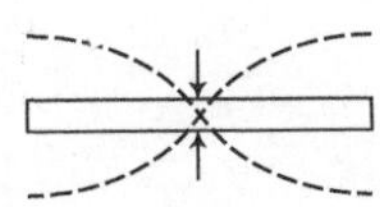

$\therefore \quad \lambda = 2(1) = 2 \text{ m}$

$v = f\lambda = 2.53 \times 2 = 5.06 \text{ kms}^{-1}$

10. (b) $v = \sqrt{\frac{Y}{\rho}} = \sqrt{\frac{1.6 \times 10^{11}}{2500}} = \frac{4}{5} \times 10^4 = 8000 \text{ ms}^{-1}$

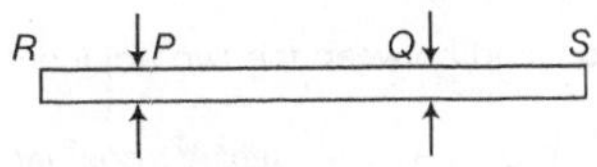

At P and Q, nodes are formed

At R and S, antinodes are formed

In the region PQ, $\quad n_1\lambda = 80 \text{ cm} \quad \ldots\text{(i)} \; (n_1 = 1, 2, 3, \cdots)$

In the region RP, $(2n_2 - 1)\frac{\lambda}{4} = 5 \text{ cm} \Rightarrow (2n_2 - 1)\lambda = 20 \text{ cm} \quad \ldots\text{(ii)}$

$(n_2 = 1, 2, 3, \cdots)$

In the region QS, $(2n_3 - 1)\frac{\lambda}{4} = 15 \text{ cm} \Rightarrow (2n_3 - 1)\lambda = 60 \text{ cm} \quad \ldots\text{(iii)}$

$(n_3 = 1, 2, 3, \cdots)$

From Eqs. (i) and (ii), we get

$$\frac{n_1}{2n_2 - 1} = 8 \quad \ldots\text{(iv)}$$

From Eqs. (i) and (iii), we get

$$\frac{n_1}{2n_3 - 1} = \frac{8}{3} \quad \ldots\text{(v)}$$

For minimum frequency, n_1, n_2 and n_3 should be least.

So, $\quad n_1 = 8, n_2 = 1, n_3 = 2$

From Eqs. (iv) and (v) for minimum frequency.

From Eq. (ii),

$$\lambda = \frac{20}{2(1) - 1} = 20 \text{ cm} \quad \text{or} \quad \lambda = 0.2 \text{ m}$$

$\therefore \quad f_{min} = \frac{v}{\lambda} = \frac{8000}{0.2} = 40 \text{ kHz}$

11. (b) The next higher values of n_1, n_2 and n_3 that satisfy Eqs. (iv) and (v) of above solution are

$$n_1 = 24, n_2 = 2 \quad \text{and} \quad n_3 = 5$$

$$\lambda = \frac{20}{2n_2 - 1} = \frac{20}{3} \text{ cm} = \frac{2}{30} \text{ m}$$

∴ Corresponding frequency is

$$f = \frac{v}{\lambda} = \frac{8000}{2} \times 30 = 120 \text{ kHz}$$

12. (c) $y = A\sin kx \cos \omega t$ (given) ...(i)

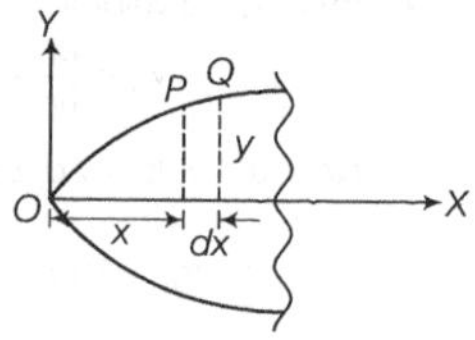

The graph of displacement *versus* position is shown in the figure.

$$\frac{dy}{dt} = v_p = -A\omega \sin kx \sin \omega t \quad \text{...(ii)}$$

$$\frac{d^2y}{dt^2} = a_p = -A\omega^2 \sin kx \cos \omega t$$

$$= -\omega^2 y \quad \text{...(iii)}$$

The kinetic energy of an elemental length dx of the rod at a distance x from the first node (assumed as origin) is given by

$$dK = \frac{1}{2}(\mu dx)\left(\frac{dy}{dt}\right)^2 = \frac{1}{2}\mu A^2\omega^2 \sin\omega t \cdot \sin^2 kx dx$$

∴ The total KE confined between consecutive nodes will be $K = \int dK$

or
$$K = \int_0^{\lambda/2} \frac{1}{2} \mu A^2\omega^2 \sin^2 \omega t \sin^2 kx dx$$

$$= \frac{\pi A^2\omega^2\mu \sin^2 \omega t}{4k} \qquad \left[\because k = \frac{2\pi}{\lambda}\right]$$

The potential energy of the elemental length, will be the negative of the work done by the internal forces, from the mean to present position.

i.e.
$$dU = -\int_0^y F\, dy = -\int_0^y (\mu dx) \frac{d^2y}{dt^2} dy$$

$$= -\int_0^y (\mu dx)(-\omega^2 y) dy \qquad \text{[From Eq. (iii)]}$$

or
$$dU = \mu\omega^2 dx \int_0^y y\, dy = \mu\omega^2 dx \left(\frac{y^2}{2}\right)$$

$$= \frac{1}{2}\mu\omega^2 A^2 \cos^2 \omega t \sin^2 kx dx$$

∴ The total potential energy confined between the two adjacent nodes, will be

$$U = \int dU = \int_0^{\lambda/2} \frac{1}{2} \mu\omega^2 A^2 \cos^2 \omega t \cdot \sin^2 kx dx = \frac{\mu\omega^2 A^2 \pi \cos^2 \omega t}{4k}$$

So, the total mechanical energy confined between any two adjacent nodes will be

$$T = K + U = \frac{\mu\omega^2 A^2 \pi}{4k}$$

13. (a,b) As, the rod is centrally clamped, there is a node at the centre. In fundamental mode there are only the antinodes at the edges.

∴ $$\frac{\lambda_0}{2} = L \Rightarrow \lambda_0 = 2L$$

∴ $$f_0 = \frac{v}{\lambda_0} = \frac{5000}{2(1.6)} = 1590 \text{ Hz}$$

In copper rod, $$f_0' = \frac{3560}{2(1.6)} \simeq 1110 \text{ Hz}$$

14. (a,b,c) From the diagram,

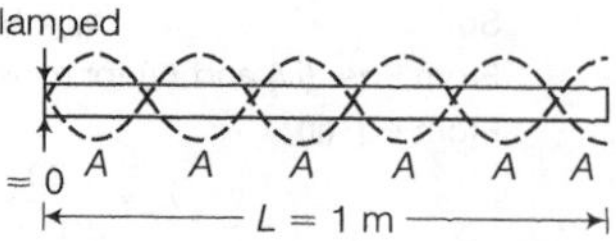

$$5\left(\frac{\lambda}{2}\right) + \frac{\lambda}{4} = 1 \text{ m}$$

$$\frac{11\lambda}{4} = 1$$

$$\Rightarrow \quad \lambda = \frac{4}{11} \text{ m}$$

or $$\lambda = 0.36 \text{ m}$$

$$v = \sqrt{\frac{Y}{\rho}} = \sqrt{\frac{6.4 \times 10^{10}}{5000}} = 3578 \text{ ms}^{-1}$$

$$f = \frac{v}{\lambda} = \frac{3578}{0.36} = 9939 \text{ Hz}$$

Amplitude at position x is $A(x) = A\sin(kx)$

$$= 4 \times 10^{-6} \sin(kx)$$

∴ $$A(x = 0.5 \text{ m}) = 4 \times 10^{-6} \sin\left(\frac{2\pi}{0.36} \times 0.5\right)$$

$$= 2.57 \times 10^{-6} \text{ m}$$

∴ $$y(\text{at } x = 0.5 \text{ m}) = 2.57 \times 10^{-6} \sin(\omega t + \delta)$$

where, $\omega = 2\pi f = 62400 \text{ s}^{-1}$

∴ $$y = 2.57 \times 10^{-6} \sin(62400\, t + \delta)$$

15. (c) Velocity of longitudinal waves in the rod is

$$v = \sqrt{\frac{Y}{\rho}} = \sqrt{\frac{2 \times 10^{11}}{8000}} = 5000 \text{ ms}^{-1}$$

The wavelength of the wave for the mode of vibration shown is given by

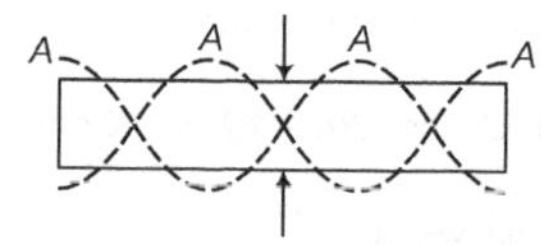

$$\frac{3\lambda}{2} = 2.5 \text{ m}$$

$$\Rightarrow \quad \lambda = \frac{5}{3} \text{ m}$$

Hence, frequency of vibration in this mode

$$\frac{v}{\lambda} = \frac{5000}{5/3} = 3000 \text{ Hz}$$

16. (a) In fundamental mode, the figure looks as shown,

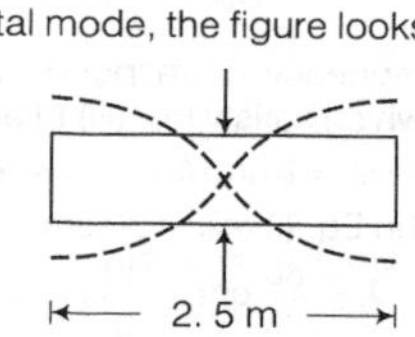

From the figure, $\lambda = 2.5 \times 2 = 5 \text{ m}$

$$f_0 = \frac{v}{\lambda_0} = \frac{5000}{5} = 1000 \text{ Hz}$$

Amplitude at antinode is $A = 2 \times 10^{-6}$ m

$$v_{max} = A\omega = A 2\pi f_0 = (2 \times 10^{-6})(2\pi)(1000)$$

$$= 12.56 \times 10^{-3} \simeq 1.25 \times 10^{-2} \text{ ms}^{-1}$$

17. (c) In this case let left end is clamped. It is given that, totally 4 antinodes are formed.

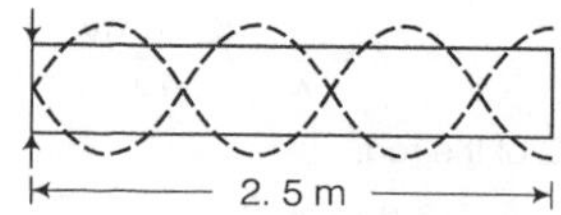

Here, $$\lambda' + \frac{3\lambda'}{4} = 2.5 \text{ m}$$

$$\frac{7\lambda'}{4} = 2.5 \text{ m}$$

$$\Rightarrow \quad \lambda' = \frac{10}{7} \text{ m}$$

Frequency f' is $$f' = \frac{v}{\lambda'} = \frac{5000}{10/7} = 3500 \text{ Hz}$$

DPP-5 Interference in Time-Beats

1. (i) In the phenomenon of beats, the waves reaching a stationary receiver (like diaphragm, eardrum etc), will have intensity variations at the position of receiver as the time elapses. Thus, the phenomenon of beats can comfortably be considered as interference in time.

(ii) No. The phenomenon of beats occurs, even if the frequency difference of waves reaching the receiver is more than 10 Hz. But our eardrum can distinguish two different sounds only, if the time gap between them is more than 0.1 s. If the time gap between them is less than 0.1 s, then it hears only one sound. Due to this limitation of our ear, beats can be 'heard' only when frequency difference is less than 10 Hz, though beats phenomenon occurs for frequency differences more than 10 Hz also.

2. (a) When a tuning fork is filed, due to decrease in its inertia, frequency increases.

Initially, beat frequency is $f_b = 4\ s^{-1} = f_1 - 256$ or $256 - f_1$

It is given that after filing T_1, beat frequency has increased i.e. becomes more than $4\ s^{-1}$.

Due to filing frequency of T_1 increases. So, f_1 should be greater than 256. Thus, $f_1 - 256 = 4\ s^{-1} \Rightarrow f_1 = 260$ Hz

3. (a) When T_1 is loaded with wax, its frequency decreases. Initially, $f_1 \sim f_2 = 4$ beats s^{-1}. Now, $f_1' < f_1$, now also $f_1' \sim f_2 = 4$ beats s^{-1}. This, happens only if f_1 is greater than f_2 by 4 Hz initially and due to loading, its frequency becomes smaller than f_2 by 4 Hz.

Thus, $f_1 = 260$ Hz and due to loading $f_1' = 252$ Hz

$\therefore$ Initially, $f_b = f_1 - f_2 = 260 - 256 = 4\ s^{-1}$

Finally, $f_b' = f_2 - f_1' = 256 - 252 = 4\ s^{-1}$

4. (c) Fundamental frequency of a stretched string is

$$f = \frac{1}{2L}\sqrt{\frac{F}{\mu}} = C\sqrt{F}$$

where, C is a constant equal to $\frac{1}{2L\sqrt{\mu}}$ and F is tension.

$$\therefore \quad \frac{\Delta f}{f} = \frac{1}{2}\frac{\Delta F}{F}$$

$$\frac{\Delta F}{F} = 2\frac{\Delta f}{f}$$

$$\Rightarrow \quad \frac{\Delta F}{F} = 2 \times \frac{5}{500} = \frac{1}{50}$$

$$\%\text{ increase in tension} = \frac{\Delta F}{F} \times 100 = \frac{1}{50} \times 100 = 2\%$$

5. (d)

Group	Beat frequency $(f' - f'')$ Hz	Beat time period $\left(\frac{1}{f' - f''}\right)$ s
S_1 and S_2	1	1
S_2 and S_3	1	1
S_1 and S_3	2	0.5

Beat time period of S_1 and S_2 is 1s. It implies that, if S_1 and S_2 are in phase at time $t = 0$, they are again in phase at $t = 1$ s. Same is the case with S_2 and S_3. But beat time period of S_1 and S_3 is 0.5 s. Therefore, beat time period for all together S_1, S_2 and S_3 will be 1s. Because, if at $t = 0$, S_1, S_2 and S_3 all are in phase, then after 1s (S_1 and S_2) and (S_2 and S_3) will again be in phase for the first time, while (S_1 and S_3) will be in phase for the second time. Or we can say all S_1, S_2 and S_3 are again in phase after 1s.

$\therefore$ beat time period is $T_b = 1$ s

$\therefore$ Beat frequency is $f_b = \frac{1}{T_b} = 1$ Hz

6. (c) Let the speed of each tuning fork (i.e. source) is v_s.

Given, $f_1 - f_2 = 3$ and $f_0 = 340$ Hz

$$\therefore \quad \left(\frac{v}{v - v_s}\right)f_0 - \left(\frac{v}{v + v_s}\right)f_0 = 3$$

or
$$\left[\left(\frac{1}{1 - \frac{v_s}{v}}\right) - \left(\frac{1}{1 + \frac{v_s}{v}}\right)\right]f_0 = 3$$

or
$$\left[\left(1 - \frac{v_s}{v}\right)^{-1} - \left(1 + \frac{v_s}{v}\right)^{-1}\right]f_0 = 3$$

The difference in apparent frequencies is very small. So, we can conclude that the speed of source $(v_s) << $ speed of sound (v). Thus, we can use binomial approximation and write.

$$\left(1 + \frac{v_s}{v}\right) - \left(1 - \frac{v_s}{v}\right) = \frac{3}{340}$$

$$\Rightarrow \quad \frac{2v_s}{v} = \frac{3}{340}$$

$$v_s = \frac{3}{340} \times \frac{v}{2} = \frac{3}{340} \times \frac{340}{2} = 1.5\ ms^{-1}$$

7. (b) $y = \left(\cos^2 \pi t - \frac{1}{4}\right)\sin(500\ \pi t)$ (given)

$$y = \left(\frac{4\cos^2 \pi t - 1}{4}\right)\sin(500\pi t)$$

$$= \left(\frac{2\cos^2 \pi t - 1 + 2\cos^2 \pi t - 1 + 1}{4}\right)\cdot\sin(500\ \pi t)$$

or
$$y = \frac{1 + 2\cos 2\pi t}{4}\sin(500\ \pi t)$$

We can express the wave function as

$$y = A_0 \sin(500\ \pi t)$$

where, $A_0 = \frac{1}{4}(1 + 2\cos 2\pi t)$ and $\omega = 500\ \pi\ s^{-1}$

$$\text{Intensity} \propto A_0^2 \Rightarrow I \propto \frac{(1 + 2\cos 2\pi t)^2}{16}$$

or
$$I = (\text{constant})\frac{(1 + 2\cos 2\pi t)^2}{16} \quad \ldots(i)$$

For maximum or minimum intensity, $\frac{dI}{dt} = 0$

$$\therefore \quad 2(1 + 2\cos 2\ \pi t)(-2\ \pi \sin 2\pi t)(2\pi) = 0$$

$$1 + 2\cos 2\pi t = 0 \quad \ldots(ii)$$

or
$$\sin 2\pi t = 0 \quad \ldots(iii)$$

From Eq. (ii),

$$\cos 2\pi t = -\frac{1}{2}$$

$$2\pi t = 2n\pi \pm \frac{2\pi}{3} \quad (n = 0, 1, 2, 3, \cdots)$$

$$t = n \pm \frac{1}{3}$$

For $t = n + \frac{1}{3}$, $\quad t = \left(\frac{1}{3}, \frac{4}{3}, \frac{7}{3}, \frac{10}{3}, \ldots\right)$

For $t = n - \frac{1}{3}$, $\quad t = \left(\frac{2}{3}, \frac{5}{3}, \frac{8}{3}, \frac{11}{3}, \ldots\right)$ $\left(\text{neglecting } -\frac{1}{3}\right)$

$$\text{Combining, } t = \left(\frac{1}{3}, \frac{2}{3}, \frac{4}{3}, \frac{5}{3}, \ldots\right) \quad \ldots(iv)$$

From Eq. (iii), $\sin 2\pi t = 0 \Rightarrow 2\pi t = n\pi$

or
$$t = \frac{n}{2}\ (n = 0, 1, 2, 3 \ldots)$$

or
$$t = \left(0, \frac{1}{2}, 1, \frac{3}{2} \ldots\right) \quad \ldots(v)$$

When Eq. (v) holds good, $\sin 2\pi t = 0$ and so $\cos 2\pi t = 1$. Thus, from Eq. (i), I is maximum.

When Eq. (iv) holds good, I is minimum.

$\therefore$ In one second, I is minimum two times and maximum two times. So, beat frequency is $2\ s^{-1}$.

8. (b) For a one end closed organ pipe, $f = \frac{(2n-1)v}{4L}$ $(n = 1, 2, 3...)$

For open organ pipe, $f = \frac{nv}{2L}$ $(n = 1, 2, 3, ...)$

Wave speed in air is $v = \sqrt{\frac{\gamma RT}{M_0}}$, we can see that $f \propto v$ and $v \propto \sqrt{T}$.

Hence, we can write $f = c\sqrt{T}$, c is a constant.

If the frequency of tuning fork is f', two expressions for beat frequency can be written as

$$f' - c\sqrt{T} = 5 \text{ or } c\sqrt{T} - f' = 5$$

With decrease in temperature, the frequency of organ pipe decreases. If we choose the first expression, as $c\sqrt{T}$ decreases, the beat frequency must increase contrary to given condition. Hence, we use second expression.

$$\therefore \quad c\sqrt{273 + 88} - f' = 5 \Rightarrow f' + 5 = c\sqrt{361} = 19c$$

$$c\sqrt{273 + 51} - f' = 1 \Rightarrow f' + 1 = c\sqrt{324} = 18c$$

$$\therefore \quad \frac{f' + 5}{f' + 1} = \frac{19}{18} \Rightarrow 18f' + 90 = 19f' + 19$$

$$f' = 71 \text{ Hz}$$

9. (d) Let the frequency of T_2 is f_2 and echo frequency it hears is f_2'. Let T_2' be the acoustic image.

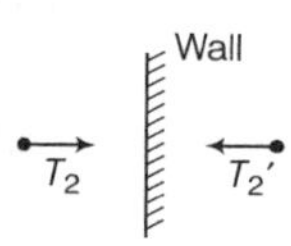

$$\therefore f_2' = \left(\frac{v + v_O}{v - v_s}\right) f_2 = \left(\frac{332 + 5}{332 - 5}\right) f_2 = 1.0306 f_2$$

$$f_2' - f_2 = 5 \text{(given)}$$

$$1.0306 f_2 - f_2 = 5$$

$$\therefore \quad f_2 = 163.4 \text{ Hz}$$

$$f_2 - f_1 = 3.5 \text{ Hz} \quad \Rightarrow \quad f_1 - f_2 = 3.5 \text{ Hz}$$

$$\therefore \quad f_1 \text{ can be} \simeq 160 \text{ Hz or } 167 \text{ Hz.}$$

It is given that when T_1 is loaded with wax, its beat frequency with T_2 increases. When T_1 is loaded, its frequency decreases. So, better value is $f_1 = 160$ Hz.

10. (b) $f_1 = \frac{1}{2L_1}\sqrt{\frac{F_1}{\pi r_1^2 \rho_1}}$ and $f_2 = \frac{1}{2L_2}\sqrt{\frac{F_2}{\pi r_2^2 \rho_2}}$

$$\frac{f_1}{f_2} = \frac{L_2}{L_1}\sqrt{\left(\frac{F_1}{F_2}\right)\left(\frac{r_2}{r_1}\right)^2\left(\frac{\rho_2}{\rho_1}\right)}$$

According to the given problem,

$$\frac{F_1}{F_2} = \frac{8}{1}, \frac{L_1}{L_2} = \frac{36}{35}, \frac{r_1}{r_2} = \frac{4}{1} \text{ and } \frac{\rho_1}{\rho_2} = \frac{1}{2}$$

$$\frac{f_1}{f_2} = \left(\frac{35}{36}\right)\sqrt{\left(\frac{8}{1}\right)\left(\frac{1}{4}\right)^2\left(\frac{2}{1}\right)} = \frac{35}{36}$$

Thus, $\quad f_2 > f_1$ and so $f_2 = 360$Hz

$$\therefore \quad \frac{f_1}{360} = \frac{35}{36} \Rightarrow f_1 = 350 \text{ Hz}$$

$$\therefore \quad f_b = 360 - 350 = 10 \text{ Hz}$$

11. (c) At 27°C, air column resonates with tuning fork. So, frequency of air column = 300 Hz. Let v_1 be the speed of sound at 27°C and L be the length of air column. Then,

$$f = \frac{v_1}{4L} = 300 \text{ Hz} \quad \ldots\text{(i)}$$

Let speed of sound at 0°C is v_0. Then,

$$\frac{v_0}{v_1} = \sqrt{\frac{273}{27 + 273}} = \sqrt{\frac{273}{300}}$$

$$\therefore \quad v_0 = (0.954)\, v_1$$

The frequency of air column at 0°C is

$$f' = \frac{v_0}{4L} = \frac{(0.954)\, v_1}{4L}$$

But from Eq. (i), $\frac{v_1}{4L} = 300$ Hz

$$\therefore \quad f' = (0.954)(300) \approx 286 \text{ Hz}$$

$\therefore$ Beat frequency is $f_b = 300 - 286 = 14$ Hz

12. (d) $y = y_1 + y_2 = A\sin(2\pi f_1 t) + A(\sin 2\pi f_2 t)$

or $\quad y = 2A \cdot \cos\left[2\pi\left(\frac{f_1 - f_2}{2}\right)t\right]\sin\left[2\pi\left(\frac{f_1 + f_2}{2}\right)t\right]$

Amplitude is

$$A' = 2A\cos\left[2\pi\left(\frac{f_1 - f_2}{2}\right)t\right] \text{ and } f = \frac{f_1 + f_2}{2}$$

If we consider A', it is clear that option (d) is correct.

13. (b) $y_1 = 4\sin[2\pi(200)t]$ and $y_2 = 3\sin(2\pi(202)t)$

$\therefore \quad$ Beat frequency $= 202 - 200 = 2\ s^{-1}$

$$\frac{I_{max}}{I_{min}} = \left(\frac{4 + 3}{4 - 3}\right)^2 = 49$$

14. (c) As motorcyclist does not hear any beats.

$$f_1' = f_2'$$

or $\quad \left(\frac{330 - v}{330 - 22}\right)176 = \left(\frac{330 + v}{330}\right)165$

$$\frac{330 - v}{330 + v} = 0.875$$

$$\Rightarrow \quad 330 - v = 288.75 + 0.875v$$

$$\Rightarrow \quad v = 22 \text{ ms}^{-1}$$

15. (a) Let f be the frequency of tuning fork. The frequency of string will be either $(f + 10)$ or $(f - 10)$. As, the tension in the string increases, its frequency increases ($f \propto \sqrt{F}$) and becomes f. This shows that the initial frequency (at $F = 129.6$ N) of string will be $(f - 10)$. Hence,

$$f - 10 = \frac{1}{2L}\sqrt{\frac{129.6}{\mu}} \text{ (intially)}$$

and $\quad f = \frac{1}{2L}\sqrt{\frac{160}{\mu}}$ (finally)

Dividing, we get

$$\frac{f - 10}{f} = \sqrt{\frac{129.6}{160}}$$

$$\Rightarrow \quad f = 100 \text{ Hz}$$

16. (a) Let the initial length of the wire is L metre.

The number of beats per second $= \frac{20}{5} = 4$

If the length decreases, frequency increases and the beats disappear.

Thus, the original frequency of the wire is less than 256 Hz by 4 Hz.

$$\therefore \quad f_0 = 256 - 4 = 252 \text{ Hz}$$

Tension in the wire is

$$F = 6.48 \times 9.8 = 63.5 \text{ N}$$

Natural frequency of wire is

$$f_0 = \frac{1}{2L}\sqrt{\frac{F}{\mu}} = \frac{1}{2L}\sqrt{\frac{F \cdot L}{m}}$$

or $\quad f_0^2 = \frac{1}{4L^2} \cdot \frac{FL}{m}$

$$\Rightarrow \quad L = \frac{F}{4m f_0^2}$$

or $\quad L = \frac{63.5}{4(0.5 \times 10^{-3})(252)^2} = 0.5 \text{ m} = 50 \text{ cm}$

17. (d) Frequency of air column is $f = \frac{v}{2L}$

$$\frac{f_1}{f_2} = \frac{v_1}{v_2} = \sqrt{\frac{T_1}{T_2}}$$

$$\frac{f_1}{f_2} = \sqrt{\frac{273+51}{273+16}} = \sqrt{\frac{324}{289}} = 1.059$$

$\therefore$ $f_2 < f_1$

As the number of beats with f_2 is smaller than the number of beats with f_1, the frequency of air column must be greater than the frequency of tuning fork. If f is the frequency of tuning fork, $f_1 = f + 4$ and $f_2 = f + 1$

$\therefore$ $\frac{f+4}{f+1} = 1.059 \Rightarrow f = 49.8$ Hz

18. (a,b) Here, only tension (F) in the vibrating string is changed. So, we write, $f_0 = \frac{1}{2L}\sqrt{\frac{F}{\mu}} = c\sqrt{F}$, where constant, $c = \frac{1}{2L\sqrt{\mu}}$ and F is tension.

According to given problem, $f - c\sqrt{F} = 5$ or $c\sqrt{F} - f = 5$. If $f - c\sqrt{F} = 5$, on decreasing the tension, beats will increase and can never be 5 again. However, if $c\sqrt{F} - f = 5$, with decrease in tension, beats will decrease, become zero (when $c\sqrt{F} = f$) and will increase $c\sqrt{F}$ becomes smaller than f. So, according to the given problem,

$$c\sqrt{100} - f = 5$$

and $f - c\sqrt{81} = 5$

or $f + 5 = 10c$ and $f - 5 = 9c$

$\therefore$ $\frac{f+5}{f-5} = \frac{10}{9} \Rightarrow f = 95$ Hz

Now, using $f + 5 = 10c$, we have $95 + 5 = 10c$

or $c = 10 \Rightarrow \frac{1}{2L\sqrt{\mu}} = 10$

$$\mu = \frac{1}{4 \times 100 \times (0.5)^2} = 0.01$$

or $\rho \times$ cross-sectional area = 0.01

$$\rho = \frac{0.01}{\pi\left(\frac{10^{-3}}{2}\right)^2}$$

or $\rho = 1.274 \times 10^4$ kg m^{-3}

19. (b,c,d) Fundamental frequency of closed organ pipe is

$$\frac{v}{4l_c} = 110 \text{ Hz} \Rightarrow l_c = \frac{330}{4 \times 110} = 0.75 \text{ m}$$

First overtone of closed organ pipe $= \frac{3v}{4l_c}$

First overtone of open organ pipe $= \frac{2v}{2l_0} = \frac{v}{l_0}$

These two produce beats of frequency 2.2 Hz, when sounded together. So,

$$\frac{3v}{4l_c} - \frac{v}{l_0} = 2.2 \quad \text{or} \quad \frac{v}{l_0} - \frac{3v}{4l_c} = 2.2$$

$$\frac{3}{4 \times 0.75} - \frac{1}{l_0} = \frac{2.2}{330} \Rightarrow l_0 = 1.0067 \text{ m}$$

$$\frac{1}{l_0} - \frac{3}{4 \times 0.75} = \frac{2.2}{330}$$

$\Rightarrow$ $l_0 = 0.9934$ m

20. (d) $f_1 = \frac{v}{4L_1} = \frac{v}{4 \times 32} = \frac{v}{128}$

and $f_2 = \frac{v}{2L_2} = \frac{v}{2 \times 66} = \frac{v}{132}$

$f_1 > f_2$

$\therefore$ $f_1 - f_2 = 8$

$\Rightarrow$ $\frac{v}{128} - \frac{v}{112} = 8$

$$v = \frac{128 \times 132 \times 8}{4} = 33792 \text{ cms}^{-1}$$

21. (d) $f_1 = \frac{33792}{4 \times 32} = 264$ Hz and $f_2 = \frac{33792}{2 \times 66} = 256$ Hz

22. (1) $f = \frac{1}{2L}\sqrt{\frac{F}{\mu}}$

$$f = c\sqrt{F}$$

where, constant $c = \frac{1}{2L\sqrt{\mu}}$

$$\frac{df}{dF} = c \cdot \frac{1}{2\sqrt{F}}$$

or $\frac{df}{dF} = \frac{1}{2L\sqrt{\mu}} \cdot \frac{1}{2\sqrt{F}} = \frac{1}{2L\sqrt{\mu}} \frac{\sqrt{F}}{2F}$

or $\frac{df}{dF} = \frac{1}{2L}\sqrt{\frac{F}{\mu}} \cdot \frac{1}{2F} = \frac{f}{2F}$

$$df = \frac{f dF}{2F} = \frac{(200)(1)}{2(100)} = 1 \text{ s}^{-1}$$

$\therefore$ Beat frequency = 1 s^{-1}

23. (8) Let the frequency of tuning fork is f Hz. Let the frequency of the note from the open organ pipe at 20°C and 30°C be f_1 and f_2 respectively. At 20°C, unison happens, So,

$$f = f_1$$

At 30°C, 5 beats per second are heard. So,

$$f_2 - f = 5$$

$$f \propto \sqrt{T} \Rightarrow \frac{f_1}{f_2} = \sqrt{\frac{T_1}{T_2}} = \sqrt{\frac{293}{303}} = 0.983$$

$\therefore$ $\frac{f}{5+f} = 0.983 \Rightarrow f = (0.983)5 + (0.983)f \Rightarrow f = 289$ Hz

$$37\beta = 289 \Rightarrow \beta = 7.8 \simeq 8$$

24. (3) Fundamental frequency of open organ pipe is

$$f_0 = \frac{v}{2L} = \frac{292}{2(0.5)} \text{ or } f_0 = 292 \text{ Hz}$$

Let L' is the stressed length of the wire. Then $L' = L + \Delta L$, where L = initial length

$\therefore$ $L' = 0.98 + 0.02 = 1$ m

The fundamental frequency of stressed wire is $f_0' = \frac{1}{2L'}\sqrt{\frac{T}{\pi r^2 \rho}}$ ...(i)

where, r = radius of the wire and ρ = density of wire. It is given that wire produces 8 beats with open pipe. Thus,

$$f_0' = f_0 \pm 8 \text{ or } f_0' = 292 \pm 8$$

So, $f_0' = 300$ Hz or $f_0' = 284$ Hz

Increasing stress f_0' increases and it is given that beat frequency decreases. Thus, $f_0' = 300$ Hz

From Eq. (i), $\frac{T}{\pi r^2} = 4 f_0'^2 (L')^2 \rho$

$$Y = \frac{T/\pi r^2}{\Delta L/L} = \frac{4f_0'^2 (L')^2 \rho}{(\Delta L/L)} = \frac{4(300)^2 (1)^2 \times 10^4}{(0.02/0.98)}$$

or $Y = 1.764 \times 10^{11}$ Nm^{-2} = $(0.588 \times 3) \times 10^{11}$ Nm^{-2} $\Rightarrow \alpha = 3$

25. (9) Let the last one has frequency = x, then the frequency of the first = $2x$.

Thus frequency of 16th tuning fork is = $2x - 15 \times 3 = x + 9 \times 3$

$\therefore$ $x = 72$Hz

$\therefore$ $f_{16} = 2(72) - 45 = 99$ Hz

or $f_{16} = 72 + 27 = 99$ Hz

$\therefore$ $y = 9$

Revisal Problems (JEE Main)

1. (b) Fundamental frequency of vibrations of string is $f_0 = \frac{1}{2L}\sqrt{\frac{F}{\mu}}$

When length is constant $f_0 \propto \sqrt{F}$ or $\frac{f_0}{\sqrt{F}}$ = constant

After increase, the tension is

$$F' = F + \frac{44}{100}F = 1.44\,F$$

$$\frac{f_0'}{\sqrt{F'}} = \frac{f_0}{\sqrt{F}}$$

$$\Rightarrow \quad f_0' = \left(\sqrt{\frac{F'}{F}}\right) f_0$$

$$f_0' = f_0 + 6 \qquad \text{(given)}$$

$$\therefore \quad f_0 + 6 = \sqrt{\frac{1.44\,F}{F}}\, f_0$$

$$\Rightarrow \quad f_0 + 6 = 1.2\, f_0$$

$$0.2 f_0 = 6 \Rightarrow f_0 = \frac{6}{0.2} = 30 \text{ Hz}$$

When tension is constant, $f_0 \propto \frac{1}{L} \Rightarrow f_0'' L'' = f_0 L$

If length increases by 20% $L'' = L + \frac{20}{100}L = 1.2\,L$

$$f_0 L = f_0'' L'' \Rightarrow f_0'' = \frac{L}{L''} f_0 = \frac{L}{1.2L} \times 30 = 25 \text{ Hz}$$

Change (decrease) in fundamental frequency = 30 − 25 = 5 Hz

2. (a) Let L_1, L_2 and L_3 be the lengths of three segments.

Given, $\quad L_1 + L_2 + L_3 = 1.1$ m

We know, $f_0 = \frac{1}{2L}\sqrt{\frac{F}{\mu}}$. When F and μ are same, $f_0 L$ = constant

$$\therefore (f_0)_1 L_1 = (f_0)_2 L_2 = (f_0)_3 L_3 \Rightarrow L_1 : L_2 : L_3 = \frac{1}{(f_0)_1} : \frac{1}{(f_0)_2} : \frac{1}{(f_0)_3}$$

$$\Rightarrow \quad L_1 : L_2 : L_3 = \frac{1}{1} : \frac{1}{2} : \frac{1}{3} = \frac{6:3:2}{6} = 6:3:2$$

So, we can write $L_1 = 6\,K$, $L_2 = 3\,K$ and $L_3 = 2\,K$, where K is a constant.

$$\therefore \quad 6K + 3K + 2K = 1.1 \Rightarrow K = 0.1$$

$$\therefore \quad L_1 = 0.6 \text{ m}, L_2 = 0.3 \text{ m and } L_3 = 0.2 \text{ m}$$

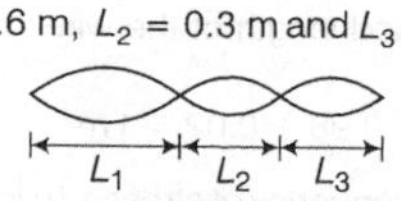

So, required distances are 0.6 m, (0.6 m + 0.3 m) = 0.9 m

or $\quad$ 0.2 m, (0.2 m + 0.3 m) = 0.5 m

So, the best answer is option (a).

3. (c)

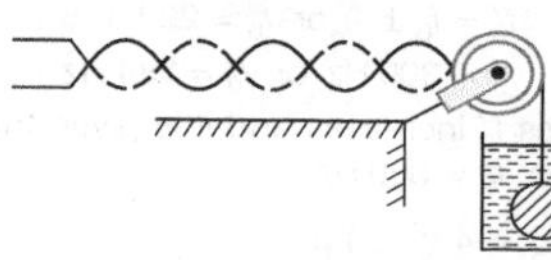

Initially FBD of sphere is

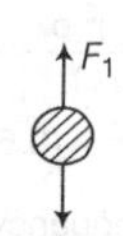

After immersion into water, FBD of sphere is

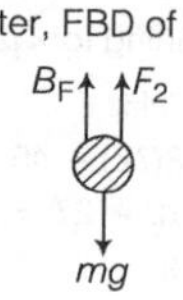

When sphere is immersed in water, buoyant force (B_F) acts upwards and reduces tension (F). The change in tension causes a change in speed of waves on the string, which in turn causes a change in the wavelength.

This altered wavelength results in the fifth harmonic. As vibrating element is same, frequency is same in both the cases.

From figures, $\quad F_1 = mg \qquad \ldots$(i)

and $\quad B_F + F_2 = mg$ or $B_F = mg - F_2 \qquad \ldots$(ii)

$$f = \frac{n_1}{2L}\sqrt{\frac{F_1}{\mu}},\; f = \frac{n_2}{2L}\sqrt{\frac{F_2}{\mu}}$$

$$\therefore \quad 1 = \frac{n_1}{n_2}\sqrt{\frac{F_1}{F_2}}$$

$$\Rightarrow \quad F_2 = \left(\frac{n_1}{n_2}\right)^2 F_1 = \left(\frac{2}{5}\right)^2 (2 \times 9.8) = 3.14 \text{ N}$$

Using this in Eq. (ii),

$$B_F = 19.6 - 3.14 = 16.46 \text{ N}$$

Also, $\quad B_F = V_{sphere}\, \rho_{water}\, g = \frac{4}{3}\pi r^3 (1000)(9.8)$

$$\therefore \quad r^3 = \frac{16.46 \times 3}{4 \times 3.14 \times 1000 \times 9.8}$$

$$\Rightarrow \quad r = 0.0738 \text{ m} = 7.38 \text{ cm}$$

Note *Only certain radii of the sphere will result In the string vibrating in a normal mode. The speed of waves on the string must be changed to a value such that the length of the string is integer multiple of half wavelengths.*

4. (d) The sound of the wind blowing across the end of the pipe contains many frequencies and the culvert responds to the sound by vibrating at the natural frequencies of air column. This problem is a relatively simple substitution problem

$$f_1 = \frac{v}{2L} = \frac{343}{2(1.23)} = 139 \text{ Hz}$$

5. (a) $f = \frac{nv}{4L} \quad (n = 1, 3, 5, \cdots)$

$$L = \frac{nv}{4f} = \frac{n(340)}{4(340)} = \frac{n}{4} \text{ m} = 0.25\, n \text{ m}$$

$$= 25\, n \text{ cm}$$

$$\therefore \quad L = 25 \text{ cm}, 75 \text{ cm}, 125 \text{ cm}$$

The length of the tube is 120 cm. So, the possible lengths of air columns are

$$L = 25 \text{ cm}, L = 75 \text{ cm}$$

If h is the height of water filled in the tube, then

$$L + h = 120 \text{ cm}$$

$$\Rightarrow \quad h = 120 - L$$

For minimum value of h, L has to be maximum. Thus,

$$h_{min} = 120 - L_{max}$$

$$= 120 - 75 = 45 \text{ cm}$$

6. (b) The wavelength of the first overtone of open pipe is

$$\lambda_0 = L + 2e = 40 + 2e$$

$$\Rightarrow \quad \frac{v}{f} = 40 + 2e$$

The wavelength of first overtone of closed pipe is

$$\frac{3}{4}\lambda_c = L' + e = 31 + e$$

$$\Rightarrow \quad \frac{3}{4}\left(\frac{v}{f}\right) = 31 + e$$

$$\therefore \quad \frac{3}{4}(40 + 2e) = 31 + e$$

$$\Rightarrow \quad e = 2 \text{ cm}$$

7. (c) Given, $\lambda_1 = \frac{90}{175}$ m, $\lambda_2 = \frac{90}{173}$ m

$$v = f_1\lambda_1 \text{ and } v = f_2\lambda_2$$

$\therefore$ $$\lambda_1 = \frac{v}{f_1} \text{ and } \lambda_2 = \frac{v}{f_2}$$

$$\lambda_2 > \lambda_1 \Rightarrow f_2 < f_1$$

If f is the frequency of the third note,

$$f_1 - f = 4 \quad \text{and } f - f_2 = 4 \Rightarrow f_1 - f_2 = 8$$

$$\frac{v}{\lambda_1} - \frac{v}{\lambda_2} = 8 \Rightarrow v\left(\frac{1}{\lambda_1} - \frac{1}{\lambda_2}\right) = 8$$

$$v\left(\frac{175}{90} - \frac{173}{90}\right) = 8 \Rightarrow v = 45 \times 8 = 360 \text{ ms}^{-1}$$

8. (d) $v = f\lambda$ and $v = \sqrt{\frac{F}{\mu}}$

$\therefore$ $$f = \sqrt{\frac{F}{\mu\lambda^2}} \Rightarrow f \propto \sqrt{F} \text{ (when } \mu \text{ and } \lambda \text{ are constant)}$$

$$\frac{f_1}{f_2} = \sqrt{\frac{F_1}{F_2}} \Rightarrow f_2 = f_1\sqrt{\frac{F_2}{F_1}} = (110)\sqrt{\frac{540}{600}} = 104.4 \text{ Hz}$$

$$f_b = f_1 - f_2 = 110 - 104.4 = 5.6 \text{ Hz}$$

9. (c) The wavelength in thin wire is given by

$$\frac{\lambda}{2} = L \Rightarrow \lambda = 2(40) = 80 \text{ cm}$$

The frequency and tension are same in both the wires.

$$f = \frac{v}{\lambda} = \frac{1}{2L}\sqrt{\frac{F}{\mu}} = \frac{1}{2(0.4)}\sqrt{\frac{4.6}{0.002}} = 59.9 \text{ Hz}$$

The linear mass density of thick wire is $\mu' = 4\mu = 0.008 \text{ kg m}^{-1}$

($\because D' = 2D$)

$$f = \frac{v'}{\lambda'} \text{ and } \frac{\lambda'}{2} = L'$$

$\therefore$ $$L' = \frac{v'}{2f} = \frac{1}{2f}\sqrt{\frac{F}{\mu'}} = \frac{1}{2(59.9)}\sqrt{\frac{4.6}{0.008}}$$

$$= 0.2 \text{ m} = 20 \text{ cm}$$

10. (a) For hot region, $\mu = \frac{v_a}{v'} = \frac{\sqrt{\frac{\gamma RT_0}{m_0}}}{\sqrt{\frac{\gamma R \cdot 4T_0}{m_0}}} = \frac{1}{2}$

μ here, is equivalent to refractive index

$\therefore$ Path difference $= |(\mu - 1)L| = \left|\left(\frac{1}{2} - 1\right)L\right| = \frac{L}{2}$

For minima, path difference $= (2n + 1)\frac{\lambda}{2}$

$\therefore$ $$\lambda = \frac{L}{2n + 1}$$

$$f = \frac{v_a}{\lambda} = \frac{v_a}{L}(2n + 1)$$

For minimum f, $2n + 1 = 1$

$\therefore$ $$f_{min} = \frac{v_a}{L}$$

11. (d) $$f_1 = f_2$$

$$f_1 = \frac{1}{2L}\sqrt{\frac{F_1}{\mu}}$$

$$f_2 = \frac{2}{2L}\sqrt{\frac{F_2}{\mu}}$$

$\therefore$ $$\sqrt{F_1} = 2\sqrt{F_2} \Rightarrow F_1 = 4F_2$$

For vertical equilibrium,

$$F_1 + F_2 = 4.8 + 1.2 = 6 \text{ kg wt} = 60 \text{ N}$$

$\therefore$ $$4F_2 + F_2 = 60 \Rightarrow F_2 = 12 \text{ N}$$

$\therefore$ $$F_1 = 48 \text{ N}$$

Taking moments about O,

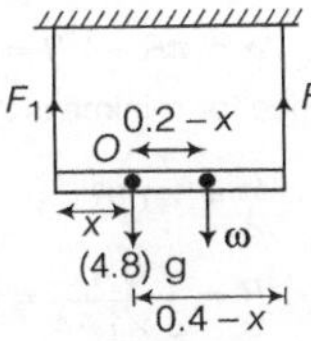

$$F_1x + W(0.2 - x) = F_2(0.4 - x)$$

$$48x + 12(0.2 - x) = 12(0.4 - x)$$

$$48x = 4.8 - 2.4 = 2.4$$

$$x = \frac{2.4}{48} = 0.05 \text{ m} = 5 \text{ cm}$$

12. (a) Frequency of string $= f_0 = \frac{1}{2L}\sqrt{\frac{F}{\mu}} = \frac{1}{2(0.2)}\sqrt{\frac{128}{10^{-3}/0.2}}$

$= 400$ Hz $=$ frequency of tuning fork

Apparent frequency heard by the observer near sonometer is

$$f' = \left(\frac{v}{v + v_s}\right)f_0 = \left(\frac{v}{v + 0.75}\right)(400)$$

Beat frequency is

$$f_b = f_0 - f' = 400 - 400\left(\frac{v}{v + 0.75}\right)$$

$$f_b = 1 \text{ (given)}$$

$\therefore$ $$400\left(1 - \frac{v}{v + 0.75}\right) = 1$$

$$1 - \frac{1}{400} = \frac{v}{v + 0.75}$$

$\Rightarrow$ $$399(v + 0.75) = 400v$$

$$v = 399(0.75) = 299.25 \text{ ms}^{-1} \approx 300 \text{ ms}^{-1}$$

13. (b) $\frac{f}{f'} = \frac{840}{1120} = \frac{3}{4}$

If fundamental frequency is f_0, then $3f_0 = 840$

$\therefore$ $$f_0 = 280 \text{ Hz}$$

$\therefore$ $$\frac{1}{2L}\sqrt{\frac{F}{\mu}} = 280$$

$\Rightarrow$ $$F = (280)^2 \times (2 \times 0.25)^2 \times (0.75 \times 10^{-3})$$

or $$F = 14.7 \text{ N}$$

14. (c) $f_0 = \frac{1}{4L} \cdot v = \frac{1}{4(1)}(320) = 80$ Hz

Even harmonics are not possible in this case. So, 320 Hz is ruled out.

15. (d) Initially, the length of water column increases, thus length of air column decreases. So, fundamental frequency increases. But when rate of inflow becomes equal to rate of out flow, length of air column becomes constant and so does the fundamental frequency.

16. (a) $\frac{\lambda_{max}}{2} = 0.4$ m

$\Rightarrow$ $$\lambda_{max} = 0.8 \text{ m}$$

$\therefore$ $$f_{min} = \frac{v}{\lambda_{max}} = \frac{5500}{0.8} = 6875 \text{ Hz}$$

17. (d) The distance between an antinode and next node is 9 cm.

$\therefore$ $$\frac{\lambda}{4} = 0.09 \text{ m}$$

$\Rightarrow$ $$\lambda = 0.36 \text{ m}$$

Speed of sound is 360 ms^{-1}.

$\therefore$ $$f = \frac{v}{\lambda} = \frac{360}{0.36} = 1000 \text{ Hz}$$

18. (c) Path difference is

$$\Delta x = \pi R - 2R = 1.14\,R$$

Minimum path difference for minima is $\Delta x = \dfrac{\lambda}{2}$

$\therefore \qquad \dfrac{\lambda}{2} = 1.14R$

$\Rightarrow \qquad R = \dfrac{0.4}{2 \times 1.14} = 0.175$ m

19. (d) When the stone is suspended in air,

$$f = \frac{1}{2L}\sqrt{\frac{W_{air}}{\mu}}$$

When the stone is suspended in water

$$f = \frac{1}{2L'}\sqrt{\frac{W_{water}}{\mu}}$$

$\therefore \qquad \dfrac{\sqrt{W_{air}}}{L} = \dfrac{\sqrt{W_{water}}}{L'}$

$\Rightarrow \qquad \dfrac{W_{air}}{W_{water}} = \dfrac{L^2}{L'^2}$

Specific gravity of material of the stone $= \dfrac{W_{air}}{W_{air} - W_{water}}$

$$\frac{1}{1 - \dfrac{W_{water}}{W_{air}}} = \frac{1}{1 - \dfrac{L'^2}{L^2}} = \frac{L^2}{L^2 - L'^2} = \frac{(40)^2}{(40)^2 - (22)^2}$$

20. (c) $f_1 = \dfrac{1}{2L} \cdot v$ and $f_2 = \dfrac{1}{2(L + x)} \cdot v$

Clearly, $f_1 > f_2$

$\therefore \qquad f_b = f_1 - f_2 = \dfrac{v}{2L} - \dfrac{v}{2(L + x)}$

$$= \frac{vL + vx - vL}{2L(L + x)}$$

$\therefore \qquad f_b = \dfrac{vx}{2L^2} \qquad (\because x << L)$

21. (a) This is a string fixed at one end and free at the other.

$\therefore$ The three first modes are as shown.

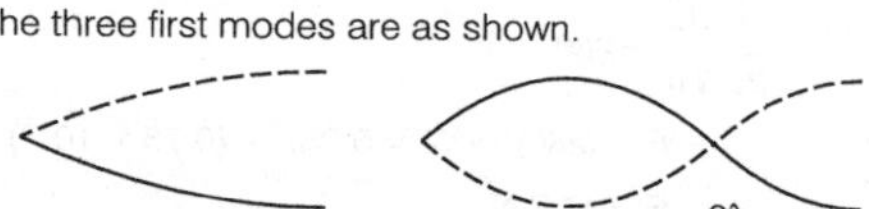

$L = \dfrac{\lambda_1}{4} \qquad L = \dfrac{3\lambda_2}{4}$

$\therefore \qquad \lambda_1 = 4 \times 2.4 = 9.6$ m

$$\lambda_2 = \frac{4 \times 2.4}{3} = 3.2 \text{ m}$$

$$\lambda_3 = \frac{4 \times 2.4}{5} = 1.92 \text{ m}$$

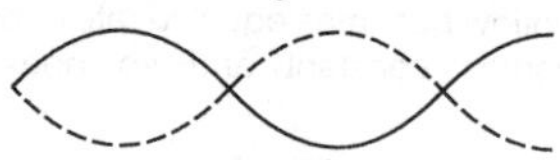

$L = \dfrac{5\lambda_3}{4}$

22. (b) Consider $y' = A \sin(\omega t + kx + \pi)$

$\therefore \qquad y' = -A \sin(\omega t + kx)$

At $x = 0$, $y = A \sin \omega t$ and $y' = -A \sin \omega t$

So, for all times, at $x = 0$, $y + y' = 0$ or x is node.

23. (b) Speed of longitudinal wave in the rod is

$$v = \sqrt{\frac{Y}{\rho}} = \sqrt{\frac{10^{11}}{10^5}} = 1000 \text{ ms}^{-1}$$

Required time is $t = \dfrac{2L}{v} = \dfrac{2(100)}{1000} = 0.2$ s

24. (a) Total path difference is

$$\Delta x = PQ + QR + \frac{\lambda}{2}$$

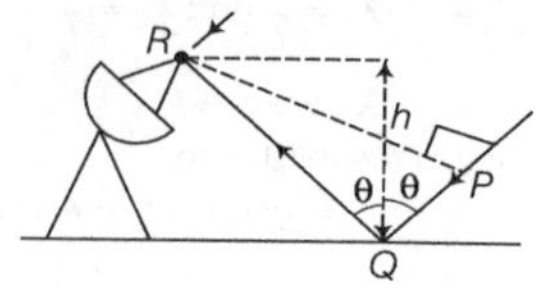

For maxima, $\qquad \cos\theta = \dfrac{h}{QR}$

$\therefore \qquad QR = h \sec\theta$

$$\cos 2\theta = \frac{PQ}{QR}$$

$\Rightarrow \qquad PQ = h \sec\theta \cos 2\theta$

For maxima, in this case is $\Delta x = \lambda$

$\therefore \qquad PQ + QR + \dfrac{\lambda}{2} = \lambda$

$\Rightarrow \qquad h \sec\theta \cos 2\theta + h \sec\theta = \dfrac{\lambda}{2}$

$$h \sec\theta (2\cos^2\theta) = \frac{\lambda}{2}$$

$\Rightarrow \qquad h = \dfrac{\lambda}{4\cos\theta}$

25. (a) Path difference is

$$\Delta x = 25 \times 2 = 50 \text{ cm}$$

For minima, $\qquad \Delta x = (2n + 1)\dfrac{\lambda}{2} \qquad (n = 0, 1, 2 \cdots)$

$$50 \text{ cm} = (2n + 1)\frac{\lambda}{2} = (2n + 1)\frac{1}{2} \cdot \frac{300}{f}$$

$$f = \frac{150}{0.5}(2n + 1) = (2n + 1)\ 300 \text{ Hz}$$

All odd multiples will be cancelled out. So, the answer is 300 Hz.

26. (d) $r = A \sin\left[\alpha\left(\dfrac{x - y}{2}\right)\right] \cos\left[\omega t - \alpha\left(\dfrac{x + y}{2}\right)\right]$

If we put $y = x + \dfrac{4\pi}{\alpha}$ in $\dfrac{\partial r}{\partial t} = -\omega A \sin\left[\alpha\left(\dfrac{x - y}{2}\right)\right]$

$$\sin\left[\omega t - \alpha\left(\frac{x + y}{2}\right)\right]$$

$\dfrac{\partial r}{\partial t}$ becomes equal to zero.

So, option (d) is correct.

27. (c) For path difference at P_1,

$$(\Delta x)_1 = 100.25\lambda + \frac{\lambda}{4} \text{ equivalent to } \frac{\lambda}{2}$$

For path difference at P_2,

$$(\Delta x)_2 = 100.25\lambda - \frac{\lambda}{4} \text{ equivalent to } 0.$$

$\therefore$ Phase difference at P_1 is π.

$\therefore \qquad A_1 = 2A - A = A$

Phase difference at P_2 is 0.

$\therefore \qquad A_2 = 2A + A = 3A$

$$I \propto A^2$$

$\Rightarrow \qquad \dfrac{I_1}{I_2} = \left(\dfrac{A}{3A}\right)^2 = \dfrac{1}{9}$

28. (d) At nodes $v = 0$ and at antinodes tension is perpendicular to velocity. So, at nodes and antinodes $P = 0$. For other points $P \neq 0$.

29. (d) Closed end is a pressure antinode, therefore pressure is not constant.

Revisal Problems (JEE Advanced)

1. (c) In case of stretched string,

$$f = \frac{1}{2L}\sqrt{\frac{F}{\mu}} = \frac{1}{2L}\sqrt{\frac{FL}{m}}$$

When the wire is stretched, the total mass of wire m remains constant.

After elongation,

$$L_1 = 1.02\,L \quad \text{and} \quad L_2 = 1.04L$$

Tension $(F) \propto$ elongation (given)

$$\therefore \quad \frac{F_1}{F_2} = \frac{2}{4} = 0.5$$

Now, $$f_1 = \frac{1}{2(1.02\,L)}\sqrt{\frac{F_1(1.02\,L)}{m}}$$

and $$f_2 = \frac{1}{2(1.04L)}\sqrt{\frac{F_2\,(1.04L)}{m}}$$

Dividing f_1 by f_2, we have

$$\frac{f_1}{f_2} = \sqrt{\frac{1.04}{1.02}\frac{F_1}{F_2}} = \sqrt{\frac{1.04}{1.02} \times 0.5}$$

$$\frac{f_1}{f_2} = 0.71$$

2. (a) When incident wave and reflected wave superimpose to produce stationary wave, the ratio of amplitudes at antinodes to at nodes is given by

$$\frac{A_{\text{max}}}{A_{\text{min}}} = \frac{A_i + A_r}{A_i - A_r}$$

This ratio is given by 1.5.

$$\therefore \quad \frac{A_i + A_r}{A_i - A_r} = \frac{3}{2}$$

$$\Rightarrow \quad \frac{1 + \dfrac{A_r}{A_i}}{1 - \dfrac{A_r}{A_i}} = \frac{3}{2} \quad \Rightarrow \quad \frac{A_r}{A_i} = \frac{1}{5}$$

$$\therefore \quad \frac{I_r}{I_i} = \left(\frac{A_r}{A_i}\right)^2 = \frac{1}{25}$$

$$\Rightarrow \quad I_r = 0.04 I_i$$

That means 4% of incident energy is reflected or 96% energy passes across the obstacle.

3. (b) We require $y_1 + y_2 = 0$

$$\Rightarrow \quad \frac{5}{(3x - 4t)^2 + 2} + \frac{-5}{(3x + 4t - 6)^2 + 2} = 0$$

$$\therefore \quad (3x - 4t)^2 = (3x + 4t - 6)^2$$

$$3x - 4t = 3x + 4t - 6$$

$$\Rightarrow \quad 8t = 6 \Rightarrow t = 0.75 \text{ s}$$

4. (a) Path $SCD = SC + CD = 2SC$

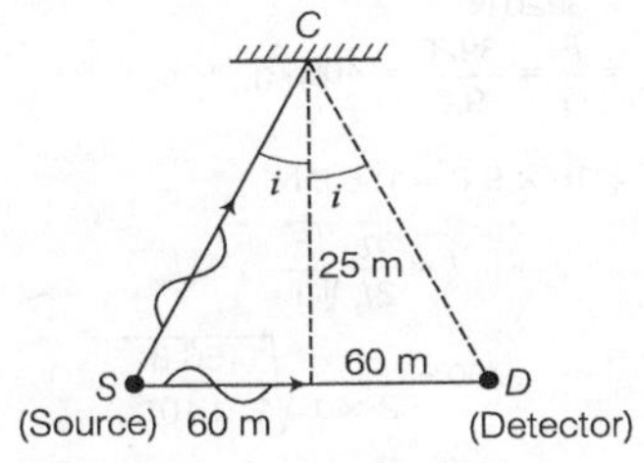

or $$\text{Path } SCD = 2\sqrt{60^2 + 25^2} = 130 \text{ m}$$

Path difference is $\Delta x = 130 - 120 = 10$ m

As one of the waves is reflected from the ceiling, it undergoes a phase change of 180°. So, conditions for constructive interference and destructive interference interchange.

$\therefore$ For minima at D, $\quad \Delta x = n\lambda$

or $\quad 10 \text{ m} = n\lambda$

$$\Rightarrow \quad \lambda = \frac{10}{n} \text{ m} \qquad (n = 1, 2, 3, \cdots)$$

$$\therefore \quad \lambda = 10 \text{ m}$$

5. (d) At point A, path difference is $\Delta x = 3\lambda$. At point 1, path difference is zero.

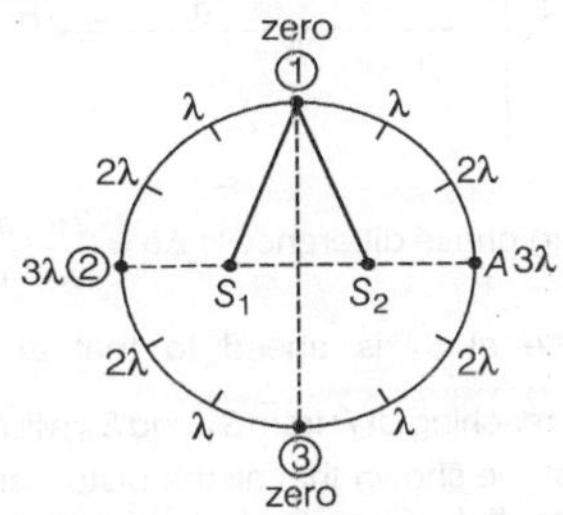

So, before reaching point (1), he hears maximum sound three times corresponding to path differences $\Delta x = 3\lambda, 2\lambda, \lambda$.

$\therefore$ Total number of times, he hears maximum sound, is 12.

6. (d) Minimum sound is heard for path difference,

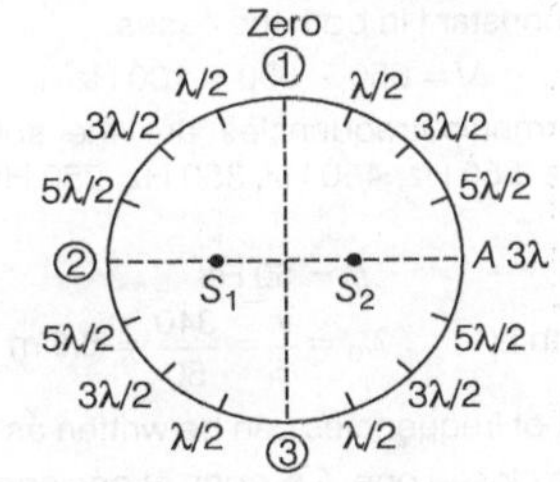

$$\Delta x = (2n + 1)\frac{1}{2} \qquad (n = 0, 1, 2 \ldots)$$

$$= \frac{\lambda}{2}, \frac{3\lambda}{2}, \frac{5\lambda}{2}, \ldots$$

Here, possible path differences for minima are $\frac{\lambda}{2}, \frac{3\lambda}{2}, \frac{5\lambda}{2}$ only.

So, number of minima = 3 × 4 = 12

7. (b) As intensities are equal, $A_1 = A_2 = A_3 = A$. So, the resultant amplitude, from the vector diagram shown, is zero.

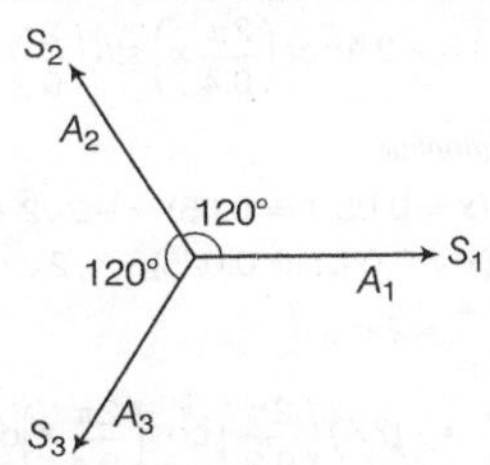

8. (c) $L_1 + L_2 = 1$ m and $L_1 - L_2 = 2$ mm

$$\Rightarrow \quad L_1 = 50.1 \text{ cm}, L_2 = 49.9 \text{ cm}, fL = \text{constant}$$

$$f(50.1) = (f + 1)\, 49.9$$

$$\Rightarrow \quad f = 249.5 \text{ Hz}$$

$$\therefore \quad f_1 = 249.5 \text{ Hz}, f_2 = 250.5 \text{ Hz}$$

9. (c) $\Delta f = 125 - 75 = 50$ Hz and this is same for any two successive frequencies. So, the possible frequencies from 175 Hz are 175, 125, 75, 25 Hz.

$\therefore f_0 = 25$ and it is a string fixed at one end and free at the other end.

$$f_0 = \frac{v}{4L} \Rightarrow L = \frac{v_0}{4f_0} = \frac{400}{4 \times 25} = 4 \text{ m}$$

10. (a) $\Delta x = a\sin\theta \simeq a\tan\theta$ or $\Delta x = a\frac{a}{b} = \frac{a^2}{b}$

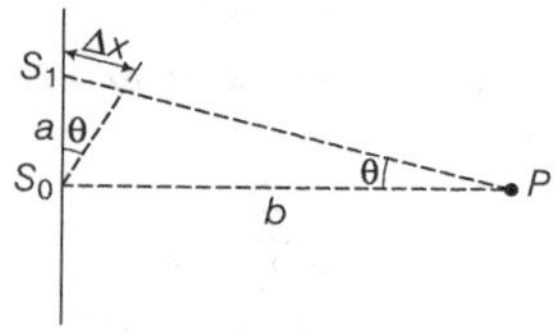

Corresponding phase difference is $\Delta\phi = \frac{2\pi}{\lambda} \cdot \frac{a^2}{b}$

As, the phase at S_1 is ahead to that at S_0 by $\frac{2\pi a^2}{b\lambda}$, the disturbances reaching at P from S_1 and S_0 will be in phase. Similarly, it can be shown that all the disturbances reaching at P will be in phase. If A_0 is amplitude of each wave, the amplitude at point P will be $7\,A_0$.

$$I \propto A^2 \Rightarrow I_P = 49\, I_0$$

11. (a,d) Harmonic frequencies are given by $f_1 \cdot n$ for open pipes and $f_1(2n-1)$ for closed pipes, the difference between adjacent harmonics is constant in both the cases.

$$\Delta f = 650 - 550 = 100 \text{ Hz}$$

Thus, the harmonic frequencies are the set of the following values. 650 Hz, 550 Hz, 450 Hz, 350 Hz, 250 Hz, 150 Hz, and 50 Hz.

$\therefore$ $f_0 = 50$ Hz

The wavelength is $\lambda_0 = \frac{v}{f} = \frac{340}{50} = 6.8$ m

The above set of frequencies can be written as $(2n - 1)$ 50 Hz.

$\therefore$ The pipe is a closed one. (i.e. open at one end and closed at the other.)

$\therefore$ $L = \frac{\lambda_0}{4} = \frac{6.8}{4} = 1.7$ m

12. (b,c,d) Distance between node and next antinode is 0.1 m (given).

$\therefore$ $\frac{\lambda}{4} = 0.1$ m

$\Rightarrow$ $\lambda = 0.4$ m

From graph $T = 0.2$ s and amplitude of standing wave is $2A = 4$ cm. Equation of standing wave is

$$y(x, t) = -2A\cos\left(\frac{2\pi}{0.4}x\right)\sin\left(\frac{2\pi}{0.2}t\right) \text{ cm}$$

Note *$x = 0$ is an antinode.*

$$y(x = 0.05, t = 0.05) = -2\sqrt{2} \text{ cm}$$

$$y(x = 0.04, t = 0.025) = -2\sqrt{2}\cos 36°$$

Speed $= \frac{\lambda}{T} = \frac{0.4}{0.2} = 2 \text{ ms}^{-1}$

$$v_p = \frac{\partial y}{\partial t} = -(2A)\left(\frac{2\pi}{0.2}\right)\cos\left(\frac{2\pi}{0.4}x\right)\cos\left(\frac{2\pi}{0.2}t\right)$$

$$v_p\left(x = \frac{1}{15}\text{m}, t = 0.1\right)$$

$$= -(4)(10\pi)\cos\left(5\pi \times \frac{1}{15}\right)\cos(10\pi \times 0.1)$$

$$= -(40\pi)\left(\frac{1}{2}\right)(-1) = 20\pi \text{ cms}^{-1}$$

13. (a) $\varepsilon = (0.1 \text{ mm})\cos\left[\frac{2\pi}{0.8}(y + 1 \text{ cm})\right]\cos(400\,t)$

End correction is 1 cm.

So, at $y = -1$ cm

$\therefore$ $\varepsilon = (0.1 \text{ mm})\cos\left[\frac{2\pi}{0.8}(-1 + 1)\right]$

$= (0.1 \text{ mm})\cos 0° \Rightarrow$ Antinode

So, upper end is open.

At the lower end, $y = 99$ cm $= 0.99$ m

$\therefore$ $\varepsilon = (0.1 \text{ mm})\cos\left[\frac{2\pi}{0.8}(0.99 + 0.01)\right]$

$= (0.1 \text{ mm})\cos\frac{5\pi}{2} = 0$

$\therefore$ Lower end is closed.

14. (b) $\frac{2\pi}{0.8} = \frac{2\pi}{\lambda} \Rightarrow \lambda = 0.8$ m

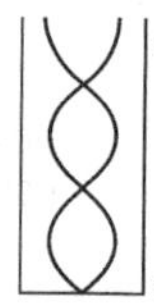

Effective length of air column $= 0.99 + 0.01 = 1$ m

$\therefore$ $\frac{L}{\lambda} = \frac{1}{0.8} = \frac{5}{4}$

$\Rightarrow$ $L = \frac{5\lambda}{4}$. So, second overtone.

15. (d) $P_{ex} = -B\frac{d\varepsilon}{dy} = -(5 \times 10^5)(0.1 \times 10^{-3})$

$$\left\{-\sin\left[\frac{2\pi}{0.8}(y + 1 \text{ cm})\right]\right\}\left(\frac{2\pi}{0.8}\right)\cos(400\,t)$$

$$= (125\pi \text{ Nm}^{-2})\sin\left\{\frac{2\pi}{0.8}(y + 1 \text{ cm})\right\}\cos(400\,t)$$

16. (d) End correction $= 0.3\,D = 1$ cm

$\Rightarrow$ $D = \frac{10}{3}$ cm

V, Volume of tube $= \left(\frac{\pi D^2}{4}\right)L = \frac{\pi}{4}\left(\frac{10}{3}\right)^2 \times 100 \text{ cm}^3$

(Taking, $L = 0.99\text{m} \simeq 1$ m)

or $V = \frac{10\pi}{36}$ litre

$\therefore$ Number of moles $= \frac{10\pi}{36 \times 22.4}$

($\because$ At NTP, 1 mole of a gas has 22.4 L volume)

17. (d) For largest mass, $n = 1$

$\therefore$ $f = \frac{n}{2L}\sqrt{\frac{F}{\mu}} \Rightarrow 700 = \frac{1}{2 \times 1}\sqrt{\frac{F}{2 \times 10^{-3}}}$

or $F = 3920$ N

$\therefore$ $M_{max} = \frac{F}{g} = \frac{3920}{9.8} = 400$ kg

18. (b) Here, $F = 16 \times 9.8 = 156.8$ N

$$f = \frac{n}{2L}\sqrt{\frac{F}{\mu}}$$

$\Rightarrow$ $700 = \frac{n}{2 \times 1}\sqrt{\frac{156.8}{2 \times 10^{-3}}}$

$\Rightarrow$ $n = 5$

19. (c) $A(x) = 6\sin\left(\frac{\pi x}{10}\right)$

At $x = 5$ cm, $\quad A(x) = 6\sin\left(\frac{\pi \times 5}{10}\right) = 6$ cm

20. (c) Incident wane in (i),

$$y_i = A\sin(kx - \omega t)$$

Reflection at free end,

$$y_r = A\sin(kx + \omega t)$$

$$y = y_1 + y_r = 2A\sin kx\cos\omega t$$

Reflection at rigid end,

$$y_r = A\sin(kx + \omega t + \pi)$$

$$= -A\sin(kx + \omega t)$$

$$y = y_i + y_r = 2A\cos kx\sin\omega t$$

Incident wave in Eq. (ii), $y_i = A\cos(kx - \omega t)$

Reflection at free end, $y_r = A\cos(kx + \omega t)$

$$y = y_i + y_r = 2A\cos kx\cos\omega t$$

Reflection at rigid end, $y_r = -A\cos(kx + \omega t)$

$$y = y_i + y_r = 2A\sin(kx)\sin(\omega t)$$

When $x = 0$ is the rigid support, then node is formed at $x = 0$. And at $x = 0$, y is zero.

When $x = 0$ is the flexible support, then antinode is formed at $x = 0$. And at $x = 0$, y_{max} is $\pm 2A$.

Hence, (i) → (p, r) (ii) → (q,s) (iii) → (q,r) (iv) → (p,s)

21. (a) In organ pipes, longitudinal waves exist. In strings, transverse waves exist. Open end is an antinode, fixed (or closed) end is a node. The least distance between node and antinode is $\frac{\lambda}{4}$ and between the nodes is $\frac{\lambda}{2}$.

Hence, (i) → (p, t) (ii) → (p,s) (iii) → (q,s) (iv) → (q,r)

22. (5) $\quad f = \frac{n}{2L}\sqrt{\frac{F}{\mu}}$

$\Rightarrow \quad 100 = \frac{n}{2(0.2)}\sqrt{\frac{0.5}{10^{-3}} \times 0.2}$

$\Rightarrow \quad n = 4$

$\therefore \quad 4 \times \frac{\lambda}{2} = 20$ cm $\Rightarrow \frac{\lambda}{2} = 5$ cm

23. (6) Velocities on Al and steel wires are different because μ is different for both of them. So, λ is also different as f is same for both.

$$\mu_1 = \mu_{steel} = \rho_1 \times \text{Area} = 7.8 \times 10^3 \times 10^{-6} = 7.8 \times 10^{-3} \text{ kg m}^{-1}$$

$$\mu_2 = \mu_{Al} = \rho_2 \times \text{Area} = 2.6 \times 10^3 \times 10^{-6} = 2.6 \times 10^{-3} \text{ kg m}^{-1}$$

$$f_1 = f_2$$

$\Rightarrow \quad \frac{n_1}{2L_1}\sqrt{\frac{F}{\mu_1}} = \frac{n_2}{2L_2}\sqrt{\frac{F}{\mu_2}} = \frac{n_1}{0.866 \times \sqrt{7.8}} = \frac{n_2}{0.6\sqrt{2.6}}$

$\Rightarrow \quad \frac{n_1}{n_2} = \frac{5}{2}$

Therefore, lowest frequency oscillation causes 2 loops in Al and 5 loops in steel.

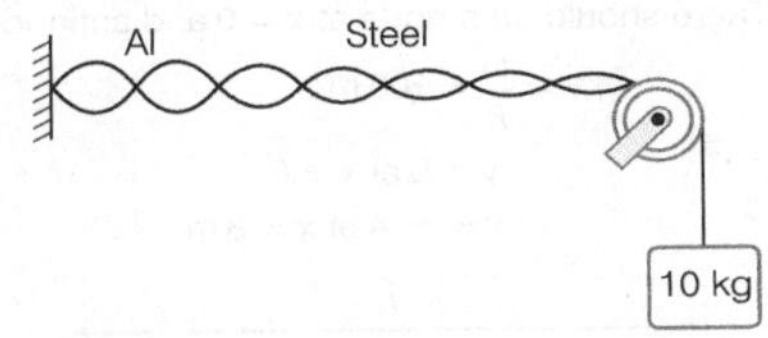

So, total no. of nodes = 8, required no. of nodes = 6

JEE Main & AIEEE Archive

1. (c) For closed organ pipe $= \frac{(2n+1)v}{4l}$ $\quad (n = 0, 1, 2, \ldots)$

$$\frac{(2n+1)v}{4l} < 1250$$

$$(2n+1) < 1250 \times \frac{4 \times 0.85}{340}$$

$$(2n+1) < 12.5$$

$\Rightarrow \quad 2n < 11.50 \quad \Rightarrow \quad n < 5.25$

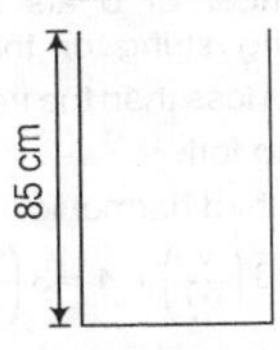

So, $\quad n = 0, 1, 2, 3, \ldots, 5$

So, we have 6 possibilities.

Alternate Method

In closed organ pipe, fundamental node

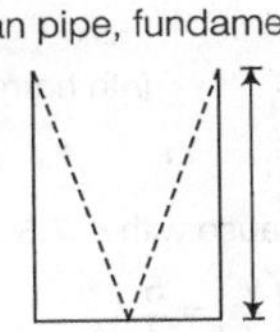

$l = 0.85 = \frac{l}{4}$

i.e. $\quad \frac{\lambda}{4} = 0.85 \Rightarrow \lambda = 4 \times 0.85$

As we know,

$$\nu = \frac{c}{\lambda} = \frac{340}{4 \times 0.85} = 100 \text{ Hz}$$

$\therefore$ Possible frequencies

= 10 Hz, 300 Hz, 500 Hz, 700 Hz, 900 Hz, 1100 Hz below 1250 Hz.

2. (d) Initially for open organ pipe, fundamental frequency,

$$\nu_o = \frac{v}{2l} = f \quad \text{(given)}$$

But when it is half dipped in water, then it becomes closed organ pipe of length $\frac{l}{2}$. In this case, fundamental frequency,

$$\nu_c = \frac{v}{4l'} = \frac{v}{4\frac{l}{2}} = \frac{v}{2l} = f$$

3. (d) As Intensity, $I = \frac{1}{2}\rho\omega^2A^2v$

Here, $\quad \rho$ = density of medium,

A = amplitude,

ω = angular frequency

and $\quad v$ = velocity of wave.

$\therefore$ Intensity depend upon amplitude, frequency as well as velocity of wave.

Also, $\quad I_1 = I_2$

4. (a)
$$y(x,t) = e^{-(ax^2 + bt^2 + 2\sqrt{ab}\,xt)}$$
$$= e^{-(\sqrt{a}\,x + \sqrt{b}\,t)^2}$$

It is a function of type
$$y = f(\omega t + kx)$$
So, $y(x,t)$ represents wave travelling along $-x$ direction.

$\therefore$ Speed of wave $= \dfrac{\omega}{k} = \dfrac{\sqrt{b}}{\sqrt{a}} = \sqrt{\dfrac{b}{a}}$

5. (a) For string fixed at both the ends, resonant frequency are given by $f = \dfrac{nv}{2L}$, where symbols have their usual meanings. It is given that 315 Hz and 420 Hz are two consecutive resonant frequencies, let these be nth and $(n+1)$th harmonics.

Then, $$315 = \frac{nv}{2L} \quad \ldots\text{(i)}$$

and $$420 = \frac{(n+1)v}{2L} \quad \ldots\text{(ii)}$$

Dividing Eq. (i) by Eq. (ii), we get
$$\frac{315}{420} = \frac{n}{n+1}$$
or $$n = 3$$
From Eq. (i), lowest resonant frequency
$$f_0 = \frac{v}{2L} = \frac{315}{3} = 105 \text{ Hz}$$

6. (b) When the string is plucked in the middle, it vibrates in one loop with nodes at fixed ends and an antinode in the middle.

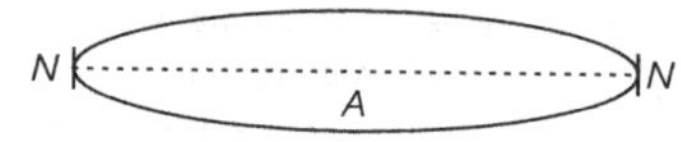

So that, $$\frac{\lambda_1}{2} = l$$

or $$\lambda_1 = 2 \times 40 = 80 \text{ cm}$$

JEE Advanced & IIT JEE Archive

1. (a,c,d) There should be a node at $x = 0$ and antinode at $x = 3$ m.

Also, $$v = \frac{\omega}{k} = 100 \text{ m/s.}$$

$\therefore$ $$y = 0 \text{ at } x = 0$$

and $$y = \pm A \text{ at } x = 3 \text{ m.}$$

2. (b.c)

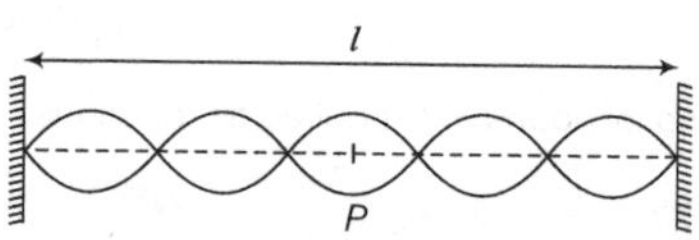

From the given equation, we can see that
$$k = \frac{2\pi}{\lambda} = 62.8 \text{ m}^{-1}$$

$\therefore$ $$\lambda = \frac{2\pi}{62.8} \text{m}$$
$$= 0.1 \text{ m}$$
$$l = \frac{5\lambda}{2} = 0.25 \text{ m}$$

The mid-point of the string is P, an antinode

$\therefore$ maximum displacement = 0.01 m
$$\omega = 2\pi f = 628 \text{ s}^{-1}$$

$\therefore$ $$f = \frac{628}{2\pi} = 100 \text{ Hz}$$

But this is fifth harmonic frequency.

$\therefore$ Fundamental frequency
$$f_0 = \frac{f}{5} = 20 \text{ Hz,}$$

3. (b) In organ pipes, longitudinal waves are formed.
In string, transverse waves are formed. Open end of pipe is displacement antinode and closed end is displacement node. In case of string fixed end of a string is node.

Further, least distance between a node and an antinode is $\dfrac{\lambda}{4}$ and between two nodes is $\dfrac{\lambda}{2}$. Keeping these points in mind answer to this question is as under

Hence, (i) $\rightarrow$ (p,t), (ii) $\rightarrow$ (p,s), (iii) $\rightarrow$ (q,s), (iv) $\rightarrow$ (q,r)

4. (b) For hollow pipe,
$$f_1 = \frac{1}{4L} \cdot v = \frac{1}{4 \times 0.8} \times 320 = 100 \text{ Hz}$$

For string, $$f_2 = \frac{2}{2L'}\sqrt{\frac{F}{\mu}}$$
$$= \frac{1}{0.5}\sqrt{\frac{50}{\mu}} = 2\sqrt{\frac{50}{\mu}}$$

Given, $$f_2 = f_1$$

$\therefore$ $$2\sqrt{\frac{50}{\mu}} = 100$$

$\Rightarrow$ $$\frac{50}{\mu} = 2500$$

$\Rightarrow$ $$\mu = \frac{1}{50} \text{ kg m}^{-1}$$

$\therefore$ $$m = \frac{1}{50} \times L'$$
$$= \frac{1}{50} \times 0.5 = \frac{1}{100} \text{ kg}$$

or $$m = 10 \text{ g}$$

5. (a) With increase in tension, frequency of vibrating string will increase. Since number of beats are decreasing. Therefore, frequency of vibrating string or third harmonic frequency of closed pipe should be less than the frequency of tuning fork by 4.

$\therefore$ Frequency of tuning fork
$$= \text{third harmonic frequency of closed pipe} + 4$$
$$= 3\left(\frac{v}{4l}\right) + 4 = 3\left(\frac{340}{4 \times 0.75}\right) + 4$$
$$= 344 \text{ Hz}$$

$\therefore$ Correct option is (a).

6. (c) $f_1 = \dfrac{v}{l}$ (2nd harmonic of open pipe)

$f_2 = n\left(\dfrac{v}{4l}\right)$ (nth harmonic of closed pipe)

Here, n is odd and $f_2 > f_1$

It is possible when $n = 5$ because with $n = 5$
$$f_2 = \frac{5}{4}\left(\frac{v}{l}\right) = \frac{5}{4} f_1$$

Last 3 Years' Questions
JEE Main & Advanced
(2016-2018)

JEE Main

2016

1. A pendulum clock loses 12 s a day if the temperature is 40°C and gains 4 s a day if the temperature is 20°C. The temperature at which the clock will show correct time, and the coefficient of linear expansion α of the metal of the pendulum shaft are, respectively **[Oscillations]**

(a) 25°C, $\alpha = 1.85 \times 10^{-5}$/°C
(b) 60°C, $\alpha = 1.85 \times 10^{-4}$/°C
(c) 30°C, $\alpha = 1.85 \times 10^{-3}$/°C
(d) 55°C, $\alpha = 1.85 \times 10^{-2}$/°C

2. A particle performs simple harmonic motion with amplitude A. Its speed is trebled at the instant that it is at a distance $\frac{2}{3}A$ from equilibrium position. The new amplitude of the motion is **[Oscillations]**

(a) $\frac{A}{3}\sqrt{41}$
(b) $3A$
(c) $A\sqrt{3}$
(d) $\frac{7}{3}A$

3. A uniform string of length 20 m is suspended from a rigid support. A short wave pulse is introduced at its lowest end. It starts moving up the string. The time taken to reach the support is (Take, $g = 10\ \text{ms}^{-2}$) **[Oscillations]**

(a) $2\pi\sqrt{2}$ s
(b) 2 s
(c) $2\sqrt{2}$ s
(d) $\sqrt{2}$ s

4. A pipe open at both ends has a fundamental frequency f in air. The pipe is dipped vertically in water, so that half of it is in water. The fundamental frequency of the air column is now **[Superposition and Standing Waves]**

(a) $\frac{f}{2}$
(b) $\frac{3f}{4}$
(c) $2f$
(d) f

2017

5. A particle is executing simple harmonic motion with a time period T. At time $t = 0$, it is at its position of equilibrium. The kinetic energy-time graph of the particle will look, like **[Oscillations]**

(a)
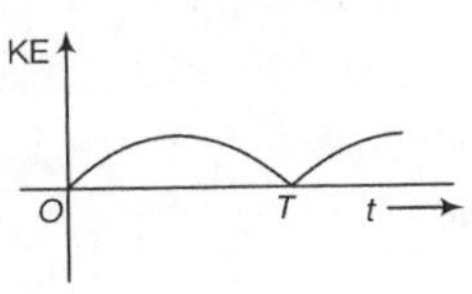

(b)
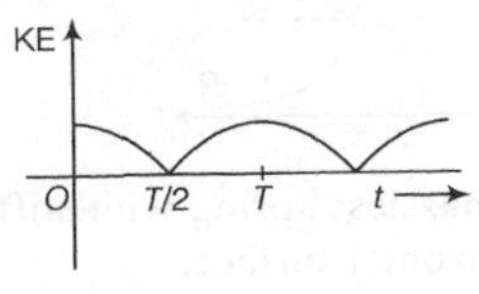

(c)
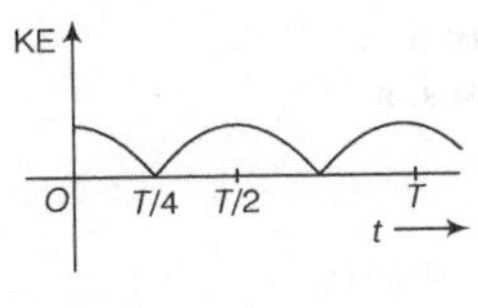

(d)
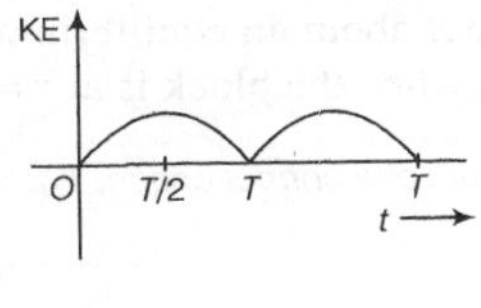

6. An observer is moving with half the speed of light towards a stationary microwave source emitting waves at frequency 10 GHz. What is the frequency of the microwave measured by the observer? (speed of light = 3×10^8 ms^{-1})

[Sound]

(a) 12.1 GHz
(b) 17.3 GHz
(c) 15.3 GHz
(d) 10.1 GHz

2018

7. A silver atom in a solid oscillates in simple harmonic motion in some direction with a frequency of 10^{12} per second. What is the force constant of the bonds connecting one atom with the other?

(Take, molecular weight of silver = 108 and Avogadro number = 6.02×10^{23} g mol^{-1})

[Oscillations]

(a) 6.4 N/m
(b) 7.1 N/m
(c) 2.2 N/m
(d) 5.5 N/m

8. A granite rod of 60 cm length is clamped at its middle point and is set into longitudinal vibrations. The density of granite is 2.7×10^3 kg/m^3 and its Young's modulus is 9.27×10^{10} Pa. What will be the fundamental frequency of the longitudinal vibrations?

[Superposition and Standing Waves]

(a) 5 kHz
(b) 2.5 kHz
(c) 10 kHz
(d) 7.5 kHz

JEE Advanced

2016

1. Two loudspeakers M and N are located 20m apart and emit sound at frequencies 118 Hz and 121 Hz, respectively. A car in initially at a point P, 1800 m away from the midpoint Q of the line MN and moves towards Q constantly at 60 km/h along the perpendicular bisector of MN. It crosses Q and eventually reaches a point R, 1800 m away from Q.

Let $v(t)$ represent the beat frequency measured by a person sitting in the car at time t. Let v_P, v_Q and v_R be the beat frequencies measured at locations P, Q and R respectively. The speed of sound in air is 330 ms^{-1}. Which of the following statement(s) is (are) true regarding the sound heard by the person? **[Superposition and Standing Waves, Type 2]**

(a) The plot below represents schematically the variation of beat frequency with time

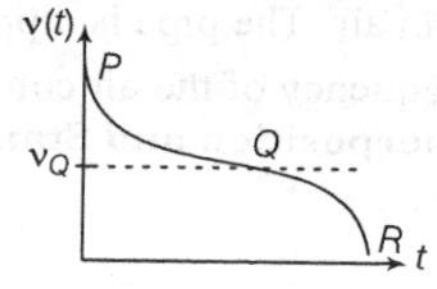

(b) The rate of change in beat frequency is maximum when the car passes through Q

(c) $v_P + v_R = 2v_Q$

(d) The plot below represents schematically the variations of beat frequency with time

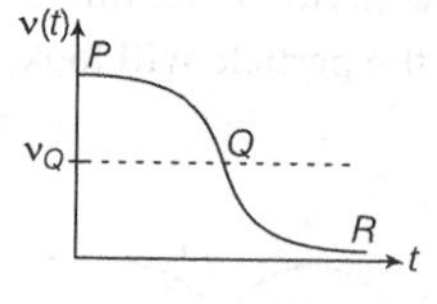

2. A block with mass M is connected by a massless spring with stiffness constant k to a rigid wall and moves without friction on a horizontal surface.

The block oscillates with small amplitude A about an equilibrium position x_0. Consider two cases : (i) when the block is at x_0 and (ii) when the block is at $x = x_0 + A$. In both the cases, a

Note ***Type-1*** *Only one correct option,* ***Type-2*** *More then one correct option,* ***Type-3*** *Single digit Answer,* ***Type-4*** *Numerical value based*

particle with mass m ($< M$) is softly placed on the block after which they stick to each other. Which of the following statement(s) is (are) true about the motion after the mass m is placed on the mass M? **[Superposition and Standing Waves, Type 2]**

(a) The amplitude of oscillation in the first case changes by a factor of $\sqrt{\frac{M}{m+M}}$, whereas in the second case it remains unchanged
(b) The final time period of oscillation in both the cases is same
(c) The total energy decreases in both the cases
(d) The instantaneous speed at x_0 of the combined masses decreases in both the cases

2017

3. A block M hangs vertically at the bottom end of a uniform rope of constant mass per unit length. The top end of the rope is attached to a fixed rigid support at O. A transverse wave pulse (Pulse 1) of wavelength λ_0 is produced at point O on the rope. The pulse takes time T_{OA} to reach point A. If the wave pulse of wavelength λ_0 is produced at point A (Pulse 2) without disturbing the position of M it takes time T_{AO} to reach point O. Which of the following options is/are correct? **[Sound, Type 2]**

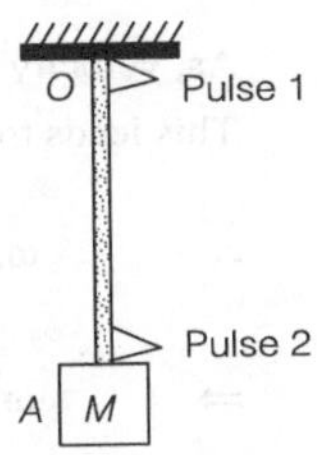

(a) The time $T_{AO} = T_{OA}$
(b) The wavelength of Pulse 1 becomes longer when it reaches point A
(c) The velocity of any pulse along the rope is independent of its frequency and wavelength
(d) The velocities of the two pulses (Pulse 1 and Pulse 2) are the same at the mid-point of rope

4. A stationary source emits sound of frequency $f_0 = 492\,\text{Hz}$. The sound is reflected by a large car approaching the source with a speed of $2\,\text{ms}^{-1}$. The reflected signal is received by the source and superposed with the original. What will be the beat frequency of the resulting signal in Hz? (Given that the speed of sound in air is $330\,\text{ms}^{-1}$ and the car reflects the sound at the frequency it has received). **[Sound, Type 3]**

2018

5. Two men are walking along a horizontal straight line in the same direction. The main in front walks at a speed $1.0\,\text{ms}^{-1}$ and the man behind walks at a speed $2.0\,\text{ms}^{-1}$. A third man is standing at a height 12 m above the same horizontal line such that all three men are in a vertical plane. The two walking men are blowing identical whistles which emit a sound of frequency 1430 Hz. The speed of sound in air $330\,\text{ms}^{-1}$. At the instant, when the moving men are 10 m apart, the stationary man is equidistant from them. The frequency of beats in Hz, heard by the stationary man at this instant, is .... **[Superposition and Standing Waves, Type 2]**

6. In an experiment to measure the speed of sound by a resonating air column, a tuning fork of frequency 500 Hz is used. The length of the air column is varied by changing the level of water in the resonance tube. Two successive resonances are heard at air columns of length 50.7 cm and 83.9 cm. Which of the following statements is (are) true? **[Progressive Waves, Type 2]**

(a) The speed of sound determined from this experiment is $332\,\text{m s}^{-1}$
(b) The end correction in this experiment is 0.9 cm
(c) The wavelength of the sound wave is 66.4 cm
(d) The resonance at 50.7 cm corresponds to the fundamental harmonic

7. A spring block system is resting on a frictionless floor as shown in the figure. The spring constant is 2.0 Nm^{-1} and the mass of the block is 2.0 kg. Ignore the mass of the spring. Initially, the spring is in an unstretched condition. Another block of mass 1.0 kg moving with a speed of 2.0 ms^{-1} collides elastically with the first block. The collision is such that the 2.0 kg block does not hit the wall. The distance, in metres, between the two blocks when the spring returns to its unstretched position for the first time after the collision is .... **[Progressive Waves, Type 4]**

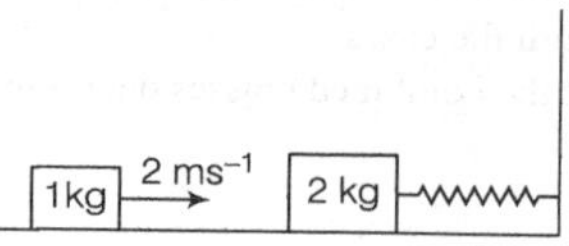

Answer *with* Explanations

JEE Main

1. *(a)* $T_0 = 2\pi\sqrt{\frac{L}{g}}$

$$T' = T_0 + \Delta T = 2\pi\sqrt{\frac{L+\Delta L}{g}}$$

$$\therefore \quad T' = T_0 + \Delta T = 2\pi\sqrt{\frac{L(1+\alpha\,\Delta\theta)}{g}}$$

$$= \left\{2\pi\sqrt{\frac{L}{g}}\right\}(1+\alpha\,\Delta\theta)^{1/2}$$

$$\approx T_0\left(1+\frac{\alpha\,\Delta\theta}{2}\right)$$

$$\Delta T = T' - T_0 = \frac{\alpha\,\Delta\theta T_0}{2} \quad \ldots(i)$$

or $$\frac{\Delta T_1}{\Delta T_2} = \frac{\alpha\,\Delta\theta_1 T_0}{\alpha\,\Delta\theta_2 T_0}$$

$$\Rightarrow \quad \frac{12}{4} = \frac{40^\circ - \theta}{\theta - 20^\circ}$$

$$\Rightarrow \quad 3(\theta - 20^\circ) = 40^\circ - \theta$$

$$\Rightarrow \quad 4\theta = 100^\circ \Rightarrow \theta = 25^\circ\,C$$

Time gained or lost is given by

$$\Delta T = \left(\frac{\Delta T}{T_0 + \Delta T}\right)t \approx \frac{\Delta t}{T_0}t$$

From Eq. (i), $\frac{\Delta T}{T_0} = \frac{\alpha\Delta\theta}{2}$

$$\therefore \quad \Delta t = \frac{\alpha\,(\Delta\theta)t}{2}$$

$$12 = \frac{\alpha\,(40^\circ - 25^\circ)(24 \times 3600)}{2}$$

$$\therefore \quad \alpha = 1.85 \times 10^{-5}/^\circ C$$

2. *(d)* $v = \omega\sqrt{A^2 - x^2}$

At, $x = \frac{2A}{3}$

$$v = \omega\sqrt{A^2 - \left(\frac{2A}{3}\right)^2} = \frac{\sqrt{5}}{3}\,\omega A$$

As, velocity is trebled, hence $v' = \sqrt{5}A\omega$

This leads to new amplitude A'

$$\therefore \quad \omega\sqrt{A'^2 - \left(\frac{2A}{3}\right)^2} = \sqrt{5}A\omega$$

$$\Rightarrow \quad \omega^2\left[A'^2 - \frac{4A^2}{9}\right] = 5A^2\omega^2$$

$$\Rightarrow \quad A'^2 = 5A^2 + \frac{4}{9}A^2$$

$$= \frac{49}{9}A^2$$

$$A' = \frac{7}{3}A$$

3. *(c)* At distance x from the bottom

$$v = \sqrt{\frac{T}{\mu}} = \sqrt{\frac{\left(\frac{mgx}{L}\right)}{\left(\frac{m}{L}\right)}} = \sqrt{gx}$$

$$\therefore \quad \frac{dx}{dt} = \sqrt{x}\sqrt{g}$$

$$\Rightarrow \quad \int_0^L x^{-1/2}dx = \sqrt{g}\int_0^t dt$$

$$\Rightarrow \quad \left[\frac{x^{1/2}}{(1/2)}\bigg|_0^L\right] = \sqrt{g}\cdot t$$

$$\Rightarrow \quad t = \frac{2\sqrt{L}}{\sqrt{g}}$$

$$\Rightarrow \quad t = 2\sqrt{\frac{20}{10}} = 2\sqrt{2}\,s$$

4. *(d)* Fundamental frequency of open pipe.

$$f = \frac{v}{2l}$$

Now, after half filled with water it becomes a closed pipe of length $\frac{l}{2}$.

Fundamental frequency of this closed pipe,

$$f' = \frac{v}{4(l/2)}$$

$$= \frac{v}{2l} = f$$

5. *(c)* KE is maximum at mean position and minimum at extreme position $\left(\text{at } t = \frac{T}{4}\right)$.

6. *(b)* As the observer is moving towards the source, so frequency of waves emitted by the source will be given by the formula

$$f_{\text{observed}} = f_{\text{actual}} \cdot \left(\frac{1 + v/c}{1 - v/c}\right)^{1/2}$$

Here, frequency $\frac{v}{c} = \frac{1}{2}$

So, $\quad f_{\text{observed}} = f_{\text{actual}} \left(\frac{3/2}{1/2}\right)^{1/2}$

$\therefore \quad f_{\text{observed}} = 10 \times \sqrt{3} = 17.3 \text{ GHz}$

7. *(b)* For a harmonic oscillator,

$$T = 2\pi\sqrt{\frac{m}{k}},$$

where k = force constant and $\quad T = \frac{1}{f}$

$$\therefore \quad k = 4\pi^2 f^2 m$$

$$= 4 \times \left(\frac{22}{7}\right)^2 \times (10^{12})^2 \times \frac{108 \times 10^{-3}}{6.02 \times 10^{23}}$$

$$\Rightarrow \quad k = 7.1 \text{ N/m}$$

8. *(a)*

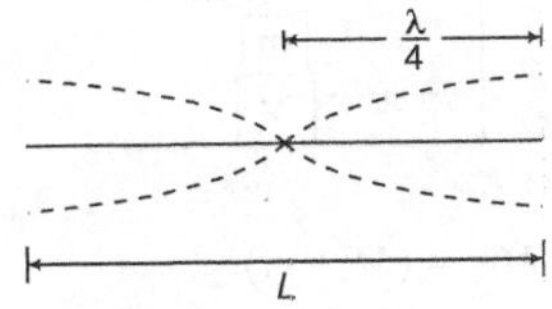

From vibration mode,

$$\frac{\lambda}{2} = L \Rightarrow \lambda = 2L$$

$\therefore$ Wave speed, $v = \sqrt{\frac{Y}{\rho}}$

So, frequency $f = \frac{v}{\lambda}$

$$\Rightarrow \quad f = \frac{1}{2L}\sqrt{\frac{Y}{\rho}}$$

$$= \frac{1}{2 \times 60 \times 10^{-2}}\sqrt{\frac{9.27 \times 10^{10}}{2.7 \times 10^3}}$$

$$\approx 5000 \text{ Hz}$$

$$f = 5 \text{ k Hz}$$

JEE Advanced

1. *(b,c,d)*

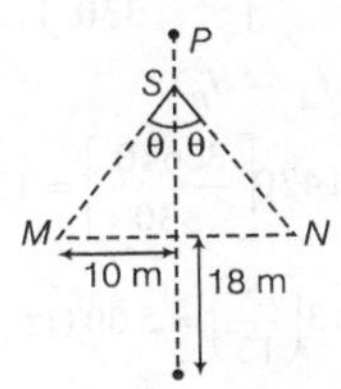

speed of car, $v = 60 \text{km/h} = \frac{500}{3}$ m/s

At a point S, between P and Q

$$v'_M = v_M\left(\frac{C + v\cos\theta}{C}\right);$$

$$v'_N = v_N\left(\frac{C + v\cos\theta}{C}\right)$$

$$\Rightarrow \quad \Delta v = (v_N - v_M)\left(1 + \frac{v\cos\theta}{C}\right)$$

Similarly, between Q and R

$$\Delta v = (v_N - v_M)\left(1 - \frac{v\cos\theta}{C}\right)$$

$$\frac{d(\Delta v)}{dt} = \pm(v_N - v_M)\frac{v}{C}\sin\theta\frac{d\theta}{dt}$$

$\theta \approx 0°$ at P and R as they are large distance apart.

$\Rightarrow$ Slope of graph is zero.

at Q, $\qquad \theta = 90°$

$\sin\theta$ is maximum also value of $\frac{d\theta}{dt}$ is maximum as $\frac{d\theta}{dt} = \frac{v}{r}$,

where v is its velocity and r is the length of the line joining P and S.

and r is minimum at Q.

$\Rightarrow$ Slope is maximum at Q.

At P, $v_P = \Delta v = (v_N - v_M)\left(1 + \frac{v}{C}\right)$

$$(\theta \approx 0°)$$

At R, $v_R = \Delta v = (v_N - v_M)\left(1 - \frac{v}{C}\right)$

$$(\theta \approx 0°)$$

At Q, $v_Q = \Delta v = (v_N - v_M)\,(\theta = 90°)$

From these equations, we can see that

$$v_P + v_R = 2v_Q$$

2. *(a,b,d)* **Case-I**

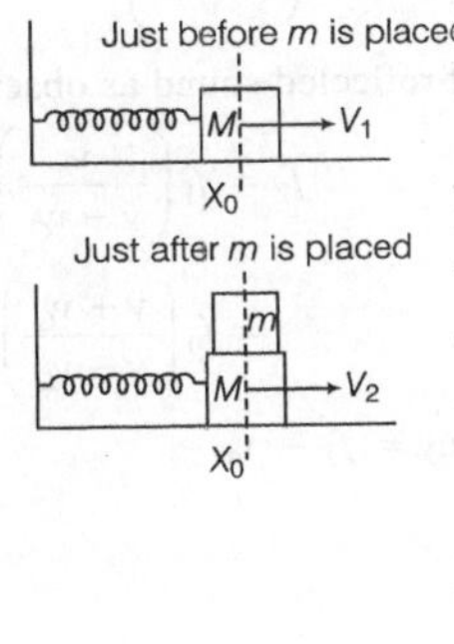

Case-II

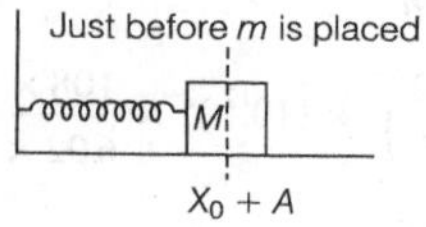

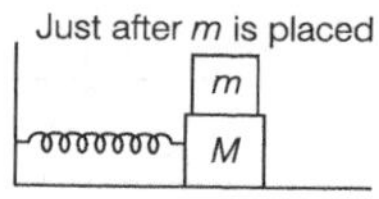

In case-1,

$$Mv_1 = (M + m)v_2$$

$$\Rightarrow \quad v_2 = \left(\frac{M}{M+m}\right)v_1$$

$$\sqrt{\frac{k}{M+m}}\, A_2 = \left(\frac{M}{M+m}\right)\sqrt{\frac{k}{M}}\, A_1$$

$$A_2 = \sqrt{\frac{k}{M+m}}\, A_1$$

In case-2, $\quad A_2 = A_1$

$$T = 2\pi\sqrt{\frac{M+m}{k}} \text{ in both cases.}$$

Total energy decreases in first case whereas remain same in 2 nd case. Instantaneous speed at x_0 decreases in both cases.

3. *(a,c,d or a,c)* $v = \sqrt{\dfrac{T}{\mu}}$, so speed at any position will be same for both pulses, therefore time taken by both pulses will be same.

$$\lambda f = v$$

$$\Rightarrow \quad \lambda = \frac{v}{f}$$

$$\Rightarrow \quad \lambda \propto v \propto T$$

since when pulse 1 reaches at A tension and hence speed decreases therefore λ decreases.

Note If we refer velocity by magnitude only, then option (a, c, d) will be correct, else only (a, c) will be correct.

4. *(6)*

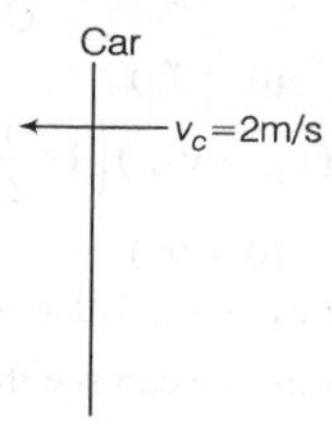

Frequency observed at car

$$f_1 = f_0\left(\frac{v + v_C}{v}\right) \qquad (v = \text{speed of sound})$$

Frequency of reflected sound as observed at the source

$$f_2 = f_1\left(\frac{v}{v - v_C}\right)$$

$$= f_0\left(\frac{v + v_C}{v - v_C}\right)$$

Beat frequency $= f_2 - f_0$

$$= f_0\left[\frac{v + v_C}{v - v_C} - 1\right]$$

$$= f_0\left[\frac{2v_C}{v - v_C}\right]$$

$$= 492 \times \frac{2 \times 2}{328} = 6 \text{ Hz}$$

5. *(5 Hz)*

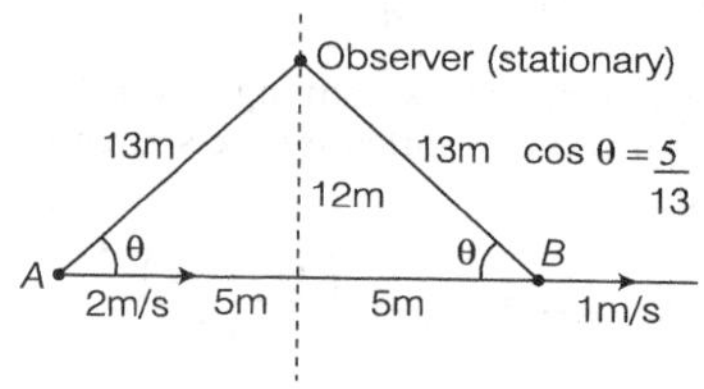

$$f_A = 1430\left[\frac{330}{330 - 2\cos\theta}\right]$$

$$= 1430\left[\frac{1}{1 - \dfrac{2\cos\theta}{330}}\right]$$

$$\approx 1430\left[1 + \frac{2\cos\theta}{330}\right]$$

[from binomial expansion]

$$f_B = 1430\left[\frac{330}{330 + 1\cos\theta}\right]$$

$$\approx 1430\left[1 - \frac{\cos\theta}{330}\right]$$

Beat frequency $= f_A - f_B$

$$= 1430\left[\frac{3\cos\theta}{330}\right] = 13\cos\theta$$

$$= 13\left(\frac{5}{13}\right) = 5.00 \text{ Hz}$$

6. *(a, c)* Let n th harmonic is corresponding to 50.7 cm and $(n+1)$th harmonic is corresponding 83.9 cm.

$\therefore$ Their difference is $\dfrac{\lambda}{2}$.

$$\therefore \quad \frac{\lambda}{2} = (83.9 - 50.7)\text{ cm}$$

$$\text{or} \quad \lambda = 66.4 \text{ cm}$$

$$\therefore \quad \frac{\lambda}{4} = 16.6 \text{ cm}$$

Length corresponding to fundamental mode must be close to $\dfrac{\lambda}{4}$ and 50.7 cm must be an odd multiple of this length.

16.6 × 3 = 49.8 cm. Therefore, 50.7 is 3rd harmonic.

If end correction is e, then

$$e + 50.7 = \frac{3\lambda}{4}$$

$$e = 49.8 - 50.7$$

$$= -0.9 \text{ cm}$$

$\therefore$ Speed of sound, $v = f\lambda$

$\Rightarrow$ $$v = 500 \times 66.4 \text{ cm/s}$$
$$= 332 \text{ m/s}$$

7. (2.09 **m**) **Just Before Collision,**

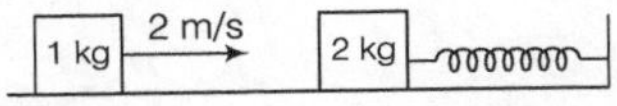

Just After Collision

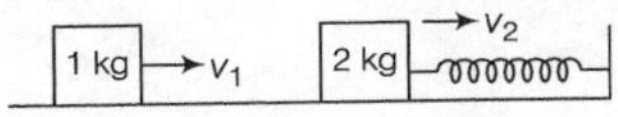

Let velocities of 1 kg and 2 kg blocks just after collision be v_1 and v_2 respectively.

From momentum conservation principle,

$$1 \times 2 = 1v_1 + 2v_2 \quad \ldots(i)$$

Collision is elastic. Hence $e = 1$ or relative velocity of separation = relation velocity of approach.

$$v_2 - v_1 = 2 \quad \ldots(ii)$$

From Eqs. (i) and (ii),

$$v_2 = \frac{4}{3} \text{ m/s},$$

$$v_1 = \frac{-2}{3} \text{ m/s}$$

2 kg block will perform SHM after collision,

$$t = \frac{T}{2} = \pi\sqrt{\frac{m}{k}} = 3.14 \text{ s}$$

$$\text{Distance} = |v_1|t = \frac{2}{3} \times 3.14$$

$$= 2.093 = 2.09 \text{ m}$$